Energy Simulation
in
Building Design

For
Kathryn,
Fiona, Karen and Andrew
with much appreciation and love

Energy Simulation
in
Building Design

2nd Edition

J A Clarke
Professor of Environmental Engineering
University of Strathclyde
Glasgow, Scotland

Routledge
Taylor & Francis Group
LONDON AND NEW YORK

First published 2001 by Butterworth-Heinemann

First published 1985
Second edition 2001

This edition published 2011 by Routledge
2 Park Square, Milton Park, Abingdon, Oxfordshire OX14 4RN
711 Third Avenue, New York, NY 10017, USA

Simultaneously published in USA and Canada

First issued in hardback 2015

Routledge is an imprint of the Taylor and Francis Group, an informa business

British Library Cataloguing in Publication Data
A catalogue record for this book is available from the British Library

Library of Congress Cataloguing in Publication Data
A catalogue record for this book is available from the Library of Congress

ISBN 13: 978-1-138-13370-9 (hbk)
ISBN 13: 978-0-7506-5082-3 (pbk)

Contents

Preface

If a system can be characterised by n parameters, each of which may assume 3 independent states, then the total number of combinations is 3^n. A major problem encountered in the design of any energy system—from a component such as a boiler, to a system such as a building—is that n is large. A building, for example, is characterised by parameters such as occupancy level, ventilation rate, degree of insulation, location of thermal capacity, glazing type, extent of HVAC provision, level of control and type of fuel, to name but a few. Even a relatively low number of parameters will give rise to a large number of combinations: $n = 10$ equates to 59,000! In short, energy systems are complex. To pretend otherwise is to design for certain failure.

Achieving a high quality indoor environment at acceptable cost has always presented a challenge for the construction industry. With aspects of sustainable development now being added to the list of requirements, and the growth in the available materials/systems that may be employed, this challenge is set to become even more formidable.

To add to the designer's problems, determining the merit of one combination over another is a non-trivial task requiring some means to translate the myriad physical interactions to information on cost and performance relating to fitness-for-purpose, energy use and environmental impact. Returning to the previous example: a building comprises several thermodynamic domains—air movement, radiation exchange, moisture flow, electrical power flow, daylight distribution etc—each one of which may interact with the others in a non-trivial manner. For example, the simple act of adjusting the position of a window shading device will have cascading effects on glare, internal daylight level, artificial lighting requirement, luminaire heat gain and space cooling, heating and electricity demand. Clearly, the construction industry has some way to go if it wishes to incorporate a rigorous life cycle analysis into its future design practice.

Given the limitations inherent in traditional design methods, is it surprising that our energy systems often fail to attain their expected performance? Many buildings stubbornly hover at around 300 kWh $m^{-2}yr^{-1}$, energy conversion and delivery systems operate at substantially less than their optimum efficiency, and human health and comfort needs are rarely fully satisfied.

Simulation represents a possible solution to the complexity dilemma by enabling comprehensive and integrated appraisals of design options under realistic operating conditions. In other words, simulation supports the emulation of future realities at the design stage. It gives practitioners the ability to appreciate the underlying behaviour of a system and, thereby, to take judicious steps to improve performance across the range of relevant criteria. To be contentious: simulation represents a paradigm shift of vast potential. It will give rise to a *cheaper*, *better* and *quicker* design process. And it will provide outcomes that better match society's aspirations for sustainable practices, environmental protection and climate change mitigation.

This book addresses the issues underlying the development, proving and use in practice of building energy simulation. While these issues are covered in a generic manner, the specific material derives from the ESP-r program, which has been under continuous development over a twenty five year period with financial support from the UK's Engineering and Physical Science Research Council and the R&D Framework Programmes of the European Commission. I am indebted to both organisations and to the many technical reviewers and project officers who recognised something of value in the ESP-r project.

The book has two complementary objectives: to cover theoretical aspects that will be of relevance to the next generation of modellers who will take the state-of-the-art to a higher plane, and to cover the use in practice aspects that will be of relevance to those practitioners who wish to adopt a computational approach to design.

I am deeply grateful to many people who, over the years, have given me guidance, support and encouragement: Professor Tom Maver of ABACUS, who made it all possible; my colleagues in ESRU, who create an intellectual environment that nurtures creativity and team working; to my many friends around the world whose daily aim, it sometimes seems, is to locate the pernicious software bugs upon which ESP-r is founded†; to the memory of my mother; and to my supportive wife and children to whom I dedicate this book.

I am especially indebted to my PhD students—Essam Aasem, Ian Beausoleil-Morrison, Francesca Born, Tin-tai Chow, Stephane Citherlet, Steven Conner, Dru Crawley, Mark Evans, Jan Hensen, Apostolis Karatolios, Nick Kelly, Amissah Patrick Ken, Jae-min Kim, Iain Macdonald, Don McLean, John MacQueen, Christoph Morbitzer, Abdul Nakhi, Cezar Negrao, Nicola Smith and Dechao Tang—who have so ably contributed to the different aspects of the integrated modelling approach, and to Jordan Denev, Milan Janak and Cor Pernot, who made important theoretical contributions.

Paul Strachan was kind enough to read the entire manuscript and give me detailed critiques, while Ian Beausoleil-Morrison, Chris Bronsdon, Andy Grant, Jon Hand, Jan Hensen, Cameron Johnstone, Nick Kelly and Lori McElroy gave me helpful advice on particular topics. Those errors and misconceptions that remain are entirely of my own making.††

Joe Clarke
Glasgow 2001

† ;-)

†† Corrections to such errors, as found, will be available at <http://www.esru.strath.ac.uk> from where further ESP-r related publications and source code downloads can also be obtained.

1

Introduction

It is not hyperbole to suggest that the better design of new buildings would result in a 50-75% reduction in their energy consumption relative to 2000 levels, and that appropriate intervention in the existing stock would readily yield a 30% reduction. Added together, this would significantly reduce a nation's energy bill, handsomely contribute to environmental impact and climate change mitigation, and help to alleviate the stressful indoor conditions experienced by many citizens. Indeed, energy efficiency may be likened to an untapped, clean energy resource of vast potential.

The barrier to accessing this resource is less to do with technological constraints—much know-how and many approaches already exist—and more to do with ineffective decision-support. This is especially true at the early design stage in the case of new build, and at the remedial options selection stage in the case of existing buildings. It is a strange paradox that we live in an information age and yet information is never in the hands of those who need it to make informed decisions.

It is in response to this deficiency that building simulation has emerged for use to appraise options for change in terms of relevant issues—from human health and comfort, through energy demand reduction, to sustainable practices. Because of the growing acceptance that simulation defines best practice, substantial attempts are being made to transfer the technology into practice. There are two main incentives for this transfer. First, buildings are complex artifacts involving 'hard' and 'soft' aspects (such as transient energy flows and stochastic occupant interactions respectively). Traditional design methods, by failing to address this complexity, fall short of best practice. Second, there is a need for rapid feedback on the cost and performance of alternative design approaches. The present system of specialist consultants, while adequate for the detailed design and final specification phases, fails to provide this *ad hoc* advice.

Such incentives provide the impetus for the growth of organisations representing the notion of 'test driving' a building using simulation: the International Building Performance Simulation Association (IBPSA 1999), the Energy Design Advice Scheme (McElroy *et al* 1997) and the Scottish Energy Systems Group (McElroy and Clarke 1999). Such organisations have brought about a better understanding of the potential of a modelling and simulation approach to building design (Howrie 1995).

Notwithstanding the advanced capabilities of contemporary simulation, there remain at least four formidable barriers to its routine and effective application in practice.

First, there are shortcomings in the user interface. These derive predominately from a conflict between the necessity for the underlying model to be comprehensive and rigorous—to adequately represent real world complexity—while also being straightforward and intuitive—to facilitate ease of user interaction. The situation is exacerbated by the divergence of the conceptual frameworks of the design-oriented program users and the technical-oriented program developers. To complete the confusion, there is the subtly different terminology of the architectural, engineering and scientific professions.

Second, there is little agreement on the data model used to define the building and its energy systems. The program specific data models that have emerged serve only to ensure that there is little commonality between the different modelling systems. This frustrates the validation process, forces applications to operate in isolation and presents a formidable barrier to collaborative design.

Third, the absence of agreed performance assessment methods has forced users to devise personalised appraisal strategies and to become expert enough to coordinate a program's operational path accordingly. Clearly, the existence of standard methods would serve to harmonise program use and make the application experience less fraught for the novice.

Fourth, it may be expected that as the rate of uptake of simulation accelerates, user expectation will grow, especially in relation to integrated modelling by which a building's multi-variate state may be appraised. Satisfying this expectation requires the integration of several complex technical domains.

This book has two principal objectives: to establish and integrate sufficiently detailed models for each technical domain comprising a building, and to elaborate and exemplify new work practices aimed at fostering a simulation-based design process. The aim is to remove the mystery surrounding simulation by concisely deriving an integrative theoretical basis and elaborating an apt mode of use.

Chapter 1 commences with an overview of building performance simulation, introduces the underlying energy flowpaths, and introduces the different possible classes of modelling method.

Chapter 2 derives the two main analytical formulations for dynamic building energy modelling—time and frequency domain response function methods—and sets out the elements for a less constrained numerical method based on conservation considerations applied to control volumes.

Chapter 3 undertakes a step-by-step formulation of the numerical method by deriving conservation equations for characteristic control volumes and structuring these equations in a manner that is the topological equivalent of the real building system.

Chapter 4 demonstrates conservation equation-set formulation for a simple building example and derives matrix partitioning protocols by which fast, variable frequency (time step), simultaneous solutions can be achieved.

Chapter 5 derives complementary approaches to the modelling of inter- and intra-room air movement and moisture flow within the building fabric. A technique for linking the flow and building models is then elaborated.

Chapter 6 applies the theory of chapter 3 to heating, ventilating and air conditioning (HVAC), renewable energy conversion (REC) and control systems and shows how the equation-sets to emerge can be solved simultaneously with the building/flow models. To support REC simulation, a numerical model of electrical power flow is introduced.

Chapter 7 introduces models for the technical sub-systems that impact upon the parameters of the conservation equations, the boundary conditions under which they must be solved and the interpretation of results: weather, non-orthogonal geometry, shading and insolation, shortwave and longwave radiation exchange, surface convection, casual heat sources, daylight illuminance distribution and mould growth.

Chapter 8 addresses use in practice with the emphasis on practical advice aimed at those readers who seek to apply simulation in the real-time, real-scale context of design practice.

Finally, chapter 9 places building energy simulation in the future context of virtual design whereby the different disciplines may collaborate in real time to ensure that buildings are acceptable in terms of their multi-variate performance and impact.

In order to retain a definite focus throughout, the book intersperses theoretical derivations relating to the different technical domains within an evolving description of the building as a complex energy system. In this way an integrated modelling system is arrived at by the book's end. This modelling system is similar in its form and content to the ESP-r system <http://www.esru.strath.ac.uk/> which, since 1974, has evolved in accordance with the software development process as espoused by Maver and Ellis (1982):

> Research into model needs, methods, algorithms and organisation. This leads to a research prototype embodying the fundamental laws governing energy flow.
>
> Development of a pilot program based on the research findings and which offers a reasonable platform for testing.
>
> Validation of the program to test the underlying models, the in-built assumptions and the various numerical schemes.
>
> Implementation trials to test the robustness, relevance and efficacy of the program when applied to practical problems.
>
> Improvement of the software and documentation with respect to commercial standards and the incorporation of the lessons learned through the validation and trial implementation studies.
>
> Commercial exploitation and the development of user training and support procedures.

It is instructive to note that the resource required at any stage is typically greater than the accumulated resource required for the preceding stages. Thus, it may be expected that the validation and implementation trial stages will be significantly more costly than the resource required to produce the initial pilot program; and that the commercial exploitation and user support stage will require a more substantial investment again.

1.1 A brief history of simulation

Design tools have traditionally been constructed by reducing the complexity of the underlying system equations in an attempt to lessen the computational load and the corresponding input burden placed on the user. Some portion of the system may be neglected (e.g. longwave radiation exchange), time invariant values may be assigned to some system parameters (e.g. material thermal properties) or simple boundary conditions may be imposed (e.g. steady state or steady cyclic). Within a simulation program such assumptions are heresy. Instead, a mathematical model is constructed to represent each possible energy flowpath and their interactions. In this sense simulation is an attempt to emulate the reality. The evolution of design tools, from traditional manual methods to contemporary simulators, is summarised in table 1.1.

Table 1.1: Evolution of design tools.

Generation	*Characteristics*	*Consequences*
1	handbook oriented simplified and piecemeal familiar to practitioners	easy to use, difficult to translate to real world, non-integrative, application limited, deficiencies hidden
2	building dynamics stressed less simplified, still piecemeal based on standard theories	| | |
3	field problem approach shift to numerical methods integrated modelling stressed graphical user interface partial interoperability enabled	increasing integrity *vis-à-vis* the real world | | |
4 and beyond	good match with reality intelligent knowledge-based fully integrated network compatible/interoperable	deficiencies overt, easy to use and interpret, predictive and multi-variate, ubiquitous and accessible

Traditionally, designers have relied on a range of disparate calculation techniques to quantify and assess building performance at the design stage. The approach is piecemeal in that, at best, only a weak coupling is evident between the various calculation steps. These calculations are based on analytical formulations that embody many simplifying assumptions to permit their formulation in the first instance. Significantly, there is no attempt to faithfully represent the energy and mass flowpaths that occur in real buildings. The intention is only to provide users with an indication of performance: a 1st generation program is consequently easy to apply but difficult to interpret since the user is required to appreciate its limitations and make appropriate allowances.

In the mid-'70s 2nd generation programs began to emerge. These stressed the temporal aspect of the problem, particularly with respect to long time constant elements such as multi-layered constructions. The underlying calculation methods remained analytical and piecemeal: time or frequency domain response factors were used to model the dynamic response of constructional elements, while HVAC system modelling was confined to the steady state.

With the advent of more powerful personal computing, 3rd generation programs began to emerged as a viable prospect in the mid-'80s. These assume that only the space and time dimensions are independent variables; all other system parameters are dependent so that no single energy transfer process can be solved in isolation. This signalled the beginning of integrated modelling whereby the thermal, visual and acoustic aspects of performance are considered together.

In the mid-'90s, the domain integration work continued apace but with the addition of program interoperability, which is essentially a data modelling issue. Also, and in response to the growing uptake by practitioners, new developments commenced concerned with knowledge-based user interfaces, application quality control and user training.

As summarised in figure 1.1a (MacCallum 1993), the use of design tools has hitherto adhered to a *tool-box* metaphor by which the designer must recognise a particular task, locate a suitable program, apply it and translate its outputs to appropriate modifications to the design hypothesis. This is an inadequate model in that the tools are decoupled from the process and require the designer to translate between data models. A more desirable approach is summarised in figure 1.1b, which shows a computer-supported design environment (CSDE). Here, the designer evolves the design hypothesis in such a way that the computer applications are able to automatically access the data describing the design and give feedback on all aspects of performance and cost in terms meaningful to the designer.

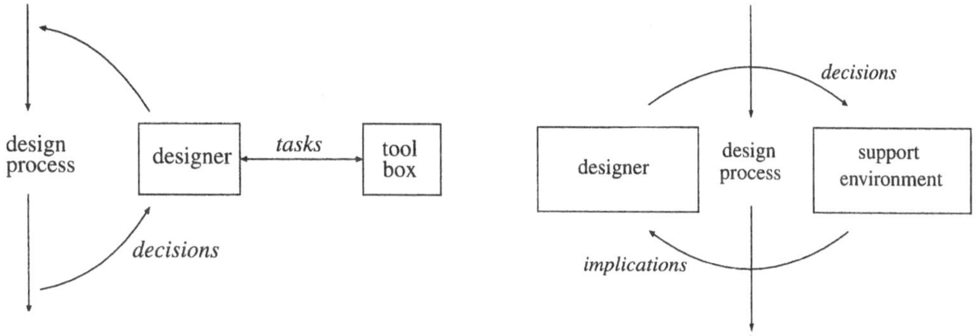

Figure 1.1a: Tool-box approach; b: CSDE approach.

The attainment of such a CSDE (see §9.1) is a non-trivial task requiring the development of a computational model of the design process in which the role of each participant, human and otherwise, is clearly defined.

1.2 Simulation overview

Consider figure 1.2, which shows the flowpaths encountered within and outwith buildings and which interact, in a dynamic manner, to dictate comfort levels and energy demands. To understand the simulation approach, it is useful to visualise such a system as an electrical network of time dependent resistances and capacitances subjected to time dependent potential differences. The currents to result in each branch of the network are then equivalent to the heat flows between the building's parts. Constructional elements, room contents, glazing systems, plant components, renewable energy devices etc may be treated as network 'nodes' and characterised by capacitance, with the inter-node connections characterised by conductance. Nodes possess 'variables of state' such as temperature and pressure (analogous to voltage). Since nodes have different capacitances, the problem is essentially dynamic: each node responding at a different rate as it competes with its neighbours to capture, store and release energy (current). It is this distributed dynamic behaviour, along with the non-trivial nature of the branch flows and network parameters, that imparts complexity to the building modelling task. The resolution of the model—that is the number of nodes—is a function of the analysis objectives. Clearly, an early design stage estimation of summertime temperatures will require a lower level of discretisation than a detailed study of indoor air quality.

From a mathematical viewpoint, several complex equation types must be solved to accurately represent such a system and, because these equations represent heat transfer processes that are highly inter-related, it is necessary to apply simultaneous solution techniques if the performance prediction is to be both accurate and preserve the spatial and temporal integrity of the modelled system.

Once established, a simulation program can be applied throughout the design process, from the early concept stage through detailed design. It is more efficient to use a single simulation

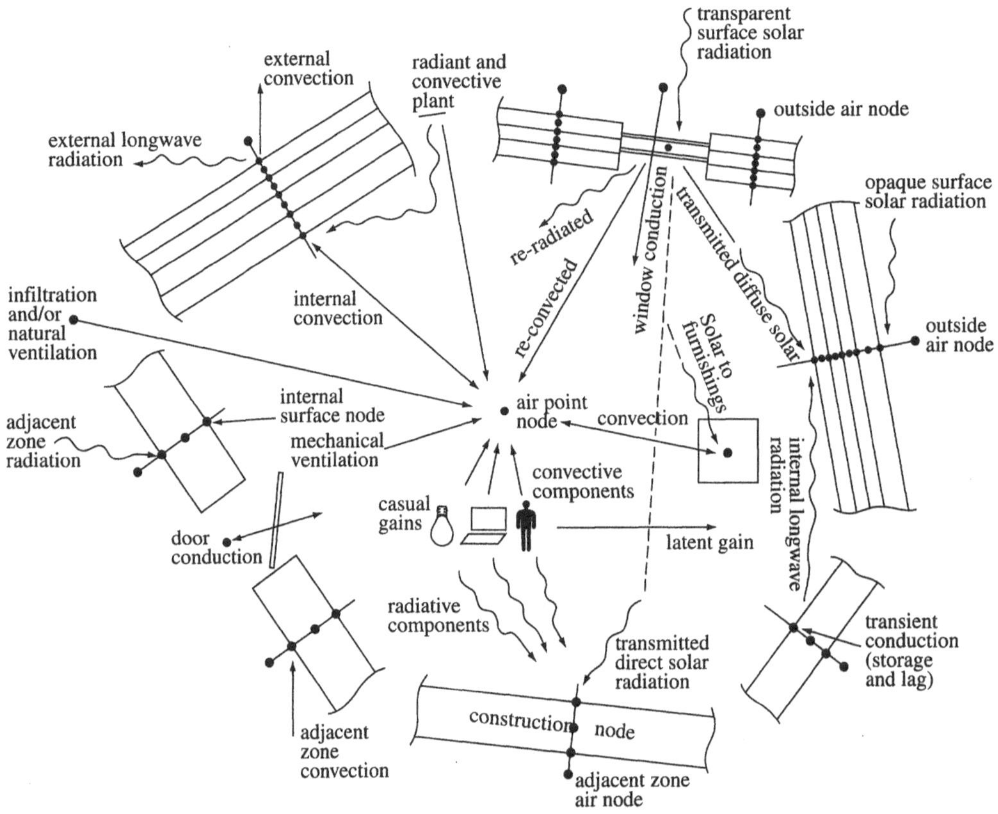

Figure 1.2: Building energy flowpaths.

program throughout the design process than to use a progression of tools—from simplified to detailed—and ignore the many theoretical discontinuities and pernicious assumptions. Consider the following application scenario.

It is possible to use simulation at an early design stage to determine the optimum combination of zone layout and constructional scheme that will provide a climate responsive solution and so minimise the need for mechanical plant. Some simulations might focus on the choice of constructional materials and their relative positioning within multi-layered constructions so that good temperature and load levelling is attained. Also, alternative daylight capture and shading strategies might be investigated to ensure glare avoidance, excess solar gain control and minimum luminaire usage.

After a fundamentally sound design has emerged, well tested in terms of its performance under a range of anticipated operating conditions, a number of alternative control scenarios can be simulated. For example, basic control studies will lead to decisions on the potential of optimum start/stop control, appropriate night set-back temperature, the efficacy of weather anticipation, the location of sensors and the interrelation of thermal and visual comfort variables. Further analysis might focus on 'smart' control by which the system is designed to respond to occupancy levels or indoor daylight illuminance. Yet further simulations might be undertaken to ensure acceptable indoor air quality or explore the feasibility of deploying local renewable energy conversion devices such as photovoltaic cells.

As the underlying relationships emerge, the designer is able to assess the benefits, or otherwise, of any given course of action before it is implemented. The appraisal permutations are essentially without limit. Simulation can be used at any design stage to address relevant questions (CIBSE 1998): what are the maximum demands for heating, cooling and lighting and where and when do they occur?; what are the main causal factors?; what will be the contribution of particular technologies—transparent insulation, advanced glazing, smart control etc—when deployed independently or jointly?; can new and renewable energy systems be used to match demand and/or operate co-operatively with the public electricity supply without loss of power quality?; what combination of energy efficiency measures will give rise to a given target saving?; is local cogeneration feasible?; and so on.

Simulation allows users to understand the interrelation between design and performance parameters, to identify potential problem areas, and so implement and test appropriate design modifications. The design to result is more energy conscious with better comfort levels and air quality attained throughout.

1.3 Integrative modelling

The aim of integrative modelling is to preserve the integrity of the entire building/plant system by simultaneously processing all energy transport paths at a level of detail commensurate with the objectives of the problem in hand and the uncertainties inherent in the describing data. To this end, a building should be regarded as being systemic (many parts make the whole), dynamic (the parts evolve at different rates), non-linear (the parameters depend on the thermodynamic state) and, above all, complex (there are myriad intra- and inter-part interactions). To achieve high modelling integrity, a simulation program aims to preserve these intrinsic characteristics.

Underlying the flowpaths of figure 1.2 is the concept of energy, mass and momentum balance which requires, in turn, a knowledge of the fundamental processes of conduction, convection, radiation and fluid flow. Many excellent texts exist that cover the fundamentals of heat transfer (e.g. Kreith 1973, Ozisik 1977, Incropera and DeWitt 1981) and no attempt is made here to provide a similar treatment of the subject. Instead, specific mathematical models are derived for each flowpath and, as the book develops, these models are combined to form a unified representation of the building and its environmental control systems. Specifically, this representation includes models of the heat, air, moisture, light and electricity flows as they occur within a building/plant system when subjected to weather factors and influenced by distributed control action and occupant interactions.

1.4 Energy flowpaths and causal effects

Before commencing the development of an integrative mathematical model, it is necessary to consider the various heat and mass transfer mechanisms and the factors that give rise to them.

Transient conduction

This lies at the heart of the building model. It is the process by which a fluctuation of heat flux at one boundary of a solid material finds its way to another boundary, being diminished in magnitude and shifted in time due to the material's thermal inertia. Within the building fabric, transient conduction is a function of the temperature and heat flux excitations at exposed surfaces, the possible generation of heat within the fabric, the temperature- and moisture-dependent (and therefore time-dependent) hygro-thermal properties of the individual materials, and the relative position of these materials. With the external weather excitations declared as known time-series data, the modelling objective is to determine the intra-fabric temperature and moisture distribution and hence the dynamic variation of heat flux at the exposed surfaces. In many situations it is important to consider heat flow in more than one direction, for example in cases where thermal bridging might be expected to occur or where corner and edge regions are large relative to the planar area of a wall.

The thermophysical properties of interest include conductivity, k ($W\ m^{-1}{}^{\circ}C^{-1}$), density, ρ ($kg\ m^{-3}$), and specific heat capacity, C ($J\ kg^{-1}{}^{\circ}C^{-1}$). These properties are time-dependent because of material temperature and/or moisture fluctuations, and may be position or direction dependent if the material is non-homogeneous or anisotropic respectively. In some applications such dependencies may be ignored and the thermophysical properties assumed constant. Appendix A lists the thermal properties of some common building materials.

Such properties can be combined to give simple indices for use at an early design stage to differentiate construction performance. The overall thermal transmittance, or U-value ($W\ m^{-2}{}^{\circ}C^{-1}$), is given by

$$U = \frac{1}{\sum_{i=1}^{N} \frac{x_i}{k_i} + R_{si} + R_{so} + R_c}$$

where N is the number of layers in the construction, x_i the thickness of layer i (m), R the combined radiative and convective thermal resistance ($m^2{}^{\circ}C\ W^{-1}$) and subscripts si, so, and c refer to the innermost surface, outermost surface and cavity respectively.

Traditionally, designers have relied on this simple steady state concept to assess the heat loss characteristics of the building fabric. In addition to ignoring the dynamic aspect of fabric behaviour, the approach does not preserve spatial integrity since different constructional arrangements will perform differently even though each may have the same U-value.

As an example, if insulation is located at the innermost position of a wall then any shortwave solar radiation penetrating windows and striking that internal surface cannot be readily stored in the construction since the insulation will act as a barrier. Instead, the solar energy will cause a surface temperature rise which, in turn, will increase the rate of energy release to the adjacent air by the process of natural convection. A space experiencing high solar energy penetration is therefore likely to overheat if cooling is not introduced. Conversely, if the insulation is relocated externally, with capacity elements exposed to the inside, then internal surface shortwave gain can access capacity to be stored. By proper design this stored energy can later be harnessed passively (rather than by mechanical means) to minimise heating requirements and avoid overheating. On the other hand, internal capacity may give rise to increased peak plant demand due to the initial high rate of transfer of energy to capacity at plant start-up in an intermittent scheme. With continuous operation, capacity can help to minimise the peaks and maximise the troughs of plant demand and so promote good load levelling. This, in turn, will give a stable environment and encourage efficient operation by allowing plant to operate consistently at or near full load. The risk of interstitial condensation is greater in the case of internally located insulation since a portion of the construction may fall below the dew point temperature of moist air permeating through the construction in the absence of an effective vapour barrier. In summary, transient conduction will affect energy requirements, load diversity, peak plant demand, load levelling, plant operating efficiency and condensation potential.

Unfortunately, there is no simple design paradigm that can be used to select an optimum construction. Table 1.2, for example, shows the effect on room cooling energy requirements, peak plant demand and load levelling for a number of combinations of construction and plant operation. From these data it is clear that there is a need to utilise dynamic models to determine the performance of alternative constructions on a case-by-case basis: whilst a U-value can be used as a simple selection index, it has no role as a predictor of fabric thermodynamic performance.

Table 1.2: Effect of construction and plant operation on fabric performance.

Construction†	*Energy (kWh)*	*Load levelling (LL) ($Q_{max} - Q_{min}$; kW)*	*Comment*
continuous operation			
40I/150C/ 40I††	410	22	LL best when more insulation to outside of capacity
5I/150C/ 75I	396	18	
75I/150C/ 5I	413	25	
75C/ 80I/ 75C	398	17	LL best when more capacity to inside of insulation
25C/ 80I/125C	411	20	
125C/ 80I/ 25C	391	16	
intermittent operation			
40I/150C/ 40I	268	37	LL best when insulation split equally either side of capacity
5I/150C/ 75I	305	39	
75I/150C/ 5I	275	43	
75C/ 80I/ 75C	330	44	LL best when more capacity to inside of insulation
25C/ 80I/125C	336	53	
125C/ 80I/ 25C	328	42	

† Each construction comprises 3 homogeneous elements each of which can be insulation (I) or capacity (C). The specifications are all inside to outside and all constructions have the same U-value.
†† Dimensions in mm.

Thermal diffusivity, $\alpha = k/\rho C$ (m^2s^{-1}), and effusivity, $\varepsilon = (k\rho C)^{\frac{1}{2}}$ ($J\ m^{-2}{}^\circ C^{-1}s^{-\frac{1}{2}}$), are useful dynamic performance indicators. Materials with high α values transmit boundary heat flux fluctuations more rapidly than do materials with correspondingly low values, while materials with high ε values will more readily absorb a surface heat flux. These parameters can be usefully applied to real constructions by reducing the multiple layers to an equivalent homogeneous layer (Mackey and Wright 1944):

$$(k\rho C)_e = \frac{1.1R_i(k\rho C)_i + \sum_m 1.1R_m(k\rho C)_m}{R_e} + \frac{(k\rho C)_o}{R_e}(R_o - 0.1R_i - \sum_m 0.1R_m)$$

where R is resistance ($m^2{}^\circ C\ W^{-1}$) and subscripts e, i, m and o refer to the equivalent, innermost, intermediate and outermost layers respectively (m = 0 for a 2 layer construction and $(k\rho C) = 0$ for an air gap). If the second term on the right-hand side of this equation becomes negative (i.e. $R_o \to 0$) then it should be neglected. The equivalent resistance is given by

$$R_e = \frac{x_o}{k_o} + \sum_m \frac{x_m}{k_m} + \frac{x_i}{k_i}$$

where, for the case of an air gap, x_m/k_m is the combined convective/radiative resistance. The equivalent (ρC) value is obtained from

$$(\rho C)_e = \frac{(k\rho C)_e R_e}{(x_o + \sum_m x_m + x_i)} \, .$$

so that the thermal diffusivity/effusivity of the equivalent layer may be determined from the equivalent k and (ρC) values. Chapter 2 describes models of transient conduction while Appendix C outlines the use of the technique to estimate a construction's time constant.

Surface convection

This is the process by which heat flux is exchanged between a surface (opaque or transparent) and the adjacent air layer. In building modelling it is usual to differentiate between external and internal exposures. In the former case, convection is usually wind induced and considered as forced whereas, with internal surfaces, natural and/or forced air movement can occur depending on the location of mechanical equipment and the flow field to result. It is normal practice to make use of time-varying, but surface-averaged, convection coefficients, h_c (W m^{-2}°C^{-1}). Several researchers (e.g. Alamdari and Hammond 1982, Halcrow 1987, Khalifa and Marshall 1990, Fisher 1995, Awbi and Hatton 1999) have addressed the specific needs of building simulation by producing h_c correlation equations for typical configurations.

Forced convection is a function of the prevailing fluid flow vector. Typically, for external building surfaces, wind speed and direction data are available for some reference height and techniques exist to estimate non-reference height values in terms of characteristic vertical velocity profiles. Forced convection estimation for internal surfaces is more problematic, requiring knowledge of the distribution and operation of air handling equipment.

Natural convection is an easier problem to study and many formulations have emerged which give convection coefficients as a function of the surface-to-air temperature difference, surface roughness, direction of heat flow and characteristic dimensions.

Chapter 7 describes approaches to the modelling of surface convection, forced and buoyant, at surface layers associated with the building fabric.

Internal surface longwave radiation exchange

In most simplified methods, surface heat transfer coefficients are treated as combinations of convection and longwave radiation although the values used are often dubious. In reality, the two processes are related by the fact that they both can raise or lower surface temperatures and so influence each other.

Inter-surface longwave radiation is a function of the prevailing surface temperatures, the surface emissivities, the extent to which the surfaces are in visual contact, referred to as the view factor, and the nature of the surface reflection (diffuse, specular or mixed). The flowpath will tend to establish surface temperature equilibrium by cooling hot, and heating cold, surfaces. It is an important flowpath where temperature asymmetry prevails, as in passive solar buildings where an attempt is made to capture solar energy at some selected surface. A standard energy efficiency measure is to upgrade windows with glazings incorporating a low emissivity coating. This increases the reflection of longwave radiation flux and so acts to break inter-surface heat exchange. The mathematical representation of the flowpath is non-linear in the temperature term and this introduces modelling complications as discussed in chapters 3 and 7.

External surface longwave radiation exchange

The exchange of energy by longwave radiation between external (opaque and transparent) surfaces and the sky vault, surrounding buildings and ground can result in a substantial lowering of surface temperatures, especially under clear sky conditions and at night. This can lead to sub-zero surface temperatures, especially with exposed roofs, and can become critical in cases of low insulation level. Conversely, the flowpath can result in a net gain of energy, although under most conditions this will be negligible.

The adequate treatment of this flowpath will require an ability to estimate several contributing factors: the effective sky temperature as a function of the prevailing cloud cover and type; the temperature of surrounding buildings; the temperature of the ground as a function of terrain conditions; the local air temperature; the surface warming effect of any incident shortwave flux; and view factor information to geometrically couple the surface with the three portions of its scene—sky, ground and surroundings.

Shortwave radiation

In most buildings, the gain of energy from the sun constitutes a significant portion of the total cooling load. The method of treatment of the shortwave flowpaths can therefore largely determine the accuracy of the overall predictions.

Some portion of the shortwave energy impinging on an external surface—arriving directly from the sun or diffusely after atmospheric scatter and terrain reflections—may, depending on subsequent temperature variations affecting transient conduction, find its way through the fabric where it will contribute to the inside surface heat flux at some later time. It is not uncommon for exposed surfaces to be as much as 15-20°C above ambient temperature. Some simplified methods utilise the 'sol-air' temperature concept to handle fabric solar gain. This corresponds to a suitably elevated ambient temperature for use in fabric conduction calculations. This is clearly inadequate on two counts:

> Unless the solar contribution to the sol-air temperature is determined on the basis of time-dependent surface properties relating to shading and convection, then a difference will prevail between actual solar absorption and that predicted.
>
> Insulation/capacity structures are often a mix of opaque, transparent and translucent materials (e.g. a transparently insulated facade). In such cases it is important to model the intra-construction shortwave absorptions.

In the case of completely transparent structures, the shortwave energy impinging on the outermost surface is partially reflected and partially transmitted. Within the glazing layers and substrates of the system many further reflections take place and some portion of the energy is absorbed within the material to raise its temperature. This temperature rise will augment the normal transient conduction process and, thereby, help to establish innerside and outerside surface temperatures which, in turn, will drive the surface convection and longwave radiation flowpaths. Thus, in effect, absorbed shortwave radiation penetrates the building via convection and longwave radiation.

The component of the incident beam which is transmitted will strike (with no perceptible time lag) some internal surface(s) where it behaves as did the external surface impingement: opaque surface absorption/reflection, transparent surface absorption/reflection and transmission (back to outside or onward to another zone), and giving rise to behind-the-surface transient conduction where the flux is stored and lagged.

Accurate solar irradiation modelling therefore requires methods for the prediction of surface position relative to the solar beam, and the assessment of the moving pattern of insolation of internal and external surfaces. The former method is a function of site latitude and longitude, time of day/year and surface geometry, while the latter method requires the existence of ray tracing techniques of the kind introduced in chapter 7.

The thermophysical properties of interest include shortwave absorptivity for opaque elements and absorptivity, transmissivity and reflectivity for transparent elements (Appendix A). The magnitude of these properties is dependent on the angle of incidence of the shortwave flux and on its spectral composition. With regard to the latter, it is common practice to accept properties that are averaged over the entire solar power spectrum.

Shading and insolation

These factors control the magnitude and point of application of solar energy and so dictate the overall accuracy of any solar processing algorithm. Both time-series require point projection or hidden line/surface techniques for their estimation, as well as access to a data structure that contains the geometry of obstruction features.

It is usual to assume that facade shading caused by remote obstructions (such as buildings and trees) will reduce the magnitude of direct insolation, leaving the diffuse beam undiminished. Conversely, shading caused by facade obstructions (such as overhangs and window recesses) should also be applied to the diffuse beam since the effective solid angle of the external scene, as subtended at the surface in question, is markedly reduced.

At any point in time the shortwave radiation directly penetrating an exposed window will be associated with one or more internal surfaces, depending on the prevailing solar angle and the internal building geometry. The receiving surface(s) may be opaque, a window in another wall (connecting the zone to another zone or back to ambient conditions), items of furniture or special surfaces included in the model to represent occupants or sensors. While it has been observed that disregarding the apportioning of window transmitted shortwave flux between the associated receiving surfaces can have a significant effect on thermal predictions, the smearing of the portion received by one surface over its entirety will have minimum effect if the surface has a uni-directional conduction heat flow representation (Robinson 1979).

Air flow

Within buildings, three air flow paths predominate: infiltration, zone-coupled flows and mechanical ventilation. These flowpaths give rise to advective (fluid-to-fluid) heat exchanges. Each are vector quantities in that only air flow into a region is considered to cause the thermal loading of that region, any loss being the driving force for a corresponding replacement to maintain a mass balance.

Infiltration is the name given to the leakage of air from outside and can be considered as comprising two components: the unavoidable movement of air through distributed leakage paths such as the small cracks around windows and doors and through the fabric itself; and the ingress of air through intentional openings (windows, vents etc) often referred to as natural ventilation. Zone coupled air flow, as with infiltration, is caused by pressure variations and by buoyancy forces resulting from the density differences associated with the temperatures of the coupled air volumes. Mechanical ventilation is the deliberate supply of air to satisfy a fresh air requirement and, perhaps, heat or cool a space.

Random occurrences, such as window/door opening, changes in the prevailing wind conditions and the intermittent use of mechanical ventilation, will influence the levels of infiltration and zone-coupled air flow. Notwithstanding the stochastic nature of these occurrences, air flow models of varying complexity can be constructed. Such models will span the spectrum from whole building predictors, based on regressions applied to measured data, to the numerical solution of equations representing the conservation of mass, momentum and energy.

At a level appropriate to building energy modelling, air movement is often represented by a nodal network in which nodes represent fluid volumes and inter-nodal connections represent the distributed leakage paths connecting these volumes and through which flow can occur. Numerical techniques are then applied to this network to establish the mass balance corresponding to any given nodal temperature field and boundary pressure condition. Such a method is well suited to the determination of the contribution of air movement to energy requirements. A more comprehensive approach involves the solution of the energy, continuity (mass) and momentum (Navier-Stokes) equations when applied to a finely discretised flow domain. In addition to supporting energy analysis, such a method will also provide information on the spatial variation of indoor air quality and thermal comfort levels. Chapter 5 describes these two approaches and elaborates a technique for their conflation with the building, HVAC and renewable energy system models derived in chapters 3 and 6 respectively.

Casual gains

In most non-domestic buildings, the effects of the heat gains from lighting installations, occupants, small power equipment, IT devices and the like can be considerable. It is important therefore to process these heat sources in as realistic a manner as possible. Typically, this will necessitate the separate processing of the heat (radiant and convective) and moisture emissions, and the provision of a mechanism to allow each casual source to change its value by prescription or via control action. It is usual to assume that the convective heat emission is experienced instantaneously as an air load whereas the radiant portion, behaving in a manner similar to shortwave radiation penetrating the building envelope, is apportioned between the internal opaque and transparent surfaces according to some distribution strategy. Because of the inherent relationship with the construction capacity, the radiant component will experience a time lag before it can contribute to the cooling load or elevate the internal air temperature.

Some casual gain sources, such as luminaires and IT equipment, will require the elaboration of a model of their electrical behaviour in order to modulate heat emission as a function of the electrical power usage. For example, this would be required in the case of daylight responsive luminaire dimming.

Heating, ventilating and air conditioning (HVAC) systems

Figure 1.3 illustrates some typical systems that may be connected to a building to service its environmental requirements.

The problem of predicting energy consumption has traditionally been divided into two distinct stages. As shown in figure 1.4, the first stage is concerned with predicting the energy requirements to satisfy the demands of the building's activities. This is found by modifying the various instantaneous heat gains and losses as a function of the distributed thermal capacities. In the second stage, these energy requirements are modified by the operating characteristics of the plant to give the energy actually consumed. The first stage is concerned with the design of the building to reduce the energy requirements, whilst the second stage is concerned with the design of the installed plant to best match these requirements and minimise consumption (and thereby the resulting gaseous emissions).

Because the building and its plant are strongly coupled, accuracy considerations dictate that they be handled simultaneously. One approach, as demonstrated in §4.2, is to incorporate plant characteristics within control statements which are then embedded within the solution of the building-side equations (or used to influence their formulation). Alternatively, and more accurately, dynamic plant system models (Lebrun 1982) can be established for solution in tandem

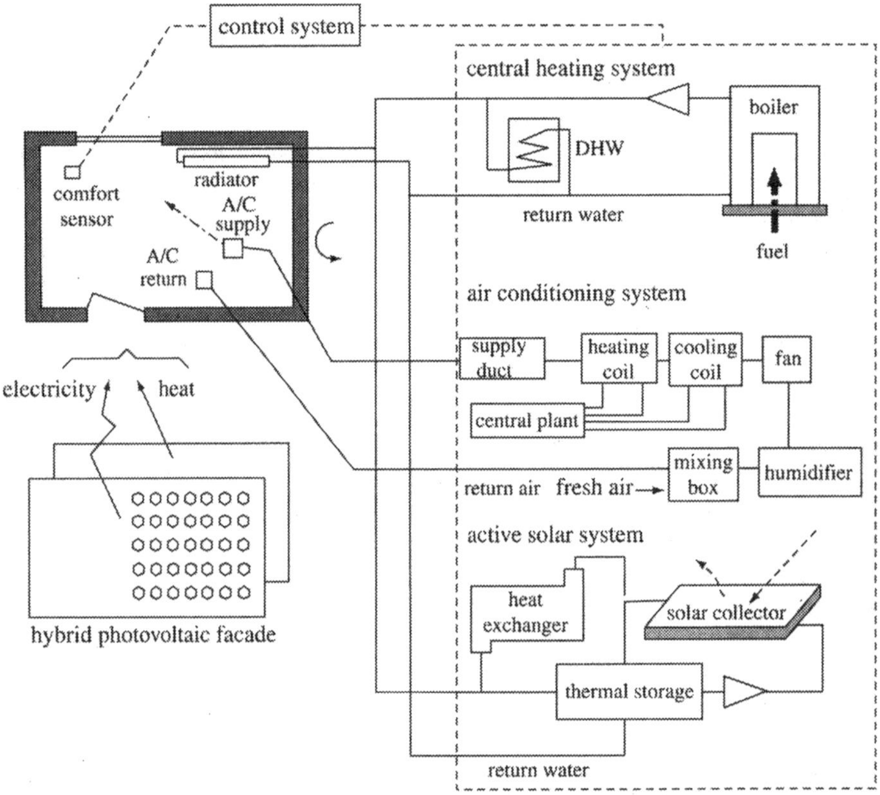

Figure 1.3: Some possible HVAC systems.

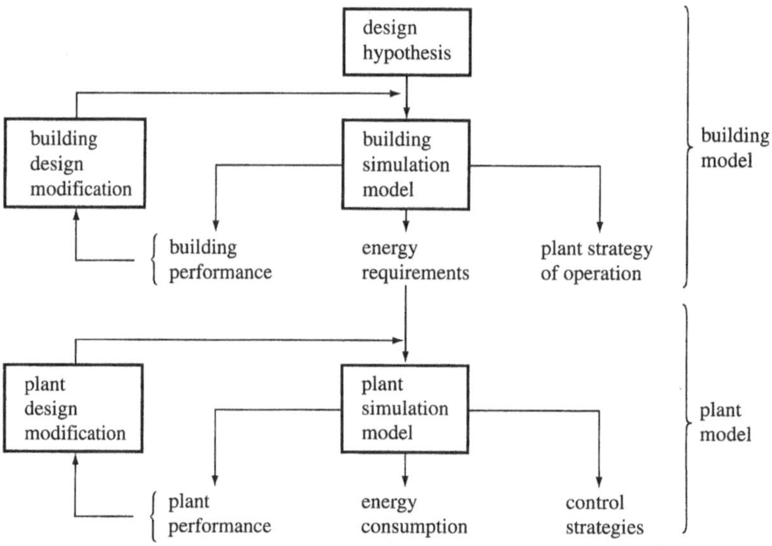

Figure 1.4: The traditional role of building and plant models.

with the building model so that the spatial and temporal interactions are fully respected. Such an approach is demonstrated in chapter 6 where selected plant systems are combined with the building model established in chapter 3.

Control

To direct the path of a simulation, the combined building and plant model is subjected to control action. This involves the establishment of several control loops, each one comprising a sensor (to measure some simulation parameter or aggregate of parameters), an actuator (to receive and act upon the controller output signal) and a regulation law (to relate the sensed condition to the actuated state). These control loops are used to regulate HVAC components and manage building-side entities, such as solar control devices, in response to departures from desired environmental conditions. Control loops can also be used to effect changes to the active model at run-time—e.g. to impose alternative heat transfer coefficients or activate a more rigorous treatment of air movement at some critical point in time. Conveniently, control loops may be used to emulate plant behaviour in terms of the location of flux addition/extraction, the prevailing radiant/convective split, and any physical constraints imposed on the installed capacity. This is a useful feature where there is insufficient information to allow a detailed plant model to be established. Section 6.4 describes alternative approaches to control system modelling.

Moisture

Dampness and mould growth are recognised as major problems affecting a significant proportion of houses throughout the world. Approximately 2.5 million UK residences are affected, with well documented cases for the rest of Europe and North America (Workshop 1992). Singh (1995) has estimated the cost of repairing the damage caused by timber decay in the UK housing stock to be approximately £400M per annum. Apart from aesthetic considerations, there is now considerable epidemiological evidence to support the view that mouldy housing has a detrimental effect on the physical and mental health of occupants (Paton 1993).

High levels of airborne spores may occur due to the growth of fungi on walls and furnishings. Data from the 1991 Scottish housing condition survey (Scottish Homes 1993) indicate that around 12.3% of Scottish houses are affected, with inadequate heating, insulation and ventilation cited as the principal causal factors.

Fluctuations in moisture levels within the building's fabric can also be problematic, leading to interstitial condensation or causing variations in a material's thermophysical properties and, thereby, adversely affecting its thermodynamic performance.

Approaches to the modelling of airborne moisture transport are described in §5.2, while §5.3 describes a procedure for the modelling of the moisture flows within porous media and §7.9 describes the conditions required for the proliferation of mould growth.

Passive solar elements

Many designers have come to favour the use of the so-called passive solar features. These act to capture and process solar radiation passively and without recourse to mechanical systems. Consider figure 1.5, which summarises the range of possible passive solar elements. In each case certain factors can be identified which, in particular, will impose technical complexity on any modelling exercise:

(a) *Non-diffusing direct gain* systems will require adequate treatment of the mapping of the solar beam onto the internal receiving surfaces or obstruction objects such as furniture.

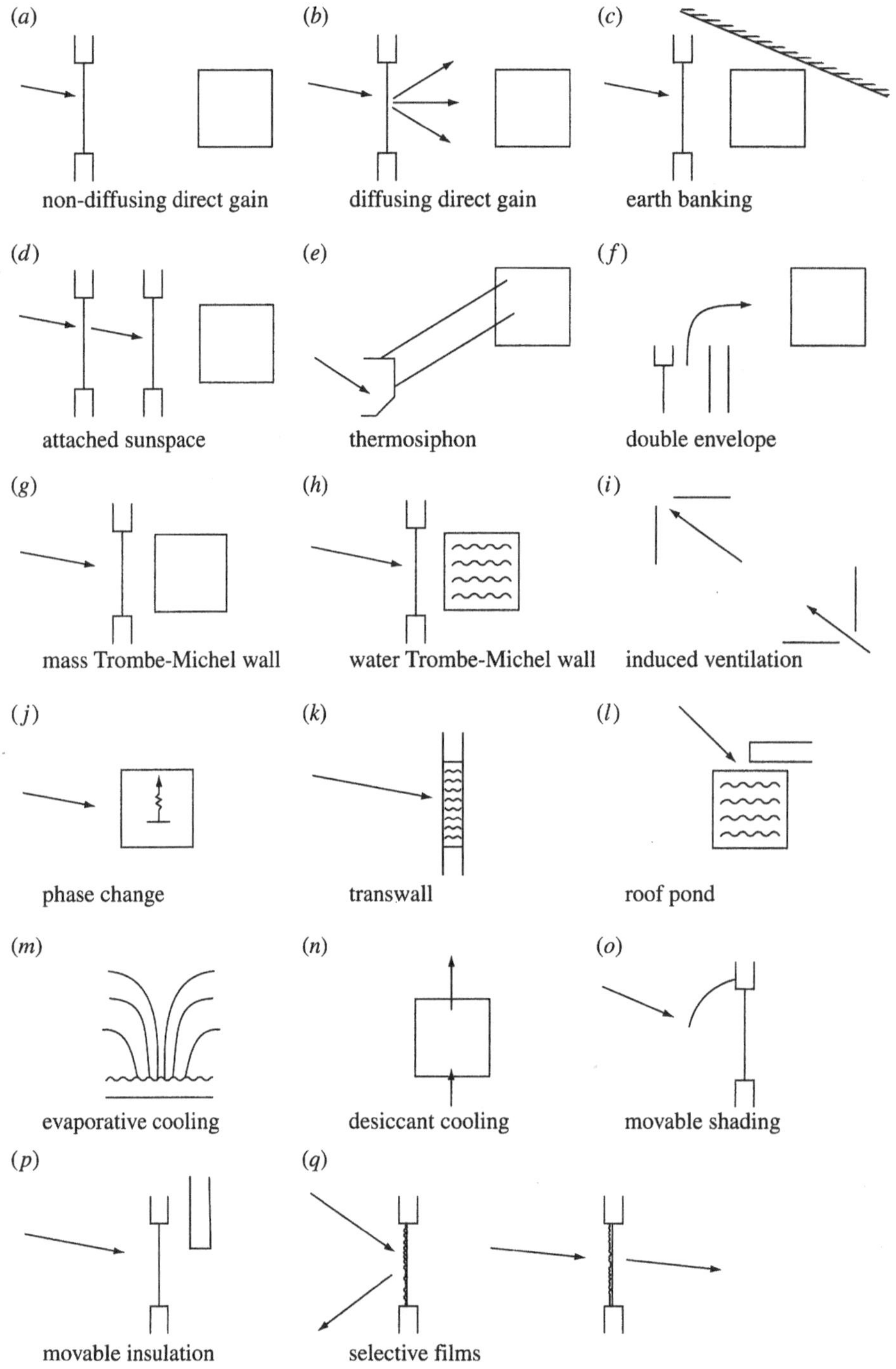

Figure 1.5: Passive solar elements in architectural design.

(b) *Diffusing direct gain* systems will require solar energy apportioning to the various internal surfaces.

(c) *Earth banking* will introduce complexity in the modelling of ground heat exchange.

(d) *Attached sunspaces* will require that the model be able to establish the level of penetration of solar radiation to interior contained zones.

(e) *Thermosiphon* systems will require buoyancy driven air flow modelling.

(f) *Double envelopes* will require sophistication on the part of solar algorithms since radiation may penetrate the external skin to cause 'deep' construction heating.

(g) *Mass Trombe-Michel walls* will require a multi-dimensional conduction model.

(h) *Water Trombe-Michel Walls* will require a detailed treatment of convective heat transfer.

(i) *Induced ventilation* schemes will require accurate modelling of buoyancy and pressure induced air flows.

(j) *Phase change materials* will necessitate the switching from sensible heating behaviour to constant temperature behaviour in the transient conduction schemes.

(k) *Transwalls* will impose demands on the solar and conduction algorithms since direct shortwave transmission and fluid motion will occur.

(l) *Roof ponds* will require accurate external longwave radiation assessment.

(m) *Evaporative cooling* systems will require combined heat and mass flow modelling.

(n) *Desiccant* materials will exhibit a change of dehumidification potential with time and require models for regeneration.

(o) *Movable shading* will require device operation modelling and sophistication on the part of the shading/insolation prediction algorithms.

(p) *Movable insulation* implies a time-dependent system definition.

(q) *Selective thin films* will require detailed spectral response modelling.

Advanced modelling systems seek to include these and other energy flowpaths while respecting the inevitable interactions and underlying complexities.

New and renewable energy systems

Future cities are likely to be characterised by a greater level of new and renewable energy (RE) systems deployment. Traditionally, such deployment has occurred at the strategic level with the grid connection of medium-to-large scale hydro stations, bio-gas plant and wind farms. The further introduction of RE systems at this scale will give rise to network balancing and power quality problems, which will limit deployment to an estimated 25% of total installed capacity (EA 1999). This limitation is due to the intermittent nature of RE sources, requiring controllable, fast responding reserve capacity to compensate for fluctuations in output, and energy storage to compensate for non-availability.

To attain higher levels of new and RE systems penetration, alternative approaches will be required, including the deployment of micro power systems at the local level (e.g. micro CHP, fuel cells and RE components). Furthermore, by utilising energy efficiency and passive solar measures, a building's energy demand may be reduced and the profile of this demand reshaped to accommodate the low power densities of RE components such as photovoltaic panels and ducted wind turbines. Any energy deficit may then be met from the public electricity supply (PES) or small scale co-generation plant operating co-operatively with the local RE systems.

To facilitate the modelling of embedded RE systems, an electrical power flow model is required. Section 6.3 describes such a model and derives related models for photovoltaic components and ducted wind turbines. Section 8.3 then demonstrates an application of the integrated modelling approach concerned with the embedding of RE components within a building in a manner that facilitates co-operative operation with the PES.

Environmental impact

Buildings will typically account for around 50% of the total energy consumption in a developed country (Harris and Elliot 1997) and a similar portion of the carbon dioxide emissions. A significant additional energy consumption is associated with the production and transportation of construction materials. In the UK, for example, this amounts to around 25% of the domestic sector energy consumption (West *et al* 1994). Associated with these consumptions are gaseous emissions that can contribute to global warming (CO_2), acidification (SO_x) and ozone depletion (NO_x). The integrated performance modelling approach espoused in this book is able to address all aspects of a building's life cycle and thereby help designers to strike a balance between energy use, indoor comfort and local/global impact.

For a succinct review of the nature of the environmental impacts associated with building construction, operation and demolition, and a simulation embedded approach to the quantification of these impacts, the reader is referred to the work of Citherlet (2001). Section 8.3 considers the use of simulation to undertake a life cycle impact assessment of a building.

Uncertainty

Since all design parameters are subject to uncertainty, programs need to be able to apply uncertainty bands to their input data and automatically use these bands to determine the impact of uncertainty on likely performance. Programs so endowed will be able to assess risk, rather than merely presenting performance data to their users. Surely it is better to give the probability of overheating than to output an operative temperature profile and trust to the user's interpretive skill. Section 8.4 considers the treatment of uncertainty within a simulation-based design process.

1.5 The need for accuracy and flexibility

It is impossible to establish, *a priori*, the optimum level of model accuracy and flexibility in the field of building energy simulation. Indeed, the trade-off between accuracy and flexibility is itself a dynamic concept that will vary according to the modelling task in hand. Nevertheless, it is important to differentiate between simplified models and comprehensive models which are capable of simplified model emulation.

In the former case a number of simplifying assumptions are applied to the underlying thermal network and/or solution scheme. Invariably, some flowpaths are crudely approximated or omitted entirely. The model to result is then valid only when applied to problems that embody the same simplifications. In the latter case, a comprehensive model is designed to operate on input data ranging from simplified to detailed. This is achieved by incorporating context specific defaults to allow the inclusion of any flowpath not explicitly addressed in the input data set. This approach is substantially more flexible, with the accuracy level changing as a function of the quality of the design information supplied. To illustrate the problems associated with the former approach, Appendix B presents the results from several methods when applied to the same problem.

As a general strategy, it would seem reasonable to aim for a high level of accuracy combined with a model structure that is capable of adapting to the information available at any design stage. It is likely that a truly simple model, as perceived by a user, will be internally comprehensive in its treatment of the energy flowpaths, relying on the proper design of the interface for its operational flexibility. This is the philosophy underlying the modelling approach promoted in this book. The contention is that accurate and flexible appraisal tools can only be achieved by an approach that achieves conservation of energy whilst including all flowpaths, ensures integrity of the mathematical model *vis-à-vis* the reality, and wins acceptability through proper interface design in conjunction with rigorous validity and applicability testing.

The three pole axiom of conservation of energy, conservation of integrity and conservation of flexibility is the essential goal of the integrated modelling approach.

1.6 Energy modelling techniques

Most contemporary simulation programs are based either on response function methods or on numerical methods in finite difference or, equivalently, finite volume form. The former method is appropriate for the solution of systems of linear differential equations possessing time invariant parameters. In use, it is usual to assume a high degree of equation decoupling. Numerical methods, on the other hand, can be used to solve time varying, non-linear equation systems without need to assume equation decoupling as a computational convenience. Numerical methods are favoured for a number of reasons. First, to ensure accuracy it is essential to preserve the spatial and temporal integrity of real energy systems by arranging that whole system partial differential equation-sets be solved simultaneously at each computational time step. Second, numerical methods, unlike their response function counterpart, can handle complex flowpath interactions. Third, time varying system parameters can be accommodated. Fourth, processing frequencies can be adapted to handle so-called 'stiff' systems in which time constants vary significantly between the different parts of the problem (building fabric, HVAC components, fluid flow domains, control system elements etc).

Chapter 2 details the response function method in its time and frequency domain forms and subsequent chapters introduce the universally applicable numerical approach. Other approaches, based on regression analysis (Claux *et al* 1982, Markus *et al* 1982, Baker *et al* 1993), stochastic modelling (Jiang and Hong 1993, Palomo and Lefebvre 1995, Hanby and Dil 1995) and neural networks/genetic algorithms (Anslett and Kreider 1993, Kreider and Haberl 1994) are described elsewhere. Gough (1999) provides a succinct overview of contemporary programs and discernible development trends.

1.7 References and further reading

Alamdari F and Hammond G P 1982 Improved Data Correlations for Buoyancy-Driven Convection in Rooms *Building Services Eng. Res. and Tech.* **4**(3) 106-12

Anslett M and Kreider J F 1993 Application of Neural Networking Models to Predict Energy Use *ASHRAE Trans.* **99**(1) 505-17

Awbi H B and Hatton A 1999 Natural Convection from Heated Room Surfaces *Energy and Buildings* **30** 233-44

Baker N, Franchiotti A and Steemers K 1993 The LT Method *Daylighting in Architecture* **10** 19-22 (London: James and James)

CIBSE 1998 *Applications Manual AM11: Building Energy and Environmental Modelling ISBN 0 900953 85 3*

Citherlet S 2001 Towards the Holistic Assessment of Building Performance Based on an Integrated Simulation Approach *PhD Thesis* (Lausanne: Swiss Federal Institute of Technology)

Claux P, Franca J P, Gilles R, Pesso A, Pouget A and Raoust M 1982 *Method 5000* (Paris: Claux Pesso Raoust)

EA 1999 *Proc Workshop on Network Connection of Photovoltaic Systems* (Capenhurst: EA Technology)

Fisher D E 1995 An Experimental Investigation of Mixed Convection Heat Transfer in a Rectangular Enclosure *ASHRAE Transactions* **103**(2) 137-48

Gough M 1999 A Review of New Techniques in Building Energy and Environmental Modelling *Report for Contract BREA-42* (Garston: Building Research Establishment)

Halcrow 1987 Heat Transfer at Internal Building Surfaces *Project Report to the Energy Technology Support Unit* (Swindon: Sir William Halcrow and Partners Ltd)

Hanby V I and Dil A J 1995 Stochastic Modelling of Building Heating and Cooling Systems *Building Services Eng. Res. and Tech.* **16**(4) 199-205

Harris D J and Elliot C J 1997 Energy Accounting for Recycled Building Components *Proc. 2nd Int. Conf. on Buildings and the Environment* CIB TG8 (Paris)

Howrie J 1995 Building Modelling: An Architect's View *BEPAC Newsletter No 12* (http://www.bepac.dmu.ac.uk/)

IBPSA 1999 (http://www.ibpsa.org/)

Incropera F P and DeWitt D P 1981 *Fundamentals of Heat Transfer* (New York: Wiley)

Jiang Y and Hong T 1993 Stochastic Analysis of Building Thermal Processes *Building and Environment* **28**(4) 509-18

Khalifa A J N and Marshall R H 1990 Validation of Heat Transfer Coefficients on Interior Building Surfaces Using a Real-sized Indoor Test Cell *Int. J. Heat and Mass Transfer* **33**(10) 2219-36

Kreider J F and Haberl J S 1994 Predicting Hourly Building Energy Use: The Great Energy Predictor Shootout - Overview and Discussion of Results *ASHRAE Trans.* **100**(2) 1104-18

Kreith F 1973 *Principles of Heat Transfer (3rd edn)* (New York: Harper & Row)

Lebrun J (ed) 1992 *Proc. Systems Simulation* (Liege: University of Liege)

Mackey C O and Wright L T 1944 Periodic Heat Flow—Homogeneous Walls or Roofs *ASHVE Trans.* **50** 293

Markus T A, Clarke J A and Morris E N 1982 Climatic Severity *Occasional Paper 82/5* (Glasgow: University of Strathclyde, Dept. of Architecture)

Maver T W and Ellis J 1982 Implementation of an Energy Model Within a Multi-disciplinary Practice *Proc. CAD82* (Brighton)

MacCallum K 1993 *Private Communication*

McElroy L B, Hand J W and Strachan P A 1997 Experience from a Design Advice Service Using Simulation *Proc. Building Simulation '97* (Prague: Technical University) ISBN 80-01-01646-3

McElroy L B and Clarke J A 1999 Embedding Simulation Within Energy Sector Businesses *Proc. Building Simulation '99, September* (Kyoto)

Ozisik M N 1977 *Basic Heat Transfer* (New York: McGraw-Hill)

Palomo E and Lefebvre G 1995 Stochastic Simulations Against Deterministic Ones: Advantages and Drawbacks *Proc. Building Simulation '95* (Madison)

Paton J 1993 Dampness, Mould Growth and Children's Health *Health and Hygiene* **14** 141-4

Robinson F 1979 Investigation of the Common Assumptions Applied to Internal Surface Insolation in Buildings *BSc Thesis* (Glasgow: University of Strathclyde, Dept. of Mechanical Eng.)

Scottish Homes 1993 *Scottish Housing Condition Survey 1991* (Edinburgh: Scottish Homes)

Singh J 1995 The Built Environment and the Developing Fungi *Building Mycology* 1-21 (London: E and F Spon)

West J, Atkinson C and Howard N 1994 Embodied Energy and Carbon Dioxide for Building Materials *Proc. 1st Int. Conf. on Buildings and Environment* CIB TG8 (Watford)

Workshop 1992 *Proc. Health Implications of Fungi in Indoor Environments* (Baarn: Centraalbureau voor Schimmelcultures)

2

Integrative modelling methods

This chapter describes the theoretical basis and development background of the much popularised response function method. Both branches of the method—time and frequency response—are derived for the case of transient conduction and intra-zone energy flowpaths and, in each case, use in practice is described.

The elements of a more flexible modelling approach, based on a finite volume conservation method, are then presented as the essential introduction to chapters 3 and 4 where a building model is formulated and solved respectively. Finally, §2.5 introduces the criteria on the basis of which an answer might be obtained to the question, 'which method is best?'.

Each of the methods provide a solution to the differential equations that govern the flow of heat in solids, heat transfer at surface layers and heat exchange between connected fluid volumes. The response function approach is usually applied to differential problems of low order with time-invariant parameters whereas the numerical method is also suited to time varying problems of high order.

2.1 Response function methods

Consider figure 2.1, which shows a homogeneous, isotropic element of thickness defined by $0 < x < l$ and, at time t, temperature $\theta(x, t)$ and heat flux $q(x, t)$. Two relationships are of interest; the change of temperature with distance and the change of flux with distance:

$$\frac{\partial \theta(x,t)}{\partial x} = -\frac{1}{k}\, q(x,t) \tag{2.1}$$

$$\frac{\partial q(x,t)}{\partial x} = -\rho C\, \frac{\partial \theta(x,t)}{\partial t}\,. \tag{2.2}$$

Combination of these equations gives the usual governing partial differential heat equation, the Fourier equation, as derived from first principles in Appendix C:

$$\frac{\partial^2 \theta(x,t)}{\partial x^2} = \frac{1}{\alpha}\,\frac{\partial \theta(x,t)}{\partial t}\,. \tag{2.3}$$

where α is the thermal diffusivity as introduced in §1.4.

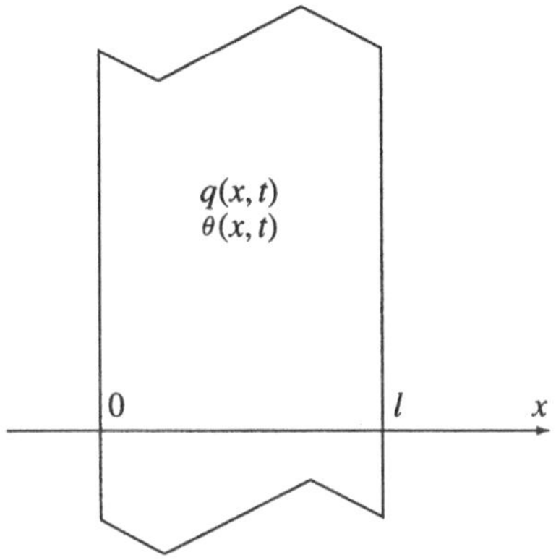

Figure 2.1: A homogeneous, isotropic element.

An analytical approach to the solution of these equations involves the use of the Laplace transformation (Carslaw and Jaeger 1959, Churchill 1958, Davies 1978). This is essentially a three stage procedure as follows.

The given equation in the time domain is transformed into a subsidiary equation in an imaginary space.

This subsidiary equation is solved by purely algebraic manipulations.

An inverse transformation is applied to this solution to obtain the solution in the time domain of the initial problem.

The interesting feature of the method is that in many cases ordinary differential equations are transformed into purely algebraic equations and partial differential equations are transformed to ordinary differential equations.

Table 2.1: Some common Laplace transform pairs.

	f(t)	f(p)
Unit impulse	$\delta(t)$	1
Unit step	$H(t)$	$1/p$
Unit ramp	t	$1/p^2$
	t^n	$n!/p^{n+1}$; n + ve integer
Delayed unit impulse	$\delta(t-\Delta)$	$e^{-p\Delta}$
Delayed unit step	$H(t-\Delta)$	$e^{-p\Delta}/p$
	e^{-at}	$1/(p+a)$
	$e^{-a(t-\Delta)}H(t-\Delta)$	$e^{-p\Delta}/(p+a)$
	te^{-at}	$1/(p+a)^2$
	$t^n e^{-at}$	$n!/(p+a)^{n+1}$
	$\sin bt$	$b/(p^2+b^2)$
	$\cos bt$	$p/(p^2+b^2)$

In practice, transforms and inverse transforms are often obtained from tables of Laplace transforms such as that shown in table 2.1 (see Healey 1967 for a more comprehensive list). With respect to the temperature variable, $\theta(x,t)$, the Laplace transform is given by

$$L[\theta(x,t)] = \theta(x,p) = \int_0^\infty e^{-pt}\theta(x,t)dt \tag{2.4}$$

where p is a complex number whose real part is positive and large enough to cause the integral to converge.

A number of theorems accompany the transform of which two are relevant here:

$$L\left[\frac{\partial\theta(x,t)}{\partial t}\right] = pL[\theta(x,t)] - \theta(x,0)$$

$$L\left[\frac{\partial^n\theta(x,t)}{\partial x^n}\right] = \frac{\partial^n L[\theta(x,t)]}{\partial x^n} .$$

The Laplacian of eqns (2.1)—(2.3) can now be written as

$$\frac{\partial\theta(x,p)}{\partial x} = -\frac{1}{k}q(x,p) \tag{2.5}$$

$$\frac{\partial q(x,p)}{\partial x} = -\rho C p\theta(x,p) + \rho C\theta(x,0) \tag{2.6}$$

$$\frac{\partial^2\theta(x,p)}{\partial x^2} = \frac{p}{\alpha}\theta(x,p) - \frac{1}{\alpha}\theta(x,0) . \tag{2.7}$$

These are the subsidiary equations which, when solved, give the Laplace transform $\theta(x,p)$ of the solution of the original equations. If $\theta(x,p)$ is found in a table of transforms then the solution for $\theta(x,t)$ may be determined immediately. If no such transform exists then $\theta(x,t)$ is determined from $\theta(x,p)$ by the inversion theorem:

$$\theta(x,t) = L^{-1}[\theta(x,p)] = \frac{1}{2\pi i}\int_{\gamma-i\infty}^{\gamma+i\infty} e^{pt}\theta(x,p)dp \tag{2.8}$$

where γ is a large number such that all the singularities of $\theta(x,p)$ lie to the left of the line $(\gamma - i\infty, \gamma + i\infty)$.

The solution of the subsidiary equations (Carslaw and Jaeger 1959) is given by

$$\theta(x,p) = \cosh[(p/\alpha)^{\frac{1}{2}}x]\theta(0,p) - \frac{\sinh[(p/\alpha)^{\frac{1}{2}}x]q(0,p)}{k(p/\alpha)^{\frac{1}{2}}}$$

$$q(x,p) = -k(p/\alpha)^{\frac{1}{2}}\sinh[(p/\alpha)^{\frac{1}{2}}x]\theta(0,p) + \cosh[(p/\alpha)^{\frac{1}{2}}x]q(0,p) .$$

It is convenient to represent these temperature and heat flux relationships in matrix notation (Pipes 1957) so that

$$\begin{bmatrix}\theta(l,p)\\ q(l,p)\end{bmatrix} = \begin{bmatrix}m_{11}(p) & m_{12}(p)\\ m_{21}(p) & m_{22}(p)\end{bmatrix} \times \begin{bmatrix}\theta(0,p)\\ q(0,p)\end{bmatrix} \tag{2.9}$$

where, by inspection, the elements of matrix **M** are given by

$$m_{11}(p) = m_{22}(p) = \cosh[(p/\alpha)^{\frac{1}{2}}l]$$

$$m_{12}(p) = -\sinh[(p/\alpha)^{\frac{1}{2}}l]/k(p/\alpha)^{\frac{1}{2}}$$

$$m_{21}(p) = -k(p/\alpha)^{\frac{1}{2}}\sinh[(p/\alpha)^{\frac{1}{2}}l]$$

and the matrix has unit determinant; that is $m_{11}m_{22} - m_{12}m_{21} = 1$.

In the terminology, **M** is the transmission matrix and its entries are the transfer functions. For composite constructions, $0 < x < L$, comprised of a number of layers in intimate contact, the formulation can be directly extended by simple matrix manipulation techniques to give

$$\begin{bmatrix} \theta(L,p) \\ q(L,p) \end{bmatrix} = \begin{bmatrix} A(p) & B(p) \\ C(p) & D(p) \end{bmatrix} \times \begin{bmatrix} \theta(0,p) \\ q(0,p) \end{bmatrix} \tag{2.10}$$

where the value of the elements A(p), B(p), C(p) and D(p) of the overall transmission matrix will depend on the properties of the component elements of the multi-layered construction and the order in which the individual element transmission matrices are combined.

For a multi-layered construction with homogeneous elements e1, e2, e3, . . , en, specified outside ($x = 0$) to inside ($x = L$), the overall transmission matrix is given by

$$\begin{bmatrix} A(p) & B(p) \\ C(p) & D(p) \end{bmatrix} = \mathbf{M}_{e1} \times \mathbf{M}_{e2} \times \mathbf{M}_{e3} \times \cdots \times \mathbf{M}_{en}$$

where, in general, $m_{11} = m_{22}$ but $A(p) \neq D(p)$.

Eqn (2.10) is the fundamental relationship underlying the time-domain and frequency-domain response function methods as described in the following sections. For the former method a rearrangement of eqn (2.10) is necessary†, which relates the flux at both surfaces to surface temperatures:

$$\begin{bmatrix} q(0,p) \\ q(L,p) \end{bmatrix} = \begin{bmatrix} D(p)/B(p) & -1/B(p) \\ 1/B(p) & -A(p)/B(p) \end{bmatrix} \times \begin{bmatrix} \theta(0,p) \\ \theta(L,p) \end{bmatrix}. \tag{2.11}$$

If two- or three-dimensional transient heat conduction is to be considered then the partial differential equation $\nabla^2\theta(x,t) = \alpha^{-1}\partial\theta(x,t)/\partial t$ can still be treated by the Laplace transform technique where the subsidiary equation to result will still be a partial differential equation but in three (space) variables instead of four (space plus time). In this way, the problem complexity is reduced.

It is at this point—the application of the inverse transform—that the time- and frequency-domain methods take on separate identities; one concerned with the response of multi-layered constructions to time-series temperature or flux pulses, the other with the response to periodic excitations of differing frequencies.

2.2 Time-domain response functions

This solution technique—concerned with the solution of eqn (2.11) in the time domain—is most commonly referred to as the response factor method. In some applications the technique can match the numerical technique although it can only be applied to an equation system which is both linear and invariable. While Mitalas (1965) has stated that such a requirement need not impose severe restrictions in a building design context, the complexity of contemporary building related issues will severely limit the method's applicability. That said, the method is

† The sign convention has first been changed by transposing L and 0 in eqn (2.10). This is done to maintain consistency with the published literature. This means that the elements of the overall transmission matrix will differ between the two equations.

capable of handling both periodic and non-periodic flux and temperature time-series and, for this reason perhaps, has enjoyed wider application, especially in North America, than the frequency-domain (or harmonic) method of §2.3.

The basic strategy is to predetermine the response of a system to some unit excitation relating to the boundary conditions anticipated in reality. A unit excitation function has a value of unity at its start and zero thereafter (i.e. $1, 0, 0, 0, \ldots$). The response of a linear, invariant equation system to this unit excitation function is termed the unit response function (URF) and the time-series representation of this URF, that is the individual terms of the series, are the response factors.

The number of URFs considered in any problem will depend on the number of combinations of excitation function (for solar radiation, dry bulb temperature, sky longwave radiation etc) and responses of interest (cooling loads, internal temperatures etc). Figure 2.2 shows a combination in which a unit change in external air temperature (the unit excitation function) produces the URF of the heat flux at the internal surface of some multi-layered construction.

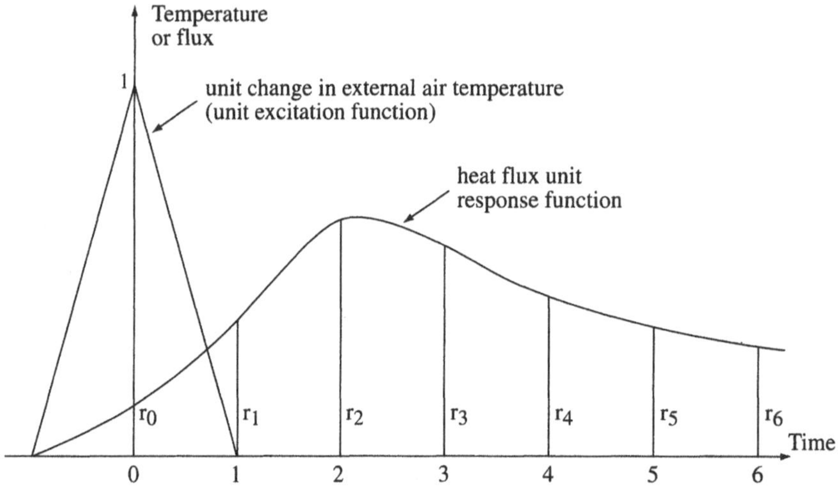

Figure 2.2: Unit excitations and response functions.

In general terms there are three steps inherent in the method after the various URFs have been determined. First, the actual excitation functions are resolved into their equivalent time-series. This can be achieved by triangular or rectangular approximation, contemporary systems favouring the former. Second, the URFs are combined with a corresponding excitation function to determine the system response. This is achieved by application of the convolution theorem, which states that the response of a linear, invariant system is given by the products of the response of the same system to a unit excitation (the URF) and the actual excitation given that the appropriate time adjustments are made. Stated mathematically:

$$R(t) = \sum_{m=0}^{\infty} RF(m\Delta)\, E(t - m\Delta) \qquad (2.12)$$

where $R(t)$ is the system response at time $t = m\Delta$ (m is an integer), $RF(m\Delta)$ is a response factor at time $m\Delta$, $E(t - m\Delta)$ is the excitation at time $(t - m\Delta)$ and Δ is the URF time-step. Finally, the individual responses from the different excitation functions are superimposed to give the overall response.

URFs are only dependent on design parameters and assumptions regarding thermophysical properties and therefore, if assumptions of time invariance are acceptable, need only be determined once for any given design. This is one of the main attractions of the method over the more generalised numerical methods where a computational exercise, equivalent to URF computation, must be implemented at each time-step as a simulation proceeds. However, should system properties vary with time, requiring that the URF be computed anew, then this computational distinction will disappear.

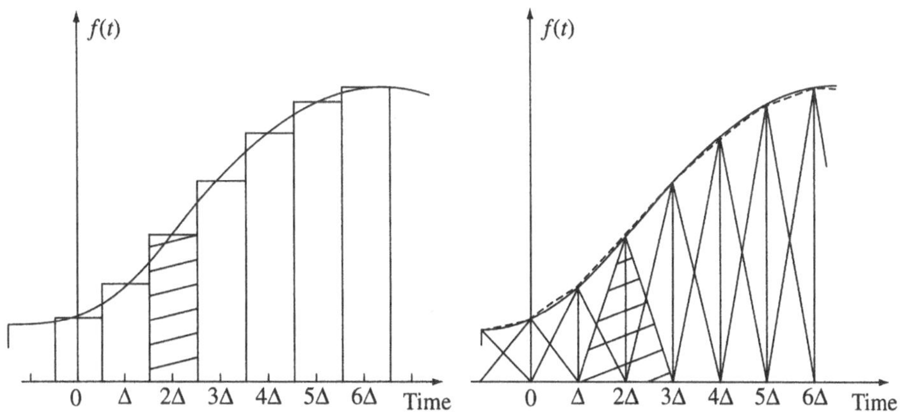

Figure 2.3: Rectangular and triangular pulse representation of a continuous function.

Pratt and Ball (1963) were among the earlier workers in the field of response function modelling. They developed a method for the calculation of room loads and temperatures using URFs derived for multi-layered constructions of up to three homogeneous elements. Stephenson and Mitalas (1967a) are largely responsible for the present day form of the method. Their formulation builds on the earlier work of Brisken and Reque (1956), who were among the first to consider response factors as a set of numbers denoting the time-series values of a URF at equally spaced intervals of time. A triangular pulse representation technique was developed in which each term in some continuous excitation function is considered as the magnitude of a triangular pulse centered at the particular time in question and with a base equal to twice the selected time-step. The summation of such overlapping triangles is equivalent to a trapezoidal approximation and represents a continuous function comprised of straight line segments. Figure 2.3 demonstrates the triangular pulse technique and, for comparative purposes, shows the rectangular pulse representation common in earlier formulations. Although for most applications triangular approximation gives a good fit, Gupta *et al* (1974) have pointed out that switched inputs such as lighting loads would be better treated by the rectangular method.

Many subsequent enhancements have been made to the response factor technique. These include the derivation of additional equations for the evaluation of interfacial temperatures and heat fluxes within multi-layered constructions (Kusuda 1969); the concept of whole building response functions (Muncey 1979); and an approach to the calculation of wall response based on an eigenfunction representation which is computationally efficient (Gough 1982).

The response factor method, like its analytical counterpart, the frequency response method, can be used to estimate the internal air temperature prevailing in an unconditioned building and the heating/cooling requirements for constant or varying internal conditions. Figure 2.4 gives the sequence of steps involved in the computation of the overall response of a building zone to variations in external and internal conditions. In most response factor implementations, the overall response is conveniently considered in two stages: the load profiles are determined relative to some fictitious and constant internal reference temperature; and any deviation of the internal temperature from the chosen reference is determined as a function of known plant operational characteristics and capacities.

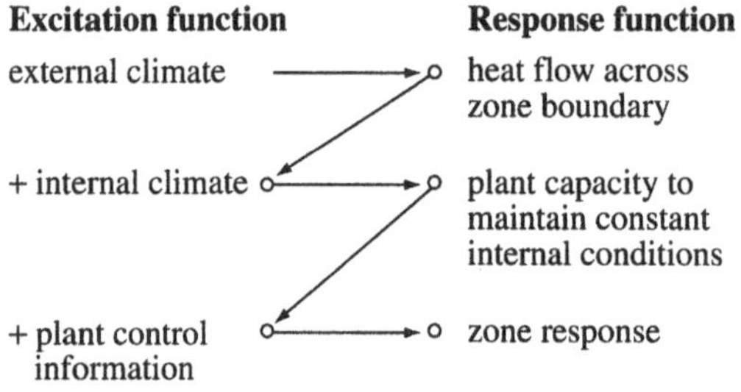

Figure 2.4: Overall zone response—sequence of steps.

In other words, the first stage is concerned with the effect of heat flow across the system boundary and subsequent internal flowpath interaction to produce a design condition plant demand. The second stage addresses the operational strategy of the installed plant.

2.2.1 Multi-layered constructions

Figure 2.5 shows the variation of heat flux at one surface of a homogeneous element due to a unit temperature pulse at either surface. The URF for such an element represents the heat flux at the innermost or outermost surfaces as caused by a unit triangular excitation applied to the other surface whilst the surface in question is held at a constant temperature.

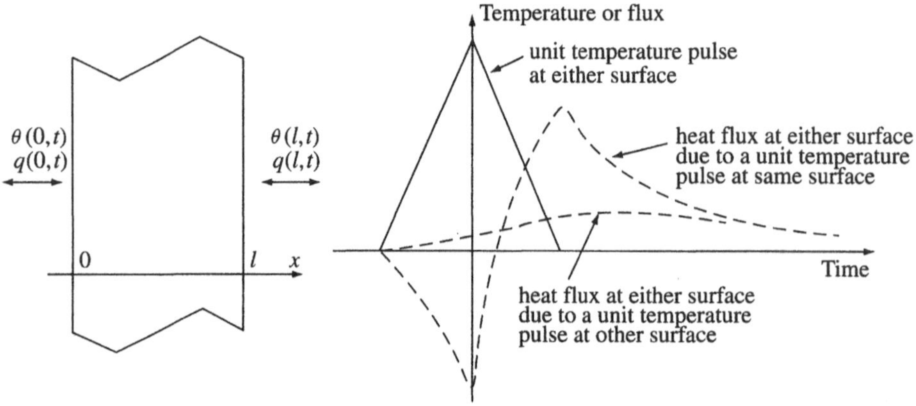

Figure 2.5: Heat flux due to a unit temperature pulse at either surface.

Recall eqn (2.11), which is the rearranged form of eqn (2.10) after L and 0 were transposed:

$$\begin{bmatrix} q(0,p) \\ q(L,p) \end{bmatrix} = \begin{bmatrix} D(p)/B(p) & -1/B(p) \\ 1/B(p) & -A(p)/B(p) \end{bmatrix} \times \begin{bmatrix} \theta(0,p) \\ \theta(L,p) \end{bmatrix}. \tag{2.13}$$

Flux unit response functions are the heat fluxes to result when first $\theta(0, p)$ then $\theta(L, p)$ is set to the Laplacian of a unit triangular pulse whilst the opposite surface temperature, $\theta(L, p)$ or $\theta(0, p)$ respectively, is held at zero. In this case three URFs will result:

$X(n\Delta)$:- the heat flux URF into the construction at $x = 0$ for a unit pulse at $x = 0$;

$Y(n\Delta)$:- the heat flux URF out of the construction at $x = 0$ or $x = L$ for a unit pulse at $x = L$ or $x = 0$ respectively;

$Z(n\Delta)$:- the heat flux URF into the construction at $x = L$ for a unit pulse at $x = L$;

where $n = 0, 1, 2, 3, 4, \ldots \infty$.

It follows from eqn (2.12) that at time $t = n\Delta$

$$q(0, n\Delta) = \sum_{m=0}^{\infty} \theta[0, (n-m)\Delta]X(m\Delta) - \sum_{m=0}^{\infty} \theta[L, (n-m)\Delta]Y(m\Delta) \tag{2.14}$$

$$q(L, n\Delta) = \sum_{m=0}^{\infty} \theta[0, (n-m)\Delta]Y(m\Delta) - \sum_{m=0}^{\infty} \theta[L, (n-m)\Delta]Z(m\Delta) \tag{2.15}$$

where $q(0, n\Delta)$ signifies heat flow into the construction at $x = 0$ and $t = n\Delta$; and $q(L, n\Delta)$ signifies heat flow out of the construction at $x = L$ and $t = n\Delta$.

In the time domain, a unit pulse can be represented by the superimposition of three ramp functions, $r(t)$, defined as

$$r(t) = \begin{cases} 0 & \text{for } t < 0 \\ t & \text{for } t \geq 0 . \end{cases}$$

Thus, a unit triangular pulse is given by

$$f(t) = r(t + \Delta) - 2r(t) + r(t - \Delta) \tag{2.16}$$

so that, for $\Delta = 1$, $f(-1) = 0$, $f(0) = 1$ and $f(1) = 0$.

The Laplacian of the ramp function $r(t) = t$ is $1/p^2$ and so, from eqn (2.13), the URF due to a ramp function at $x = 0$ with $\theta(L, p) = 0$ gives

$$q(0, p) = D(p)/p^2B(p)$$

$$q(L, p) = 1/p^2B(p)$$

and establishing the ramp function at $x = L$ with $\theta(0, p) = 0$ gives

$$q(0, p) = -\, 1/p^2B(p)$$

$$q(L, p) = -\, A(p)/p^2B(p) .$$

Application of the inverse transform of eqn (2.8) gives the heat flux time-series in the time domain: these (with appropriate signs) are the URFs $X(m\Delta)$, $Y(m\Delta)$ and $Z(m\Delta)$; that is

$$X(m\Delta) = L^{-1}[D(p)/p^2B(p)]$$

$$Y(m\Delta) = L^{-1}[1/p^2B(p)]$$

$$Z(m\Delta) = L^{-1}[A(p)/p^2B(p)] .$$

The method of residues can now be used to achieve the integration of the inverse transform. Noticing that the expressions for the URFs can each be represented by $\xi(p) = R(p)/p^2S(p)$, the residue theorem gives

$$\xi(t) = L^{-1}[\xi(p)] = \frac{R(0)}{S(0)} t + \frac{d}{dp}\left(\frac{R(p)}{S(p)}\right)\Bigg|_{p=0} + \sum_{j=1}^{\infty} \frac{R(a_j)}{a_j^2 S'(a_j)} e^{a_j t}$$

where $R(a_j)/S'(a_j)$ is the residue of the transfer function R/S at its jth singularity (or pole), a_j, and S' indicates the derivative. The poles of R/S are the zeros of S and so the task in hand is to determine the roots of $S(p) = 0$ where, in this case, $S(p) = B(p)$ as determined from the matrix analysis applied to the multi-layered construction in question (eqn (2.13)). Thus, for $X(m\Delta)$

$$\xi(t) = \frac{D(0)}{B(0)} t + \frac{d}{dp}\left(\frac{D(p)}{B(p)}\right)\Bigg|_{p=0} + \sum_{j=1}^{\infty} \frac{D(a_j)}{a_j^2 B'(a_j)} e^{a_j t} \tag{2.17}$$

with the response factor terms given, from eqn (2.16), by

$$\begin{aligned} X(0) &= \xi(\Delta) \\ X(\Delta) &= \xi(2\Delta) - 2\xi(\Delta) \\ X(m\Delta) &= \xi[(m+1)\Delta] - 2\xi(m\Delta) + \xi[(m-1)\Delta] \; ; \; m = 2, 3, 4, \ldots \end{aligned} \tag{2.18}$$

with similar expressions emerging for $Y(m\Delta)$ and $Z(m\Delta)$.

With most multi-layered constructions the elements of the overall transmission matrix are complex hyperbolic functions and determination of the roots of $B(p) = 0$ is achieved by numerical search procedures which attempt to locate a sign change in $B(p)$ as p is incremented in small steps. Hittle (1981) has proposed an efficient root finding procedure which allows larger search increments whilst still ensuring root location. This method makes use of an observation that the roots of $B(p) = 0$ are bracketed by the roots of $A(p) = 0$. Gough (1982) proposed an alternative root-finding algorithm which makes use of the fact that $B(p)$ can be expressed as a product expansion in terms of its zeros a_j.

For a homogeneous element, from eqn (2.9):

$$p^2 B(p) = p^2 \sinh[(p/\alpha)^{\frac{1}{2}} l]/[k(p/\alpha)^{\frac{1}{2}}]$$

and the roots of $\sinh[(p/\alpha)^{\frac{1}{2}} l]$ are given by

$$a_j = -j^2 \pi^2 \alpha / l^2 \, .$$

Stephenson and Mitalas (1967a) have shown that, from eqn (2.18):

$$X(0) = -kl/\alpha\Delta\left[-1/3 - \alpha\Delta/l^2 - 2/\pi^2 \sum_{j=1}^{\infty} \gamma_j/j^2\right]$$

$$X(1) = -kl/\alpha\Delta\left[1/3 + 2/\pi^2 \sum_{j=1}^{\infty} (\gamma_j^2 - 2\gamma_j)/j^2\right]$$

$$X(n) = -2kl/\alpha\Delta\pi^2 \sum_{j=1}^{\infty} [(\gamma_{j(n+1)} - 2\gamma_{j(n)} + \gamma_{j(n-1)})/j^2]; \text{ for } n \geq 2$$

$$Y(0) = -\,kl/\alpha\Delta\left[1/6 - \alpha\Delta/l^2 + 2/\pi^2 \sum_{j=1}^{\infty}(-1)^j\gamma_j/j^2\right]$$

$$Y(1) = -\,kl/\alpha\Delta\left[-1/6 + 2/\pi^2 \sum_{j=1}^{\infty}(-1)^j(\gamma_j^2 - 2\gamma_j)/j^2\right]$$

$$Y(n) = -\,2kl/\alpha\Delta\pi^2 \sum_{j=1}^{\infty}[(-1)^j(\gamma_{j(n+1)} - 2\gamma_{j(n)} + \gamma_{j(n-1)})/j^2];\ \text{for } n \geq 2 \tag{2.19}$$

where $\gamma_{j(\xi)} = \exp(-j^2\pi^2\alpha\Delta/l^2)$ evaluated at time-row ξ. Note that for a homogeneous element $A(p) = D(p)$ and so $X(m\Delta) = Z(m\Delta)$.

These unit response functions relate to some time-step Δ and can be converted to the values corresponding to a time-step of 2Δ using

$$r_{2\Delta(n)} = 0.5r_{\Delta(2n-1)} + r_{\Delta(2n)} + 0.5r_{\Delta(2n+1)}$$

where $r_{2\Delta(n)}$ is the response factor for a 2Δ time-step and at time n, and $r_{\Delta(\xi)}$ is the response factor at a Δ time-step and at time ξ.

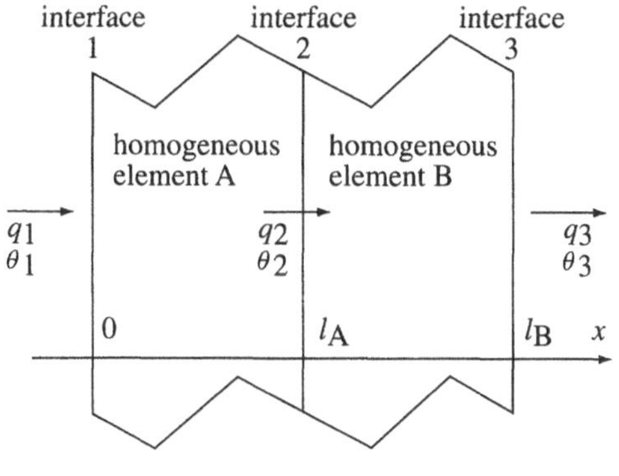

Figure 2.6: Combination of response factors.

The URFs for a multi-layered construction are found by computation of the overall transmission matrix to determine the transfer functions for use in eqn (2.17). Alternatively, the factors given by eqn (2.19) can be combined for each homogeneous element in turn. As proof of this, and with reference to figure 2.6, the heat fluxes at interfaces 1, 2 and 3 are given by

$$q_1 = X_A\theta_1 - Y_A\theta_2 \tag{2.20}$$

$$q_2 = Y_A\theta_1 - Z_A\theta_2 \tag{2.21}$$

$$= X_B\theta_2 - Y_B\theta_3 \tag{2.22}$$

$$q_3 = Y_B\theta_2 - Z_B\theta_3 \tag{2.23}$$

and from eqns (2.21) and (2.22):

$$\theta_2 = (Y_A\theta_1 + Y_B\theta_3)/(Z_A + X_B)$$

so that substitution in eqns (2.20) and (2.23) gives

$$q_1 = \left(X_A - \frac{Y_A^2}{Z_A + X_B}\right)\theta_1 - \left(\frac{Y_A Y_B}{Z_A + X_B}\right)\theta_3 = U\theta_1 - V\theta_3$$

$$q_3 = \left(\frac{Y_A Y_B}{Z_A + X_B}\right)\theta_1 - \left(Z_B - \frac{Y_B^2}{Z_A + X_B}\right)\theta_3 = V\theta_1 - W\theta_3$$

where U, V and W are the response factor time-series for the two layer composite. Continued combination in the same manner for the remaining layers gives the overall URF.

Gough (1984) has stated that this procedure is slightly suspect as it embodies an assumption of linear interpolation between sample values at every layer boundary. The errors so introduced may not be negligible for a wall composed of several layers.

Concerning the infinite series of eqns (2.17) and (2.19), on evaluation it is observed that after a finite number of terms all subsequent terms decay with a common ratio and that each successive term is given by

$$X(m\Delta) = X(m\Delta - 1)e^{a_j} .$$

In the early seventies Stephenson and Mitalas (1971) introduced a further transform, known as the z-transform from a method by Jury (1964). This allows the boundary conditions to be represented by a train of samples, with linear (or some other form of) interpolation used to convert the samples to a continuous function as with the response factor method. The results from the z-transform method are identical to those from the response factor method, for identical interpolation assumptions, but substantially quicker to obtain. In this method the heat flux at the inside surface of a multi-layered construction is given as a weighted finite sum of both input and output excitation histories:

$$q(l, t) = \sum_{m=0}^{N} E(m\Delta)\theta(0, t - m\Delta) - \sum_{m=1}^{M} H(m\Delta)q(l, t - m\Delta)$$

where E and H are z-transfer functions derived from consideration of the ratio of input and output time-series. They are related to the p-transfer functions $X(m\Delta)$, $Y(m\Delta)$ and $Z(m\Delta)$ by

$$E(m\Delta) = \sum_{j=0}^{N} X(j\Delta)H(m\Delta - j\Delta)$$

$$F(m\Delta) = \sum_{j=0}^{N} Y(j\Delta)H(m\Delta - j\Delta)$$

$$G(m\Delta) = \sum_{j=0}^{N} Z(j\Delta)H(m\Delta - j\Delta)$$

and $H(m\Delta)$ $(m = 1, 2, 3, \ldots)$ is obtained as the coefficient of Z^{-m} in the expansion

$$\sum_{m=0}^{\infty} H(m\Delta)Z^{-m} = \prod_{m=1}^{\infty}(1 - Z^{-m}e^{a_j\Delta})$$

where a_j are the poles of B(p) as in the formulation of eqn (2.17).

2.2.2 Zone energy balance

For a given zone, in order to determine the overall URF under any unit excitation, it is necessary to consider the entire energy field as an integrated system because of the inter-relationship of the various heat flowpaths. This can be done by formulating energy balance equations for each major region within the zone, to effectively link all regions over space and time. The simultaneous solution of this equation-set, when subjected to some unit excitation function with all other excitations set at zero, gives the corresponding URF. Repetition of this procedure for each unit excitation gives the complete URF set from which the total zone response is determined.

Consider figure 2.7 which shows an elementary zone comprised of six opaque multi-layered constructions, one single glazed window and one fluid volume representing the thoroughly mixed internal air. Surface 1 is an external construction with internal surface denoted by 1i and an external surface by 1e. Since the thermal storage effects of the glazing (surface 7) can be assumed negligible, and since the glass resistance contributes little to the overall resistance (glass + boundary layers), then the inside and outside surface temperatures are here considered to be equal and so only a one surface representation is required. This restriction is easy to remove as demonstrated in chapter 3. The remaining external surfaces, 2e – 6e, are assumed to make contact with adjacent zones of known temperature θ_{A2} through θ_{A6} respectively. Explicit matrix equation representation for systems including furnishings, multiple glazings, multi-zone geometries etc are given in chapters 3 and 4.

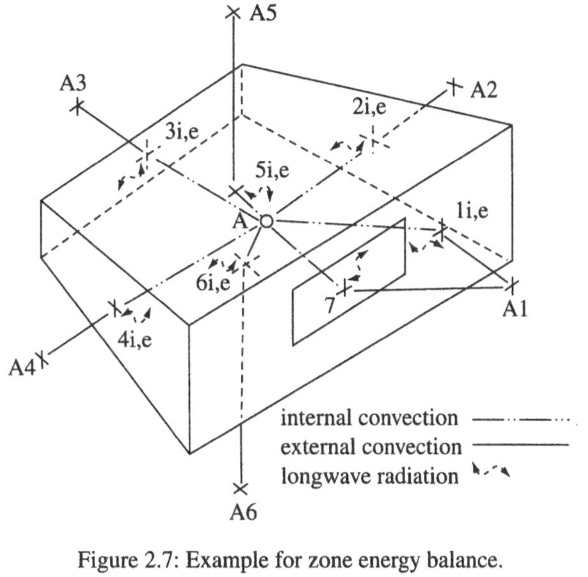

Figure 2.7: Example for zone energy balance.

Assuming that internal surfaces have no associated thermal capacity, since all structural capacity effects are included in the response functions of §2.2.1, then the surface heat balance is given by

$$q_c(t) + q_R(t) + q_K(t) + q_S(t) + q_M(t) = 0 \tag{2.24}$$

where $q_c(t)$ is the surface convection, $q_R(t)$ is the longwave radiation gain from surrounding surfaces, $q_K(t)$ is the uni-directional conduction flux at the surface, $q_S(t)$ is the shortwave radiation excitation and $q_M(t)$ is the summation of the miscellaneous radiant excitations from lights, appliances etc. All flux terms have units of Wm^{-2} and t is time.

The convective flux is given by

$$q_c(t) = h_{cs}[\theta_x(t) - \theta_s(t)] \tag{2.25}$$

where h_{cs} is the time invariant (or time averaged) convective heat transfer coefficient for surface s (W m^{-2}°C^{-1}), $\theta_x(t)$ the adjacent air temperature (°C) (equal to $\theta_A(t)$ for an inside surface and $\theta_{Aj}(t)$ for outermost surface j), and $\theta_s(t)$ is the surface temperature (°C).

The longwave radiation gain is given by

$$q_R(t) = \sum_{j=1}^{N} h_{rj,s}[\theta_j(t) - \theta_s(t)] \tag{2.26}$$

where N is the total number of surfaces in thermal communication, $h_{rj,s}$ is the linearised radiation coefficient between surface j and the surface s under consideration (W m^{-2}°C^{-1}), and $\theta_j(t)$ is the temperature of the communicating surface (°C).

For outermost surfaces it is convenient here to combine radiative and convective exchanges into a single term relating to the adjacent air layer. This is necessary since, in this formulation, no external or adjacent zone surface temperature information is available. The heat flux for an outermost surface is then given by

$$q_c(t) + q_R(t) = h_{Ts}[\theta_{Aj}(t) - \theta_s(t)] \tag{2.27}$$

where h_{Ts} is the combined heat transfer coefficient related to the air temperature (W m^{-2}°C^{-1}).

Sections 7.5 and 7.6 cover the evaluation of time dependent h_r and h_c values respectively for use in the numerical method formulated in chapter 3, and chapter 4 describes a matrix equation organisation, partitioning and processing technique which allows treatment of combined multi-zone problems and permits the removal of the assumption of linearity applied to the non-linear longwave radiation flowpath.

The conduction heat flux at an inside surface is dependent on the history of temperature and heat flux variations at the exposed surfaces of the multi-layered construction and the thermophysical properties of the individual layers. Section 2.2.1 demonstrated URF formulation for transient conduction within multi-layered constructions which, when combined with the surface temperature time-series, gives the conduction heat flux. Thus, from eqn (2.15):

$$q_K(t) = -\sum_{m=0}^{\infty} \theta_{si}(t - m\Delta)Z(m\Delta) + \sum_{m=0}^{\infty} \theta_{se}(t - m\Delta)Y(m\Delta) \tag{2.28}$$

where m is the time-step index measured from commencement (m = 0) of the application of some unit excitation, $Z(m\Delta)$ represents the response factor terms of the inside surface URF due to a unit temperature excitation at the inside surface (W m^{-2}°C^{-1}), $Y(m\Delta)$ are the response factor terms of the inside surface URF due to a unit temperature excitation at the outside surface, and si, se denote inside and outermost surfaces respectively.

Because of the dependence of heat flux on the relative position of component elements within composite constructions, the flux at an outermost surface is given, from eqn (2.14), by

$$q_K(t) = \sum_{m=0}^{\infty} \theta_{se}(t - m\Delta)X(m\Delta) - \sum_{m=0}^{\infty} \theta_{si}(t - m\Delta)Y(m\Delta) \tag{2.29}$$

where $X(m\Delta)$ are the response factor terms of the outermost surface URF due to a unit temperature excitation at the outermost surface and $Y(m\Delta)$ are the response factor terms of the outermost surface URF due to a unit temperature excitation at the inside surface (both W m^{-2}°C^{-1}).

Section 7.4 describes a technique for the estimation of the internal (and external) shortwave radiation distribution, $q_S(t)$, and §7.7 discusses the treatment of the convective and radiant portions of casual heat sources, $q_M(t)$.

Substituting eqns (2.25)-(2.29) into eqn (2.24) and collecting all unknown terms on the equation left-hand side gives, for an internal surface

$$-\left(h_{cs} + \sum_{j=1}^{N} h_{rj,s} + Z(0)\right)\theta_{si}(t) + Y(0)\theta_{se}(t) + \sum_{j=1}^{N} h_{rj,s}\theta_j(t)$$

$$= -\, h_{cs}\theta_A(t) - q_s(t) - q_M(t) + \sum_{m=1}^{\infty} \theta_{si}(t - m\Delta)Z(m\Delta)$$

$$- \sum_{m=1}^{\infty} \theta_{si}(t - m\Delta)Y(m\Delta) \qquad (2.30)$$

and for an outermost surface

$$-\,[h_{Ts} - x(0)]\theta_{se}(t) - Y(0)\theta_{si}(t)$$

$$= -\, h_{Ts}\theta_{As}(t) - q_S(t) - q_M(t) - \sum_{m=1}^{\infty} \theta_{se}(t - m\Delta)X(m\Delta)$$

$$+ \sum_{m=1}^{\infty} \theta_{si}(t - m\Delta)Y(m\Delta)\,. \qquad (2.31)$$

In forming these relationships the assumption is made that shortwave distribution and miscellaneous radiant gains can be independently assessed for all time-rows. Also, at this stage in the procedure, the internal air temperature is held constant at some design value and so is prescribed for all time. Heat balance applied to the internal air volume yields

$$\sum_{j=1}^{N} A_j h_{cj}[\theta_{sj}(t) - \theta_A(t)] + \frac{\rho CV}{\delta t}[\theta_A(t) - \theta_A(t - \delta t)] + q_p(t) + q_L(t) + q_I(t) = 0 \qquad (2.32)$$

where δt is the calculation time-step (s), A_j the area of surface j (m^2), ρ the air density (kg m^{-3}), C the air specific heat (J $kg^{-1}{}^{\circ}C^{-1}$), V the enclosed volume (m^3), q_p the convective plant requirement to maintain the design temperature (W), $q_L(t)$ the convective heat gain from casual sources such as lights (W), and $q_I(t)$ the advective load due to infiltration and zone coupled air flow (W).

The heat balance relationships of eqns (2.30) and (2.31) can now be applied to each internal and external surface in turn and the complete equation-set (including the window heat balance equation) represented in standard matrix notation (figure 2.8) as

$$\mathbf{A} \times \theta = \mathbf{B} \qquad (2.33)$$

where **A** is a non-homogeneous matrix of future time-row coefficients of the surface temperature terms as represented by the left-hand side of eqns (2.30) and (2.31), θ is a column matrix (vector) of future time-row (t) surface temperatures, and **B** is a column matrix formulated from the terms that are dependent on the excitation components at time t and the entire surface temperature history from the start of the solution stream to the previous time-step. Note that in practice only a finite number of terms need be considered (see §2.2.3). The contents of the **B** matrix is defined by the right-hand side of eqns (2.30) and (2.31).

The solution of eqn (2.33) is now obtained as

$$\theta = \mathbf{A}^{-1} \times \mathbf{B} \qquad (2.34)$$

where $\mathbf{A}^{-1}$ is unchanged with time and so only the **B** matrix need be reformulated at each new time-step using the latest values determined at the previous time-row. The objective then is to obtain the surface temperature URFs from eqn (2.34) and hence the cooling and/or heating

$$
\mathbf{A}
$$

$$
\begin{bmatrix}
-h_{c1}-\sum_{j=1}^{7} h_{rj,1}-z_1(0) & h_{r2,1} & h_{r3,1} & h_{r4,1} & h_{r5,1} & h_{r6,1} & h_{r7,1} \\
h_{r1,2} & -h_{c2}-\sum_{j=1}^{7} h_{rj,2}-z_2(0) & h_{r3,2} & h_{r4,2} & h_{r5,2} & h_{r6,2} & h_{r7,2} \\
h_{r1,3} & h_{r2,3} & -h_{c3}-\sum_{j=1}^{7} h_{rj,3}-z_3(0) & h_{r4,3} & h_{r5,3} & h_{r6,3} & h_{r7,3} \\
h_{r1,4} & h_{r2,4} & h_{r3,4} & -h_{c4}-\sum_{j=1}^{7} h_{rj,4}-z_4(0) & h_{r3,4} & h_{r6,4} & h_{r7,4} \\
h_{r1,5} & h_{r2,5} & h_{r3,5} & h_{r4,5} & -h_{c5}-\sum_{j=1}^{7} h_{rj,5}-z_5(0) & h_{r6,5} & h_{r7,5} \\
h_{r1,6} & h_{r2,6} & h_{r3,6} & h_{r4,6} & h_{r5,6} & -h_{c6}-\sum_{j=1}^{7} h_{rj,6}-z_6(0) & h_{r7,6} \\
h_{r1,7} & h_{r2,7} & h_{r3,7} & h_{r4,7} & h_{r5,7} & h_{r6,7} & -h_{c7}-\sum_{j=1}^{7} h_{rj,7}-z_7(0) \\
-y_1(0) & & & & & & \\
 & -y_2(0) & & & & & \\
 & & -y_3(0) & & & & \\
 & & & -y_4(0) & & & \\
 & & & & -y_5(0) & & \\
 & & & & & -y_6(0) &
\end{bmatrix}
$$

(continued)

Figure 2.8: The matrix equation $\mathbf{A}\theta = \mathbf{B}$ (first part).

$$
\underbrace{\begin{bmatrix}
-y_1(0) & & & & & \\
 & -y_2(0) & & & & \\
 & & -y_3(0) & & & \\
 & & & -y_4(0) & & \\
 & & & & -y_5(0) & \\
 & & & & & -y_6(0) \\
 & & & & & \\
-h_{r1}+x_1(0) & & & & & \\
 & -h_{r2}+x_2(0) & & & & \\
 & & -h_{r3}+x_3(0) & & & \\
 & & & -h_{r4}+x_4(0) & & \\
 & & & & -h_{r5}+x_5(0) & \\
 & & & & & -h_{r6}+x_6(0)
\end{bmatrix}}_{\mathbf{A}\text{ (continued)}}
\mathbf{X}
\underbrace{\begin{bmatrix}
\theta_{1i} \\ \theta_{2i} \\ \theta_{3i} \\ \theta_{4i} \\ \theta_{5i} \\ \theta_{6i} \\ \theta_{7i} \\ \theta_{1e} \\ \theta_{2e} \\ \theta_{3e} \\ \theta_{4e} \\ \theta_{5e} \\ \theta_{6e}
\end{bmatrix}}_{\theta}
=
\underbrace{\begin{bmatrix}
-h_{c1}\theta_A(t) + \sum_{m=1}^{\infty} \theta_{1i}(t-m)z_1(m) - \sum_{m=1}^{\infty} \theta_{1e}(t-m)y_1(m) - q_{E,1i}(t) \\
-h_{c2}\theta_A(t) + \sum_{m=1}^{\infty} \theta_{2i}(t-m)z_2(m) - \sum_{m=1}^{\infty} \theta_{2e}(t-m)y_2(m) - q_{E,2i}(t) \\
-h_{c3}\theta_A(t) + \sum_{m=1}^{\infty} \theta_{3i}(t-m)z_3(m) - \sum_{m=1}^{\infty} \theta_{3e}(t-m)y_3(m) - q_{E,3i}(t) \\
-h_{c4}\theta_A(t) + \sum_{m=1}^{\infty} \theta_{4i}(t-m)z_4(m) - \sum_{m=1}^{\infty} \theta_{4e}(t-m)y_4(m) - q_{E,4i}(t) \\
-h_{c5}\theta_A(t) + \sum_{m=1}^{\infty} \theta_{5i}(t-m)z_5(m) - \sum_{m=1}^{\infty} \theta_{5e}(t-m)y_5(m) - q_{E,5i}(t) \\
-h_{c6}\theta_A(t) + \sum_{m=1}^{\infty} \theta_{6i}(t-m)z_6(m) - \sum_{m=1}^{\infty} \theta_{6e}(t-m)y_6(m) - q_{E,6i}(t) \\
-h_{c7}\theta_A(t) + q_{E,7}(t) \\
+\sum_{m=1}^{\infty} \theta_{1e}(t-m)x_1(m) + \sum_{m=1}^{\infty} \theta_{1i}(t-m)y_1(m) - q_{E.1e}(t) \\
+\sum_{m=1}^{\infty} \theta_{2e}(t-m)x_2(m) + \sum_{m=1}^{\infty} \theta_{2i}(t-m)y_2(m) - q_{E.2e}(t) \\
+\sum_{m=1}^{\infty} \theta_{3e}(t-m)x_3(m) + \sum_{m=1}^{\infty} \theta_{3i}(t-m)y_3(m) - q_{E.3e}(t) \\
+\sum_{m=1}^{\infty} \theta_{4e}(t-m)x_4(m) + \sum_{m=1}^{\infty} \theta_{4i}(t-m)y_4(m) - q_{E.4e}(t) \\
+\sum_{m=1}^{\infty} \theta_{5e}(t-m)x_5(m) + \sum_{m=1}^{\infty} \theta_{5i}(t-m)y_5(m) - q_{E.5e}(t) \\
+\sum_{m=1}^{\infty} \theta_{6e}(t-m)x_6(m) + \sum_{m=1}^{\infty} \theta_{6i}(t-m)y_6(m) - q_{E.6e}(t)
\end{bmatrix}}_{\mathbf{B}}
$$

Figure 2.8: The matrix equation $\mathbf{A}\theta = \mathbf{B}$ (second part).

URFs from eqn (2.32).

Returning to the single zone problem, figure 2.8 gives the contents of the **A**, θ and **B** matrices where

$q_{E,si}(t)$ = internal surface excitation at time t = $q_S(t) + q_M(t)$

$q_{E,7}(t)$ = window surface excitation at time t = $q_S(t) + q_M(t) + h_{T7}\theta_{A1}$

$q_{E,se}(t)$ = outermost surface excitation at time t = $q_S(t) + q_M(t) + h_{Ts}\theta_{As}$.

Once established, eqn (2.34) is solved at each successive time-step by re-establishing the **B** matrix from previously computed temperature values. This gives the solution for $\theta_s(t)$, the surface temperature response factors for any given unit excitation function. The strategy is to let one excitation component assume a unit time-series variation (1,0,0,0,...) whilst all other excitations are held at zero (0,0,0,0,...) and the zone air temperature is held constant at the reference temperature. The contents of the **B** matrix are then evaluated. Subsequent post-multiplication of the (constant) inverted **A** matrix by the **B** matrix gives the surface temperatures prevailing at time t = 0. The surface temperature history—at this stage for the first time-step only—is then used to re-evaluate the **B** matrix at time t = 1 and the surface temperatures prevailing at the next time-step determined. The process then continues until the ratio of successive terms in the respective surface temperature time-series becomes constant. Continuation of this geometric progression, for as many terms as desired, gives the surface temperature URF. The entire process is then repeated for all other unit excitations of interest. Plant capacity response factors corresponding to any unit excitation function are then given by:

$$P_r(t) = -\sum_{j=1}^{N} A_j h_{cj} [S_{rj}(t) - \theta_A(t)]$$

where $P_r(t)$ is the plant capacity response factor at time t and corresponding to some unit excitation function (W), and $S_{rj}(t)$ the surface temperature response factor at time t for surface j and corresponding to some unit excitation (°C).

Note that there are as many plant capacity URFs as unit excitations of interest—e.g. external air temperature and solar radiation. When combined with the appropriate surface excitation time-series, the plant capacity response factors yield the heat transfer to or from the zone. This is the amount of heat that must be removed or added to the air to maintain constant temperature levels.

In practice, zone air temperatures will deviate from the design value due to intermittent plant operation and the unavoidable temporal mismatch between load and extraction (or addition) rates caused by control system action and equipment inertia.

For an internal air volume, eqn (2.32) can be rewritten as

$$\sum_{j=1}^{N} A_j h_{cj} \theta_{sj}(t) - \sum_{j=1}^{N} A_j h_{cj} \theta_A(t) + \frac{\rho C V}{\delta t} \theta_A(t)$$

$$= \frac{\rho C V}{\delta t} \theta_A(t - \delta t) - q_p(t) - q_L(t) - q_I(t)$$

This equation can now be incorporated in the matrix equation of figure 2.8 so that, for each unit excitation considered, a corresponding air temperature URF will emerge. Alternatively, the equation system can be subjected to a unit time-series variation in internal air temperature with all other excitations held at zero. In each case the surface temperature URFs give the plant capacity response factors. These are the zone weighting factors W(mΔ).

2.2.3 Response function application

The total heating or cooling load, Q(t), is determined by modifying the individual heat loss and gain flowpaths by the corresponding weighting factors:

$$Q(t) = \sum_{m=0}^{\infty} \sum_{j=1}^{N} W_j(m\Delta) H_j(t - m\Delta) \tag{2.35}$$

where $W_j(m\Delta)$ is the weighting factor for flowpath j and $H_j(t - m\Delta)$ is the heat gain or loss due to flowpath j (W). Kimura (1977) gives a detailed description of the algorithmic counterpart of this process as embodied in the SHASE program (Saito and Kimura 1974).

In applying construction response or zone weighting factors, it is necessary to operate with finite summations rather than with the infinite summations of §2.2.1 and §2.2.2. Eqn (2.35) can be rewritten (for any flowpath j) to give

$$Q(t) = \sum_{m=0}^{k} W(m\Delta) H(t - m\Delta) + \sum_{m=1}^{\infty} C^{m\Delta} W(k) H(t - k - m\Delta) \tag{2.36}$$

where C is the common ratio of $W(m\Delta)$ so that $C^{m\Delta}W(k)$ defines the time-series after the kth term where the geometric progression is assumed to commence. Also, from eqn (2.36), the expression for $Q(t - 1)$ is given by

$$Q(t-1) = \sum_{m=0}^{k} W(m\Delta) H(t - 1 - m\Delta) + \sum_{m=1}^{\infty} C^{m\Delta} W(k) H(t - 1 - k - m\Delta)\,. \tag{2.37}$$

Multiplying eqn (2.37) by C and subtracting from eqn (2.36) gives the expression containing only a finite summation:

$$Q(t) = CQ(t-1) + W(0)H(t) + \sum_{m=1}^{k} W(m\Delta - 1) H(t - \Delta)\,. \tag{2.38}$$

The final zone air temperature will depend on the design temperature, the cooling or heating load corresponding to this design temperature and the actual heat extraction or addition rate. For a cooling unit with a simple proportional control system characterised by

$$E = C_d + D\delta(t)$$

where E is the actual rate of heat extraction (W), C_d the plant capacity when operating at the design temperature (W) and D the change in plant potential caused by a one degree rise in air temperature (W °C^{-1}), Stephenson and Mitalas (1967b) have shown that the deviation of the actual air temperature from the design value can be found from

$$\delta(t) = \frac{L_d(t) - C_d + \sum_{m=1}^{k} W_1(m\Delta)\delta(t - m\Delta) - \sum_{m=1}^{k} W_2(m\Delta)[E(t - m\Delta) - L_d(t - m\Delta)]}{[D - W_1(0)]}$$

where $\delta(t)$ is the deviation of the air temperature from the assumed design value (°C), $L_d(t)$ the capacity demand relative to the design temperature (W), and W_1, W_2 are appropriate weighting factors.

Kimura (1977) gives an alternative formulation for the calculation of zone temperature during plant OFF times:

$$\theta(t) = \frac{-1}{W_T(0) + K(t)} \left(\sum_{m=1}^{\infty} W_T(m\Delta)\theta(t - m\Delta) + L_d(t) \right)$$

where W_T is the total weighting function ($= \sum W_j$), and K(t) is the infiltration conductance coefficient.

Several computer modelling systems exist that are based on the conduction transfer and zone weighting factor approach and the interested reader is directed elsewhere for further details on technique implementation (ASHRAE 1975, Mitalas and Arseneault 1970, Hittle 1979, York and Tucker 1980).

2.3 Frequency domain response functions

The fundamental assumption underlying the frequency domain (or harmonic) method is that weather time-series can be represented by a series of periodic cycles. In this way the weather's influence can be represented by a steady state term accompanied by a number of sine wave harmonics with, in general, increasing frequency and reducing amplitude. The division of weather time-series into component sinusoidal variations about a mean condition is achieved by Fourier series representation by which a continuous function, f(t), can be approximated by a series of sine and cosine functions:

$$f(t) = a_0 + \sum_{m=1}^{k} a_m \sin(2\pi mt/L) + \sum_{m=1}^{k} b_m \cos(2\pi mt/L)$$

where $1/L$ is the fundamental frequency (Hz) and t is time (s).

Each selected harmonic can then be processed separately and modified by thermal response factors appropriate to its frequency—factors that can be mathematically determined from eqn (2.10) only because the boundary condition has been prescribed as a sine wave in the first instance. The principle of superimposition (Stephenson and Mitalas 1967b) is then invoked to allow the system response to be obtained by summing the individual effects of the separate harmonics with respect to the mean condition.

For convenience, the frequency of the fundamental harmonic is often set at 24 hours with the remaining harmonics having diminishing periods such as 12, 6, 3, 1.5 hours and so on. As a function of harmonic frequency, the thermal factors can be determined and applied, as operators, to the individual terms of the corresponding weather harmonic.

The method can handle many of the energy transfers within buildings although some may be crudely approximated. For example, longwave radiation exchanges are normally handled by the environmental temperature method (Danter 1973); casual gains are subject to Gibbs' Phenomenon (Gower and Baker 1974) since their usually square profile will be 'clipped' when represented by a finite family of sine waves; and internal solar models are typically crude. Also, by accepting the principle of superimposition, it is difficult to preserve the integrity of strongly coupled domains such as the building and its HVAC systems.

Much of the early work on harmonic prediction techniques (Alford *et al* 1939, Mackey and Wright 1944, 1946) concentrated on the estimation of fabric heat flow under assumptions of constant internal temperatures. External air temperatures were combined with the prevailing shortwave solar radiation to give a single sol-air temperature term, and internal radiant loads (solar and radiant casual) were considered as instantaneous fluxes applied directly at the air point—an assumption no longer deemed valid since no account is taken of the interaction with fabric thermal capacity.

Later workers removed the limitation imposed by the assumption of constant internal air temperature (Nottage and Parmelee 1955, Muncey 1953) and increased the accuracy of internal surface radiative and convective modelling (Pipes 1957, Gupta 1964). To render the method amenable to manual use, Danter (1960) developed the 'means and swings' variant. In its simplest form, this was adopted by the Chartered Institute of Building Services Engineers (CIBSE) and is commonly referred to as the *admittance method* (Loudon 1968). The method allows the estimation of flux transfers under steady cyclic (periodic) conditions, where external flux or temperature variations are repeated over a period of time. Significantly, it relies only on thermal response factors relating to some fundamental harmonic (usually taken as the 24 hour period harmonic) but applied to the actual temperature or flux excitation.

2.3.1 Multi-layered constructions

Recall the matrix relationship of eqn (2.10), which expresses the Laplace transforms of the temperature and flux at one extreme boundary of a multi-layered construction in terms of the corresponding transforms at the other extreme boundary:

$$\begin{bmatrix} \theta(L,p) \\ q(L,p) \end{bmatrix} = \begin{bmatrix} A(p) & B(p) \\ C(p) & D(p) \end{bmatrix} \times \begin{bmatrix} \theta(0,p) \\ q(0,p) \end{bmatrix}. \tag{2.39}$$

Expanding the overall transmission matrix to include boundary layers of combined convective/radiative resistance, R_L and R_0 ($m^2 °C\ W^{-1}$), gives

$$\begin{bmatrix} E(p) & F(p) \\ G(p) & H(p) \end{bmatrix} = \begin{bmatrix} 1 & R_L \\ 0 & 1 \end{bmatrix} \times \begin{bmatrix} A(p) & B(p) \\ C(p) & D(p) \end{bmatrix} \times \begin{bmatrix} 1 & R_0 \\ 0 & 1 \end{bmatrix} \tag{2.40}$$

since, as $\rho C \to 0$

$$\begin{bmatrix} m_{11} & m_{12} \\ m_{21} & m_{22} \end{bmatrix} = \begin{bmatrix} 1 & R \\ 0 & 1 \end{bmatrix}.$$

Eqn (2.39) now becomes

$$\begin{bmatrix} \theta(L,p) \\ q(L,p) \end{bmatrix} = \begin{bmatrix} A(p)+R_L C(p) & R_0 A(p)+R_L R_0 C(p)+B(p)+R_L D(p) \\ C(p) & R_0 C(p)+D(p) \end{bmatrix} \times \begin{bmatrix} \theta(0,p) \\ q(0,p) \end{bmatrix}$$

$$= \begin{bmatrix} E(p) & F(p) \\ G(p) & H(p) \end{bmatrix} \times \begin{bmatrix} \theta(0,p) \\ q(0,p) \end{bmatrix}. \tag{2.41}$$

By similar reasoning the transforms of the temperature and heat flux at $x = 0$ are given by

$$\begin{bmatrix} \theta(0,p) \\ q(0,p) \end{bmatrix} = \begin{bmatrix} A(p)+R_0 C(p) & R_L A(p)+R_0 R_L C(p)+B(p)+R_0 D(p) \\ C(p) & R_L C(p)+D(p) \end{bmatrix} \times \begin{bmatrix} \theta(L,p) \\ q(L,p) \end{bmatrix}$$

$$= \begin{bmatrix} I(p) & J(p) \\ K(p) & L(p) \end{bmatrix} \times \begin{bmatrix} \theta(L,p) \\ q(L,p) \end{bmatrix}. \tag{2.42}$$

Both matrix equations involve four unknown values and, in order to permit solution, some assumptions must be made concerning the problem boundary conditions. Setting one boundary surface temperature harmonic to zero while the other surface is subjected to a periodic temperature state defined by $\sin(\omega t)$, where $\omega = 2\pi/L$ is the angular frequency of the temperature variation, gives the frequency response function of the multi-layered construction. That is, for $\theta(0, p) = 0$, eqn (2.41) yields

$$\theta(L, p) = F(p)q(0, p)$$

$$q(L, p) = H(p)q(0, p) \tag{2.43}$$

and for $\theta(L, p) = 0$, eqn (2.42) yields

$$\theta(0, p) = J(p)q(L, p)$$

$$q(0, p) = L(p)q(L, p). \tag{2.44}$$

Now if $p = j\omega$, where j is the imaginary operator defined by $j^2 = -1$, then the Laplace transform becomes the Fourier transform, for which the inverse transforms of eqns (2.43) and (2.44) are

$$\theta_h(L, t) = F(j\omega)q_h(0, t) \tag{2.45}$$

$$q_h(L, t) = H(j\omega)q_h(0, t) \tag{2.46}$$

$$\theta_h(0, t) = J(j\omega)q_h(L, t) \tag{2.47}$$

$$q_h(0, t) = L(j\omega)q_h(L, t) \tag{2.48}$$

where θ_h and q_h are the temperature and flux harmonics. The matrix elements F, H, J and L are determined from the basic homogeneous element relationships with $p = j\omega$; that is, for a homogeneous element under steady periodic conditions, eqn (2.9) can be rewritten as

$$\begin{bmatrix} \theta_h(l, t) \\ q_h(l, t) \end{bmatrix} = \begin{bmatrix} m_{11}(j\omega) & m_{12}(j\omega) \\ m_{21}(j\omega) & m_{22}(j\omega) \end{bmatrix} \times \begin{bmatrix} \theta_h(0, t) \\ q_h(0, t) \end{bmatrix}$$

where

$$m_{11}(j\omega) = m_{22}(j\omega) = \cosh[(j\omega/\alpha)^{\frac{1}{2}}l]$$

$$m_{12}(j\omega) = -\sinh[(j\omega/\alpha)^{\frac{1}{2}}l]/[k(j\omega/\alpha)^{\frac{1}{2}}]$$

$$m_{21}(j\omega) = -k(j\omega/\alpha)^{\frac{1}{2}}\sinh[(j\omega/\alpha)^{\frac{1}{2}}l]$$

and the overall transmission matrix can be established by the usual matrix multiplication process.

Recalling the earlier assumption that $\theta(x, p) = 0$, it should be appreciated that this corresponds to setting the surface (or adjacent air) temperature harmonic to zero. This implies that the temperature is held constant but not necessarily at zero.

Thus, from eqn (2.45), the flux at surface $x = 0$ due to a steady periodic condition $\theta_h(L, t) = \sin \omega t$ is given by

$$q_h(0, t) = \sin \omega t/F(j\omega)$$

and, at surface $x = L$ due to $\theta_h(0, t) = \sin \omega t$, from eqn (2.47), by

$$q_h(L, t) = \sin \omega t/J(j\omega)$$

where F and J are complex numbers and $q_h(0, t)$ and $q_h(L, t)$ are the heat fluxes at the outermost and innermost boundaries, corresponding to unit sinusoidal, opposite boundary, temperature variations of angular frequency ω.

In the 'means and swings' technique three separate response factors have been identified:

decrement response, surface response and admittance response. Each of these factors possess a corresponding phase angle that determines the response time of the process the factor addresses; that is, the time delay between cause and effect (where the phase angle is positive).

The *decrement response factor* is defined as the ratio of the cyclic flux transmission to the steady state flux transmission and is applied to fluctuations (about the mean) in external temperature or flux impinging on exposed opaque surfaces. This gives the related fluctuation within a building at some later point in time depending on the decrement factor time lag. Figure 2.9 shows a building element exposed to a sinusoidal external air temperature or solar radiation time-series. The corresponding cyclic heat flux at the inside surface (x = 0) is also shown after time-series modification by the decrement factor of the intermediate element (or series of elements) and application of the appropriate time shift.

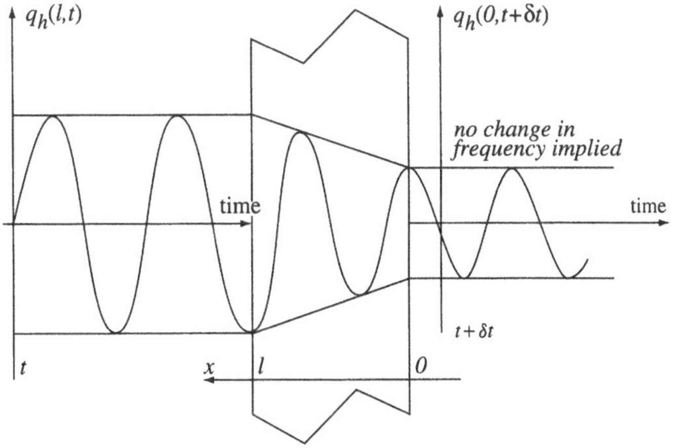

Figure 2.9: Effect of decrement factor and time lag.

The *surface response factor* defines the portion of the heat flux at an internal surface which is re-admitted to the internal environmental point when temperatures are held constant. The factor is applied to cyclic energy inputs at an internal surface to give the corresponding cyclic energy variations at the environmental point; the concept is analogous to time shifted reflections and is illustrated in figure 2.10. Typical applications include the modification of the transmitted component of solar radiation through windows and the radiant component of casual gains.

The *admittance response factor* is defined as the amount of energy entering a surface for each degree of temperature swing at the environmental point. It is used to represent enclosure response and give the equivalent swing in temperature about the mean value due to a cyclic heat load on the enclosure.

The final temperature or flux prediction is obtained by summing the cyclic contribution from each harmonic and expressing the result with respect to the contribution made by the steady state term. To simplify the procedure, Danter (1960) proposed that the factors for the fundamental 24 hour cycle be applied to the actual weather parameter profile. While Milbank and Harrington-Lynn (1974) have stated that this is not an unreasonable approximation, Gupta *et al* (1974) have advised that the fundamental frequency should be greater than twice the thermal time constant and not merely 24 hours if a steady periodic regime is required. The three factors are now defined mathematically.

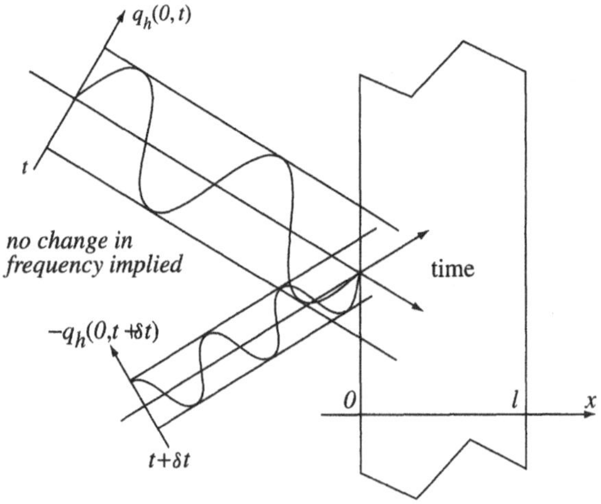

Figure 2.10: Effect of surface factor and time lag.

Decrement response factor

Recalling that the decrement factor, d, is the ratio of the cyclic flux transmission (usually from outside to inside) to the steady state flux transmission:

$$d_{0,L} = q_h(L,t)/U\theta_h(0,t)$$

where U is the overall thermal transmittance (W m^{-2}°C^{-1}).

Thus, from eqn (2.47), it follows that

$$d_{0,L} = 1/J(j\omega)U$$

$$= 1/[R_L A(j\omega) + R_L R_0 C(j\omega) + B(j\omega) + R_0 D(j\omega)]U \;.$$

This expression will have both real and imaginary parts, r_1 and v_1 respectively, so that the magnitude of the decrement factor is

$$d_{o,L} = | r_1 + jv_1 | = (r_1^2 + v_1^2)^{\frac{1}{2}}$$

with a corresponding time lag (in radians) of

$$\phi_d = \tan^{-1}(v_1/r_1) \;.$$

Surface response factor

This is defined as the ratio of the heat flux re-admitted from a surface (usually internal) to the total flux absorbed. It is equivalent to the ratio of the overall impedance, minus the impedance of the surface layer, to the overall impedance. For an internal surface, and with reference to figure 2.11, the surface response factor, s, is given by

$$s_{L,0} = [Z(L,t) - R_L]/Z(L,t)$$

$$= 1 - R_L/Z(L,t)$$

where Z(L, t) is the overall impedance from $x = 0$ to $x = L$, and equals $\theta_h(L,t)/q_h(L,t)$

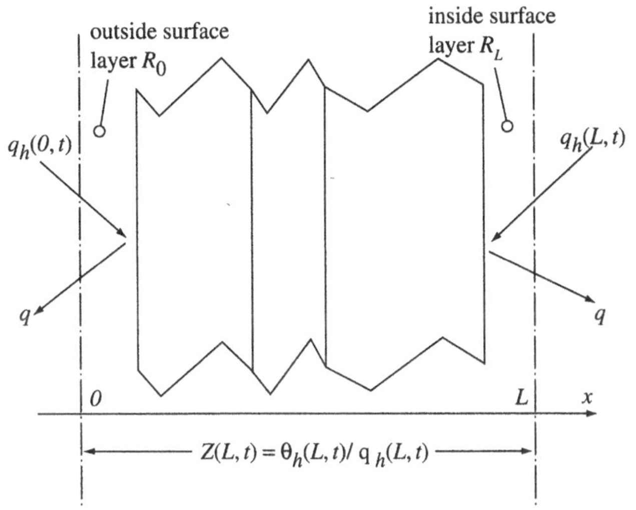

Figure 2.11: The overall thermal impedance of a multi-layered construction.

($m^2K\ W^{-1}$). This implies that

$$s_{L,0} = 1 - \frac{R_L q_h(L, t)}{\theta_h(L, t)}$$

and, from eqns (2.45) and (2.46),

$$s_{L,0} = 1 - \frac{R_L H(j\omega)}{F(j\omega)}$$

$$= \frac{R_0 A(j\omega) + B(j\omega)}{R_0 A(j\omega) + R_0 R_L C(j\omega) + B(j\omega) + R_L D(j\omega)} .$$

As before, this expression will have real and imaginary parts, r_2 and v_2, so that

$$s_{L,0} = | r_2 + jv_2 | = (r_2^2 + v_2^2)^{\frac{1}{2}}$$

with a time lag (radians) of

$$\phi_s = \tan^{-1}(v_2/r_2).$$

Admittance response factor

The admittance response factor, a, is defined as the ratio of the energy entering an internal surface to the corresponding temperature swing:

$$a_{L,0} = q_h(L, t)/\theta_h(L, t)$$

$$= \frac{R_0 C(j\omega) + D(j\omega)}{R_0 A(j\omega) + R_0 R_L C(j\omega) + B(j\omega) + R_L D(j\omega)} = r_3 + jv_3 .$$

Thus:

$$a_{L,0} = (r_3^2 + v_3^2)^{\frac{1}{2}}$$

with the temperature swing leading the absorption by

$$\phi_a = \tan^{-1}(v_3/r_3) \, .$$

Similar reasoning gives rise to the response factors for the reverse ordered cases:

$$d_{L,0} = 1/[R_0A(j\omega) + R_0R_LC(j\omega) + B(j\omega) + R_LD(j\omega)]U$$

$$s_{0,L} = [R_LA(j\omega) + B(j\omega)]/[R_LA(j\omega) + R_0R_LC(j\omega) + B(j\omega) + R_0D(j\omega)]$$

$$a_{0,L} = [R_LC(j\omega) + D(j\omega)]/[R_LA(j\omega) + R_0R_LC(j\omega) + B(j\omega) + R_0D(j\omega)] \, .$$

2.3.2 Zone energy balance

In determining the overall transmission matrix of eqns (2.41) and (2.42) surface boundary layers are represented by combined surface resistances determined from

$$R = 1/(h_c + 1.2\varepsilon h_r)$$

where h_c is the surface convection coefficient, h_r the radiative heat transfer coefficient (both W m^{-2}°C^{-1}), ε the surface emissivity and the 1.2 modulus is a required adjustment to account for the use of environmental temperature (Clarke 1977).

This treatment assumes that the inter-surface longwave radiation exchange can be linearised and evaluated as a function of some weighting of zone air and mean radiant temperatures. Environmental temperature (Danter 1973) is often used for this purpose. Since all heat flows through this fictitious temperature point, the inter-surface radiation exchange need not be separately considered.

2.3.3 Response function application

The factors derived in §2.3.1 can now be applied to the component harmonics of some real weather excitation to determine the building's response. This section demonstrates the use of the factors to determine internal temperatures, heating/cooling requirements and the effects of variable ventilation and intermittent plant operation.

Internal temperature prediction

The magnitude of the various load fluctuations at each point in time, for each harmonic considered and about the mean condition, are determined and modified by the decrement and surface factors appropriate to their type and frequency. This gives the energy fluctuation imposed on the enclosure at the environmental point at each frequency and experienced at some time after the initial excitation; a delay which depends on the time lag related to the factor used. The total fluctuating load at each point in time and at each frequency is obtained by summing the individual load fluctuations realised at each time-row over the interval in question. These fluctuations are then modified by the appropriate admittance factor to give the contribution of each harmonic to the fluctuation about the mean of internal environmental temperature at some later time depending on the lag associated with the admittance response factor.

The mean internal environmental temperature is determined from

$$\bar{\theta}_{ei} = \bar{\theta}_{ao} + \bar{Q}_T/(\sum_{i=1}^{c} A_iU_i + \bar{C}_v) \qquad (2.49)$$

where $\bar{\theta}_{ei}$ is the daily mean internal environmental temperature (°C), $\bar{\theta}_{ao}$ the mean external air temperature (°C), $\bar{Q}_T$ the mean total heat flux from all sources (W), $\bar{C}_v$ the mean ventilation conductance (W °C^{-1}), $\sum A_i U_i$ represents the sum of the product of areas and overall thermal transmittance values (W °C^{-1}) and c is the number of constructions.

The mean total heat flux is given by

$$\bar{Q}_T = \bar{Q}_{fs} + \bar{Q}_s + \bar{Q}_c$$

where $\bar{Q}_{fs}$ is the mean solar gain through opaque surfaces, $\bar{Q}_s$ the mean solar gain through transparent surfaces, and $\bar{Q}_c$ the mean gain from casual sources (all measured in W). These terms are now considered in turn.

$$\bar{Q}_{fs} = \sum_{i=1}^{N}[A_i U_i R_{oi}(\alpha_i \bar{I}_{soi} - \varepsilon_i I_{oi})]$$

where N is the total number of exposed opaque surfaces, α_i the surface shortwave absorptivity, $\varepsilon_i I_{oi}$ the longwave radiation exchange with the surroundings (W m^{-2}), R_{oi} the combined surface resistance (m^2°C W^{-1}) and $\bar{I}_{soi}$ the mean solar flux incident on the opaque surface:

$$\sum_{t=1}^{m} I_{soi}(t)/m$$

where m = 24 for the daily mean and $I_{soi}(t)$ the instantaneous solar flux impinging on opaque surface i (W m^{-2}).

The mean solar gain through transparent surfaces is found from

$$\bar{Q}_s = \sum_{t=1}^{m} Q_s(t)/m$$

where $Q_s(t)$ is the instantaneous solar gain at time t (W).

Section 7.4 details a solar modelling procedure that gives $I_{so}(t)$ and $Q_s(t)$ as a function of location, time of year/day, building geometry, site obstructions and thermo-optical properties. As a simplification, the use of 'solar gain factors' has been proposed (Loudon 1968). These are predetermined values that depend on solar incidence angles and the properties of window/blind arrangements. They are used to determine the portion of the incident solar flux that penetrates the boundary:

$$Q_s(t) = \sum_{i=1}^{L} A_i S_i(t) I_{soi}(t)$$

where L is the total number of transparent surfaces and $S_i(t)$ the solar gain factor at time t.

$$Q_c(t) = \sum_{j=1}^{K} Q_{cj}(t)$$

where K is the number of casual sources and Q_{cj} the magnitude of any source (W).

The contribution of each harmonic to the swing in internal environmental temperature about the mean value is then computed from

$$\tilde{\theta}_{ei}(t) = \frac{\tilde{Q}_t(t - \phi_a)}{\sum Aa + C_v} \qquad (2.50)$$

where $\tilde{\theta}_{ei}(t)$ is the fluctuation in internal environmental temperature about the mean at some time t (°C), $\tilde{Q}_t(t - \phi_a)$ the total fluctuating load at the environmental point at time $(t - \phi_a)$ for any harmonic (W); ϕ_a the time lag associated with the admittance factor (s), and $\sum Aa$ the sum of the product of area and admittance, $a_{L,o}$, for all internal surfaces (W °C^{-1}).

The total fluctuating energy gain at the environmental point due to a given excitation frequency is given by

$$\tilde{Q}_t(t) = \tilde{Q}_{fs}(t) + \tilde{Q}_s(t) + \tilde{Q}_c(t) + \tilde{Q}_{fc}(t) + \tilde{Q}_{gc}(t) + \tilde{Q}_v(t)$$

where $\tilde{Q}_{fs}(t)$ is the opaque surface solar gain fluctuation at time t, $\tilde{Q}_s(t)$ the transparent surface solar gain fluctuation, $\tilde{Q}_c(t)$ the casual gain fluctuation, $\tilde{Q}_{fc}(t)$ the opaque surface conduction gain fluctuation, $\tilde{Q}_{gc}(t)$ the transparent surface conduction gain fluctuation, and $\tilde{Q}_v(t)$ the ventilation or infiltration fluctuation (all measured in W). Each load fluctuation is now considered in turn.

$$\tilde{Q}_{fs}(t) = \sum_{i=1}^{N}[A_i U_i R_{oi} d_i \alpha_i \tilde{I}_{so}(t - \phi_d)]$$

where d_i is the decrement factor ($d_{L,0}$) for the layers behind surface i, ϕ_d the associated time lag, $\tilde{I}_{so}(t - \phi_d)$ the fluctuation about the mean of the solar intensity incident on opaque surfaces at some time $(t - \phi_d)$, measured in W m^{-2} and equal to $I_{so}(t - \phi_d) - \bar{I}_{so}$.

$$\tilde{Q}_s(t) = \tilde{Q}_{s1}(t - \phi_s) + \tilde{Q}_{s2}(t)$$

where $\tilde{Q}_{s1}(t - \phi_s)$ is the fluctuation due to the directly transmitted (time lagged) component of solar radiation through transparent surfaces and $\tilde{Q}_{s2}(t)$ the fluctuation due to the absorbed component of the incident solar radiation which is retransmitted (with no time lag) inwards to the environmental point (via the combined convective and radiative resistance), both measured in W. Section 7.4 derives models to calculate these terms.

The simplified admittance procedure proposes the use of 'alternating solar gain' factors, which allow the fluctuation in energy at the environmental point due to the fluctuation in solar gain through transparent surfaces to be determined from

$$\tilde{Q}_s(t) = \sum_{i=1}^{L} A_i \tilde{S}_i(t - \phi_s)\tilde{I}_{si}(t - \phi_s)$$

where $\tilde{S}_i(t - \phi_s)$ is the alternating solar gain factor which includes the effect of the surface response factor.

$$\tilde{Q}_c(t) = Q_c(t) - \bar{Q}_c$$

where $\tilde{Q}_c(t)$ is the total instantaneous casual load (W).

With the use of environmental temperature, internally generated casual gains are only immediately realised as cooling loads if the heat input is in the proportion of two-thirds radiation to one-third convection. In many cases casual gains will not conform to this split and so they must be separated into their convective and radiant components. The convective component can then be assumed to act immediately at the air point with the radiant component treated in a manner similar to solar gains and modified by the internal mass of the enclosure. In general terms

$$\tilde{Q}_c(t) = s_{L,0}(t - \phi_s)\tilde{Q}_R(t - \phi_s) + L_c\tilde{Q}_c(t)$$

where $\tilde{Q}_R(t - \phi_s)$ is the fluctuating radiant portion of the casual load (W), $\tilde{Q}_c(t)$ the fluctuating convective portion (W), L_c the convective load correction factor, which adjusts this component to account for injection at the environmental point rather than at the air point, given by $h\sum A/(h\sum A + C_v)$, and h is the hypothetical conductance associated with the environmental point ($W\ m^{-2}{}^{\circ}C^{-1}$).

$$\tilde{Q}_{fc}(t) = \sum_{i=1}^{o} A_i U_i d_i \tilde{\theta}_{Ao}(t - \phi_d)$$

where $\tilde{\theta}_{Ao}(t)$ is the fluctuation in outside air temperature (°C) and o the number of opaque constructions.

$$\tilde{Q}_{gc}(t) = \sum_{i=1}^{T} A_i U_i \tilde{\theta}_{Ao}(t)$$

where T is the number of transparent constructions. It is usual practice to assume that window conduction processes undergo negligible time delay.

$$\tilde{Q}_v(t) = C_v(t)\tilde{\theta}_{Ao}(t)$$

where, again, it is usual to assume a zero time lag.

Appendix D gives a worked example relating to the estimation of peak summertime temperatures in an unconditioned space.

Heating/cooling requirements

The fluctuation in energy at the environmental point due to the various temperature and energy fluctuations are computed relative to some fixed internal temperature. This is done by reversing the procedure for internal temperature prediction to compute the plant capacity to maintain specified temperature conditions.

Variable ventilation

It is possible to combine the foregoing equations in such a way that the effects of known variations in the infiltration/ventilation rate can be assessed. This is of particular use in the analysis of the intermittent operation of ventilation plant where the ventilation rate, although variable, is known. Such a formulation allows an assessment of the effects of increasing or reducing the ventilation rate relative to some design value in order to gain an insight into the sensitivity of a system under real conditions.

The variable ventilation harmonic method (Harrington-Lynn 1974a) is obtained by combining the heat balance equations involving the mean and fluctuating energy gains at the environmental point. The method allows the internal temperature to be established for any known or hypothetical ventilation scheme.

Combining eqns (2.49) and (2.50) gives

$$\bar{Q}_T + \tilde{Q}_T(t) = (\sum AU + C_v)(\bar{\theta}_{ei} - \bar{\theta}_{Ao}) + (\sum Aa + C_v)\tilde{\theta}_{ei}(t)$$

and assuming that the ventilation conductance is time-dependent:

$$Q_x(t) = \sum AU(\bar{\theta}_{ei} - \bar{\theta}_{Ao}) + \sum Aa\tilde{\theta}_{ei}(t) + C_v(t)[\tilde{\theta}_{ei}(t) + \bar{\theta}_{ei} - \bar{\theta}_{Ao}] \qquad (2.51)$$

where

$$Q_x(t) = \bar{Q}_T(t) + \tilde{Q}_T(t)\ .$$

Now, for $\Delta\theta_{ei} = \bar{\theta}_{ei} - \bar{\theta}_{Ao}$,

$$Q_{x(t)} = \sum AU\Delta\theta_{ei} + \sum Aa\tilde{\theta}_{ei}(t) + C_v(t)[\tilde{\theta}_{ei}(t) + \Delta\theta_{ei}]\ .$$

For a daily analysis, taking hourly values, there are 24 equations of this type. These equations contain 25 unknown quantities ($\tilde{\theta}_{ei}(t)$ at each hour and the daily value of $\Delta\theta_{ei}$) and consequently another equation is required. Summation of the 24 equations gives

$$\sum_{t=1}^{24} Q_x(t) = 24 \sum AU\Delta\theta_{ei} + \sum_{t=1}^{24} \sum Aa\tilde{\theta}_{ei}(t) + \sum_{t=1}^{24} C_v(t)[\tilde{\theta}_{ei}(t) + \Delta\theta_{ei}]$$

and since

$$\frac{1}{24} \sum_{t=1}^{24} \sum Aa\tilde{\theta}_{ei}(t) = 0$$

this reduces to

$$\bar{Q}_x = \Delta\theta_{ei}(\sum AU + \bar{C}_v) + \frac{1}{24} \sum_{t=1}^{24} C_v(t)\tilde{\theta}_{ei}(t)$$

where

$$\bar{Q}_x = \frac{1}{24} \sum_{t=1}^{24} Q_x(t) \tag{2.52}$$

$$\bar{C}_v = \frac{1}{24} \sum_{t=1}^{24} C_v(t) .$$

Intermittent plant operation

Harmonic methods can also be used to analyse pre-defined plant operational schemes (Harrington-Lynn 1974b). Consider eqn (2.51) which relates to the variable ventilation case. With the addition of a time-dependent plant input or extract term this equation becomes

$$Q_x(t) + Q_p(t) = \sum AU\Delta\theta_{ei} + \sum Aa\tilde{\theta}_{ei}(t) + C_v(t)[\tilde{\theta}_{ei}(t) + \Delta\theta_{ei}] \tag{2.53}$$

where $Q_p(t)$ is the plant exchange with the environmental point (W).

For any given ON period, the environmental temperature control state is given by

$$\theta_c = \tilde{\theta}_{ei} + \bar{\theta}_{ei} = \tilde{\theta}_{ei} + \Delta\theta_{ei} + \bar{\theta}_{Ao} .$$

Substitution in eqn (2.53) gives, for an ON period,

$$Q_x(t) + Q_p(t) = \sum AU\Delta\theta_{ei} + \sum Aa(\theta_c - \Delta\theta_{ei} - \bar{\theta}_{Ao}) + C_v(t)(\theta_c - \bar{\theta}_{Ao}) \tag{2.54}$$

and, for an OFF period

$$Q_x(t) = \sum AU\Delta\theta_{ei} + \sum Aa\tilde{\theta}_{ei}(t) + C_v(t)[\theta_{ei}(t) + \Delta\theta_{ei}] . \tag{2.55}$$

Again, consider a 24 hour period and assume that the plant is ON for X hours, giving X equations of type (2.54) and (24 − X) equations of type (2.55). As before, the 24 equations incorporate 25 unknowns ($\tilde{\theta}_{ei}(t)$ at each OFF hour, $Q_p(t)$ at each ON hour, and the daily value of $\Delta\theta_{ei}$) and another equation is required. Summation of the 24 equations gives

$$\sum_{ON} Q_p(t) + \sum_{t=1}^{24} Q_x(t) = 24 \sum AU\Delta\theta_{ei} + \sum_{OFF} \sum Aa\tilde{\theta}_{ei}(t) + \sum_{OFF} C_v(t)[\tilde{\theta}_{ei}(t) + \Delta\theta_{ei}]$$

$$+ \sum_{ON} \sum Aa(\theta_c - \Delta\theta_{ei} - \bar{\theta}_{Ao}) + \sum_{ON} C_v(t)(\theta_c - \bar{\theta}_{Ao}) . \tag{2.56}$$

Now, from eqn (2.52)

$$\sum_{t=1}^{24} Q_x(t) = 24\bar{Q}_x$$

and by definition

$$\sum_{OFF} \sum Aa\tilde{\theta}_{ei}(t) + \sum_{ON} \sum Aa(\theta_c - \Delta\theta_{ei} - \bar{\theta}_{Ao}) = 0$$

and so eqn (2.56) becomes

$$\sum_{ON} Q_p(t) + 24\bar{Q}_x = 24 \sum AU\Delta\theta_{ei} + \sum_{OFF} C_v(t)[\theta_{ei}(t) + \Delta\theta_{ei}] + \sum_{ON} C_v(t)(\theta_c - \bar{\theta}_{Ao}) \quad (2.57)$$

and eqns (2.54), (2.55) and (2.57) can be solved to obtain the plant capacity to maintain any ON period environmental temperature θ_c.

2.4 Numerical methods

The principle of superimposition, which underpins the response function method, has long been used by modellers to determine the response of a system to a set of excitations. This entails summing the responses, determined independently, of the system's component parts. If these parts are strongly interacting, as in the case of buildings and their HVAC systems, then this will lead to an inherent inaccuracy because the parts are decoupled. The practice of assuming model parameters (e.g. surface heat transfer coefficients, fabric conductivity etc) to be time invariant has the effect of decoupling parts and thereby rendering the principle of superimposition acceptable.

Figure 2.12 summarises some of the domains that are candidates for coupling within any advanced simulation program (Clarke 2001, Citherlet *et al* 2001). First order couplings include, but are not limited to, building thermal processes and natural illuminance distribution; building and plant thermal processes and distributed fluid flow; building thermal processes and intra-room air movement; building distributed air flow and intra-room air movement; electrical demand and micro power systems (renewable energy based or otherwise); and construction heat and moisture flow.

Unlike the response function method, numerical techniques are well suited to this integration challenge because they can be used to handle problems of almost any degree of complexity. Whereas analytical solutions will result in equations of limited applicability, but which permit the calculation of temperature and heat flux at any point within the modelled system, numerical methods are generally applicable but relate to only preselected points. These points possess properties that may be considered representative of some relatively small region, e.g. a part of a wall, room, window, HVAC component or renewable energy conversion system.

The construction of a numerical model is essentially a three stage process: system discretisation is followed by the establishment of a nodal equation-set, which is then solved simultaneously to obtain the distribution of the state variables. For the analysis of the heat and inter-zone air flow within a ten zone building, such a model might typically comprise 3500 equations—and substantially more where intra-zone air movement is included.

Numerical methods are based on an approximation of some governing partial differential equation, such as the Fourier heat equation, which is applicable to regions where conduction prevails, or the Navier-Stokes momentum equation relating to fluid flow. This approximation is usually achieved by truncated Taylor series expansion or by application of conservation principles to small control volumes.

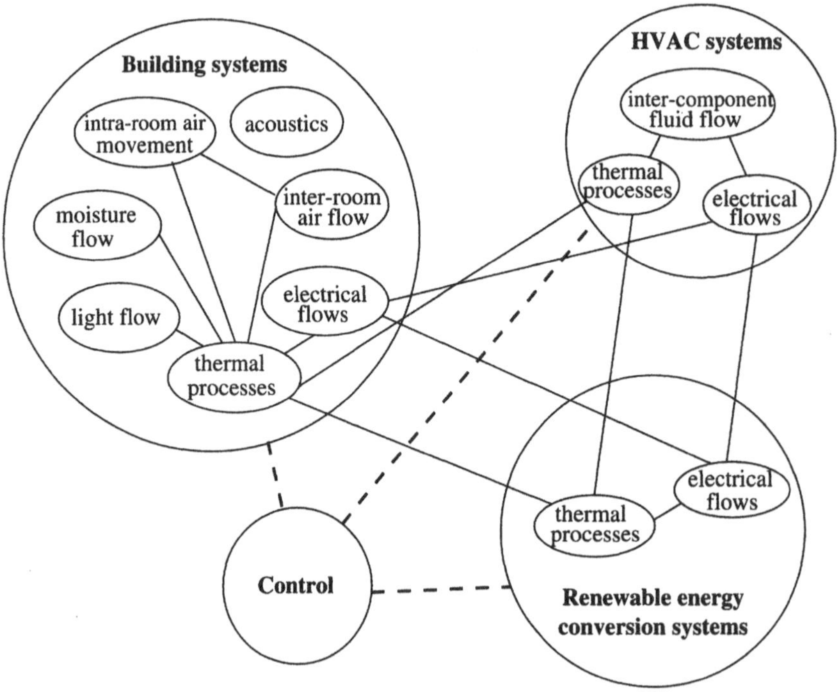

Figure 2.12: Examples of coupled domains.

2.4.1 Taylor series expansion

Consider figure 2.13, which shows a continuous function $f(\gamma)$ over the range $(\gamma - \delta\gamma) \leq \gamma \leq (\gamma + \delta\gamma)$. The replacement of the derivatives of $f(\gamma)$ by finite differences involves expressing these derivatives in terms of a truncated Taylor series expansion. Taylor's theorem applied to this function gives

$$f(\gamma + \delta\gamma) = f(\gamma) + \delta\gamma f^1(\gamma) + \frac{(\delta\gamma)^2}{2} f^2(\gamma) + \frac{(\delta\gamma)^3}{6} f^3(\gamma) + \cdots \tag{2.58}$$

and

$$f(\gamma - \delta\gamma) = f(\gamma) - \delta\gamma f^1(\gamma) + \frac{(\delta\gamma)^2}{2} f^2(\gamma) - \frac{(\delta\gamma)^3}{6} f^3(\gamma) + \cdots \tag{2.59}$$

where $f^n(\gamma)$ is $d^n f(\gamma)/d\gamma^n$. Adding these equations through those terms involving $(\delta\gamma)^3$ gives

$$f^2(\gamma) = \frac{f(\gamma + \delta\gamma) - 2f(\gamma) + f(\gamma - \delta\gamma)}{(\delta\gamma)^2} + \varepsilon[(\delta\gamma)^2] \tag{2.60}$$

where $\varepsilon[(\delta\gamma)^2]$ indicates that the truncation error resulting from the approximate representation of the second order derivative is of order $(\delta\gamma)^2$.

Subtracting eqn (2.59) from eqn (2.58) through those terms involving $(\delta\gamma)^2$ gives

$$f^1(\gamma) = \frac{f(\gamma + \delta\gamma) - f(\gamma - \delta\gamma)}{2\delta\gamma} + \varepsilon[(\delta\gamma)^2]. \tag{2.61}$$

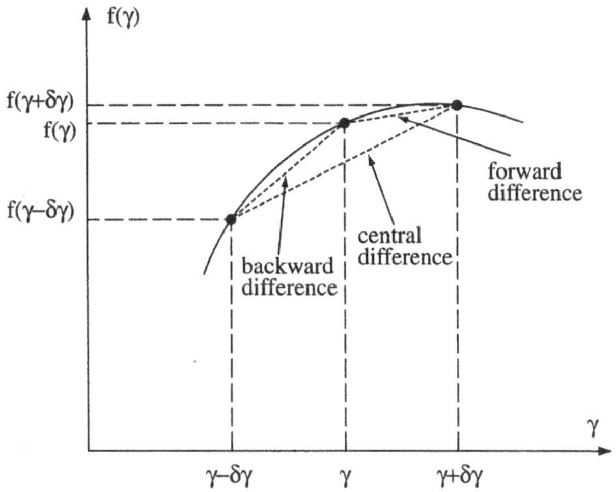

Figure 2.13: A continuous function of γ.

Eqns (2.60) and (2.61) are termed central difference approximations to the second and first order derivatives respectively. Truncation of eqn (2.58) after the term involving $\delta\gamma$ gives the first forward difference representation:

$$f^1(\gamma) = \frac{f(\gamma + \delta\gamma) - f(\gamma)}{\delta\gamma} + \varepsilon[(\delta\gamma)] \tag{2.62}$$

and application of a similar truncation to eqn (2.59) gives the first backward difference representation:

$$f^1(\gamma) = \frac{f(\gamma) - f(\gamma - \delta\gamma)}{\delta\gamma} + \varepsilon[(\delta\gamma)]\ . \tag{2.63}$$

The truncation error in eqns (2.62) and (2.63) is of order $\delta\gamma$, that is halving the discretisation step will only approximately half the error. On the other hand, the central difference approximation has a truncation error of order $(\delta\gamma)^2$ and so halving the discretisation step will approximately quarter the error.

Alternative mixes of these approximation schemes can be employed, leading to explicit and implicit difference formulations. Consider the Fourier heat equation in one space variable and with heat generation as derived in Appendix C:

$$\frac{\partial^2\theta(x,t)}{\partial x^2} = \frac{1}{\alpha}\frac{\partial\theta(x,t)}{\partial t} - \frac{q}{k}\ . \tag{2.64}$$

Figure 2.14 shows a homogeneous material layer arbitrarily located within a larger homogeneous region. Two finite difference schemes are possible: explicit and implicit. A scheme of explicit enumeration is obtained by representing the second order derivative of eqn (2.64) in the central difference form of eqn (2.60) and the first order derivative in first forward difference form as given by eqn (2.62). Ignoring the error term and assuming (for the present) that thermophysical properties are time independent gives, for node I at time t

$$\frac{\theta(I+1,t) - 2\theta(I,t) + \theta(I-1,t)}{(\delta x)^2} = \frac{1}{\alpha}\frac{\theta(I,t+\delta t) - \theta(I,t)}{\delta t} - \frac{q(I,t)}{k}$$

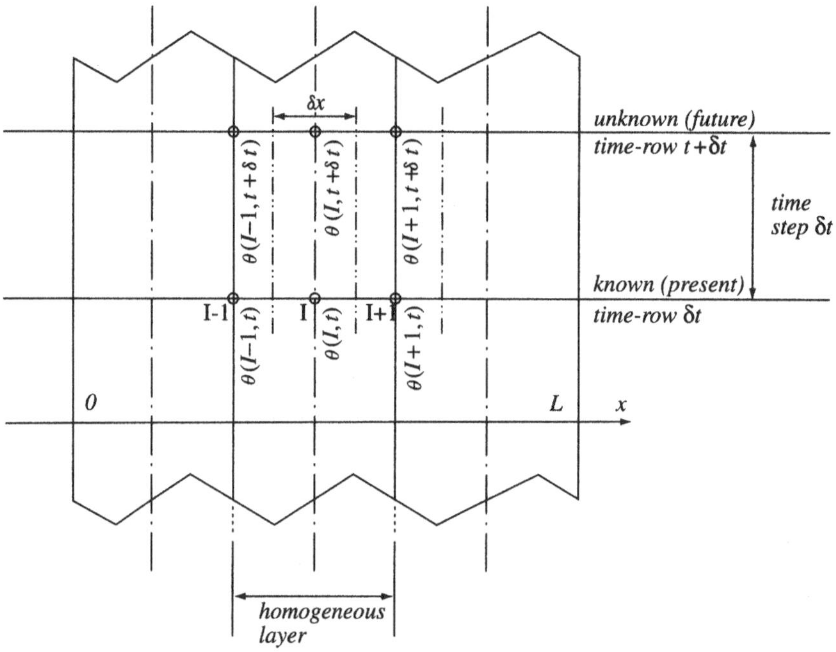

Figure 2.14: Homogeneous layer: space and time nodal scheme.

which, on rearrangement, gives

$$\theta(\mathrm{I},\mathrm{t}+\delta\mathrm{t}) = \frac{\mathrm{k}\delta\mathrm{t}}{\rho\mathrm{C}(\delta\mathrm{x})^2}\,\theta(\mathrm{I}+1,\mathrm{t}) + \left(1 - \frac{2\mathrm{k}\delta\mathrm{t}}{\rho\mathrm{C}(\delta\mathrm{x})^2}\right)\theta(\mathrm{I},\mathrm{t})$$
$$+ \frac{\mathrm{k}\delta\mathrm{t}}{\rho\mathrm{C}(\delta\mathrm{x})^2}\,\theta(\mathrm{I}-1,\mathrm{t}) + \frac{\mathrm{q}(\mathrm{I},\mathrm{t})\delta\mathrm{t}}{\rho\mathrm{C}}\,. \tag{2.65}$$

Note that the sum of the present time-row temperature term coefficients is unity. This implies that, in the absence of heat generation, the future time-row nodal temperature of any region is a weighted average of the present time-row temperature in the region and the temperatures in adjacent regions in thermal contact. If an equation of this form can be written for every region within a system then, given initial and boundary conditions, the discrete temperature history over any required period can be determined. Explicit schemes of this type are relatively easy to formulate and solve but can become unstable in certain circumstances. Consider the coefficient of the present time-row temperature term of some region I. If this coefficient should become negative, that is

$$1 - \frac{2\mathrm{k}\delta\mathrm{t}}{\rho\mathrm{C}(\delta\mathrm{x})^2} < 0$$

this would imply that the warmer the region I is now, the colder it will be after some time-step δt. To avoid such an absurdity, a stability criterion is introduced:

$$\frac{2\mathrm{k}\delta\mathrm{t}}{\rho\mathrm{C}(\delta\mathrm{x})^2} \leq 1$$

more usually expressed as

$$\frac{\alpha\delta t}{(\delta x)^2} \leq \tfrac{1}{2} \quad \text{or} \quad F \leq \tfrac{1}{2}$$

where F is the Fourier number, which defines the ratio of the heat conduction rate to the heat storage rate. High values represent good conductors with relatively poor storage potential; low values represent the converse: poor conductors with relatively good storage potential.

Simple operations performed on eqn (2.65) allow the formulation of rudimentary graphical techniques for the assessment of transient conduction in homogeneous systems. Ignoring the heat source term, eqn (2.65) can be rewritten to give

$$\theta(I, t+\delta t) = F[\theta(I+1, t) + (1/F - 2)\theta(I, t) + \theta(I-1, t)] \,.$$

Setting $F = \frac{1}{2}$, which from the stability criteria is the largest value permitted, gives

$$\theta(I, t+\delta t) = \tfrac{1}{2}\,[\theta(I+1, t) + \theta(I-1, t)] \,.$$

This is the basis of the Binder-Schmidt graphical method (Simonson 1967).

A scheme of implicit enumeration is one in which the unknown temperature $\theta(I, t+\delta t)$ is expressed in terms of both future and present time-row temperatures prevailing in all regions in thermal contact. Any given system will therefore be represented by a connected series of algebraic equations, which must be solved simultaneously for each time-step. The second order derivative of eqn (2.64) is replaced by the central difference formulation of eqn (2.60) but using the unknown temperature values at the future time-row rather than the known values at the present time-row as in the explicit formulation. The first order derivative is expressed in the first backward formulation of eqn (2.63). Again ignoring the error term and assuming constant thermophysical properties gives

$$\frac{\theta(I+1, t+\delta t) - 2\theta(I, t+\delta t) + \theta(I-1, t+\delta t)}{(\delta x)^2} = \frac{1}{\alpha}\,\frac{\theta(I, t+\delta t) - \theta(I, t)}{\delta t} - \frac{q(I, t+\delta t)}{k}$$

which, on rearrangement, gives

$$\left(1 + \frac{2k\delta t}{\rho C(\delta x)^2}\right)\theta(I, t+\delta t) = \theta(I, t) + \frac{k\delta t}{\rho C(\delta x)^2}\,[\theta(I+1, t+\delta t) + \theta(I-1, t+\delta t)] + \frac{q(I, t+\delta t)\delta t}{\rho C} \,. \qquad (2.66)$$

Implicit formulations are unconditionally stable for all space and time discretisation schemes although large space or time steps will result in excessive discretisation error. Discretisation is discussed in §3.1.

A weighted average of eqns (2.65) and (2.66) can now be taken to construct a generalised formulation. Multiplying eqn (2.65) by (1 − W) and adding the result to eqn (2.66) when multiplied by (W) gives

$$(1 + 2WF)\theta(I, t+\delta t) = WF[\theta(I+1, t+\delta t) + \theta(I-1, t+\delta t)] + [1 - 2F(1-W)]\theta(I, t) + (1-W)F[\theta(I+1, t) + \theta(I-1, t)] + \frac{\delta t}{\rho C}\,[Wq(I, t+\delta t) + (1-W)q(I, t)] \,. \qquad (2.67)$$

Setting $W < 0.5$ gives an explicit scheme for which the stability criterion is

$$F \le \frac{1}{2}\left(\frac{1}{1-W}\right)$$

and setting $W \ge 0.5$ gives an implicit scheme, with $W = 0.5$ resulting in the commonly used Crank-Nicolson formulation much favoured because of its stability combined with accuracy.

By similar reasoning it is possible to devise finite difference formulations for more than one space dimension (see chapter 3), and for explicit methods corresponding stability criteria will emerge (Croft and Lilley 1977).

2.4.2 Control volume heat balance

Differencing by Taylor series expansion is the formal method of establishing a finite difference scheme from some known partial differential equation. Unfortunately the technique can prove cumbersome and difficult to apply in all but simple problems. In a building design context several complications arise as follows.

The simultaneous presence of multiple heat transfer processes (conduction, convection, radiation, advection and heat generation).

The time and positional dependency of heat generation due to solar radiation, mechanical plant etc.

The use of a discretisation method that leads to non-homogeneous, anisotropic finite volumes.

The presence of multi-dimensional effects.

An alternative approach is to directly apply conservation principles to small control volumes established to represent the physical system. This ensures that the resulting solution satisfies the conservation laws even if the number of control volumes is small (although in such a case discretisation errors might well dominate and/or the underlying physical models become invalid).

Consider figure 2.15, which shows a control volume, I, communicating thermodynamically with four surrounding regions via the processes of conduction, convection, radiation and mass exchange (e.g. air and/or moisture flow). Internal heat generation is also considered to take place within the region I. Assuming (for the present purpose only) that the inter-region heat exchange can be represented as a linear function of the temperature difference, each flowpath may be represented by

$$Q_{J,I} = K_{J,I}(\theta_J - \theta_I)\ ;\ J = 1, 2, 3, 4$$

where $K_{J,I}$ is a linearised heat flow conductance (W °C^{-1}).

The conductance term is not considered further here since it is treated, in detail, in the equation derivations of chapter 3. Also, the assumption of linearity is made here for convenience only and chapter 4 demonstrates the treatment of non-linear systems.

The rate of heat generation within the region I is, in the present formulation, considered to be substantially independent of region temperature—perhaps originating outside the region as with shortwave solar flux—and is denoted simply by Q_I (W).

The rate of heat storage (W) within the region over some finite time interval δt is given by

$$Q_s = \frac{\rho_I(\xi)C_I(\xi)\delta V_I(\xi)}{\delta t}\,[\theta(I, t + \delta t) - \theta(I, t)]$$

where $\rho_I(\xi)$ is the characteristic density of region I at time ξ (kg m^{-3}), $C_I(\xi)$ the characteristic

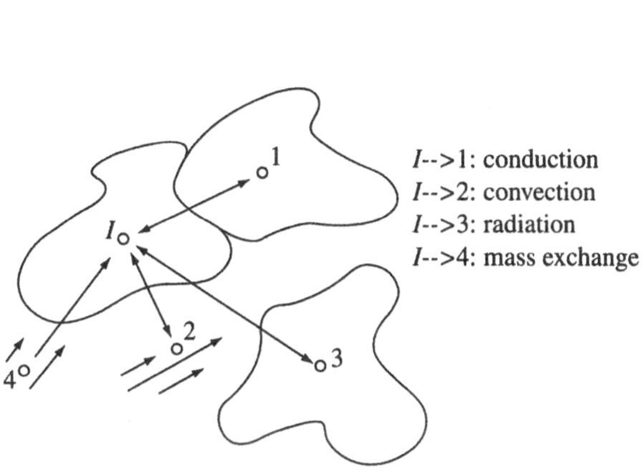

Figure 2.15: Heat exchanges in a physical system.

specific heat capacity (J $kg^{-1}{}^{\circ}C^{-1}$) and $\delta V_I(\xi)$ the region volume (m^3).

Now, in the limit, the rate at which heat is being stored within the region I can be equated to the net rate of heat flow to the region and so, for the system of figure 2.15, heat balance considerations yield

$$\frac{\rho_I(\xi)C_I(\xi)\delta V_I(\xi)}{\delta t}\,[\theta(I, t+\delta t) - \theta(I, t)] = \sum_{j=1}^{N} K_{j,I}(\theta_j - \theta_I)\Big|_{t=\xi} + q_I\Big|_{t=\xi} \quad . \tag{2.68}$$

Evaluation of the heat flux and generation terms at the present time-row, $\xi = t$, gives the explicit formulation, while evaluation at the future time-row, $\xi = t + \delta t$, gives the fully implicit formulation. As before, any explicit/implicit mix can be obtained as a weighted summation of both schemes.

From a mathematical classification viewpoint there are three types of differential equation required to describe energy exchanges in building and plant systems: first order ordinary differential equations used to represent physical regions possessing 'lumped' thermophysical properties; second order parabolic partial differential equations used to describe insulation/capacity regions requiring detailed modelling; and elliptic/hyperbolic partial differential equations used to describe fluid flow and convective coupling.

Eqn (2.68) will become identical to the first type as the space and time increments approach their limit. It can also be proved (Lambert 1973) that eqn (2.68) will become identical to types 2 and 3 if the so called semi-discretisation is applied to the space variables of these partial differential equations.

Chapter 3 demonstrates the application of eqn (2.68) to multi-zone building systems when subjected to weather excitations causing heat flow transients and time-dependent system properties.

2.4.3 Numerical solution techniques

Explicit schemes result in a set of independent equations—one for each node—for which solution is relatively straightforward. At time $t = 0$ the nodal temperature field is prescribed and the solution consists of establishing the future time-row values at each consecutive time-step from the independent nodal equations.

With implicit schemes, each nodal equation will contain present and future time-row temperature terms relating to the node in question and all surrounding nodes in thermodynamic contact. Such a system of equations must be solved simultaneously at each time-step. Two main approaches exist to achieve this: direct and iterative. A direct method yields a solution in a finite number of computational steps that can be determined in advance. For the case of a linear equation-set, the solution is complete for the current time-step. Where the equation-set is non-linear, some property being temperature dependent for example, the solution process must be repeated after the equation coefficients have been updated. For such a case, an iterative approach is preferred. An iterative method will generally commence with some guessed solution for the state variables which, when inserted into the equation-set, will give rise to a residue. This residue is then used to modify the initial guess in such a way as to cause convergence on the true solution. The number of computational steps is therefore dependent on the convergence criteria used and on the level of accuracy required.

Iterative methods are often applied to equation structures that are sparse or which are known to converge rapidly, since solution by direct means would demand, in the former case, high storage space and, in the latter case, unnecessary computation. The overall equation-set to emerge from the application of numerical techniques to buildings is invariably sparse (see chapter 4) and so iterative methods would seem appropriate. However, this sparseness can be removed by matrix partitioning techniques, allowing each partitioned sub-matrix to be processed, by a customised direct method, at a frequency depending on the building component to which it relates and the degree to which its contents (the nodal conservation equation coefficients) change with time. Thus, the greater part of the matrix processing can be achieved by rapid direct techniques with, in some cases, iteration superimposed. This is the strategy underlying the matrix processing techniques of chapter 4 where the partitioning and mixed solution scheme is demonstrated. For this reason, and since detailed descriptions are given elsewhere (e.g. Kreyszig 1979), only a brief summary of the apposite solution approaches is presented here.

Direct Methods

Consider the system of I simultaneous linear equations in J unknown temperature terms (θ) as shown in figure 2.16. Also shown is the equivalent matrix representation $\mathbf{A}\theta = \mathbf{B}$. Simultaneous solution is basically a five stage process:

Step 1: Denoting the first equation as the primary and the second equation as the secondary, each of the terms of the primary are multiplied by the ratio of the first existing coefficient in the secondary (which shares a temperature term with the primary) to the corresponding coefficient in the primary (the pivotal coefficient); that is a_{21}/a_{11}. Note that if a pivotal coefficient is small, the subtracted number can become large causing numerical problems or amplifying any uncertainty in the coefficients. It is therefore important to ensure that such coefficients are constrained by rescaling or reordering the equations as necessary; the latter process is known as pivoting.

Step 2: The primary is now subtracted from the secondary to eliminate the temperature term (from the secondary) for which the ratio was established.

Step 3: The process of steps 1 and 2 is now repeated for all primary/secondary pairs. In fact there are

$$\sum_{k=1}^{I-1}(I-k)$$

possible pairings: the first equation with $2, 3, 4, \ldots, I$; the second equation with $3, 4, 5, \ldots, I$; up to equation $I - 1$ with I.

Step 4: The last secondary to be processed will contain only one temperature term (but only if $I \geq J$) and so this temperature value can be determined.

Step 5: The complete solution is given by backward substitution of this known temperature in the equation in two temperature variables and so on.

$$
\begin{array}{lllllll}
a_{11}\ \theta_1 & + a_{12}\ \theta_2 & + & \cdots\cdots & + a_{1J}\ \theta_J & = b_1 & \\
a_{21}\ \theta_1 & + a_{22}\ \theta_2 & + & \cdots\cdots & + a_{2J}\ \theta_J & = b_2 & \\
\vdots & \vdots & & & \vdots & \vdots & \equiv \\
a_{I1}\ \theta_1 & + a_{I2}\ \theta_2 & + & \cdots\cdots & + a_{IJ}\ \theta_J & = b_I &
\end{array}
$$

$$
\underset{\mathbf{A}}{\begin{bmatrix} a_{11} & a_{12} & \cdots\cdots & a_{1J} \\ a_{21} & a_{22} & \cdots\cdots & a_{2J} \\ \vdots & \vdots & & \vdots \\ a_{I1} & a_{I2} & \cdots\cdots & a_{IJ} \end{bmatrix}} \times \underset{\boldsymbol{\theta}}{\begin{bmatrix} \theta_1 \\ \theta_2 \\ \vdots \\ \theta_J \end{bmatrix}} = \underset{\mathbf{B}}{\begin{bmatrix} b_1 \\ b_2 \\ \vdots \\ b_I \end{bmatrix}}
$$

Figure 2.16: Matrix representation of I simultaneous linear equations in J unknowns.

Figure 2.17 demonstrates this process for an example equation-set. Important points to note are: the primary/secondary combinations can be handled in any order; in any secondary equation, any temperature term can be selected for elimination as long as the corresponding term exists in the primary; all non-eliminated terms are then carried through to the other equations; and the final equation in the forward reduction process ($18z = 18$ in this case) embodies the characteristics of the entire system. The matrix processing techniques of chapter 4 and 6 employ these devices to achieve the computationally efficient solution of sparse equation systems representing multi-zone building systems and environmental control systems respectively.

The technique, known as Gaussian elimination, consists of organised multiplication, division and subtraction operations. For a system with characteristic matrix **A** square and order N, the number of such operations is of order $N^3/3$. A number of variants exist. These include the Gauss-Jordan and Cholesky methods as described in the literature (Press *et al* 1986).

Iterative methods

A number of iterative methods exist each with the same underlying technique: a guess is made of the nodal state variables (e.g. temperature), the equations are evaluated to obtain updated state variable values and these updates are used to repeat the process until subsequent updates differ only slightly from the previous iteration. In general, alternative methods can be differentiated by the stage at which an updated value is incorporated in subsequent equation evaluation.

The Gauss-Seidel technique makes use of newly computed values as soon as they become available and for this reason is known as a method of successive corrections. The Jacobi technique is termed a method of simultaneous correction because no newly computed value is used until each of the equations have been processed in a particular iteration step. These iterative methods are subject to convergence criteria relating to the eigenvalues (or latent roots) of a

$$\begin{aligned} 4x - \;\; y + 2z &= 15 \quad (a) \\ -x + 2y + 3z &= \;\; 5 \quad (b) \\ 5x - 7y + 9z &= \;\; 8 \quad (c) \end{aligned}$$

Step 1 *and* 2: (*b*) - (a) x -1/4

$$\begin{aligned} -x + 0.25y - 0.5z &= -3.75 \\ 1.75y + 3.5z &= \;\; 8.75 \\ 5x - 7.00y + 9.0z &= \;\; 8.00 \end{aligned}$$

Step 3 (1): (*c*) - (a) x −5

$$\begin{aligned} 5x - 1.25y + 2.5z &= \;\; 18.75 \\ 1.75y + 3.5z &= \;\;\; 8.75 \\ -5.75y + 6.5z &= -10.75 \end{aligned}$$

Step 3 (2): (*c*) - (b) x −5.75/1.75

$$\begin{aligned} 5x - 1.25y + \;\; 2.5z &= \;\; 18.75 \quad (d) \\ -\,5.75y - 11.5z &= -28.75 \quad (e) \\ 18.0z &= \;\; 18.00 \quad (f) \end{aligned}$$

Step 4: from (f) $z = 1$

Step 5: from (e) $y = 3$
from (d) $x = 4$

Figure 2.17: Gaussian elimination method.

corresponding 'iteration matrix' (Kreyszig 1979).

In practice, iteration methods often employ the method of over-relaxation to improve convergence. When the guessed state variable values are applied to any particular nodal equation a residual will result since the values will not, in general, represent the actual solution. The objective of over-relaxation is to so adjust the newly computed temperature values that any new residual is not set to zero but changed in sign in anticipation that subsequent operations on neighbouring nodal equations will have a favourable effect on the over-relaxed residual. Other relaxation techniques, such as block and group relaxation, can also be used to improve convergence. These techniques are described elsewhere (Croft and Lilley 1977).

In building physics applications, variations of the Newton-Raphson method are often employed. This method is described in chapters 5 and 6 where it is applied to the non-linear equation-sets representing distributed fluid and electrical power flow networks.

2.5 Which method?

In the context of design tools intended to provide an early indication of performance trends, the response function and numerical modelling approaches are equally apt. Both can handle the dynamic interactions occurring within buildings, with the linearity and invariability assumptions of the former method being largely acceptable in terms of tool purpose.

It is when this purpose changes to that of emulating reality that a clear distinction emerges. The response function method is a specific analytical technique, mathematically elegant and the outcome of many years of accumulated research and development. However, it is a technique which essentially emerged in response to the need to introduce dynamic considerations into manual methods. Numerical methods, on the other hand, evolved as a result of the dramatic inflation in computing power. The generality of these methods allow their direct application to the spectrum of target domains—building heat transfer, HVAC psychrometric processes, control, indoor air quality, electrical power flow, renewable energy conversion etc—and, more significantly, to the integration of these domains.

Programs based on response function methods do have one distinct advantage: they are often easier to validate. Consider the imposition of an adiabatic plane at an inside or outside surface of a multi-layered construction. With a program based on response functions, this is achieved by simply setting $q_K(t) = 0$ in eqn (2.24). The inquiring reader is invited to consider the problems of imposing the same condition on a program based on a numerical method (§4.1.5 considers some options). The point is that, because the different parts of a numerical model are inter-dependent, it is difficult to impose constraints as required by a particular validation test. It is likely that this has resulted in theoretically inferior analytical programs performing better than their theoretically superior numerical counterpart, when the latter is made to operate with an inappropriately configured model. An action that is incongruous in the case of a numerical method is to attempt to reassign a state variable's value at run-time. Because the state variables are the output from the numerical solution process, any attempt to assign a particular value will violate energy balance. Instead, it is necessary to adjust the source term of the nodal equation in question to effect the required state.

When it comes to the realistic testing of design prototypes, the numerical method has no master: this, at any rate, is the thesis underlying the remainder of this book.

2.6 References and further reading

ASHRAE 1975 *Procedures for Determining Heating and Cooling Loads for Computerised Energy Calculations* (ASHRAE: Task Group on Energy Requirements for Heating and Cooling of Buildings)

Alford J S, Ryan J E and Urban F O 1939 Effects of Heat Storage and Variation in Outdoor Temperature and Solar Intensity on Heat Transfer Through Walls *ASHVE Trans.* (1123) 369-96

Brisken W R and Reque S G 1956 Heat Load Calculations by Thermal Response *ASHVE Trans.* (1123) **62** 391

Carslaw H S and Jaeger J C 1959 *Conduction of Heat in Solids* (2nd edn) (Oxford: Oxford University Press)

Churchill R V 1958 *Operational Mathematics* (New York: McGraw-Hill)

Citherlet S, Clarke J A and Hand J 2001 Integration in Building Physics Simulation *Energy and Building* **33**(5) 451-61

Clarke J A 1977 Environmental Systems Performance (*PhD Thesis* (University of Strathclyde)

Clarke J A 2001 Domain Integration in Building Simulation Simulation *Energy and Buildings* **33** 303-8

Croft D R and Lilley D G 1977 *Heat Transfer Calculations Using Finite Difference Equations* (London: Applied Science Ltd.)

Danter E 1960 Periodic Heat Flow Characteristics of Simple Walls and Roofs *J. IHVE (now CIBSE)* **28** 136-46

— 1973 Heat Exchanges in a Room and the Definition of Room Temperature *Proc. IHVE (now CIBSE) Symp.* (June)

Davies B 1978 Integral *Transforms and Their Applications* (New York: Springer-Verlag)

Gough M 1982 Modelling Heat Flow in Buildings: An Eigenfunction Approach *PhD Thesis* (University of Cambridge)

— 1984 *Private Communication*

Gower N W and Baker J E 1974 *Fourier Series* (London: Chatto and Windus and Collins)

Gupta C L 1964 Matrix Method for Predicting the Thermal Response of Unconditioned Buildings *J. IHVE (now CIBSE)* **32** 159

Gupta C L, Spencer J W and Muncey R W R 1974 A Conceptual Survey of Computer-Oriented Thermal Calculation Methods *Proc. 2nd Symp. Use of Computers for Environ. Eng. Related to Build.*

Harrington-Lynn J 1974a The Admittance Procedure: Variable Ventilation *J. IHVE (now CIBSE)* **42** 199-200

— 1974b The Admittance Procedure: Intermittent Plant Operation *J. IHVE (now CIBSE)* **42** 219-21

Healey M 1967 *Tables of Laplace, Heaviside, Fourier and Z transforms* (Edinburgh: Chambers)

Hittle D C 1979 Building Loads Analysis and System Thermodynamics (BLAST) Users Manual Version 2 *Technical Report E-153* (Champaign, Ill: US Army Construction Eng. Research Laboratory)

— 1981 An Improved Root-Finding Procedure for Use in Calculating Transient Heat Flow Through Multilayered Slabs *Preprint* (Champaign, Ill: US Army Construction Eng. Research Laboratory)

Jury E I 1964 *Theory and Application of the Z-Transform Method* (New York: Wiley)

Kimura K 1977 *Scientific Basis of Air Conditioning* (London: Applied Science)

Kreyszig E 1979 *Advanced Engineering Mathematics* (New York: Wiley)

Kusuda T 1969 Thermal Response Factors for Multi-Layer Structures of Various Heat Conduction Systems *ASHRAE Trans.* **75** 246

Lambert J D 1973 *Computational Methods in Ordinary Differential Equations* (New York: Wiley)

Loudon A G 1968 Summertime Temperatures in Buildings Without Air Conditioning *BRS CP 46* (Garston: Building Research Establishment)

Mackey C O and Wright L T 1944 Periodic Heat Flow—Homogeneous Walls or Roofs *ASHVE Trans.* **50** 293

— 1946 Periodic Heat Flow—Composite Walls or Roofs *Heating, Piping and Air Conditioning* **18**(6) 107-10

Milbank N O and Harrington-Lynn J 1974 Thermal Response and the Admittance Procedure *BRS CP 61* (Garston: Building Research Establishment)

Mitalas G P 1965 An Assessment of Common Assumptions in Estimating Cooling Loads and Space Temperatures *ASHRAE Trans.* **71**(2) 72

Mitalas G P and Arseneault J G 1970 Z Transfer Functions for the Calculation of Transient Heat Transfer through Walls and Roofs *Proc. 1st Symp. Use of Computers for Environ. Eng. Related to Build.*

Muncey R W R 1953 The Calculation of Temperatures Inside Buildings Having Variable External Conditions *J. Appl. Sci.* **4** 189

— 1979 Heat Transfer Calculations for Buildings (Barking: Appl. Sci.)

Nottage H B and Parmelee G V 1955 Circuit Analysis Applied to Load Estimating (pt 2) *ASHRAE Trans.* **61** 125

Pipes L A 1957 Matrix Analysis of Heat Transfer Problems *J. Franklin Institute* **263** 195

Pratt A W and Ball E F 1963 Transient Cooling of a Heated Enclosure *J. Heat and Mass Trans.* **6** 703-18

Press W H, Flannery B P, Teukolsky S A and Vettlering W T 1986 *Numerical Recipes: the Art of Scientific Computing* (Cambridge University Press)

Saito H and Kimura K 1974 Computerised Calculation Procedures of Dynamic Air Conditioning Load Developed by SHASE of Japan *Proc. 2nd Symp. Use of Computers for Environ. Eng. Related to Buildings*

Simonson J R 1967 *An Introduction to Engineering Heat Transfer* (London: McGraw Hill)

Stephenson D G and Mitalas G P 1967a Room Thermal Response Factors *ASHVE Trans.* **73** (2019)

— 1967b Cooling Load Calculations by Thermal Response Factor Method *ASHVE Trans.* **73** (2018)

— 1971 Calculation of Heat Conduction Transfer Functions for Multilayer Slabs *ASHVE Trans.* **2** 117-26

York D A and Tucker E F (eds) 1980 *DOE-2 Reference Manual Version 2.1* (Rep. LA 7689 M) (Los Alamos: Los Alamos Scientific Laboratory)

3

Building simulation

The previous chapter demonstrated the two principal methods by which some governing partial differential equation can be solved analytically or by numerical approximation. This chapter utilises the latter method to construct an energy model that is capable of simulating any building/plant system whilst preserving its spatial and temporal integrity. Model formulation is essentially a three stage process as follows.

> The continuous building/plant system is made discrete by the placement of 'nodes' at preselected points of interest. These nodes represent homogeneous or non-homogeneous physical volumes corresponding to room air, opaque and transparent boundary surfaces, constructional elements, plant component parts, renewable energy components, room contents and so on.
>
> For each node in turn, and in terms of all surrounding nodes representing regions deemed to be in thermodynamic contact, conservation equations are developed to represent the nodal condition and the inter-nodal transfers of energy, mass and momentum.
>
> The entire equation-set is solved simultaneously for successive time steps to obtain the future time-row nodal state variables as a function of the present time-row states and prevailing boundary conditions at both time-rows.

This chapter applies the above process to formulate a building-side energy model, while chapter 6 extends this model to include HVAC, renewable energy conversion and control systems. In order to enumerate the inter-nodal mass transfers, which appear as parameters in the building- and plant-side conservation equations, chapter 5 develops models for inter-zone air flow, intra-zone air/vapour movement and intra-construction moisture flow. Before embarking on the building-side model formulation, there are four prerequisite issues that must be considered.

First, it is impossible to prescribe an optimum spatial discretisation scheme for each part of the building to be included in the model. Clearly, different parts will require different treatments: some demanding fine subdivision (many nodes), others requiring only low resolution (few nodes). One way to address this issue is to utilise the developed model within a parametric study to establish a set of context-dependent defaults.

Second, the discretised conservation equations will have a variable number of coefficients depending on the node type to which they relate. This, in turn, will require a carefully designed matrix coefficient indexing scheme to facilitate efficient equation solution.

Third, because different system parts will have different time constants and coupling strengths, equation processing must be structured to allow these effects to be reconciled whilst not enforcing a lowest common denominator processing frequency.

Last, since different domain equations possess different characteristics—for example, some are highly non-linear—an approach which depends on several co-operating solvers will be more computationally efficient than an approach that attempts to coerce the disparate equation-sets into a form suitable for a single solver type.

The model formulation commenced in this chapter and continued in chapters 4 through 7, is designed to accommodate these issues: it supports any spatial discretisation strategy—from one- to three-dimensional heat flow, and from low to high resolution discretisation; it separates the tasks of equation-set structuring and solution; and it allows the application of multiple solvers and variable frequency solution techniques.

This chapter has three objectives: to discuss system discretisation; to derive the simulation equations for the various 'primitive parts' from which a building model may be constructed; and to demonstrate an approach to equation structuring that derives its form from considerations of building topology. Chapter 4 then describes conservation equation-set formulation in relation to a specific example and details a method for the variable frequency (time step) solution of this equation-set when subjected to weather boundary conditions.

The derivations that follow are elaborated in detail in order to reveal the intricacies of numerical model formulation. This is done because, as ever, the devil is in the detail! Many excellent text exist that give detailed insights into the application of finite differencing techniques (e.g. Richtmyer 1957, Levy 1959, Dusinberre 1961, Hildebrand 1968, Mitchell 1969, John 1982, Lambert 1983).

3.1 System discretisation

There are two main types of error associated with finite differencing schemes: rounding errors and discretisation errors. The former occur in cases where computations include an insufficient number of significant figures. Any tendency towards an accumulation of such errors can rapidly become critical, especially in large numerical schemes involving many computational operations. Fortunately, errors of this type can be reduced to insignificance by the careful design of the numerical scheme and by operating, where appropriate, in double precision.

Discretisation errors result from the replacement of derivatives by finite differences. Although unavoidable, such errors can usually be minimised by reducing the space and time increments (see §2.4). Whilst accuracy considerations dictate that such increments be small, considerations of computational speed require that they be made relatively large. Although it is impossible to predetermine the space and time increments for a given accuracy level, optimum values can be ascertained from simple parametric studies using the developed model. This, of course, implies that a model must first be developed against the assumption that any increment is possible. This greatly promotes the use of implicit formulations because they are unconditionally stable and, if well designed, consistent with the original partial differential equation-set.

Such a parametric study (Clarke 1977) was conducted with an early version of the ESP-r system, which adheres to the theory presented in this chapter. Figure 3.1 shows the

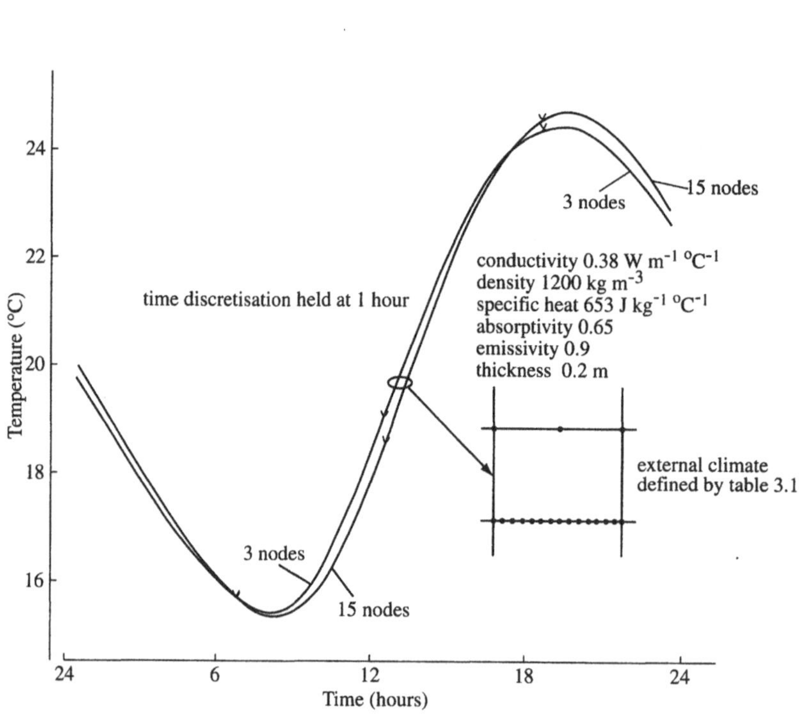

Figure 3.1: Effect of space discretisation.

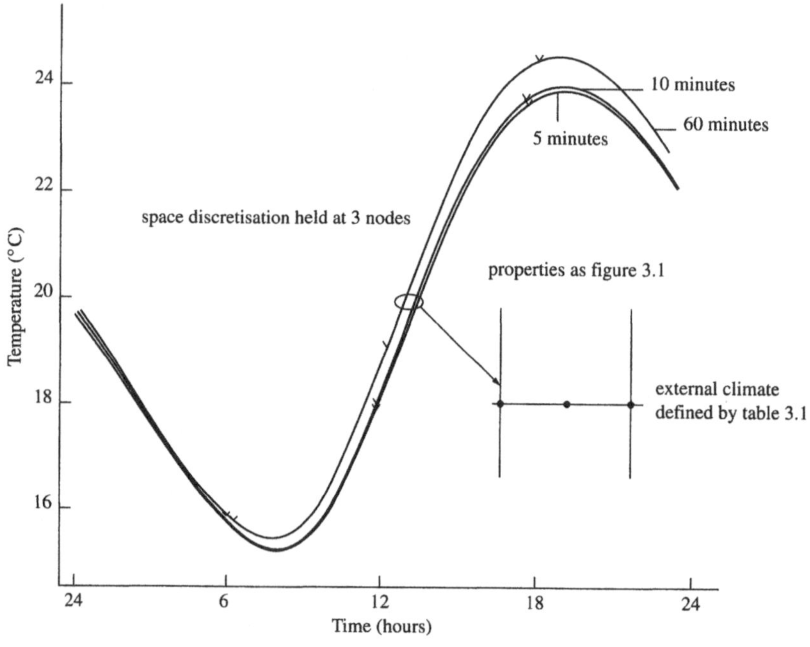

Figure 3.2: Effect of time discretisation.

temperature variations at the internal surface of a construction for the case of uni-directional transient conduction as the number of thermally uniform regions (nodes) is varied (i.e. the space step in the differencing scheme is varied while the time step is held constant). Figure 3.2 shows the corresponding variations as the time step is varied with the space step held constant. To facilitate result reproduction, table 3.1 gives the weather data used in the study.

Table 3.1: External climate definition for figures 3.1 and 3.2.

Hour	*d.b. temp.* °C	*dir. n. rad.* W m^{-2}	*dif. h. rad.* W m^{-2}	*wind sp.* m s^{-1}	*wind dir.* ° *from N*	*r.h.* %
1	16.3	0.	0.	0.0	000	81
2	16.2	0.	0.	0.0	000	85
3	15.2	0.	0.	0.3	045	86
4	15.9	0.	0.	0.8	085	81
5	15.2	1.	10.	1.0	095	81
6	15.9	7.	41.	1.3	110	80
7	18.2	140.	77.	2.2	130	78
8	20.6	405.	95.	4.4	155	68
9	22.3	575.	105.	5.4	165	64
10	23.6	622.	130.	5.7	170	60
11	25.0	634.	158.	6.4	165	55
12	26.2	605.	217.	7.0	160	50
13	26.7	557.	241.	7.2	165	48
14	27.1	568.	214.	7.2	170	48
15	28.0	610.	224.	7.2	170	45
16	28.7	585.	218.	7.0	160	43
17	28.2	475.	172.	6.7	155	45
18	27.5	390.	123.	6.2	155	45
19	26.7	235.	81.	4.9	150	49
20	25.8	49.	40.	3.1	150	52
21	23.8	4.	8.	2.1	160	62
22	23.6	0.	0.	2.2	190	64
23	22.5	0.	0.	2.1	210	69
24	21.8	0.	0.	2.1	225	70

The results from the study suggested that a spatial discretisation scheme equal to or exceeding 3 nodes per homogeneous element will, in most practical situations, be consistent with acceptable accuracy. Any node situated at the boundary between different homogeneous elements will represent mixed thermal property regions, while nodes situated at extreme surfaces—undergoing convective, conductive and radiative heat exchange—will have an associated thermal capacity equal to some fraction of the capacity of the next-to-surface element. It is also evident that computational time steps in excess of one hour should be avoided with no lower limit imposed.

There are alternative nodal placement strategies that attempt to subdivide multi-layered constructions as a function of thermal rather than geometrical criteria and so improve accuracy or reduce nodal subdivision to minimise processing. Two constructs are relevant here: the Biot Number (β) and the dwell time (t_d). For the innermost or outermost layer in a multi-layered construction, the former is given by

$$\beta = \frac{0.5hx}{k}$$

where h is the surface convection coefficient (W m^{-2}°C^{-1}), x is the layer thickness (m) and k the conductivity (W m^{-1}°C^{-1}). If β is much less than 1 then the layer may be represented by the lumped capacitance method. Where this condition is not met, spatial effects are important and the layer should be discretised. Hensen and Nakhi (1994) have reported on the relationship between Biot Number and conduction modelling accuracy in the context of building performance prediction. The second construct, dwell time, is defined by

$$t_d = \frac{(x^2 \rho C)}{k} .$$

A multi-layered construction comprising N layers may be restructured such that the square root of the dwell time across each new layer is the same:

$$t_d^{\frac{1}{2}} = \frac{1}{N}\left(\sum_{i=1}^{N} \frac{x_i (\rho C)_i^{\frac{1}{2}}}{k_i^{\frac{1}{2}}}\right).$$

Commencing at layer 1, the individual dwell time square roots are summed until

$$\sum_i \frac{x_i (\rho C)_i^{\frac{1}{2}}}{k_i^{\frac{1}{2}}} \geq t_d^{\frac{1}{2}} .$$

When this condition occurs, the dwell time square root, for the last layer to be included in the summation, is subtracted and the layer subdivided until the average dwell time square root value is obtained. The dwell time summation is then set to zero and the process continues from the current location, proceeding until the entire construction has been processed. After restructuring there may be more layers present than existed in the original construction but now the layers will more closely match the distribution of capacity.

In transient conduction schemes involving more than one space dimension it is not possible to prescribe the nodal placements since this will depend on such factors as internal and external surface insolation, the existence of localised convection, the presence of corner effects and thermal bridges, and the shape of the capacity/insulation system being modelled: all factors causing position dependent transient effects. Nevertheless, in many applications n-dimensional transient conduction schemes will become necessary, with mixed-dimensional schemes proving useful. Figure 3.3, for example, gives some mixed schemes and their corresponding application. In the following derivations the full 3-dimensional scheme is assumed for transient conduction nodes, with the reduction to lower dimensions demonstrated where appropriate.

Likewise, it is not possible to prescribe the spatial subdivision of fluid volumes (room air, boiler combustion chamber, wall cavity etc) although a number of general points can be made.

It is usually desirable to subdivide the volume vertically to include the buoyancy effects of density variations resulting in stratification.

Local, fine discretisation will be required adjacent to bounding surfaces—to allow the effects of solar patch movement to be studied or to support a link between the building fabric and an adjacent computational fluid dynamics domain.

Global, fine discretisation will be required where intra-space air movement, comfort distribution and indoor air quality are the issues to be studied.

In general terms, the subdivision criteria will depend on the expected variations of fundamental thermophysical properties and heat fluxes throughout the system, on the extent to which distinct regions will be subjected to control action, and the ultimate simulation objectives. Particular discretisation schemes are given in §3.3, where the equations derived in §3.2 are combined to demonstrate the construction of an equation-set for an example problem.

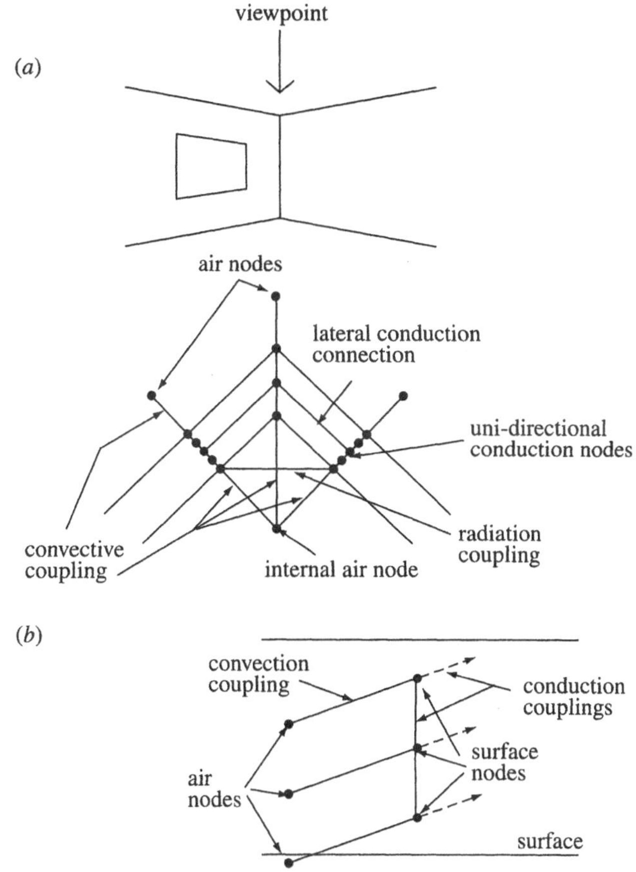

Figure 3.3: Some mixed nodal schemes and their typical applications. (a) Corner effects—a combined one- and two-dimensional scheme. (b) Surface temperature gradient—a two-dimensional scheme.

3.2 Finite volume energy equation formulation

This section applies the finite volume heat balance method introduced in chapter 2 to the characteristic regions encountered within buildings. For each region type, or primitive part, an energy conservation equation is derived to link the region with other regions that are in thermodynamic contact by conduction, convection, radiation and fluid flow.

In preparation for the matrix processing method introduced in chapter 4, note that the equations to emerge have an underlying similarity: the terms grouped on the left-hand side of the equality relate to the future (as yet unknown) time-row of some arbitrary time step; those on the right-hand side address the present (known) time-row. In the terminology, the coefficient of the state variable to which the equation applies (the target) is termed the 'self-coupling' coefficient. The remaining equation coefficients are termed 'cross-coupling' since they link the target node with coupled regions. All terms relating to boundary condition excitations are gathered on the right-hand side since they are known for all time regardless of the time-row to which they relate. It is this pattern of similarity that permits the matrix partitioning technique of chapter 4 and the efficient solution method that follows.

Figure 3.4 shows the various energy flowpaths occurring within buildings and so candidates for inclusion within a simulation model.

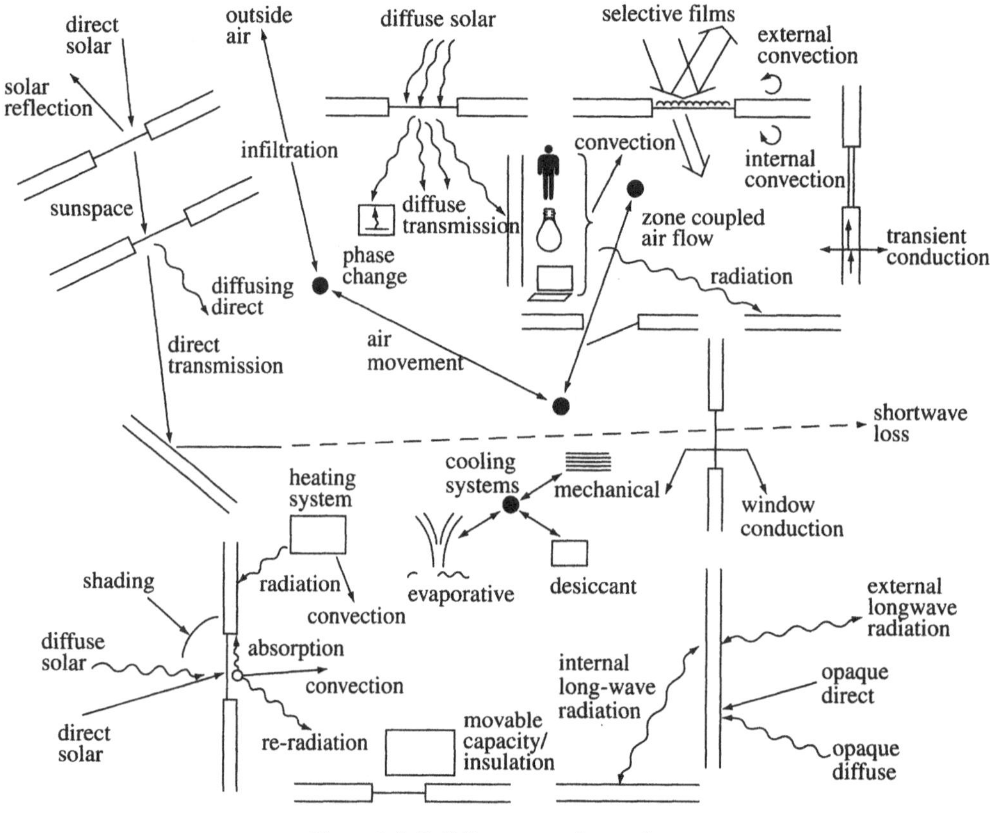

Figure 3.4: Building energy flowpaths.

Recalling the energy balance relationship of §2.4.2 and assuming that the thermal properties of region I are time dependent:

$$\frac{\rho_I(\xi)C_I(\xi)\delta V_I(\xi)}{\delta t}[\theta(I, t+\delta t) - \theta(I, t)] = \sum_{i=1}^{N} K_{i,I}[\theta(i, \xi) - \theta(I, \xi)] + q_I(\xi) + \varepsilon \qquad (3.1)$$

where $\rho_I(\xi)$ is the volume-averaged or otherwise representative density of region I at some time ξ (kg m^{-3}), $C_I(\xi)$ the representative specific heat capacity of the region (J kg^{-1}°C^{-1}), $\delta V_I(\xi)$ the region volume (m^3), δt the discretisation time step (s), $\theta(I, \xi)$ the representative temperature of region I at time ξ (°C), and $K_{i,I}(\xi)$ the heat flow conductance between region i and I (W °C^{-1}). The heat generation within region I is denoted by $q_I(\xi)$ (W), ε is the error resulting from the evaluation over finite (as opposed to infinitesimal) space- and time-increments, N the number of energy exchange flowpaths between region I and surrounding regions, t the present time-row, and $t + \delta t$ the future time-row.

Evaluation of eqn (3.1) at the present (known) time-row, $\xi = t$, gives the fully explicit scheme in which all nodal equations are independent—since they contain only present values of all coupled nodes—and so can be solved directly. Evaluation at the future (unknown) time-row, $\xi = t + \delta t$, gives the fully implicit scheme in which all nodes are linked at the future time-row and so the entire equation-set must be solved simultaneously.

Chapter 2 discussed the main advantages and disadvantages of both formulations and outlined the concatenation of implicit and explicit schemes to provide a method that combines best accuracy with unconditional stability. Such a method, based on eqn (3.1), can now be applied to the characteristic node types that represent the different portions of a buildings.

Nodes that represent the energy balance of regions located within capacity/insulation systems such as the material comprising the building fabric and room contents.

Nodes that represent the energy balance at bounding surfaces such as indoor finishes and exposed roofs.

Nodes that represent the energy balance within fluid volumes such as portions of room air.

The derived equations are general and can be applied equally to building and plant components; indeed chapter 6 applies the same equations to HVAC and renewable energy system components.

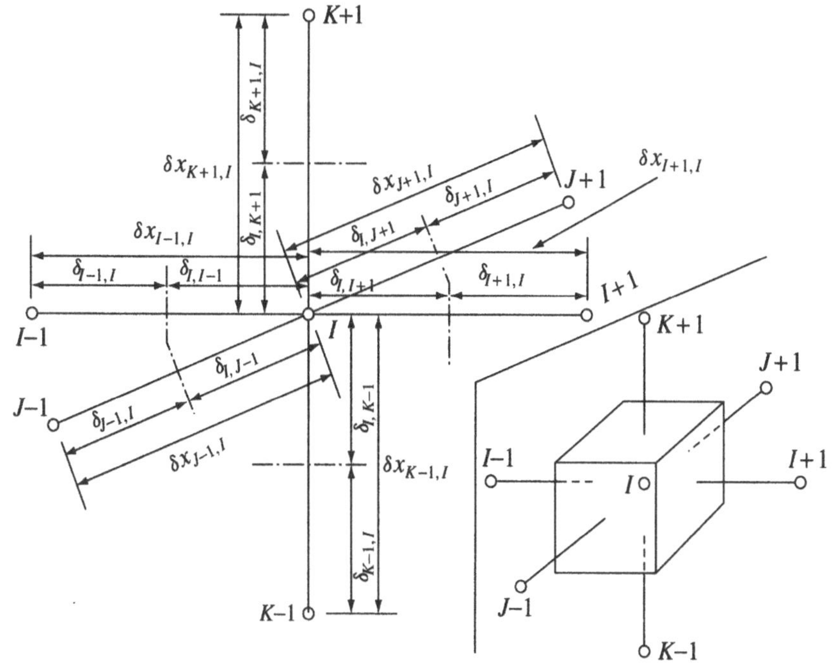

Figure 3.5: Nodes for transient conduction.

3.2.1 Capacity/insulation systems

Consider figure 3.5, which shows a number of discrete regions, denoted I, I − 1, I + 1, J − 1 etc, in conductive communication. Within this scheme node I represents the discrete finite volume given by

$$(\delta_{I,I-1} + \delta_{I,I+1})(\delta_{I,J-1} + \delta_{I,J+1})(\delta_{I,K-1} + \delta_{I,K+1}) .$$

The heat flux by conduction towards node I is given by

$$q_{I-1,I} = k'_{I-1,I}(\delta_{I,J-1} + \delta_{I,J+1})(\delta_{I,K-1} + \delta_{I,K+1})(\theta_{I-1} - \theta_I)/\delta x_{I-1,I}$$

$$q_{I+1,I} = k'_{I+1,I}(\delta_{I,J-1} + \delta_{I,J+1})(\delta_{I,K-1} + \delta_{I,K+1})(\theta_{I+1} - \theta_I)/\delta x_{I+1,I}$$

$$q_{J-1,I} = k'_{J-1,I}(\delta_{I,I-1} + \delta_{I,I+1})(\delta_{I,K-1} + \delta_{I,K+1})(\theta_{J-1} - \theta_I)/\delta x_{J-1,I}$$

$$q_{J+1,I} = k'_{J+1,I}(\delta_{I,I-1} + \delta_{I,I+1})(\delta_{I,K-1} + \delta_{I,K+1})(\theta_{J+1} - \theta_I)/\delta x_{J+1,I}$$

$$q_{K-1,I} = k'_{K-1,I}(\delta_{I,I-1} + \delta_{I,I+1})(\delta_{I,J-1} + \delta_{I,J+1})(\theta_{K-1} - \theta_I)/\delta x_{K-1,I}$$

$$q_{K+1,I} = k'_{K+1,I}(\delta_{I,I-1} + \delta_{I,I+1})(\delta_{I,J-1} + \delta_{I,J+1})(\theta_{K+1} - \theta_I)/\delta x_{K+1,I}$$

where k' is the average inter-nodal conductivity (W m^{-1}°C^{-1}), δx the inter-nodal distance in the direction of heat flow (m), and $\delta_{i\pm1,I} = \delta_{I,i\pm1} = \delta x_{i\pm1,I}/2$; i = I, J, K.

The average conductivity value is necessary to account for the possibility that the inter-nodal connections may not be homogeneous but comprised of different materials. Also, since region I may not be homogeneous and/or isotropic, it is necessary to express the other thermophysical properties as a volumetric weighting of the properties of the different materials that comprise the region. It follows from eqn (3.1) that the 3-dimensional energy balance relationship for a node undergoing transient conduction with a potential for heat generation is given by

$$[W_I(t+\delta t)\theta(I,t+\delta t) - W_I(t)\theta(I,t)](\delta_{I,I-1} + \delta_{I,I+1})(\delta_{I,J-1} + \delta_{I,J+1})(\delta_{I,K-1} + \delta_{I,K+1})/\delta t$$

$$= k'_{I-1,I}(\xi)(\delta_{I,J-1} + \delta_{I,J+1})(\delta_{I,K-1} + \delta_{I,K+1})[\theta(I-1,\xi) - \theta(I,\xi)]/\delta x_{I-1,I}$$

$$+ k'_{I+1,I}(\xi)(\delta_{I,J-1} + \delta_{I,J+1})(\delta_{I,K-1} + \delta_{I,K+1})[\theta(I+1,\xi) - \theta(I,\xi)]/\delta x_{I+1,I}$$

$$+ k'_{J-1,I}(\xi)(\delta_{I,I-1} + \delta_{I,I+1})(\delta_{I,K-1} + \delta_{I,K+1})[\theta(J-1,\xi) - \theta(I,\xi)]/\delta x_{J-1,I}$$

$$+ k'_{J+1,I}(\xi)(\delta_{I,I-1} + \delta_{I,I+1})(\delta_{I,K-1} + \delta_{I,K+1})[\theta(J+1,\xi) - \theta(I,\xi)]/\delta x_{J+1,I}$$

$$+ k'_{K-1,I}(\xi)(\delta_{I,I-1} + \delta_{I,I+1})(\delta_{I,J-1} + \delta_{I,J+1})[\theta(K-1,\xi) - \theta(I,\xi)]/\delta x_{K-1,I}$$

$$+ k'_{K+1,I}(\xi)(\delta_{I,I-1} + \delta_{I,I+1})(\delta_{I,J-1} + \delta_{I,J+1})[\theta(K+1,\xi) - \theta(I,\xi)]/\delta x_{K+1,I} + q_I(\xi) + \varepsilon \tag{3.2}$$

where W_I is the volume weighted product of density and specific heat capacity of region I (J m^{-3}°C^{-1}); that is

$$\frac{\sum_{i=1}^{N} \rho_i C_i \delta V_i}{\sum_{i=1}^{N} \delta V_i}$$

and N is the number of different materials comprising region I.

This is the fundamental 3-dimensional relationship for transient conduction within capacity/insulation systems. As the space and time steps (δ, δx and δt) approach zero, the resulting partial differential heat equation is the Fourier Field Equation with heat generation:

$$\rho_I C_I \frac{\partial \theta_I}{\partial t} = \frac{\partial}{\partial x_I}\left(k_{I-1,I+1}\frac{\partial \theta_I}{\partial x_I}\right) + \frac{\partial}{\partial x_J}\left(k_{J-1,J+1}\frac{\partial \theta_I}{\partial x_J}\right) + \frac{\partial}{\partial x_K}\left(k_{K-1,K+1}\frac{\partial \theta_I}{\partial x_K}\right) + q'_I \tag{3.3}$$

where q_I' is the heat generation per unit volume within region I (W m^{-3}). For an isotropic, homogeneous material this equation reduces to

$$\frac{1}{\alpha}\frac{\partial \theta_I}{\partial t} = \nabla^2 \theta_I + q_I'/k_I$$

where α is the thermal diffusivity (m^{2}s^{-1}).

Eqn (3.2) is now used to obtain the general form of the simulation equation for transient conduction nodes. Evaluation of the equation at the present time-row, $\xi = t$, yields, after rearrangement, the temperature explicit formulation:

$$\begin{aligned}
\theta(I, t+\delta t) = & \left(\frac{W_I(t)}{W_I(t+\delta t)} - \frac{k_{I-1,I}'(t)}{S_{I-1,I}(t+\delta t)(\delta_{I,I-1}+\delta_{I,I+1})} \right. \\
& - \frac{k_{I+1,I}'(t)}{S_{I+1,I}(t+\delta t)(\delta_{I,I-1}+\delta_{I,I+1})} \\
& - \frac{k_{J-1,I}'(t)}{S_{J-1,I}(t+\delta t)(\delta_{I,J-1}+\delta_{I,J+1})} - \frac{k_{J+1,I}'(t)}{S_{J+1,I}(t+\delta t)(\delta_{I,J-1}+\delta_{I,J+1})} \\
& \left. - \frac{k_{K-1,I}'(t)}{S_{K-1,I}(t+\delta t)(\delta_{I,K-1}+\delta_{I,K+1})} - \frac{k_{K+1,I}'(t)}{S_{K+1,I}(t+\delta t)(\delta_{I,K-1}+\delta_{I,K+1})} \right) \theta(I, t) \\
& + \frac{k_{I-1,I}'(t)\theta(I-1, t)}{S_{I-1,I}(t+\delta t)(\delta_{I,I-1}+\delta_{I,I+1})} + \frac{k_{I+1,I}'(t)\theta(I+1, t)}{S_{I+1,I}(t+\delta t)(\delta_{I,I-1}+\delta_{I,I+1})} \\
& + \frac{k_{J-1,I}'(t)\theta(J-1, t)}{S_{J-1,I}(t+\delta t)(\delta_{I,J-1}+\delta_{I,J+1})} + \frac{k_{J+1,I}'(t)\theta(J+1, t)}{S_{J+1,I}(t+\delta t)(\delta_{I,J-1}+\delta_{I,J+1})} \\
& + \frac{k_{K-1,I}'(t)\theta(K-1, t)}{S_{K-1,I}(t+\delta t)(\delta_{I,K-1}+\delta_{I,K+1})} + \frac{k_{K+1,I}'(t)\theta(K+1, t)}{S_{K+1,I}(t+\delta t)(\delta_{I,K-1}+\delta_{I,K+1})} \\
& + \frac{\delta t\, q_I(t)}{W_I(t+\delta t)(\delta_{I,I-1}+\delta_{I,I+1})(\delta_{I,J-1}+\delta_{I,J+1})(\delta_{I,K-1}+\delta_{I,K+1})} + \varepsilon
\end{aligned}$$

where $S_{i\pm1,I}(t+\delta t) = W_I(t+\delta t)\delta x_{i\pm1,I}/\delta t$; $i = I, J, K$. Note that, although the formulation is explicit in the temperature variable, each present time-row temperature coefficient contains region thermophysical properties that must be evaluated at the future time-row. The expression will, however, become fully explicit if the assumption is made that region properties are invariant in the time dimension.

If ξ is now set to $t + \delta t$ in eqn (3.2) then the fully implicit formulation is obtained:

$$\begin{aligned}
\theta(I, t+\delta t) = & \frac{W_I(t)}{W_I(t+\delta t)}\theta(I, t) - \left(\frac{k_{I-1,I}'(t+\delta t)}{S_{I-1,I}(t+\delta t)(\delta_{I,I-1}+\delta_{I,I+1})} \right. \\
& + \frac{k_{I+1,I}'(t+\delta t)}{S_{I+1,I}(t+\delta t)(\delta_{I,I-1}+\delta_{I,I+1})} + \frac{k_{J-1,I}'(t+\delta t)}{S_{J-1,I}(t+\delta t)(\delta_{I,J-1}+\delta_{I,J+1})} \\
& + \frac{k_{J+1,I}'(t+\delta t)}{S_{J+1,I}(t+\delta t)(\delta_{I,J-1}+\delta_{I,J+1})} + \frac{k_{K-1,I}'(t+\delta t)}{S_{K-1,I}(t+\delta t)(\delta_{I,K-1}+\delta_{I,K+1})} \\
& \left. + \frac{k_{K+1,I}'(t+\delta t)}{S_{K+1,I}(t+\delta t)(\delta_{I,K-1}+\delta_{I,K+1})} \right) \theta(I, t+\delta t) \\
& + \frac{k_{I-1,I}'(t+\delta t)\theta(I-1, t+\delta t)}{S_{I-1,I}(t+\delta t)(\delta_{I,I-1}+\delta_{I,I+1})} + \frac{k_{I+1,I}'(t+\delta t)\theta(I+1, t+\delta t)}{S_{I+1,I}(t+\delta t)(\delta_{I,I-1}+\delta_{I,I+1})}
\end{aligned}$$

$$+\frac{k'_{J-1,I}(t+\delta t)\theta(J-1,t+\delta t)}{S_{J-1,I}(t+\delta t)(\delta_{I,J-1}+\delta_{I,J+1})}+\frac{k'_{J+1,I}(t+\delta t)\theta(J+1,t+\delta t)}{S_{J+1,I}(t+\delta t)(\delta_{I,J-1}+\delta_{I,J+1})}$$
$$+\frac{k'_{K-1,I}(t+\delta t)\theta(K-1,t+\delta t)}{S_{K-1,I}(t+\delta t)(\delta_{I,K-1}+\delta_{I,K+1})}+\frac{k'_{K+1,I}(t+\delta t)\theta(K+1,t+\delta t)}{S_{K+1,I}(t+\delta t)(\delta_{I,K-1}+\delta_{I,K+1})}$$
$$+\frac{\delta t\, q_I(t+\delta t)}{W_I(t+\delta t)(\delta_{I,I-1}+\delta_{I,I+1})(\delta_{I,J-1}+\delta_{I,J+1})(\delta_{I,K-1}+\delta_{I,K+1})}+\varepsilon\,.$$

Adding the explicit and implicit formulations, and grouping future time-row temperature terms on the equation left-hand side, gives

$$\left(2+\frac{k'_{I-1,I}(t+\delta t)}{S_{I-1,I}(t+\delta t)(\delta_{I,I-1}+\delta_{I,I+1})}+\frac{k'_{I+1,I}(t+\delta t)}{S_{I+1,I}(t+\delta t)(\delta_{I,I-1}+\delta_{I,I+1})}\right.$$
$$+\frac{k'_{J-1,I}(t+\delta t)}{S_{J-1,I}(t+\delta t)(\delta_{I,J-1}+\delta_{I,J+1})}+\frac{k'_{J+1,I}(t+\delta t)}{S_{J+1,I}(t+\delta t)(\delta_{I,J-1}+\delta_{I,J+1})}$$
$$\left.+\frac{k'_{K-1,I}(t+\delta t)}{S_{K-1,I}(t+\delta t)(\delta_{I,K-1}+\delta_{I,K+1})}+\frac{k'_{K+1,I}(t+\delta t)}{S_{K+1,I}(t+\delta t)(\delta_{I,K-1}+\delta_{I,K+1})}\right)\theta(I,t+\delta t)$$
$$-\frac{k'_{I-1,I}(t+\delta t)\theta(I-1,t+\delta t)}{S_{I-1,I}(t+\delta t)(\delta_{I,I-1}+\delta_{I,I+1})}-\frac{k'_{I+1,I}(t+\delta t)\theta(I+1,t+\delta t)}{S_{I+1,I}(t+\delta t)(\delta_{I,I-1}+\delta_{I,I+1})}$$
$$-\frac{k'_{J-1,I}(t+\delta t)\theta(J-1,t+\delta t)}{S_{J-1,I}(t+\delta t)(\delta_{I,J-1}+\delta_{I,J+1})}-\frac{k'_{J+1,I}(t+\delta t)\theta(J+1,t+\delta t)}{S_{J+1,I}(t+\delta t)(\delta_{I,J-1}+\delta_{I,J+1})}$$
$$-\frac{k'_{K-1,I}(t+\delta t)\theta(K-1,t+\delta t)}{S_{K-1,I}(t+\delta t)(\delta_{I,K-1}+\delta_{I,K+1})}-\frac{k'_{K+1,I}(t+\delta t)\theta(K+1,t+\delta t)}{S_{K+1,I}(t+\delta t)(\delta_{I,K-1}+\delta_{I,K+1})}$$
$$-\frac{\delta t\, q_I(t+\delta t)}{W_I(t+\delta t)(\delta_{I,I-1}+\delta_{I,I+1})(\delta_{I,J-1}+\delta_{I,J+1})(\delta_{I,K-1}+\delta_{I,K+1})}$$
$$=\left(\frac{2W_I(t)}{W_I(t+\delta t)}-\frac{k'_{I-1,I}(t)}{S_{I-1,I}(t+\delta t)(\delta_{I,I-1}+\delta_{I,I+1})}-\frac{k'_{I+1,I}(t)}{S_{I+1,I}(t+\delta t)(\delta_{I,I-1}+\delta_{I,I+1})}\right.$$
$$-\frac{k'_{J-1,I}(t)}{S_{J-1,I}(t+\delta t)(\delta_{I,J-1}+\delta_{I,J+1})}-\frac{k'_{J+1,I}(t)}{S_{J+1,I}(t+\delta t)(\delta_{I,J-1}+\delta_{I,J+1})}$$
$$\left.-\frac{k'_{K-1,I}(t)}{S_{K-1,I}(t+\delta t)(\delta_{I,K-1}+\delta_{I,K+1})}-\frac{k'_{K+1,I}(t)}{S_{K+1,I}(t+\delta t)(\delta_{I,K-1}+\delta_{I,K+1})}\right)\theta(I,t)$$
$$+\frac{k'_{I-1,I}(t+\delta t)\theta(I-1,t)}{S_{I-1,I}(t+\delta t)(\delta_{I,I-1}+\delta_{I,I+1})}+\frac{k'_{I+1,I}(t+\delta t)\theta(I+1,t)}{S_{I+1,I}(t+\delta t)(\delta_{I,I-1}+\delta_{I,I+1})}$$
$$+\frac{k'_{J-1,I}(t+\delta t)\theta(J-1,t)}{S_{J-1,I}(t+\delta t)(\delta_{I,J-1}+\delta_{I,J+1})}+\frac{k'_{J+1,I}(t+\delta t)\theta(J+1,t)}{S_{J+1,I}(t+\delta t)(\delta_{I,J-1}+\delta_{I,J+1})}$$
$$+\frac{k'_{K-1,I}(t+\delta t)\theta(K-1,t)}{S_{K-1,I}(t+\delta t)(\delta_{I,K-1}+\delta_{I,K+1})}+\frac{k'_{K+1,I}(t+\delta t)\theta(K+1,t)}{S_{K+1,I}(t+\delta t)(\delta_{I,K-1}+\delta_{I,K+1})}$$
$$+\frac{\delta t\, q_I(t)}{W_I(t+\delta t)(\delta_{I,I-1}+\delta_{I,I+1})(\delta_{I,J-1}+\delta_{I,J+1})(\delta_{I,K-1}+\delta_{I,K+1})}+\varepsilon\,.$$

It should be noted that the coefficients of the temperature variables of this equation are dimensionless Fourier numbers.

If a model based on this and later equations is to be used for the investigation of zero capacity (i.e. steady state) systems, then the capacity term of the denominator will cause problems due to division by zero. This difficulty can be overcome by multiplying throughout by the common W_I term present in S to give

$$\left(2W_I(t+\delta t) + \frac{\delta t\, k'_{I-1,I}(t+\delta t)}{\delta x_{I-1,I}(\delta_{I,I-1}+\delta_{I,I+1})} + \frac{\delta t\, k'_{I+1,I}(t+\delta t)}{\delta x_{I+1,I}(\delta_{I,I-1}+\delta_{I,I+1})}\right.$$
$$+ \frac{\delta t\, k'_{J-1,I}(t+\delta t)}{\delta x_{J-1,I}(\delta_{I,J-1}+\delta_{I,J+1})} + \frac{\delta t\, k'_{J+1,I}(t+\delta t)}{\delta x_{J+1,I}(\delta_{I,J-1}+\delta_{I,J+1})} + \frac{\delta t\, k'_{K-1,I}(t+\delta t)}{\delta x_{K-1,I}(\delta_{I,K-1}+\delta_{I,K+1})}$$
$$\left.+ \frac{\delta t\, k'_{K+1,I}(t+\delta t)}{\delta x_{K+1,I}(\delta_{I,K-1}+\delta_{I,K+1})}\right)\theta(I,t+\delta t)$$
$$- \frac{\delta t\, k'_{I-1,I}(t+\delta t)\theta(I-1,t+\delta t)}{\delta x_{I-1,I}(\delta_{I,I-1}+\delta_{I,I+1})} - \frac{\delta t\, k'_{I+1,I}(t+\delta t)\theta(I+1,t+\delta t)}{\delta x_{I+1,I}(\delta_{I,I-1}+\delta_{I,I+1})}$$
$$- \frac{\delta t\, k'_{J-1,I}(t+\delta t)\theta(J-1,t+\delta t)}{\delta x_{J-1,I}(\delta_{I,J-1}+\delta_{I,J+1})} - \frac{\delta t\, k'_{J+1,I}(t+\delta t)\theta(J+1,t+\delta t)}{\delta x_{J+1,I}(\delta_{I,J-1}+\delta_{I,J+1})}$$
$$- \frac{\delta t\, k'_{K-1,I}(t+\delta t)\theta(K-1,t+\delta t)}{\delta x_{K-1,I}(\delta_{I,K-1}+\delta_{I,K+1})} - \frac{\delta t\, k'_{K+1,I}(t+\delta t)\theta(K+1,t+\delta t)}{\delta x_{K+1,I}(\delta_{I,K-1}+\delta_{I,K+1})}$$
$$- \frac{\delta t\, q_I(t+\delta t)}{(\delta_{I,I-1}+\delta_{I,I+1})(\delta_{I,J-1}+\delta_{I,J+1})(\delta_{I,K-1}+\delta_{I,K+1})}$$
$$= \left(2W_I(t) - \frac{\delta t\, k'_{I-1,I}(t)}{\delta x_{I-1,I}(\delta_{I,I-1}+\delta_{I,I+1})} - \frac{\delta t\, k'_{I+1,I}(t)}{\delta x_{I+1,I}(\delta_{I,I-1}+\delta_{I,I+1})}\right.$$
$$- \frac{\delta t\, k'_{J-1,I}(t)}{\delta x_{J-1,I}(\delta_{I,J-1}+\delta_{I,J+1})} - \frac{\delta t\, k'_{J+1,I}(t)}{\delta x_{J+1,I}(\delta_{I,J-1}+\delta_{I,J+1})}$$
$$\left.- \frac{\delta t\, k'_{K-1,I}(t)}{\delta x_{K-1,I}(\delta_{I,K-1}+\delta_{I,K+1})} - \frac{\delta t\, k'_{K+1,I}(t)}{\delta x_{K+1,I}(\delta_{I,K-1}+\delta_{I,K+1})}\right)\theta(I,t)$$
$$+ \frac{\delta t\, k'_{I-1,I}(t)\theta(I-1,t)}{\delta x_{I-1,I}(\delta_{I,I-1}+\delta_{I,I+1})} + \frac{\delta t\, k'_{I+1,I}(t)\theta(I+1,t)}{\delta x_{I+1,I}(\delta_{I,I-1}+\delta_{I,I+1})} + \frac{\delta t\, k'_{J-1,I}(t)\theta(J-1,t)}{\delta x_{J-1,I}(\delta_{I,J-1}+\delta_{I,J+1})}$$
$$+ \frac{\delta t\, k'_{J+1,I}(t)\theta(J+1,t)}{\delta x_{J+1,I}(\delta_{I,J-1}+\delta_{I,J+1})} + \frac{\delta t\, k'_{K-1,I}(t)\theta(K-1,t)}{\delta x_{K-1,I}(\delta_{I,K-1}+\delta_{I,K+1})} + \frac{\delta t\, k'_{K+1,I}(t)\theta(K+1,t)}{\delta x_{K+1,I}(\delta_{I,K-1}+\delta_{I,K+1})}$$
$$+ \frac{\delta t\, q_I(t)}{(\delta_{I,I-1}+\delta_{I,I+1})(\delta_{I,J-1}+\delta_{I,J+1})(\delta_{I,K-1}+\delta_{I,K+1})} + \varepsilon\,. \qquad (3.4)$$

This equation defines the general transient conduction node case and is equivalent to a Crank-Nicolson difference formulation: a weighted average of the fully explicit scheme—in which the second order space derivative of eqn (3.3) is expressed in central difference form with the first order time derivative expressed as a forward difference—and the implicit scheme—in which the first order time derivative is expressed as a backward difference with the second order space derivative in central difference form. This method can be shown to be consistent, convergent and stable, providing the possibility of variable time stepping and well adapted for the solution of so-called 'stiff' problems (Mitchell 1969) in which time constants vary by more than an order of magnitude.

For a homogeneous or non-homogeneous region, I, eqn (3.4) gives

$$C_s(t+\delta t)\theta(I,t+\delta t) - \sum_{i=1}^{N} C_{ci}(t+\delta t)\theta(i,t+\delta t) - \frac{\delta t\, q_I(t+\delta t)}{\delta V_I}$$

$$= C_s(t)\theta(I,t) + \sum_{i=1}^{N} C_{ci}(t)\theta(i,t) + \frac{\delta t\, q_I(t)}{\delta V_I} + \varepsilon$$

where $C_s(\xi)$ is the *self-coupling* coefficient at time ξ, $C_c(\xi)$ the *cross-coupling* coefficient, and N the number of inter-nodal contacts.

In a large building/plant model, economy of nodal discretisation is often required in order to obtain acceptable run-times. In this regard it is important to distinguish between plant and building components with respect to transient conduction modelling.

With plant components it is usually the flow processes and surface heat transfers that are of prime concern and the formulations of §§3.2.2 and 3.2.3 are relevant. It is therefore necessary to use only a few nodes to adequately represent the transient conduction within the material comprising such components. Chapter 6 demonstrates the application of eqn (3.4) to represent the material of typical plant components.

With building components (walls, windows, furniture and the range of passive solar features such as mass walls, phase change materials etc), the detailed modelling of transient conduction is crucial to an accurate simulation of the overall system. It is important, therefore, to devise a mathematical model of transient conduction which is sensitive to the relative positions of the constituent elements. For this reason eqn (3.4) is now used to derive node specific formulations.

Consider figure 3.6 which shows an arbitrary multi-layered construction with a three-dimensional nodal mesh imposed.

Opaque homogeneous element nodes

Assume node I is situated at the centre plane of an opaque homogeneous element located within some multi-layered construction (wall, ceiling, floor, furniture, window system etc). The distance $I-1 \rightarrow I+1$ defines the thickness of the element. For this case

$$\delta x_{I-1,I} = \delta x_{I+1,I} = \delta x_I$$

$$\delta x_{J-1,I} = \delta x_{J+1,I} = \delta x_J$$

$$\delta x_{K-1,I} = \delta x_{K+1,I} = \delta x_K \;.$$

Noting that

$$k'_{I-1,I} = k'_{I+1,I} = k_I$$

$$k'_{J-1,I} = k'_{J+1,I} = k_J$$

$$k'_{K-1,I} = k'_{K+1,I} = k_K$$

$$W_I = \rho_I C_I$$

$$\delta x_{i\pm1}(\delta_{I,i-1} + \delta_{I,i+1}) = \delta x_i^2 \;;\; i = I, J, K$$

then, from eqn (3.4), it follows that

$$\left(2\rho_I(t+\delta t)C_I(t+\delta t) + \frac{2\delta t\, k_I(t+\delta t)}{\delta x_I^2} + \frac{2\delta t\, k_J(t+\delta t)}{\delta x_J^2} + \frac{2\delta t\, k_K(t+\delta t)}{\delta x_K^2}\right)\theta(I,t+\delta t)$$
$$- \frac{\delta t\, k_I(t+\delta t)}{\delta x_I^2}\,\theta(I-1,t+\delta t) - \frac{\delta t\, k_I(t+\delta t)}{\delta x_I^2}\,\theta(I+1,t+\delta t)$$

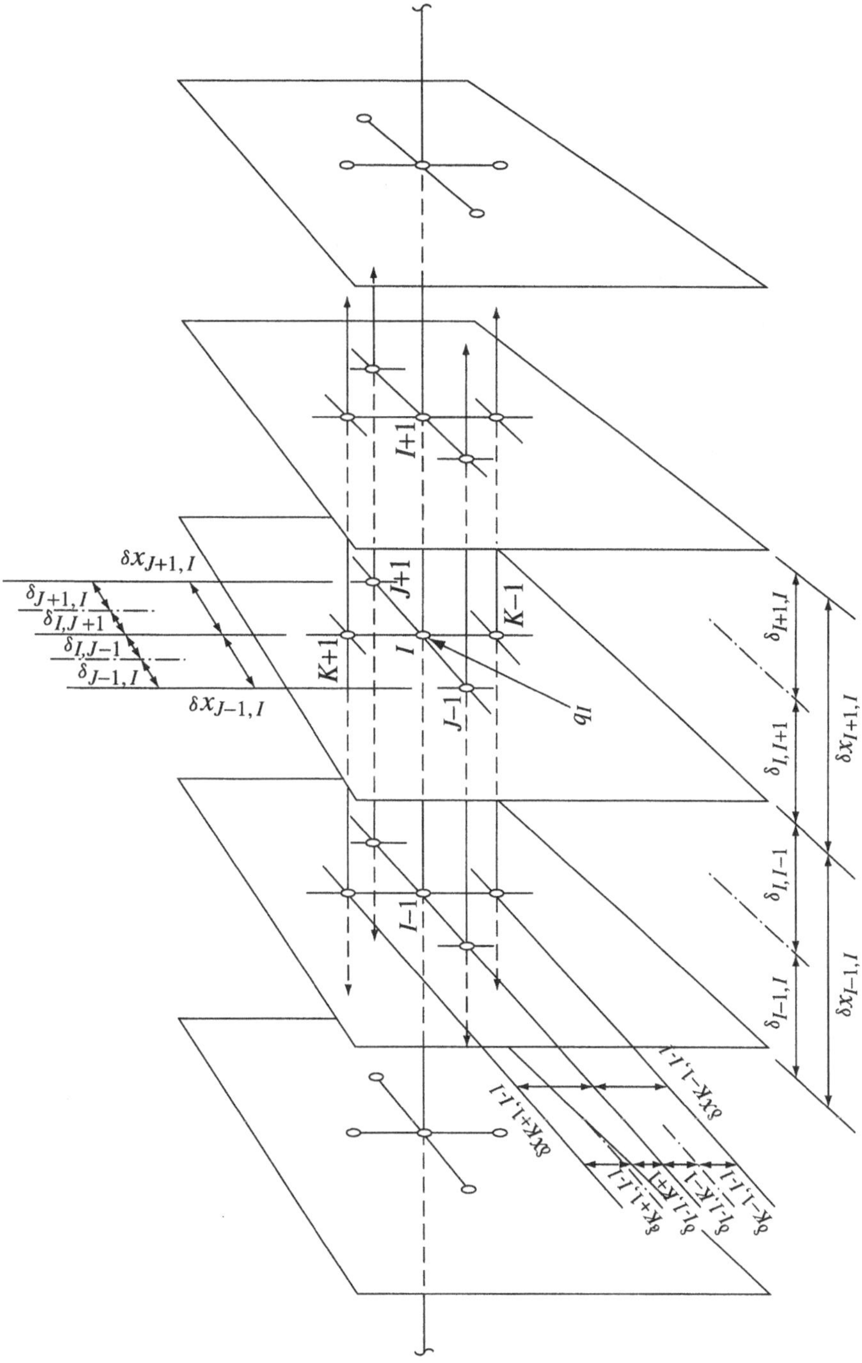

Figure 3.6: Multi-layered construction discretisation.

$$
\begin{aligned}
&- \frac{\delta t\, k_J(t+\delta t)}{\delta x_J^2}\theta(J-1,t+\delta t) - \frac{\delta t\, k_J(t+\delta t)}{\delta x_J^2}\theta(J+1,t+\delta t) \\
&- \frac{\delta t\, k_K(t+\delta t)}{\delta x_K^2}\theta(K-1,t+\delta t) - \frac{\delta t\, k_K(t+\delta t)}{\delta x_K^2}\theta(K+1,t+\delta t) - \frac{\delta t\, q_I(t+\delta t)}{\delta x_I \delta x_J \delta x_K} \\
&= \left(2\rho_I(t)C_I(t) - \frac{2\delta t\, k_I(t)}{\delta x_I^2} - \frac{2\delta t\, k_J(t)}{\delta x_J^2} - \frac{2\delta t\, k_K(t)}{\delta x_K^2}\right)\theta(I,t) \\
&+ \frac{\delta t\, k_I(t)}{\delta x_I^2}\theta(I-1,t) + \frac{\delta t\, k_I(t)}{\delta x_I^2}\theta(I+1,t) + \frac{\delta t\, k_J(t)}{\delta x_J^2}\theta(J-1,t) \\
&+ \frac{\delta t\, k_J(t)}{\delta x_J^2}\theta(J+1,t) + \frac{\delta t\, k_K(t)}{\delta x_K^2}\theta(K-1,t) + \frac{\delta t\, k_K(t)}{\delta x_K^2}\theta(K+1,t) \\
&+ \frac{\delta t\, q_I(t)}{\delta x_I \delta x_J \delta x_K}\,. \qquad (3.5)
\end{aligned}
$$

If the assumption of isotropic, homogeneous behaviour is made, then

$$k_I = k_J = k_K = k\,.$$

Eqn (3.5) can be utilised to represent the conduction within any homogeneous medium by dividing the medium into a number of finite volumes. In the case of a wall construction, for example, the accuracy level can be improved by simple element subdivision as demonstrated in figure 3.7. The heat generation term permits the direct nodal injection or extraction of heat to model, for example, an under-floor heating system or an electrical storage heater. Note that the $q_I(t + \delta t)$ term may be used by a controller (see §4.2 and §6.4) to emulate the behaviour of an HVAC system or, alternatively, it may provide a run-time link between building- and plant-side models running in tandem.

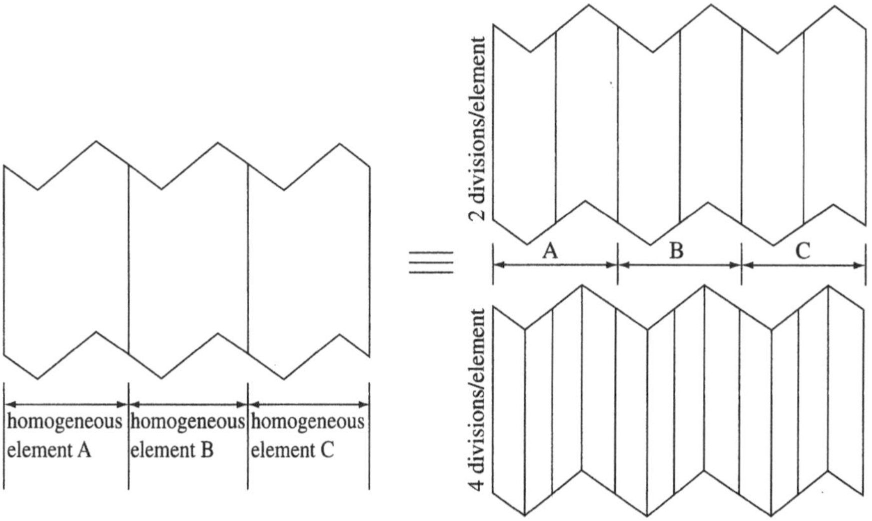

Figure 3.7: Homogeneous element subdivision to improve accuracy.

The uni-directional counterpart of eqn (3.5) can now be established:

$$\left(2\rho_I(t+\delta t)C_I(t+\delta t) + \frac{2\delta t\, k(t+\delta t)}{\delta x_I^2}\right)\theta(I,t+\delta t)$$

$$- \frac{\delta t\, k(t+\delta t)}{\delta x_I^2}\theta(I-1,t+\delta t) - \frac{\delta t\, k(t+\delta t)}{\delta x_I^2}\theta(I+1,t+\delta t) - \frac{\delta t\, q_I(t+\delta t)}{\delta x_I \delta x_J \delta x_K}$$
$$= \left(2\rho_I(t)C_I(t) - \frac{2\delta t\, k(t)}{\delta x_I^2}\right)\theta(I,t)$$
$$+ \frac{\delta t\, k(t)}{\delta x_I^2}\theta(I-1,t) + \frac{\delta t\, k(t)}{\delta x_I^2}\theta(I+1,t) + \frac{\delta t\, q_I(t)}{\delta x_I \delta x_J \delta x_K}. \qquad (3.6)$$

Transparent homogeneous element nodes

Eqn (3.5) also holds for a node representing a glazing element, but here the heat generation term will also include the absorption of shortwave energy as it travels through the transparent medium. The derivation of an algorithm for shortwave absorption is undertaken in §7.4.

Phase change material nodes

Eqn (3.5) may also be utilised to model materials which undergo a change of phase, to absorb or release the latent heat of vaporisation or fusion at constant temperature. When the temperature of the phase change is reached, the heat generation term can be employed to maintain a constant node temperature until that quantity of energy has been absorbed (or released) at which temperature change will recommence. A simple counter mechanism can be established to maintain a record of the latent energy in 'store' at any time as a function of an algorithm representing material behaviour.

Boundary nodes separating two homogeneous elements

Referring back to figure 3.6, assume node I is situated at the boundary between two homogeneous elements (opaque and/or transparent) of different thermophysical properties. The distance $I-1 \rightarrow I$ defines the half thickness of one element denoted by suffix A (node $I-1$ is located at the centre plane of this element) and $I \rightarrow I+1$ defines the half thickness of the second element denoted by suffix B. Note that if elements A and B have undergone subdivision as shown in figure 3.7, then nodes $I-1$ and $I+1$ will be relocated closer to their common interface.

For boundary nodes it is necessary to implement a volumetric weighting to establish representative thermophysical properties for each inter-nodal flowpath and the region represented by node I. A contact resistance, R_c, is also introduced to impose an additional resistance to heat flow at the interface to emulate the case of non-perfect mechanical contact. Noting that

$$k'_{I-1,I} = k_A$$

$$k'_{I+1,I} = k_B$$

$$k'_{J-1,I} = k'_{J+1,I} = k'_{K-1,I} = k'_{K+1,I} = (\delta_{I-1,I}k_A + \delta_{I+1,I}k_B)/(\delta_{I-1,I} + \delta_{I+1,I})$$

$$= k_{AB} \text{ (assuming isotropic behaviour)}$$

$$W_I = (\rho_A C_A \delta V_A + \rho_B C_B \delta V_B)/(\delta V_A + \delta V_B)$$

$$(\delta_{i-1,I} + \delta_{i+1,I}) = \delta_{I-1,I+1} \;;\; i = I, J, K$$

then eqn (3.4) gives

$$\left(2W_I(t+\delta t) + \frac{\delta t[k_A(t+\delta t)R_c(t+\delta t) + 2\delta x_{I-1,I}]}{\delta x_{I-1,I}R_c(t+\delta t)\delta_{I-1,I+1}} + \frac{\delta t[k_B(t+\delta t)R_c(t+\delta t) + 2\delta x_{I+1,I}]}{\delta x_{I+1,I}R_c(t+\delta t)\delta_{I-1,I+1}} + \frac{\delta t\, k_{AB}(t+\delta t)}{\delta x_{J-1,I}\delta_{J-1,J+1}} + \frac{\delta t\, k_{AB}(t+\delta t)}{\delta x_{J+1,I}\delta_{J-1,J+1}} + \frac{\delta t\, k_{AB}(t+\delta t)}{\delta x_{K-1,I}\delta_{K-1,K+1}} + \frac{\delta t\, k_{AB}(t+\delta t)}{\delta x_{K+1,I}\delta_{K-1,K+1}}\right)\theta(I,t+\delta t)$$
$$- \frac{\delta t[k_A(t+\delta t)R_c(t+\delta t) + 2\delta x_{I-1,I}]}{\delta x_{I-1,I}R_c(t+\delta t)\delta_{I-1,I+1}}\theta(I-1,t+\delta t) - \frac{\delta t[k_B(t+\delta t)R_c(t+\delta t) + 2\delta x_{I+1,I}]}{\delta x_{I+1,I}R_c(t+\delta t)\delta_{I-1,I+1}}\theta(I+1,t+\delta t)$$
$$- \frac{\delta t\, k_{AB}(t+\delta t)}{\delta x_{J-1,I}\delta_{J-1,J+1}}\theta(J-1,t+\delta t) - \frac{\delta t\, k_{AB}(t+\delta t)}{\delta x_{J+1,I}\delta_{J-1,J+1}}\theta(J+1,t+\delta t) - \frac{\delta t\, k_{AB}(t+\delta t)}{\delta x_{K-1,I}\delta_{K-1,K+1}}\theta(K-1,t+\delta t) - \frac{\delta t\, k_{AB}(t+\delta t)}{\delta x_{K+1,I}\delta_{K-1,K+1}}\theta(K+1,t+\delta t) - \frac{\delta t\, q_I(t+\delta t)}{\delta_{I-1,I+1}\delta_{J-1,J+1}\delta_{K-1,K+1}}$$
$$= \left(2W_I(t) - \frac{\delta t[k_A(t)R_c(t) + 2\delta x_{I-1,I}]}{\delta x_{I-1,I}R_c(t)\delta_{I-1,I+1}} - \frac{\delta t[k_B(t)R_c(t) + 2\delta x_{I+1,I}]}{\delta x_{I+1,I}R_c(t)\delta_{I-1,I+1}} - \frac{\delta t\, k_{AB}(t)}{\delta x_{J-1,I}\delta_{J-1,J+1}} - \frac{\delta t\, k_{AB}(t)}{\delta x_{J+1,I}\delta_{J-1,J+1}} - \frac{\delta t\, k_{AB}(t)}{\delta x_{K-1,I}\delta_{K-1,K+1}} - \frac{\delta t\, k_{AB}(t)}{\delta x_{K+1,I}\delta_{K-1,K+1}}\right)\theta(I,t)$$
$$+ \frac{\delta t[k_A(t)R_c(t) + 2\delta x_{I-1,I}]}{\delta x_{I-1,I}R_c(t)\delta_{I-1,I+1}}\theta(I-1,t) + \frac{\delta t[k_B(t)R_c(t) + 2\delta x_{I+1,I}]}{\delta x_{I+1,I}R_c(t)\delta_{I-1,I+1}}\theta(I+1,t) + \frac{\delta t\, k_{AB}(t)}{\delta x_{J-1,I}\delta_{J-1,J+1}}\theta(J-1,t) + \frac{\delta t\, k_{AB}(t)}{\delta x_{J+1,I}\delta_{J-1,J+1}}\theta(J+1,t)$$
$$+ \frac{\delta t\, k_{AB}(t)}{\delta x_{K-1,I}\delta_{K-1,K+1}}\theta(K-1,t) + \frac{\delta t\, k_{AB}(t)}{\delta x_{K+1,I}\delta_{K-1,K+1}}\theta(K+1,t) + \frac{\delta t\, q_I(t)}{\delta_{I-1,I+1}\delta_{J-1,J+1}\delta_{K-1,K+1}}. \tag{3.7}$$

The heat generation term, q_I, will permit, in addition to plant interaction potential, the absorption of shortwave radiant energy if either element is transparent and exposed to solar flux as, for example, in a window/blind system.

The uni-directional form of eqn (3.7) is given by

$$\left(2W_I(t+\delta t) + \frac{\delta t[k_A(t+\delta t)R_c(t+\delta t) + 2\delta x_{I-1,I}]}{\delta x_{I-1,I}R_c(t+\delta t)\delta_{I-1,I+1}} + \frac{\delta t[k_B(t+\delta t)R_c(t+\delta t) + 2\delta x_{I+1,I}]}{\delta x_{I+1,I}R_c(t+\delta t)\delta_{I-1,I+1}}\right)\theta(I,t+\delta t)$$
$$- \frac{\delta t[k_A(t+\delta t)R_c(t+\delta t) + 2\delta x_{I-1,I}]}{\delta x_{I-1,I}R_c(t+\delta t)\delta_{I-1,I+1}}\theta(I-1,t+\delta t) - \frac{\delta t[k_B(t+\delta t)R_c(t+\delta t) + 2\delta x_{I+1,I}]}{\delta x_{I+1,I}R_c(t+\delta t)\delta_{I-1,I+1}}\theta(I+1,t+\delta t) - \frac{\delta t\, q_I(t+\delta t)}{\delta_{I-1,I+1}\delta_{J-1,J+1}\delta_{K-1,K+1}}$$
$$= \left(2W_I(t) - \frac{\delta t[k_A(t)R_c(t) + 2\delta x_{I-1,I}]}{\delta x_{I-1,I}R_c(t)\delta_{I-1,I+1}} - \frac{\delta t[k_B(t)R_c(t) + 2\delta x_{I+1,I}]}{\delta x_{I+1,I}R_c(t)\delta_{I-1,I+1}}\right)\theta(I,t)$$

$$+ \frac{\delta t[k_A(t)R_c(t) + 2\delta x_{I-1,I}]}{\delta x_{I-1,I}R_c(t)\delta_{I-1,I+1}} \theta(I-1,t) + \frac{\delta t[k_B(t)R_c(t) + 2\delta x_{I+1,I}]}{\delta x_{I+1,I}R_c(t)\delta_{I-1,I+1}} \theta(I+1,t)$$
$$+ \frac{\delta t\, q_I(t)}{\delta_{I-1,I+1}\delta_{J-1,J+1}\delta_{K-1,K+1}} . \tag{3.8}$$

Lumped material nodes

When applying eqn (3.4) to capacity regions contained by some environment, as opposed to separating two environments—such as room contents, structural beams, columns or an underground thermal store—it is convenient to use a modified form of eqn (3.5) given by

$$\left(2W_I(t+\delta t) + \sum_{i=1}^{N} \frac{\delta t \delta A_{i,I}\, k'_{i,I}(t+\delta t)}{\delta V_I \delta x_{i,I}} \right) \theta(I, t+\delta t)$$
$$- \sum_{i=1}^{N} \frac{\delta t \delta A_{i,I}\, k'_{i,I}(t+\delta t)}{\delta V_I \delta x_{i,I}} \theta(i, t+\delta t) - \frac{\delta t\, q_I(t+\delta t)}{\delta V_I}$$
$$= \left(2W_I(t) - \sum_{i=1}^{N} \frac{\delta t \delta A_{i,I}\, k'_{i,I}(t)}{\delta V_I \delta x_{i,I}} \right) \theta(I,t)$$
$$+ \sum_{i=1}^{N} \frac{\delta t \delta A_{i,I}\, k'_{i,I}(t)}{\delta V_I \delta x_{i,I}} \theta(i,t) + \frac{\delta t\, q_I(t)}{\delta V_I} \tag{3.9}$$

where $\delta A_{i,I}$ is the area normal to the direction of heat flow between nodes i and I (m^2), δV_I the volume of the region represented by node I (m^3), and N the number of inter-node thermal connections. This equation can be used to establish a rudimentary nodal network to represent capacity and so introduce additional inertia to a model. Figure 3.8 shows two nodal schemes established to represent furnishings and thermal storage. With such systems it is usual to maximise the quantity $\delta A/\delta V$ to represent the high ratio of exposed area to contained volume typical of such components.

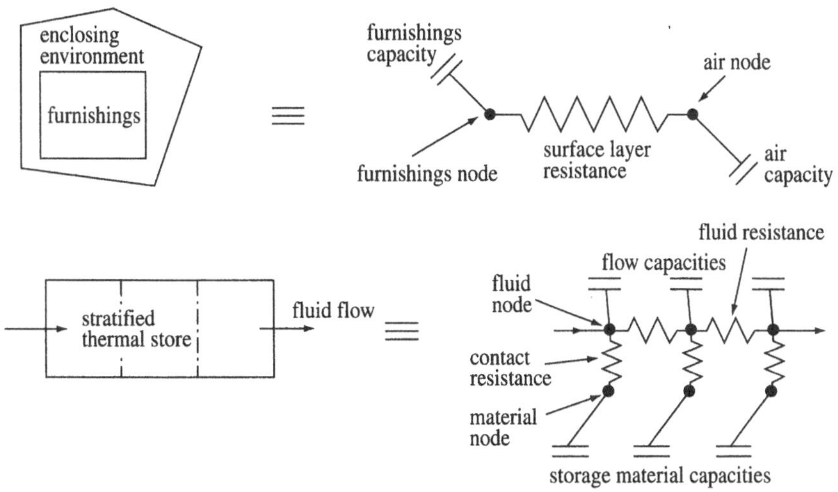

Figure 3.8: Nodal schemes for furnishings and thermal storage.

Taken together eqns (3.4) through (3.9) allow the construction of a discrete nodal network representing the transient energy flows within a capacity/insulation system containing opaque and/or transparent parts. Section 3.3 demonstrates the technique of structuring these equations (and those that follow) to represent real building configurations.

3.2.2 Exposed surface layers

Consider figure 3.9, which shows node I now located at the exposed surface of a multi-layered construction. (Note that this could equally well be the surface of a radiator.) Node I − 1 is the adjacent node buried within the material of the next-to-surface layer, while node I + 1 represents the adjacent fluid volume. Assuming that the boundary layer thickness is negligible, then the volume of the finite volume represented by I is given by

$$\delta_{I,I-1}(\delta_{I,J-1} + \delta_{I,J+1})(\delta_{I,K-1} + \delta_{I,K+1}) \; .$$

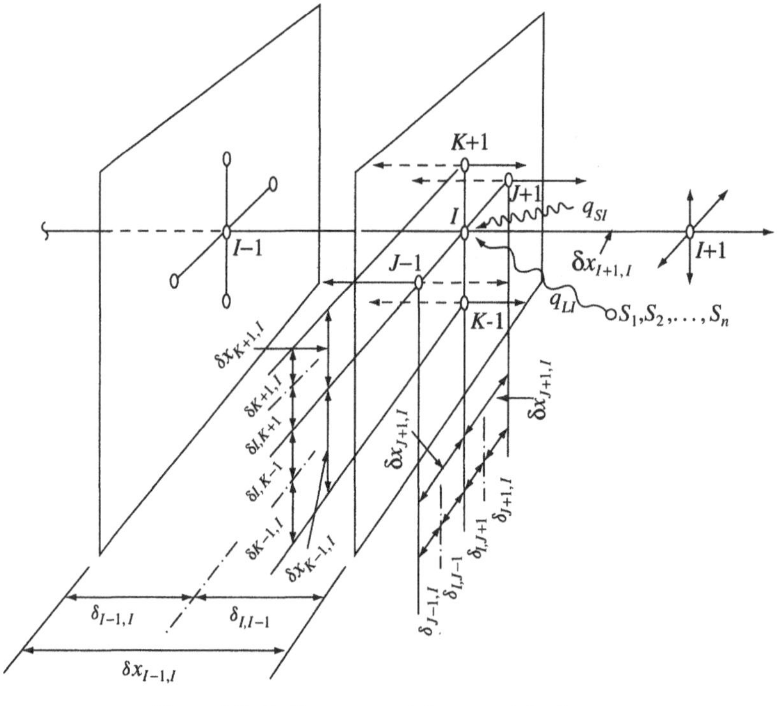

Figure 3.9: Surface energy balance nodal scheme.

For each inter-nodal flowpath, the heat flux is given by

$$q_{I-1,I} = k'_{I-1,I}(\delta_{I,J-1} + \delta_{I,J+1})(\delta_{I,K-1} + \delta_{I,K+1})(\theta_{I-1} - \theta_I)/\delta x_{I-1,I}$$

$$q_{I+1,I} = h_{cI+1,I}(\delta_{I,J-1} + \delta_{I,J+1})(\delta_{I,K-1} + \delta_{I,K+1})(\theta_{I+1} - \theta_I)$$

$$q_{J-1,I} = k'_{J-1,I}\delta_{I,I-1}(\delta_{I,K-1} + \delta_{I,K+1})(\theta_{J-1} - \theta_I)/\delta x_{J-1,I}$$

$$q_{J+1,I} = k'_{J+1,I}\delta_{I,I-1}(\delta_{I,K-1} + \delta_{I,K+1})(\theta_{J+1} - \theta_I)/\delta x_{J+1,I}$$

$$q_{K-1,I} = k'_{K-1,I}\delta_{I,I-1}(\delta_{I,J-1} + \delta_{I,J+1})(\theta_{K-1} + \theta_I)/\delta x_{K-1,I}$$

$$q_{K+1,I} = k'_{K+1,I}\delta_{I,I-1}(\delta_{I,J-1} + \delta_{I,J+1})(\theta_{K+1} + \theta_I)/\delta x_{K+1,I}$$

and noting that for this node type

$$q_I = q_{SI} + q_{LI} + q_{RI} + q_{PI}$$

$$q_{LI} = \sum_{s=1}^{N} h_{rs,I}(\delta_{I,J-1} + \delta_{I,J+1})(\delta_{I,K-1} + \delta_{I,K+1})(\theta_s - \theta_I)$$

where q_{SI} is the shortwave energy absorption, q_{LI} the longwave radiation exchange with the surroundings, q_{RI} the radiant energy from casual sources and q_{PI} the radiant component of plant input (all measured in W); h_r is the inter-surface radiation coefficient (W $m^{-2}{}^{\circ}C^{-1}$), N the number of surfaces in longwave contact, and h_c the surface convection coefficient (W $m^{-2}{}^{\circ}C^{-1}$). Note that the longwave radiation term has been linearised. Chapter 4 discusses the impact of this treatment and describes alternative approaches to the handling of non-linearity.

Substituting these heat fluxes in eqn (3.1) gives the exposed surface energy balance relationship:

$$\begin{aligned}
&[W_I(t+\delta t)\theta(I,t+\delta t) - W_I(t)\theta(I,t)]\delta_{I,I-1}(\delta_{I,J-1} + \delta_{I,J+1})(\delta_{I,K-1} + \delta_{I,K+1})/\delta t \\
&= k'_{I-1,I}(\xi)(\delta_{I,J-1} + \delta_{I,J+1})(\delta_{I,K-1} + \delta_{I,K+1})[\theta(I-1,\xi) - \theta(I,\xi)]/\delta x_{I-1,I} \\
&\quad + h_{cI+1,I}(\xi)(\delta_{I,J-1} + \delta_{I,J+1})(\delta_{I,K-1} + \delta_{I,K+1})[\theta(I+1,\xi) - \theta(I,\xi)] \\
&\quad + k'_{J-1,I}(\xi)\delta_{I,I-1}(\delta_{I,K-1} + \delta_{I,K+1})[\theta(J-1,\xi) - \theta(I,\xi)]/\delta x_{J-1,I} \\
&\quad + k'_{J+1,I}(\xi)\delta_{I,I-1}(\delta_{I,K-1} + \delta_{I,K+1})[\theta(J+1,\xi) - \theta(I,\xi)]/\delta x_{J+1,I} \\
&\quad + k'_{K-1,I}(\xi)\delta_{I,I-1}(\delta_{I,J-1} + \delta_{I,J+1})[\theta(K-1,\xi) - \theta(I,\xi)]/\delta x_{K-1,I} \\
&\quad + k'_{K+1,I}(\xi)\delta_{I,I-1}(\delta_{I,J-1} + \delta_{I,J+1})[\theta(K+1,\xi) - \theta(I,\xi)]/\delta x_{K+1,I} \\
&\quad + \sum_{s=1}^{N} h_{rs,I}(\xi)(\delta_{I,J-1} + \delta_{I,J+1})(\delta_{I,K-1} + \delta_{I,K+1})[\theta(s,\xi) - \theta(I,\xi)] \\
&\quad + q_{SI}(\xi) + q_{RI}(\xi) + q_{PI}(\xi)\,.
\end{aligned}$$

As with the transient conduction formulation, this equation can be evaluated at some present time-row, t, some future time-row, $t + \delta t$, and the two formulations weighted to give the characteristic energy conservation equation for a surface exposed to an adjacent fluid region. The temperature explicit formulation is

$$\begin{aligned}
\theta(I,t+\delta t) = &\left(\frac{W_I(t)}{W_I(t+\delta t)} - \frac{k'_{I-1,I}(t)}{S_{I-1,I}(t+\delta t)\delta_{I,I-1}} - \frac{\delta t\, h_{cI+1,I}(t)}{W_I(t+\delta t)\delta_{I,I-1}} \right. \\
&- \frac{k'_{J-1,I}(t)}{S_{J-1,I}(t+\delta t)\delta_{J-1,J+1}} - \frac{k'_{J+1,I}(t)}{S_{J+1,I}(t+\delta t)\delta_{J-1,J+1}} - \frac{k'_{K-1,I}(t)}{S_{K-1,I}(t+\delta t)\delta_{K-1,K+1}} \\
&\left. - \frac{k'_{K+1,I}(t)}{S_{K+1,I}(t+\delta t)\delta_{K-1,K+1}} - \frac{\delta t \sum_{s=1}^{N} h_{rs,I}(t)}{W_I(t+\delta t)\delta_{I,I-1}} \right) \theta(I,t) \\
&+ \frac{k'_{I-1,I}(t)}{S_{I-1,I}(t+\delta t)\delta_{I,I-1}}\theta(I-1,t) + \frac{\delta t\, h_{cI+1,I}(t)}{W_I(t+\delta t)\delta_{I,I-1}}\theta(I+1,t) \\
&+ \frac{k'_{J-1,I}(t)}{S_{J-1,I}(t+\delta t)\delta_{J-1,J+1}}\theta(J-1,t) + \frac{k'_{J+1,I}(t)}{S_{J+1,I}(t+\delta t)\delta_{J-1,J+1}}\theta(J+1,t)
\end{aligned}$$

$$+\frac{k'_{K-1,I}(t)}{S_{K-1,I}(t+\delta t)\delta_{K-1,K+1}}\theta(K-1,t)+\frac{k'_{K+1,I}(t)}{S_{K+1,I}(t+\delta t)\delta_{K-1,K+1}}\theta(K+1,t)$$

$$+\frac{\delta t\sum_{s=1}^{N}h_{rs,I}(t)\theta(s,t)}{W_I(t+\delta t)\delta_{I,I-1}}+\frac{\delta t[q_{sI}(t)+q_{RI}(t)+q_{PI}(t)]}{W_I(t+\delta t)\delta_{I,I-1}\delta_{J-1,J+1}\delta_{K-1,K+1}}+\varepsilon$$

where S is as defined in §3.2.1. The implicit formulation is

$$\theta(I,t+\delta t)=\frac{W_I(t)}{W_I(t+\delta t)}\theta(I,t)-\left(\frac{k'_{I-1,I}(t+\delta t)}{S_{I-1,I}(t+\delta t)\delta_{I,I-1}}+\frac{\delta t\, h_{cI+1,I}(t+\delta t)}{W_I(t+\delta t)\delta_{I,I-1}}\right.$$
$$+\frac{k'_{J-1,I}(t+\delta t)}{S_{J-1,I}(t+\delta t)\delta_{J-1,J+1}}+\frac{k'_{J+1,I}(t+\delta t)}{S_{J+1,I}(t+\delta t)\delta_{J-1,J+1}}+\frac{k'_{K-1,I}(t+\delta t)}{S_{K-1,I}(t+\delta t)\delta_{K-1,K+1}}$$
$$\left.+\frac{k'_{K+1,I}(t+\delta t)}{S_{K+1,I}(t+\delta t)\delta_{K-1,K+1}}+\frac{\delta t\sum_{s=1}^{N}h_{rs,I}(t+\delta t)}{W_I(t+\delta t)\delta_{I,I-1}}\right)\theta(I,t+\delta t)$$
$$+\frac{k'_{I-1,I}(t+\delta t)}{S_{I-1,I}(t+\delta t)\delta_{I,I-1}}\theta(I-1,t+\delta t)+\frac{\delta t\, h_{cI+1,I}(t+\delta t)}{W_I(t+\delta t)\delta_{I,I-1}}\theta(I+1,t+\delta t)$$
$$+\frac{k'_{J-1,I}(t+\delta t)}{S_{J-1,I}(t+\delta t)\delta_{J-1,J+1}}\theta(J-1,t+\delta t)+\frac{k'_{J+1,I}(t+\delta t)}{S_{J+1,I}(t+\delta t)\delta_{J-1,J+1}}\theta(J+1,t+\delta t)$$
$$+\frac{k'_{K-1,I}(t+\delta t)}{S_{K-1,I}(t+\delta t)\delta_{K-1,K+1}}\theta(K-1,t+\delta t)+\frac{k'_{K+1,I}(t+\delta t)}{S_{K+1,I}(t+\delta t)\delta_{K-1,K+1}}\theta(K+1,t+\delta t)$$
$$+\frac{\delta t\sum_{s=1}^{N}h_{rs,I}(t+\delta t)\theta(s,t+\delta t)}{W_I(t+\delta t)\delta_{I,I-1}}+\frac{\delta t[q_{sI}(t+\delta t)+q_{RI}(t+\delta t)+q_{PI}(t+\delta t)]}{W_I(t+\delta t)\delta_{I,I-1}\delta_{J-1,J+1}\delta_{K-1,K+1}}+\varepsilon\,.$$

An equal weighting is now performed and all unknown future time-row terms grouped on the equation left-hand side. Known future time-row quantities, such as boundary condition solar and casual radiant flux injections, are removed to the equation right-hand side.

$$\left(2W_I(t+\delta t)+\frac{\delta t\, k'_{I-1,I}(t+\delta t)}{\delta x_{I-1,I}\delta_{I,I-1}}+\frac{\delta t\, h_{cI+1,I}(t+\delta t)}{\delta_{I,I-1}}\right.$$
$$+\frac{\delta t\, k'_{J-1,I}(t+\delta t)}{\delta x_{J-1,I}\delta_{J-1,J+1}}+\frac{\delta t\, k'_{J+1,I}(t+\delta t)}{\delta x_{J+1,I}\delta_{J-1,J+1}}$$
$$\left.+\frac{\delta t\, k'_{K-1,I}(t+\delta t)}{\delta x_{K-1,I}\delta_{K-1,K+1}}+\frac{\delta t\, k'_{K+1,I}(t+\delta t)}{\delta x_{K+1,I}\delta_{K-1,K+1}}+\frac{\delta t\sum_{s=1}^{N}h_{rs,I}(t+\delta t)}{\delta_{I,I-1}}\right)\theta(I,t+\delta t)$$
$$-\frac{\delta t\, k'_{I-1,I}(t+\delta t)}{\delta x_{I-1,I}\delta_{I,I-1}}\theta(I-1,t+\delta t)-\frac{\delta t\, h_{cI+1,I}(t+\delta t)}{\delta_{I,I-1}}\theta(I+1,t+\delta t)$$
$$-\frac{\delta t\, k'_{J-1,I}(t+\delta t)}{\delta x_{J-1,I}\delta_{J-1,J+1}}\theta(J-1,t+\delta t)-\frac{\delta t\, k'_{J+1,I}(t+\delta t)}{\delta x_{J+1,I}\delta_{J-1,J+1}}\theta(J+1,t+\delta t)$$
$$-\frac{\delta t\, k'_{K-1,I}(t+\delta t)}{\delta x_{K-1,I}\delta_{K-1,K+1}}\theta(K-1,t+\delta t)-\frac{\delta t\, k'_{K+1,I}(t+\delta t)}{\delta x_{K+1,I}\delta_{K-1,K+1}}\theta(K+1,t+\delta t)$$

$$
- \frac{\delta t \sum_{s=1}^{N} h_{rs,I}(t+\delta t)\theta(s,t+\delta t)}{\delta_{I,I-1}} - \frac{\delta t\, q_{PI}(t+\delta t)}{\delta_{I,I-1}\delta_{J-1,J+1}\delta_{K-1,K+1}}
$$

$$
= \left(2W_I(t) - \frac{\delta t\, k'_{I-1,I}(t)}{\delta x_{I-1,I}\delta_{I,I-1}} - \frac{\delta t\, h_{cI+1,I}(t)}{\delta_{I,I-1}} - \frac{\delta t\, k'_{J-1,I}(t)}{\delta x_{J-1,I}\delta_{J-1,J+1}} - \frac{\delta t\, k'_{J+1,I}(t)}{\delta x_{J+1,I}\delta_{J-1,J+1}} \right.
$$

$$
\left. - \frac{\delta t\, k'_{K-1,I}(t)}{\delta x_{K-1,I}\delta_{K-1,K+1}} - \frac{\delta t\, k'_{K+1,I}(t)}{\delta x_{K+1,I}\delta_{K-1,K+1}} - \frac{\delta t \sum_{s=1}^{N} h_{rs,I}(t)}{\delta_{I,I-1}} \right) \theta(I,t)
$$

$$
+ \frac{\delta t\, k'_{I-1,I}(t)}{\delta x_{I-1,I}\delta_{I,I-1}} \theta(I-1,t) + \frac{\delta t\, h_{cI+1,I}(t)}{\delta_{I,I-1}} \theta(I+1,t)
$$

$$
+ \frac{\delta t\, k'_{J-1,I}(t)}{\delta x_{J-1,I}\delta_{J-1,J+1}} \theta(J-1,t) + \frac{\delta t\, k'_{J+1,I}(t)}{\delta x_{J+1,I}\delta_{J-1,J+1}} \theta(J+1,t)
$$

$$
+ \frac{\delta t\, k'_{K-1,I}(t)}{\delta x_{K-1,I}\delta_{K-1,K+1}} \theta(K-1,t) + \frac{\delta t\, k'_{K+1,I}(t)}{\delta x_{K+1,I}\delta_{K-1,K+1}} \theta(K+1,t) + \frac{\delta t \sum_{s=1}^{N} h_{rs,I}(t)\theta(s,t)}{\delta_{I,I-1}}
$$

$$
+ \frac{\delta t[q_{PI}(t) + q_{SI}(t) + q_{RI}(t)]}{\delta_{I,I-1}\delta_{J-1,J+1}\delta_{K-1,K+1}} + \frac{\delta t[q_{SI}(t+\delta t) + q_{RI}(t+\delta t)]}{\delta_{I,I-1}\delta_{J-1,J+1}\delta_{K-1,K+1}} + \varepsilon \,. \tag{3.10}
$$

This equation corresponds to nodes located at exposed opaque or transparent surfaces experiencing heat exchange by conduction within the next-to-surface material, by convection with the adjacent fluid layer, by longwave radiation exchanges with surrounding surfaces, and by absorption of radiation from solar or other sources. Chapter 7 introduces models for the determination of the parameters of this equation, including the surface convection and radiation coefficients. The uni-directional formulation of eqn (3.10) is given by

$$
\left(2W_I(t+\delta t) + \frac{\delta t\, k'_{I-1,I}(t+\delta t)}{\delta x_{I-1,I}\delta_{I,I-1}} + \frac{\delta t\, h_{cI+1,I}(t+\delta t)}{\delta_{I,I-1}} + \frac{\delta t \sum_{s=1}^{N} h_{rs,I}(t+\delta t)}{\delta_{I,I-1}} \right) \theta(I,t+\delta t)
$$

$$
- \frac{\delta t\, k'_{I-1,I}(t+\delta t)}{\delta x_{I-1,I}\delta_{I,I-1}} \theta(I-1,t+\delta t) - \frac{\delta t\, h_{cI+1,I}(t+\delta t)}{\delta_{I,I-1}} \theta(I+1,t+\delta t)
$$

$$
- \frac{\delta t \sum_{s=1}^{N} h_{rs,I}(t+\delta t)\theta(s,t+\delta t)}{\delta_{I,I-1}} - \frac{\delta t\, q_{PI}(t+\delta t)}{\delta_{I,I-1}\delta_{J-1,J+1}\delta_{K-1,K+1}}
$$

$$
= \left(2W_I(t) - \frac{\delta t\, k'_{I-1,I}(t)}{\delta x_{I-1,I}\delta_{I,I-1}} - \frac{\delta t\, h_{cI+1,I}(t)}{\delta_{I,I-1}} - \frac{\delta t \sum_{s=1}^{N} h_{rs,I}(t)}{\delta_{I,I-1}} \right) \theta(I,t)
$$

$$
+ \frac{\delta t\, k'_{I-1,I}(t)}{\delta x_{I-1,I}\delta_{I,I-1}} \theta(I-1,t) + \frac{\delta t\, h_{cI+1,I}(t)}{\delta_{I,I-1}} \theta(I+1,t) + \frac{\delta t \sum_{s=1}^{N} h_{rs,I}(t)\theta(s,t)}{\delta_{I,I-1}}
$$

$$+\frac{\delta t[q_{PI}(t)+q_{SI}(t)+q_{RI}(t)+q_{SI}(t+\delta t)+q_{RI}(t+\delta t)]}{\delta_{I,I-1}\delta_{J-1,J+1}\delta_{K-1,K+1}}+\varepsilon\,. \quad (3.11)$$

Some special applications of eqn (3.10) are now described.

Multi-layered construction surface nodes

In this case, node I represents the exposed surface of the construction and node I + 1 represents the adjacent zone air, either internally (same zone or adjacent zone) or externally located. Where externally located, the I + 1 node term can be relocated to the equation right-hand side since it is an ambient boundary condition and therefore known for all time. Where I + 1 is located within an adjacent zone, it may also be relocated to the equation right-hand side and the related air temperature treated as a known quantity. This decouples zones and allows their energy balance matrix equations to be assigned to separate processors. An additional mechanism would then be required to ensure that the internal air temperature solution to emerge from each calculation stream is imposed as a boundary condition on other related streams.

Ground contact nodes

For the special case where node I is located at the outermost surface of a floor slab and is in contact with the ground, node I + 1 is associated with a finite volume representing the ground. This situation can be handled by one of two mechanisms: the ground temperature can be treated as known with (say) seasonal variation and the I + 1 nodal term relocated to the equation right-hand side as before, or node I + 1 can be treated as unknown and connected to a subterranean nodal network (using the techniques of figure 3.3) which links node I + 1 through to ambient conditions and to a depth where temperatures become relatively stable and substantially independent of ambient fluctuations.

Air gap boundary nodes

If node I is situated at the interface of an air gap and an adjacent material layer then eqn (3.10) can be used to represent the usual processes of conduction, convection and inter-surface longwave radiation exchange. One simplification can then be introduced without significant loss of accuracy in many cases. By assuming a combined convective/radiative boundary layer given by

$$R=\frac{1}{h_c+\varepsilon h_r}$$

where ε is the surface emissivity, then the convection coefficient, h_c, of eqn (3.10) can be replaced with a combined coefficient, h_T ($= h_c + \varepsilon h_r$), and the longwave radiation terms discarded. Such a trade-off between accuracy and flexibility has the useful effect of ensuring that the number of cross-coupling coefficients is the same as for the transient conduction nodes of §3.2.1: a device which will substantially reduce matrix processing requirements as explained in chapter 4.

3.2.3 Fluid volumes

Figure 3.10 shows the case where node I is located at the centroid of some fluid volume bounded by a collection of fictitious and/or real faces comprising the interfaces with adjacent fluid volumes, exposed surface layers, ambient conditions, plant components etc. The heat flow conductances of eqn (3.1) will depend on the type of heat flow connection between region

I and its neighbours. Two possibilities exist: fluid-to-fluid exchange due to pressure, temperature and/or mechanical driving forces; and fluid-to-surface exchange due to convective effects.

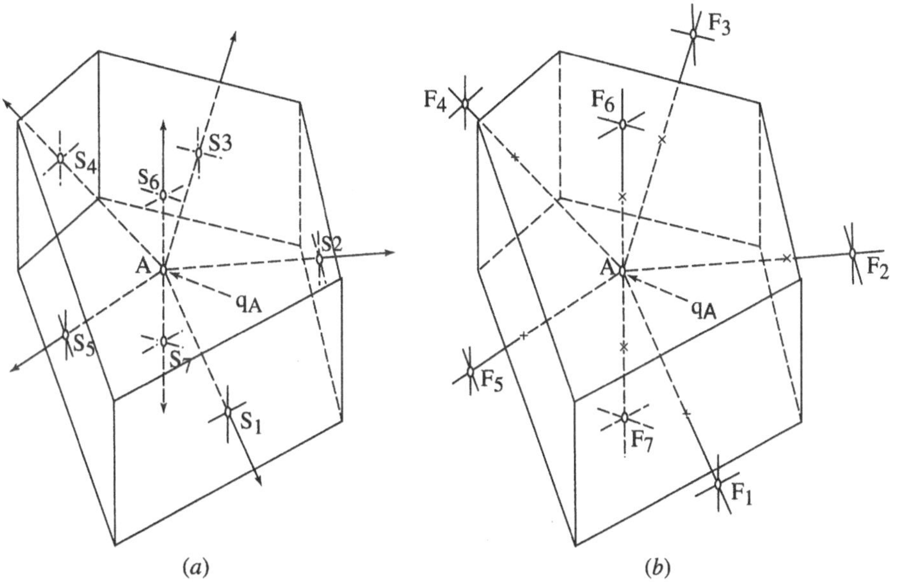

Figure 3.10: Fluid volume energy balance nodal scheme. (a) Fluid volume contained by real surfaces; convective heat transfer. (b) Fluid volumes contained by fictitious surfaces; advective heat transfer.

For the case of fluid-to-fluid exchange the conductance is given by

$$K_{i,I}(\xi) = v_{i,I}(\xi)\bar{\rho}_{i,I}(\xi)\bar{C}_{i,I}(\xi)$$

where $v_{i,I}$ is the inter-region volume flow rate relative to volume I (m^3s^{-1}), and $\bar{\rho}_{i,I}$, $\bar{C}_{i,I}$ are the density and specific heat capacity of the fluid evaluated at the temperature of the sending region ($kg\ m^{-3}$ and $J\ kg^{-1}{}^{\circ}C^{-1}$ respectively).

Note that the flow rate term is a vector quantity and therefore fluid leaving region I will not impose a load, only fluid entering the region being significant as it is raised or lowered to the region's temperature.

For fluid-to-surface exchange, the conductance is

$$K_{i,I}(\xi) = h_{ci,I}(\xi)\delta A_{i,I}$$

where $\delta A_{i,I}$ is the area normal to the flowpath (m^2).

Combining the evaluations of eqn (3.1) at the present and future time-rows gives

$$\left(2W_I(t+\delta t) + \frac{\delta t \sum_{i=1}^{N} h_{ci,I}(t+\delta t)\delta A_{i,I}}{\delta V_I} + \frac{\delta t \sum_{j=1}^{M} v_{j,I}(t+\delta t)\bar{\rho}_{j,I}(t+\delta t)\bar{C}_{j,I}(t+\delta t)}{\delta V_I} \right) \theta(I, t+\delta t)$$

$$
-\frac{\delta t\sum_{i=1}^{N} h_{ci,I}(t+\delta t)\delta A_{i,I}\theta(i,t+\delta t)}{\delta V_I} - \frac{\delta t\sum_{j=1}^{M} v_{j,I}(t+\delta t)\bar{\rho}_{j,I}(t+\delta t)\bar{C}_{j,I}(t+\delta t)\theta(j,t+\delta t)}{\delta V_I}
$$

$$
-\frac{\delta t\, q_I(t+\delta t)}{\delta V_I} = \left(2W_I(t) - \frac{\delta t\sum_{i=1}^{N} h_{ci,I}(t)\delta A_{i,I}}{\delta V_I} - \frac{\delta t\sum_{j=1}^{M} v_{j,I}(t)\bar{\rho}_{j,I}(t)\bar{C}_{j,I}(t)}{\delta V_I}\right)\theta(I,t)
$$

$$
+\frac{\delta t\sum_{i=1}^{N} h_{ci,I}(t)\delta A_{i,I}\theta(i,t)}{\delta V_I} + \frac{\delta t\sum_{j=1}^{M} v_{j,I}(t)\bar{\rho}_{j,I}(t)\bar{C}_{j,I}(t)\theta(j,t)}{\delta V_I} + \frac{\delta t\, q_I(t)}{\delta V_I} + \varepsilon \qquad (3.12)
$$

where N is the number of flowpaths linking fluid node I with its boundary surface nodes, and M the number of fluid flow streams converging on node I.

Chapters 5 and 7 introduce the theoretical considerations underlying the advective and convective terms respectively. Special instances of this equation are now considered.

Air gap nodes

Referring to figure 3.6, and for the case where node I is located at the centre plane of an air gap, with nodes $I-1$ and $I+1$ located at the air gap boundaries, the heat flows are

$$
q_{I-1,I} = h_{cI-1,I}(\delta_{I,J-1} + \delta_{I,J+1})(\delta_{I,K-1} + \delta_{I,K+1})(\theta_{I-1} - \theta_I)
$$

$$
q_{I+1,I} = h_{cI+1,I}(\delta_{I,J-1} + \delta_{I,J+1})(\delta_{I,K-1} + \delta_{I,K+1})(\theta_{I+1} - \theta_I)
$$

$$
q_{J-1,I} = v_{J-1,I}\bar{\rho}_{J-1,I}\bar{C}_{J-1,I}(\theta_{J-1} - \theta_I)
$$

$$
q_{J+1,I} = v_{J+1,I}\bar{\rho}_{J+1,I}\bar{C}_{J+1,I}(\theta_{J+1} - \theta_I)
$$

$$
q_{K-1,I} = v_{K-1,I}\bar{\rho}_{K-1,I}\bar{C}_{K-1,I}(\theta_{K-1} - \theta_I)
$$

$$
q_{K+1,I} = v_{K+1,I}\bar{\rho}_{K+1,I}\bar{C}_{K+1,I}(\theta_{K+1} - \theta_I)
$$

and therefore eqn (3.12) becomes

$$
\left(2W_I(t+\delta t) + \frac{\delta t\, h_{cI-1,I}(t+\delta t)}{\delta_{I-1,I+1}} + \frac{\delta t\, h_{cI+1,I}(t+\delta t)}{\delta_{I-1,I+1}} + \frac{\delta t\sum_i v_{i\pm 1,I}(t+\delta t)\bar{\rho}_{i\pm 1,I}(t+\delta t)\bar{C}_{i\pm 1,I}(t+\delta t)}{\delta_{I-1,I+1}\delta_{J-1,J+1}\delta_{K-1,K+1}}\right)\theta(I,t+\delta t)
$$

$$
-\frac{\delta t\, h_{cI-1,I}(t+\delta t)}{\delta_{I-1,I+1}}\theta(I-1,t+\delta t) - \delta t\,\frac{h_{cI+1,I}(t+\delta t)}{\delta_{I-1,I+1}}\theta(I+1,t+\delta t)
$$

$$
-\frac{\delta t\sum_i v_{i\pm 1,I}(t+\delta t)\bar{\rho}_{i\pm 1,I}(t+\delta t)\bar{C}_{i\pm 1,I}(t+\delta t)\theta(i\pm 1,t+\delta t)}{\delta_{I-1,I+1}\delta_{J-1,J+1}\delta_{K-1,K+1}} - \frac{\delta t\, q_I(t+\delta t)}{\delta_{I-1,I+1}\delta_{J-1,J+1}\delta_{K-1,K+1}}
$$

$$
= \left(2W_I(t) - \frac{\delta t\, h_{cI-1,I}(t)}{\delta_{I-1,I+1}} - \frac{\delta t\, h_{cI+1,I}(t)}{\delta_{I-1,I+1}} - \frac{\delta t\sum_i v_{i\pm 1,I}(t)\bar{\rho}_{i\pm 1,I}(t)\bar{C}_{i\pm 1,I}(t)}{\delta_{I-1,I+1}\delta_{J-1,J+1}\delta_{K-1,K+1}}\right)\theta(I,t)
$$

$$+ \frac{\delta t\, h_{cI-1,I}(t)}{\delta_{I-1,I+1}} \theta(I-1,t) + \frac{\delta t\, h_{cI+1,I}(t)}{\delta_{I-1,I+1}} \theta(I+1,t)$$
$$+ \frac{\delta t \sum_i v_{i\pm1,I}(t)\bar{\rho}_{i\pm1,I}(t)\bar{C}_{i\pm1,I}(t)\theta(i\pm1,t)}{\delta_{I-1,I+1}\delta_{J-1,J+1}\delta_{K-1,K+1}}$$
$$+ \frac{\delta t\, q_I(t)}{\delta_{I-1,I+1}\delta_{J-1,J+1}\delta_{K-1,K+1}} \;;\; i = J, K. \tag{3.13}$$

Note that any or all of the coupled nodes (J − 1, J + 1, K − 1, K + 1) can be located external to the air gap and, in this way, time-dependent cavity ventilation can be introduced. The uni-directional formulation of eqn (3.13) with cavity ventilation, q_{vI} (W m^{-2}), is

$$\left(2W_I(t+\delta t) + \frac{\delta t\, h_{cI-1,I}(t+\delta t)}{\delta_{I-1,I+1}} + \frac{\delta t\, h_{cI+1,I}(t+\delta t)}{\delta_{I-1,I+1}}\right)\theta(I,t+\delta t)$$
$$- \frac{\delta t\, h_{cI-1,I}(t+\delta t)}{\delta_{I-1,I+1}}\theta(I-1,t+\delta t) - \frac{\delta t\, h_{cI+1,I}(t+\delta t)}{\delta_{I-1,I+1}}\theta(I+1,t+\delta t)$$
$$- \frac{\delta t\, q_I(t+\delta t)}{\delta_{I-1,I+1}}$$
$$= \left(2W_I(t) - \frac{\delta t\, h_{cI-1,I}(t)}{\delta_{I-1,I+1}} - \frac{\delta t\, h_{cI+1,I}(t)}{\delta_{I-1,I+1}}\right)\theta(I,t)$$
$$+ \frac{\delta t\, h_{cI-1,I}(t)}{\delta_{I-1,I+1}}\theta(I-1,t) + \frac{\delta t\, h_{cI+1,I}(t)}{\delta_{I-1,I+1}}\theta(I+1,t)$$
$$+ \frac{\delta t\,[q_I(t) + q_{VI}(t) + q_{VI}(t+\delta t)]}{\delta_{I-1,I+1}}\,. \tag{3.14}$$

Fluid flow in ducts and pipes

Assuming uni-directional flow as indicated in figure 3.11, then the heat flows are given by

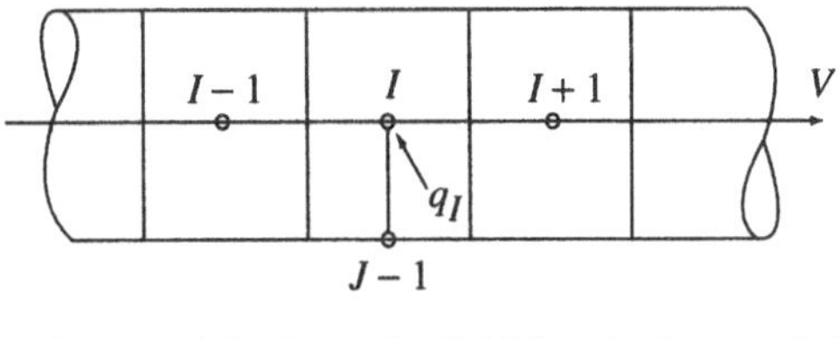

Figure 3.11: Nodal scheme for fluid flow in ducts and pipes.

$$q_{I-1,I} = v_{I-1,I}\bar{\rho}_{I-1,I}\bar{C}_{I-1,I}(\theta_{I-1} - \theta_I)$$

$$q_{I+1,I} = v_{I+1,I}\bar{\rho}_{I+1,I}\bar{C}_{I+1,I}(\theta_{I+1} - \theta_I)$$

$$q_{J-1,I} = h_{fcJ-1,I}A_{J-1,I}(\theta_{J-1} - \theta_I)$$

$$q_{J+1,I} = q_{K-1,I} = q_{K+1,I} = 0$$

where h_{fc} is the forced convection coefficient (W m^{-2}°C^{-1}) and A_{J-1} is the surface area (m^2) of the duct/pipe walls exposed to the fluid volume represented by node I. Substitution of these heat flow quantities in eqn (3.12) gives

$$\left(2W_I(t+\delta t) + \frac{A_{J-1}\delta t\, h_{fcJ-1,I}(t+\delta t)}{\delta V_I} + \frac{\delta t\, v_{I-1,I}(t+\delta t)\bar{\rho}_{I-1,I}(t+\delta t)\bar{C}_{I-1,I}(t+\delta t)}{\delta V_I} + \frac{\delta t\, v_{I+1,I}(t+\delta t)\bar{\rho}_{I+1,I}(t+\delta t)\bar{C}_{I+1,I}(t+\delta t)}{\delta V_I}\right)\theta(I,t+\delta t)$$
$$- \frac{A_{J-1}\delta t\, h_{fcJ-1,I}(t+\delta t)}{\delta V_I}\theta(J-1,t+\delta t)$$
$$- \frac{\delta t\, v_{I-1,I}(t+\delta t)\bar{\rho}_{I-1,I}(t+\delta t)\bar{C}_{I-1,I}(t+\delta t)}{\delta V_I}\theta(I-1,t+\delta t)$$
$$- \frac{\delta t\, v_{I+1,I}(t+\delta t)\bar{\rho}_{I+1,I}(t+\delta t)\bar{C}_{I+1,I}(t+\delta t)}{\delta V_I}\theta(I+1,t+\delta t) - \frac{\delta t\, q_I(t+\delta t)}{\delta V_I}$$
$$= \left(2W_I(t) - \frac{A_{J-1}\delta t\, h_{fcJ-1,I}(t)}{\delta V_I} - \frac{\delta t\, v_{I-1,I}(t)\bar{\rho}_{I-1,I}(t)\bar{C}_{I-1,I}(t)}{\delta V_I} - \frac{\delta t\, v_{I+1,I}(t)\bar{\rho}_{I+1,I}(t)\bar{C}_{I+1,I}(t)}{\delta V_I}\right)\theta(I,t)$$
$$+ \frac{A_{J-1}\delta t\, h_{fcJ-1,I}(t)}{\delta V_I}\theta(J-1,t) + \frac{\delta t\, v_{I-1,I}(t)\bar{\rho}_{I-1,I}(t)\bar{C}_{I-1,I}(t)}{\delta V_I}\theta(I-1,t)$$
$$+ \frac{\delta t\, v_{I+1,I}(t)\bar{\rho}_{I+1,I}(t)\bar{C}_{I+1,I}(t)}{\delta V_I}\theta(I+1,t) + \frac{\delta t\, q_I(t)}{\delta V_I}. \qquad (3.15)$$

In use, equations of this form will normally have control considerations superimposed to dictate the value of v as a function of pump or fan status.

Isolated fluid volumes

In many applications it is necessary to establish further volume subdivision in order to obtain information about temperature distribution. In this case many of the discrete fluid volumes will be coupled only to other fluid volumes, with no direct flowpath to opaque or transparent surfaces. For such a case eqn (3.12) reduces to

$$\left(2W_I(t+\delta t) + \frac{\delta t \sum_{j=1}^{M} v_{j,I}(t+\delta t)\bar{\rho}_{j,I}(t+\delta t)\bar{C}_{j,I}(t+\delta t)}{\delta V_I}\right)\theta(I,t+\delta t)$$
$$- \frac{\delta t \sum_{j=1}^{M} v_{j,I}(t+\delta t)\bar{\rho}_{j,I}(t+\delta t)\bar{C}_{j,I}(t+\delta t)\theta(j,t+\delta t)}{\delta V_I} - \frac{\delta t\, q_I(t+\delta t)}{\delta V_I}$$
$$= \left(2W_I(t) - \frac{\delta t \sum_{j=1}^{M} v_{j,I}(t)\bar{\rho}_{j,I}(t)\bar{C}_{j,I}(t)}{\delta V_I}\right)\theta(I,t) + \frac{\delta t \sum_{j=1}^{M} v_{j,I}(t)\bar{\rho}_{j,I}(t)\bar{C}_{j,I}(t)\theta(j,t)}{\delta V_I}$$
$$+ \frac{\delta t\, q_I(t)}{\delta V_I}. \qquad (3.16)$$

It is possible to establish a mesh of nodes, in two or three dimensions, in such a way that, by repeated application of eqn (3.16), and the more general form of eqn (3.12), the energy flowpaths within the fluid volumes comprising a zone can be interconnected to define all flowpaths up to the internal surface boundary of the zone. The equations of sections §§3.2.1 and 3.2.2 are then employed to extend the equation structure to some external boundary condition. Models for the determination of the v terms are described in chapter 5.

3.3 Equation structuring

Utilising the formulations of §3.2, it is possible to construct a mathematical model of a real scale building. The procedure involves devising a suitable discretisation scheme, generating the corresponding conservation equations, and arranging for equation-set solution when parts of the problem are constrained by control action.

The purpose of this section is merely to demonstrate equation structuring by presenting some example problems. The procedure is then fully elaborated in chapter 4 where an equation-set is established, step-by-step, and a solution technique described in detail. No attempt is made here to elaborate individual equations but, instead, to demonstrate the form of the overall matrix equation to result for some typical cases.

Single zone

Figure 3.12 shows a single zone consisting of 6 multi-layered constructions, 6 surface layers and 2 fluid volumes. A rudimentary nodal scheme is imposed in which only uni-directional transient conduction is permitted, with the enclosed air being divided into a lower and upper portion. In this case the entire system is represented by 50 nodes and so there will be 50 simultaneous equations each comprising one self- and several cross-coupling coefficients relating to the present and future time-rows of the computational time steps.

Figure 3.13 shows the form of the **A** coefficients matrix of the overall equation-set

$$\mathbf{A}\theta_{n+1} = \mathbf{B}\theta_n + \mathbf{C}. \tag{3.17}$$

The coefficient matrix **A** is not square since coefficients relating to future time-row boundary nodes are retained on the left-hand side (future time-row) of each equation even though, in any given problem, their numerical value is known. This is necessary to allow matrix interlocking when other components or building zones are added to effectively remove the problem boundary elsewhere. Even when boundary condition terms are removed to the equation right-hand side, the **A** matrix will remain non-square for all systems in which some nodal heat injection/extraction is to be determined as a function of user-specified control constraints. The solution of eqn (3.17) can then be achieved by the inclusion of negative feedback or feedforward control action as demonstrated in chapter 4. Alternatively, the coefficient entry relating to the heat interaction terms can be replaced by a plant matrix equation which is interlocked with the various building zones. The descriptive examples that follow are elaborated in chapter 6 where models of representative environmental control systems are introduced.

The two-dimensional arrays **A** and **B** have the same number of rows as system equations and any existing element a_{ij} or b_{ij} $(i \neq j+1)$ is a coefficient which links two nodal regions at the future (a_{ij}) and present (b_{ij}) time-row of the time step for which the matrix was established. Any zero valued a_{ij} or b_{ij} indicates that no coupling exists between the nodal regions in question. Elements a_{ii+1} and b_{ii+1} are self-coupling coefficients, which represent the storage potential of the region represented by equation i.

The column matrices θ_{n+1} and θ_n contain the nodal temperature terms and heat injection/extractions at the future and present time-rows respectively. The column matrix **C** contains the known boundary condition excitations due to the temperature and heat flux fluctuations that act on selected nodes to cause energy flow and so 'drive' a simulation.

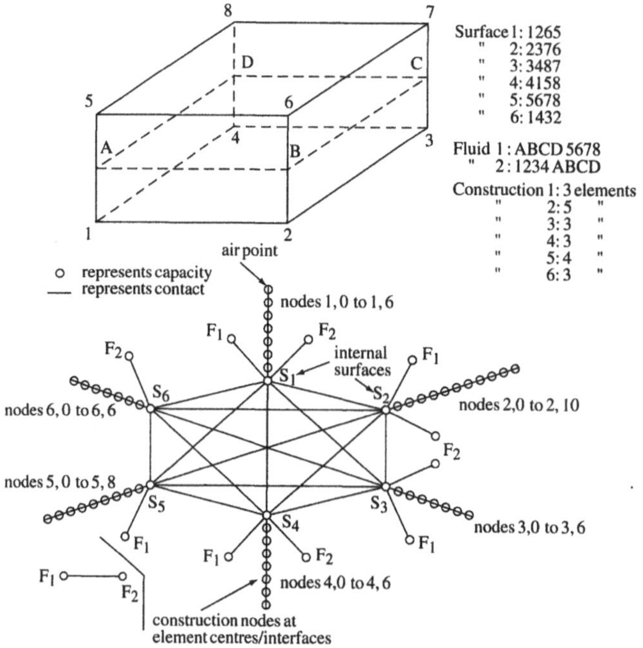

Figure 3.12: A single zone system and equivalent nodal scheme.

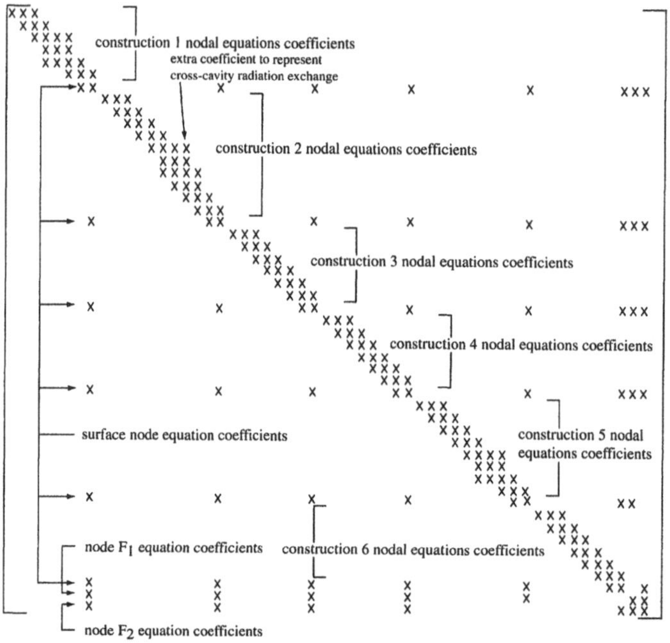

Figure 3.13: The future time-row coefficients matrix (**A**) of the single zone matrix equation $\mathbf{A}\theta_{n+1} = \mathbf{B}\theta_n + \mathbf{C}$.

Since all terms on the right-hand side of eqn (3.17) relate either to the known present time-row ($\mathbf{B}$, θ_n) or are known boundary terms ($\mathbf{C}$), it is appropriate to generate a column matrix $\mathbf{Z}$ where

$$\mathbf{Z} = \mathbf{B}\theta_n + \mathbf{C} . \tag{3.18}$$

This allows eqn (3.17) to be re-expressed as

$$\mathbf{A}\theta_{n+1} = \mathbf{Z}$$

and the solution that follows by

$$\theta_{n+1} = \mathbf{A}^{-1}\mathbf{Z} . \tag{3.19}$$

Various techniques exist to achieve this solution as discussed in chapters 4 through 6. In any event, the outputs from such solutions will support an assessment of energy requirement and comfort (surface temperatures, temperature gradients etc) and will provide causal energy breakdowns that quantify the impacts of infiltration, ventilation, shortwave and longwave radiation, casual gains and so on. Such outputs would not support the study of localised convective phenomena, thermal bridging, indoor air quality or the assessment of plant performance to predict energy consumption. For this purpose the nodal network must be extended.

Single zone with contents and room-side appliances

Figure 3.14 shows the single zone of figure 3.12 but with contents and heating system radiators added. In this case the uni-directional construction conduction model is retained, but a two-dimensional radiator model is added to represent the water and metal volumes at the top, middle and bottom of the radiator. Contents are represented by a single node and, in this model, no radiative coupling is allowed between the contents and radiator surface nodes, and the zone surface nodes. The equivalent equation structure is as shown in figure 3.15. As before, time-step solution is given by eqn (3.19).

This formulation will allow an assessment of energy requirements, surface temperatures and energy flowpath causes/interactions as with the previous single zone model. In addition, radiator surface and water temperature distribution can be predicted and related to the zone air temperature evolution. For this equation structure, boundary conditions consist of time-series temperatures and heat fluxes at the outermost surface of each multi-layered construction of the zone as well as radiator inlet conditions. It is therefore possible to predict zone energy requirements—for any desired environmental condition—as a function of radiator operational characteristics. However, it is still not possible with such a model to determine system energy consumption to meet these requirements. This would require a further extension of the model to include HVAC system detail. Note that radiator control can be applied as a function of any single or aggregate nodal condition to emulate anticipated controller characteristics. For example, to model the effect of a thermostatic radiator valve, the radiator flow rate can be adjusted as a function of some sensed temperature calculated as a weighting between zone air and radiant temperatures.

Multi-zone with central plant

Figure 3.16 shows several single zones combined to form a small multi-zone system with a central boiler and distributed radiator system superimposed. Again, transient conduction is predominantly uni-directional (for clarity) but in one zone a multi-directional scheme has been incorporated to facilitate the study (say) of corner effects and thermal bridges. A simple nodal

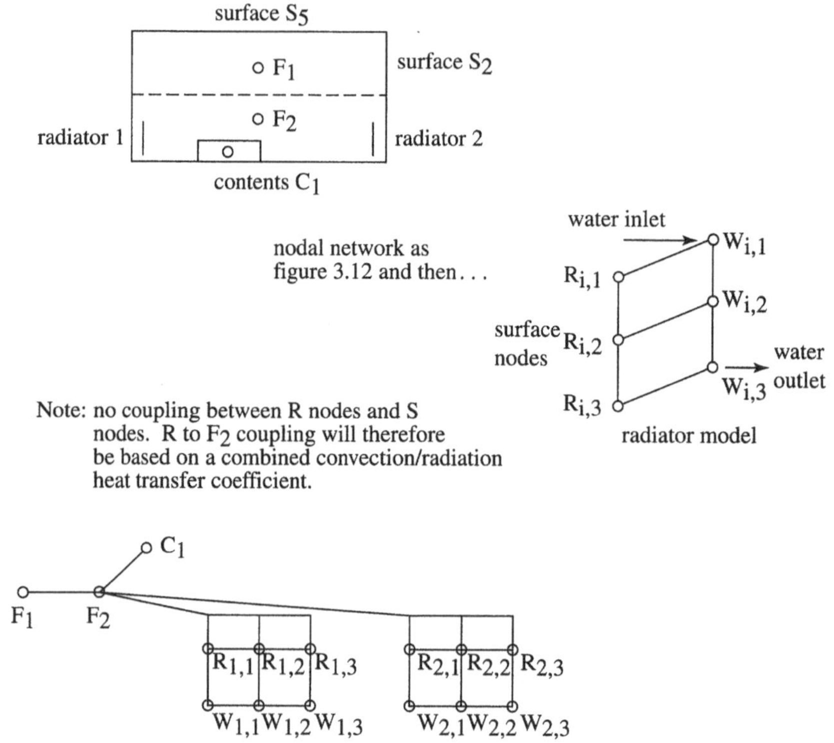

Figure 3.14: The single zone of figure 3.12 with room contents and radiators added.

Figure 3.15: The coefficients matrix of figure 3.13 with zone contents and radiators added.

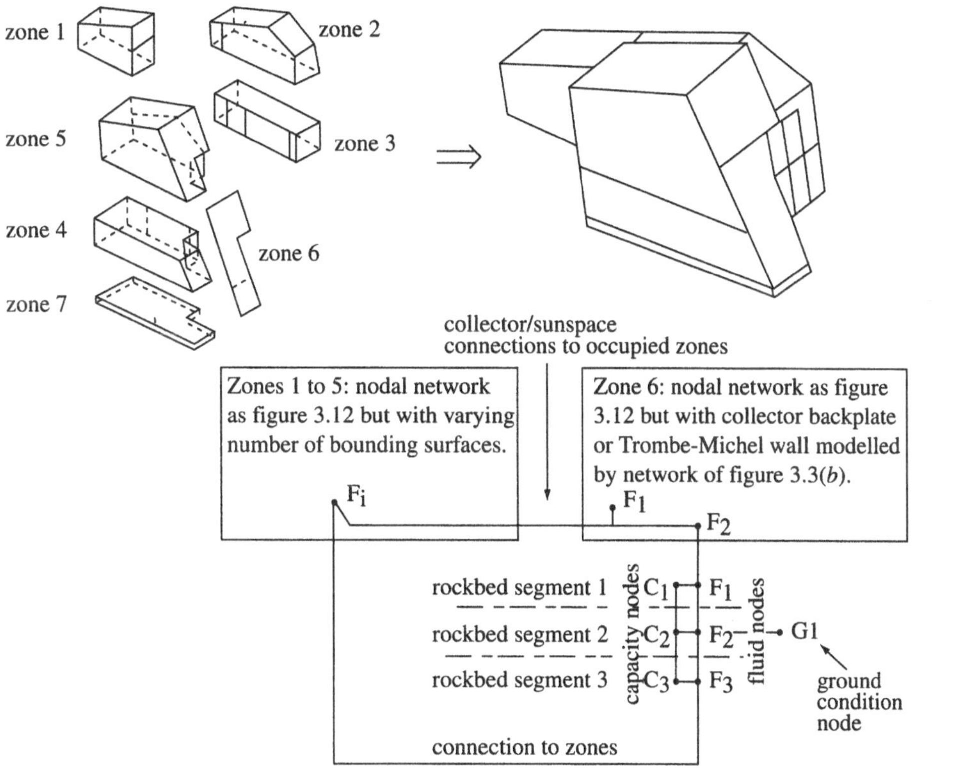

Figure 3.18: An passive/ active solar system.

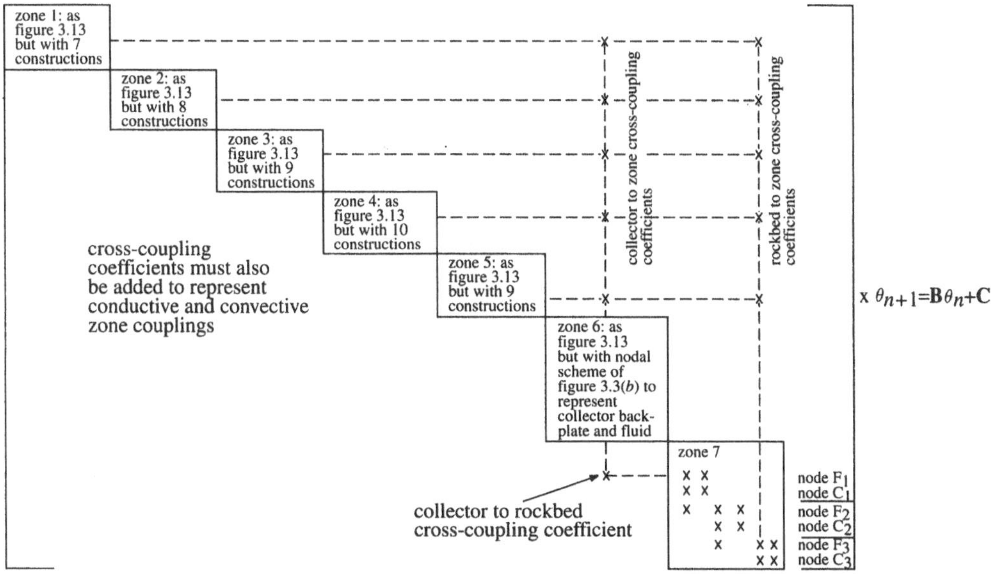

Figure 3.19: The energy balance matrix equation for the system of figure 3.18.

Active/passive solar system with remote storage

Figure 3.18 shows a 7 zone low energy house in which zones 1 through 4 are the bedroom and living zones, zone 5 is an attic space, zone 6 is an active solar collector (in the form of a Trombe-Michel wall) and zone 7 is a remote rockbed thermal store. A detailed nodal scheme has been introduced to the collector to facilitate accurate modelling of the asymmetric heating likely to be experienced due to solar pick-up at the collector back wall, and to permit realistic inlet and outlet air couplings between the collector and the occupied zones or rockbed. This model will allow the storage of the collected energy and predict the variation of surface temperature over the collector back wall.

Figure 3.19 gives the corresponding matrix equation structure, which can be subjected to several control statements governing active or passive energy collection, utilisation or storage, and solved simultaneously at each time-step with the computed future time-row nodal values becoming the basis for matrix reformulation at the next time-step.

This model allows the optimisation of the elements of active and passive solar collection and utilisation as well as the study of comfort conditions in terms of temperature gradients and air movement.

Clearly, the usefulness of the numerical approach is dependent on the ease with which models, such as those described above, can be established at the design stage. One way to achieve this is to automate the discretisation procedure so that a problem, when expressed in terms of its geometry and construction, may be efficiently translated to its numerical counterpart. This is the issue addressed in the next chapter.

3.4 References and further reading

Clarke J A 1977 Environmental Systems Performance *PhD Thesis* (Glasgow: University of Strathclyde)

Dusinberre G M 1961 *Heat Transfer Calculations by Finite Differences* (International Textbook Co)

Hensen J L M and Nakhi A E 1994 Fourier and Biot Numbers and the Accuracy of Conduction Modelling *Proc. Building Environmental Performance '94* (York)

Hildebrand F B 1968 *Finite Difference Equations and Simulations* (New Jersey: Prentice-Hall)

John F 1982 *Partial Differential Equations* (New York: Springer-Verlag)

Lambert J D 1973 *Computational Methods in Ordinary Differential Equations* (New York: Wiley)

Levy H 1959 *Finite Difference Equations* (London: Pitman)

Mitchell A R 1969 *Computational Methods in Partial Differential Equations* (New York: Wiley)

Richtmyer R D 1957 *Difference Methods for Initial-Value Problems* (New York: Interscience)

4

Processing the building energy equations

Chapter 3 derived the energy conservation equations for the characteristic building node types. These equations represent the discretised state of the building over some relatively small interval of time. In matrix notation, the system of equations can be expressed as

$$\mathbf{A}\theta_{n+1} = \mathbf{B}\theta_n + \mathbf{C} = \mathbf{Z} \tag{4.1}$$

where **A** is a sparse matrix of future time-row coefficients of the nodal temperature or heat injection terms of the conservation equations, **B** the corresponding matrix established at the present time-row, **C** a column matrix of known boundary excitations relating to the present and future time-rows, θ a column matrix of nodal temperatures and heat injections, $n+1$ refers to the future time-row, n the present time-row, and **Z** is a column matrix. Initial conditions are given by $\theta(0) = \theta_0$.

Because the system of equations contain both present and future time-row terms, they must be solved simultaneously. And because the entire system is sparse and populated by clusters of equations relating to components with different time constants, a specially adapted solver will be required to minimise the computational effort.

As described in §2.4.3 there are two main solution techniques: iterative and direct. In general terms, direct methods require more storage space than do iterative methods and may prove inefficient when applied to large, sparse equation systems as is the case here. However, direct methods may be used as the building block of an efficient solver for sparse equation systems where different equation clusters are processed at a different frequency depending on the time constant of the corresponding physical system. This allows control decisions to be made, and thermophysical properties recomputed, more frequently for an item of plant requiring a computational time-step of, say, 1 minute, than for a heavyweight construction requiring, say, 60 minutes. Such an approach to solution ensures that only the actual physical scheme is addressed by partitioning the overall (and sparse) system matrix into a number of discrete sub-matrices so that solution can be achieved in the lowest number of computational steps.† Each partitioned

† Because the capabilities and features of simulation programs are constantly evolving, computation times are still problematic despite advances in processor technology.

matrix can then be processed as far as possible by a direct method, and at any frequency, with the inter-component information brought together to permit the global solution stream to continue. As circumstances allow, any sub-matrix need not be reprocessed until its contents (the equation coefficients) have changed by an appreciable amount. Taken to an extreme, where no sub-matrix is reprocessed after the first time-step, the approach will resemble the operation of the time domain response function method as described in chapter 2.

This chapter is concerned with the derivation of an efficient solver based on the direct solution technique while chapters 5 and 6 introduce a specific iterative method in relation to fluid and power flow simulation. Initial sections establish a multi-zone matrix equation and discuss the problems inherent in establishing the time-dependent matrix coefficients at the future time-row. Later sections introduce matrix partitioning in combination with a direct reduction technique suitable for the provision of sufficiently accurate solutions. A mixed direct/iterative technique is then introduced for use where exact solutions are sought, where non-linear processes dominate, or where the solution is to be determined on the basis of multiple node conditions to accommodate aspects of sensor location and response.

Section 2.4.3 gives the salient features of the commonly used solution methods; this, and other relevant texts (e.g. Churchill 1958, Hartree 1958, Collatz 1960, Willoughby and Rose 1972, Stewart 1973, George and Lui 1981, Miranker 1981, Press *et al* 1986), is essential reading for those uninitiated in matrix algebra.

4.1 Establishing the energy matrix equation

The first step is to establish the interrelated equation-set, representing the discrete nodal network, by gathering together the conservation equations, one for each node, in some ordered manner subject to a linking protocol. Equation-sets can be established for any distributed system consisting of multiple zones when serviced by heating, cooling, ventilation and power delivery systems of arbitrary complexity. However, as discussed in §4.1.4, the treatment of time dependent quantities will require special care.

4.1.1 Single zone formulation

Recall figure 3.12 and assume uni-directional conduction within the constructions detailed in table 4.1. Nodes 1,0 through 6,0 are boundary conditions and, as such, will not have an associated equation: for each multi-layered construction, node i,1 (where i is the construction number) is the starting point. Node 1,1 is an outside surface node as represented by eqn (3.11). This equation is now located within the system matrix of figure 4.1 as row 1 in the **A** and **B** matrices (hereafter referred to as the 'row n' equation) where

$$a_{1,1}\theta_{1,0}(t+\delta t) + a_{1,2}\theta_{1,1}(t+\delta t) + a_{1,3}\theta_{1,2}(t+\delta t)$$

$$= b_{1,1}\theta_{1,0}(t) + b_{1,2}\theta_{1,1}(t) + b_{1,3}\theta_{1,2}(t) + c_1 = z_1$$

where

$$a_{1,1} = -\frac{4\delta t\ h_c(t+\delta t)}{\delta x_{11}} \qquad\qquad b_{1,1} = a_{1,1}(t)^\dagger$$

† Her $a_{1,1}(t) = a_{1,1}$ but with $t + \delta t$ replaced by t.

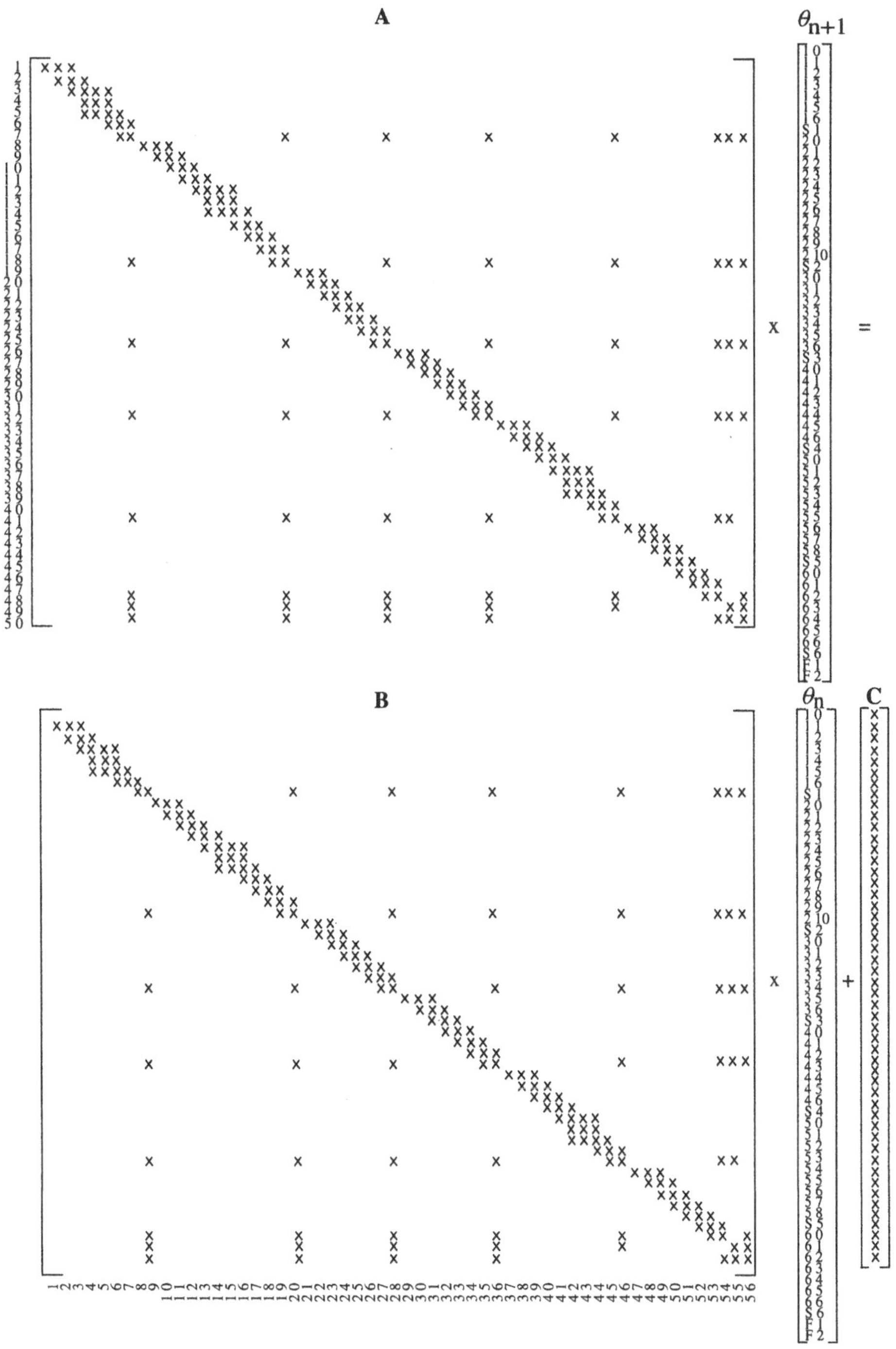

Figure 4.1: The single zone matrix equation $\mathbf{A}\theta_{n+1} = \mathbf{B}\theta_n + \mathbf{C}$.

$$a_{1,2} = 2\rho_{11}(t+\delta t)C_{11}(t+\delta t) + \frac{8\delta t\, k_{11}(t+\delta t)}{\delta x_{11}^2} + \frac{4\delta t\, h_c(t+\delta t)}{\delta x_{11}} + \frac{4\delta t\, h_{rs}(t+\delta t)}{\delta x_{11}} + \frac{4\delta t\, h_{rb}(t+\delta t)}{\delta x_{11}} + \frac{4\delta t\, h_{rg}(t+\delta t)}{\delta x_{11}}$$

$$b_{1,2} = -\,a_{1,2}(t) + 4\rho_{11}(t)C_{11}(t)$$

$$a_{1,3} = -\frac{8\delta t\, k_{11}(t+\delta t)}{\delta x_{11}^2}$$

$$b_{1,3} = -\,a_{1,3}(t)$$

$$c_1 = \{4\delta t[h_{rs}(t+\delta t)\theta_s(t+\delta t) + h_{rb}(t+\delta t)\theta_b(t+\delta t) + h_{rg}(t+\delta t)\theta_g(t+\delta t) + h_{rs}(t)\theta_s(t) + h_{rb}(t)\theta_b(t) + h_{rg}(t)\theta_g(t) + q'_{SI}(t+\delta t) + q'_{SI}(t)]\}/\delta x_{11}\ .$$

Here δx_{11} is the layer thickness (m), k_{11} the layer conductivity (W m^{-1}°C^{-1}), q'_{SI} the shortwave radiation absorption per unit area (W m^{-2}); h_{rs}, h_{rb} and h_{rg} are the longwave radiation coefficients between the surface and the sky, surroundings and ground respectively (all W m^{-2}°C^{-1}); and θ_s, θ_b and θ_g are the temperatures of the sky, surroundings and ground respectively (°C). Section 7.5 describes procedures for the evaluation of these temperatures.

Table 4.1: Constructional details for the single zone example of figure 3.12.

Construction	*Element*†	*Thermophysical properties subscript*
1: External window	Glass	11
	Air gap	12
	Glass	13
2: External wall	Rendering	21
	Brick	22
	Cavity (ventilated)	23
	Brick	24
	Light plaster	25
3: Internal wall	Light plaster	31
	Brick	32
	Light plaster	33
4: Internal wall	Light plaster	41
	Brick	42
	Light plaster	43
5: External roof	Concrete tile	51
	Sarking	52
	Air gap	53
	Insulation	54
6: Ground floor	Dense concrete	61
	Concrete screed	62
	Plastic tile	63
Fluid volume 1		1
Fluid volume 2		2

† Constructions specified from outermost to innermost element relative to the contained volume.

Note that no plant interaction is allowed at this node and so $q'_{\Pi} = 0$ at both time-rows. Thus, row 1 of the **A** matrix will have 3 coefficient entries occupying the first 3 column positions.

Node 1,2 is a homogeneous conduction node of the transparent type as represented by eqn (3.6) and is located in row 2 of figure 4.1 in such a manner that

$$a_{2,2}\theta_{1,1}(t+\delta t) + a_{2,3}\theta_{1,2}(t+\delta t) + a_{2,4}\theta_{1,3}(t+\delta t)$$

$$= b_{2,2}\theta_{1,1}(t) + b_{2,3}\theta_{1,2}(t) + b_{2,4}\theta_{1,3}(t) + c_2 = z_2$$

where

$$a_{2,2} = -\frac{4\delta t\, k_{11}(t+\delta t)}{\delta x_{11}^2} \qquad b_{2,2} = -a_{2,2}(t)$$

$$a_{2,3} = 2\rho_{11}(t+\delta t)C_{11}(t+\delta t) + \frac{8\delta t\, k_{11}(t+\delta t)}{\delta x_{11}^2} \qquad b_{2,3} = -a_{2,3}(t) + 4\rho_{11}(t)C_{11}(t)$$

$$a_{2,4} = -\frac{4\delta t\, k_{1,1}(t+\delta t)}{\delta x_{11}^2} \qquad b_{2,4} = -a_{2,4}(t)$$

$$c_2 = \frac{2\delta t[q'_{SI}(t+\delta t) + q'_{SI}(t)]}{\delta x_{11}}\ .$$

As with node 1,1 no plant interaction is allowed but, since the element is transparent, short-wave solar absorption, q'_{SI}, occurs, which is a known quantity at both time-rows (see §7.4) and so the future time-row term is relocated to the equation right-hand side.

Row 2 of the **A** matrix also has 3 coefficients but in this case there is a column offset to preserve the spatial relationship.

Node 1,3 is located at the interface between the outer glass element and the inter-element air gap and is represented by eqn (3.11) so that

$$a_{3,3}\theta_{1,2}(t+\delta t) + a_{3,4}\theta_{1,3}(t+\delta t) + a_{3,5}\theta_{1,4}(t+\delta t) + a_{3,6}\theta_{1,5}(t+\delta t)$$

$$= b_{3,3}\theta_{1,2}(t) + b_{3,4}\theta_{1,3}(t) + b_{3,5}\theta_{1,4}(t) + b_{3,6}\theta_{1,5}(t) + c_3 = z_3$$

where

$$a_{3,3} = -\frac{8\delta t\, k_{11}(t+\delta t)}{\delta x_{11}^2} \qquad b_{3,3} = -a_{3,3}(t)$$

$$a_{3,4} = 2\rho_{11}(t+\delta t)C_{11}(t+\delta t) + \frac{8\delta t\, k_{11}(t+\delta t)}{\delta x_{11}{}^2} + \frac{4\delta t\, h_c(t+\delta t)}{\delta x_{11}} + \frac{4\delta t\, h_{rA}(t+\delta t)}{\delta x_{11}}$$

$$b_{3,4} = -a_{3,4}(t) + 4\rho_{11}(t)C_{11}(t)$$

$$a_{3,5} = -\frac{4\delta t\, h_c(t+\delta t)}{\delta x_{11}} \qquad b_{3,5} = -a_{3,5}(t)$$

$$a_{3,6} = -\frac{4\delta t\, h_{rA}(t+\delta t)}{\delta x_{11}} \qquad b_{3,6} = -a_{3,6}(t)$$

$$c_3 = \frac{4\delta t[q'_{SI}(t+\delta t) + q'_{SI}(t)]}{\delta x_{11}}$$

and h_{rA} is the longwave radiation coefficient representing the cavity inter-surface exchange (see §7.5).

Again, shortwave solar absorption is possible within the region represented by $\theta_{1,3}$. Note that an additional coefficient is introduced to link nodes 1,3 and 1,5 to represent the longwave radiation exchange.

Node 1,4 is located at the centre plane of the cavity and is represented by eqn (3.14) so that

$$a_{4,4}\theta_{1,3}(t+\delta t) + a_{4,5}\theta_{1,4}(t+\delta t) + a_{4,6}\theta_{1,5}(t+\delta t)$$

$$= b_{4,4}\theta_{1,3}(t) + b_{4,5}\theta_{1,4}(t) + b_{4,6}\theta_{1,5}(t) + c_4 = z_4$$

where

$$a_{4,4} = -\frac{2\delta t\, h_{co}(t+\delta t)}{\delta x_{12}} \qquad b_{4,4} = -a_{4,4}(t)$$

$$a_{4,5} = 2\rho_{12}(t+\delta t)C_{12}(t+\delta t) + \frac{2\delta t\, h_{co}(t+\delta t)}{\delta x_{12}} + \frac{2\delta t\, h_{ci}(t+\delta t)}{\delta x_{12}}$$

$$b_{4,5} = -a_{4,5}(t) + 4\rho_{12}(t)C_{12}(t)$$

$$a_{4,6} = -\frac{2h_{ci}(t+\delta t)\delta t}{\delta x_{12}} \qquad b_{4,6} = -a_{4,6}(t)$$

$$c_4 = \frac{2\delta t[v_o(t)\bar{\rho}_A(t)\bar{C}_A(t)\theta_o(t) + v_o(t+\delta t)\bar{\rho}_A(t+\delta t)\bar{C}_A(t+\delta t)\theta_o(t+\delta t)]}{\delta x_{12}A_c}\,.$$

Here, h_{co} is the convection coefficient for the outermost cavity boundary (W $m^{-2}{}^{\circ}C^{-1}$), h_{ci} the convection coefficient for the innermost cavity boundary, v_o the cavity ventilation rate (m^3s^{-1}) and θ_o the temperature of the ventilating air (°C).

Note that the cavity air is here assumed to be non-absorbing to shortwave radiation and so the heat generation term $q_I = 0$. Also, the cavity ventilation rate is assumed known at the future time-row and so is moved to the equation right-hand side. During a simulation the quantity v_o could, of course, be determined from a separate flow model established using the theory of chapter 6. It is appropriate to set $v_o = 0$ for the case of a sealed, non-ventilated cavity.

Node 1,5 is primarily a repeat of node 1,3 but with the fourth nodal coefficient handling cavity longwave exchange by cross-linking to node 1,3. The row 5 equation therefore utilises matrix elements $a_{5,4}$, $a_{5,5}$, $a_{5,6}$, $a_{5,7}$, $b_{5,4}$, $b_{5,5}$, $b_{5,6}$, $b_{5,7}$ and c_5 where, as before, the c_5 term will contain shortwave heat generation because the multi-layered system is transparent and therefore exposed to solar flux.

Node 1,6 is a repeat of node 1,2 utilising matrix elements $a_{6,6}$, $a_{6,7}$, $a_{6,8}$, $b_{6,6}$, $b_{6,7}$, $b_{6,8}$ and c_6.

The nodal progression has now arrived at the inside surface node, S1, associated with the first multi-layered construction. The relevant conservation equation is given by eqn (3.11) so that

$$a_{7,7}\theta_{1,6}(t+\delta t) + a_{7,8}\theta_{S1}(t+\delta t) + a_{7,j}\theta_f(t+\delta t) + a_{7,20}\theta_{S2}(t+\delta t)$$

$$+ a_{7,28}\theta_{S3}(t+\delta t) + a_{7,36}\theta_{S4}(t+\delta t) + a_{7,46}\theta_{S5}(t+\delta t) + a_{7,54}\theta_{S6}(t+\delta t)$$

$$= b_{7,7}\theta_{1,6}(t) + b_{7,8}\theta_{S1}(t) + b_{7,j}\theta_f(t) + b_{7,20}\theta_{S2}(t) + b_{7,28}\theta_{S3}(t) + b_{7,36}\theta_{S4}(t)$$

$$+ b_{7,46}\theta_{S5}(t) + b_{7,54}\theta_{S6}(t) + c_7 = z_7$$

where

$$a_{7,7} = -\frac{8\delta t\, k_{13}(t+\delta t)}{\delta x_{13}^2} \qquad b_{7,7} = -a_{7,7}(t)$$

$$a_{7,8} = 2\rho_{13}(t+\delta t)C_{13}(t+\delta t) + \frac{8\delta t\, k_{13}(t+\delta t)}{\delta x_{13}^2} + \frac{4\delta t\, h_c(t+\delta t)}{\delta x_{13}} + \frac{4\delta t\, h_{r2}(t+\delta t)}{\delta x_{13}}$$

$$+ \frac{4\delta t\, h_{r3}(t+\delta t)}{\delta x_{13}} + \frac{4\delta t\, h_{r4}(t+\delta t)}{\delta x_{13}} + \frac{4\delta t\, h_{r5}(t+\delta t)}{\delta x_{13}} + \frac{4\delta t\, h_{r6}(t+\delta t)}{\delta x_{13}}$$

$$b_{7,8} = -a_{7,8}(t) + 4\rho_{13}(t)C_{13}(t)$$

$$a_{7,j} = -\frac{4\delta t\, h_c(t+\delta t)}{\delta x_{13}} \qquad b_{7,j} = -a_{7,j}(t)$$

$$a_{7,20} = -\frac{4\delta t\, h_{r2}(t+\delta t)}{\delta x_{13}} \qquad b_{7,20} = -\, a_{7,20}(t)$$

$$a_{7,28} = -\frac{4\delta t\, h_{r3}(t+\delta t)}{\delta x_{13}} \qquad b_{7,28} = -\, a_{7,28}(t)$$

$$a_{7,36} = -\frac{4\delta t\, h_{r4}(t+\delta t)}{\delta x_{13}} \qquad b_{7,36} = -\, a_{7,36}(t)$$

$$a_{7,46} = -\frac{4\delta t\, h_{r5}(t+\delta t)}{\delta x_{13}} \qquad b_{7,46} = -\, a_{7,46}(t)$$

$$a_{7,54} = -\frac{4\delta t\, h_{r6}(t+\delta t)}{\delta x_{13}} \qquad b_{7,54} = -\, a_{7,54}(t)$$

$$c_7 = \frac{4\delta t[q_{SI}(t) + q_{SI}(t+\delta t) + q_{RI}(t) + q_{RI}(t+\delta t)]}{\delta x_{13} A_{S1}} .$$

Here h_{ri} is the radiation coefficient between surface i and the surface being processed (W $m^{-2}{}^{\circ}C^{-1}$) and A_{S1} the area of surface 1. The determination of h_{ri} is covered in §7.5 where a specific linearisation method is applied.

As before, plant interaction potential has been removed and all solar and casual gain terms treated as known for all time since they can be independently determined.

The terms $a_{7,j}\theta_f(t+\delta t)$ and $b_{7,j}\theta_f(t)$ relate the surface node, S1, to the adjacent air node at temperature θ_f. It is possible to subdivide surface 1 vertically to introduce two nodal schemes and allow independent connections between one surface subdivision and air node F1, and between the other surface subdivision and air node F2. The choice here is to connect S1 to some contact area weighting of nodes F1 and F2 such that

$$\theta_f = \frac{A_{S1f1}\theta_{f1} + A_{S1f2}\theta_{f2}}{A_{S1f1} + A_{S1f2}}$$

where A_{S1f1} is the contact area between surface 1 and the volume represented by F1, and A_{S1f2} is the contact area between surface 1 and the volume represented by F2. This allows the previous equation to be expanded to give

$$a_{7,7}\theta_{1,6}(t+\delta t) + a_{7,8}\theta_{S1}(t+\delta t) + a_{7,55}\theta_{f1}(t+\delta t) + a_{7,56}\theta_{f2}(t+\delta t)$$
$$+ a_{7,20}\theta_{S2}(t+\delta t) + a_{7,28}\theta_{S3}(t+\delta t) + a_{7,36}\theta_{S4}(t+\delta t) + a_{7,46}\theta_{S5}(t+\delta t)$$
$$+ a_{7,54}\theta_{S6}(t+\delta t) = b_{7,7}\theta_{1,6}(t) + b_{7,8}\theta_{S1}(t) + b_{7,55}\theta_{f1}(t) + b_{7,56}\theta_{f2}(t)$$
$$+ b_{7,20}\theta_{S2}(t) + b_{7,28}\theta_{S3}(t) + b_{7,36}\theta_{S4}(t) + b_{7,46}\theta_{S5}(t) + b_{7,54}\theta_{S6}(t) + c_7 = z_7$$

where

$$a_{7,55} = -\frac{4A_{S1f1}\delta t\, h_c(t+\delta t)}{\delta x_{13}(A_{S1f1} + A_{S1f2})} \qquad b_{7,55} = -\, a_{7,55}(t)$$

$$a_{7,56} = -\frac{4A_{S1f2}\delta t\, h_c(t+\delta t)}{\delta x_{13}(A_{S1f1} + A_{S1f2})} \qquad b_{7,56} = -\, a_{7,56}(t)$$

and so couplings have been achieved between surface 1 and both adjacent fluid volumes. The magnitude of the convection coefficient, h_c, will be some averaged value for the entire surface although it will be determined on the basis of the weighted fluid temperature θ_f. Should this assumption prove unacceptable then surface subdivision would be required to create a number of distinct surface-to-fluid flowpaths, each with an independent h_c value determined from fundamental considerations as described in §7.6.

The indexing scheme of this surface node equation requires additional comment. As can be seen, the $a_{7,7}$ and $a_{7,8}$ elements (similar reasoning holds for the b elements) are determined on the basis of simple diagonal progression. The inter-surface coupling coefficients $a_{7,20}$, $a_{7,28}$, $a_{7,36}$, $a_{7,46}$ and $a_{7,56}$ are determined as a function of the number of nodes representing the next highest numbered multi-layered construction. In this way it can be predetermined that internal surface 1 will always be referenced by a coefficient in column n + 1, where n is the number of nodes representing the associated multi-layered construction. The value of n is therefore related to the number of homogeneous elements. In a 3 node per element scheme (with the nodes located at the boundaries and centre of the element), $n = 2 \times ne + 1$, where ne is the number of elements. In general, the column location of surface coefficients is given by

$$\sum_{i=1}^{N}(2ne_i + 2)$$

where N is the total number of surfaces up to and including the one of interest.

It is also convenient to locate coefficients that link to fluid nodes at the highest available column number in the matrix system. This forms a matrix pattern that is easy to interpret.

Matrix construction is now largely repetitive as each multi-layered construction node is processed in turn until a fluid node is reached. Nevertheless, some specific problems will be encountered. Node 2,3, for example, is located at the boundary between two elements and so represents a mixed property region for which the relevant equation is eqn (3.8), which can be rewritten as

$$a_{10,11}\theta_{2,2}(t+\delta t) + a_{10,12}\theta_{2,3}(t+\delta t) + a_{10,13}\theta_{2,4}(t+\delta t)$$

$$= b_{10,11}\theta_{2,2}(t) + b_{10,12}\theta_{10,12}(t) + b_{10,13}\theta_{2,4}(t) + c_{10} = z_{10}$$

where

$$a_{10,11} = -\frac{8\delta t[k_{21}(t+\delta t)R_c(t+\delta t) + \delta x_{21}]}{\delta x_{21}R_c(t+\delta t)} \qquad b_{10,11} = -a_{10,11}(t)$$

$$a_{10,12} = \frac{2[\delta x_{21}\rho_{21}(t+\delta t)C_{21}(t+\delta t) + \delta x_{22}\rho_{22}(t+\delta t)C_{22}(t+\delta t)]}{(\delta x_{21} + \delta x_{22})}$$

$$+ \frac{8\delta t[k_{21}(t+\delta t)R_c(t+\delta t) + \delta x_{21}]}{\delta x_{21}R_c(t+\delta t)} + \frac{8\delta t[k_{22}(t+\delta t)R_c(t+\delta t) + \delta x_{22}]}{\delta x_{22}R_c(t+\delta t)}$$

$$b_{10,12} = -a_{10,12}(t) + \frac{4[\delta x_{21}\rho_{21}(t)C_{21}(t) + \delta x_{22}\rho_{22}(t)C_{22}(t)]}{(\delta x_{21} + \delta x_{22})}$$

$$a_{10,13} = -\frac{8\delta t[k_{22}(t+\delta t)R_c(t+\delta t) + \delta x_{22}]}{\delta x_{22}R_c(t+\delta t)} \qquad b_{10,13} = -a_{10,13}(t)$$

$$c_{10} = 0\,.$$

For this nodal case it has been necessary to determine region storage potential on the basis of a weighting of the temperature dependent thermal properties of the component elements.

Each of surfaces 1-4 have equations which involve both fluid nodes. Surfaces 5 and 6, however, in keeping with the physical scheme, have only one fluid node connection.

Rows 49 and 50 hold the equations for fluid nodes F1 and F2 for which cases eqn (3.12) can be located as follows for F1

$$a_{49,8}\theta_{S1}(t+\delta t) + a_{49,20}\theta_{S2}(t+\delta t) + a_{49,28}\theta_{S3}(t+\delta t) + a_{49,36}\theta_{S4}(t+\delta t)$$

$$a_{49,46}\theta_{S5}(t+\delta t) + a_{49,55}\theta_{f1}(t+\delta t) + a_{49,56}\theta_{f2}(t+\delta t) = b_{49,8}\theta_{S1}(t) + b_{49,20}\theta_{S2}(t)$$

$$+ b_{49,28}\theta_{S3}(t) + b_{49,36}\theta_{S4}(t) + b_{49,46}\theta_{S5}(t) + b_{49,55}\theta_{f1}(t) + b_{49,56}\theta_{f2}(t) + c_{49} = z_{49}$$

where

$$a_{49,8} = -\frac{\delta A_{S1}\delta t\ h_{cf1S1}(t+\delta t)}{\delta V_{f1}} \qquad b_{49,8} = -a_{49,8}(t)$$

$$a_{49,20} = -\frac{\delta A_{S2}\delta t\ h_{cf1S2}(t+\delta t)}{\delta V_{f1}} \qquad b_{49,20} = -a_{49,20}(t)$$

$$a_{49,28} = -\frac{\delta A_{S3}\delta t\ h_{cf1S3}(t+\delta t)}{\delta V_{f1}} \qquad b_{49,28} = -a_{49,28}(t)$$

$$a_{49,36} = -\frac{\delta A_{S4}\delta t\ h_{cf1S4}(t+\delta t)}{\delta V_{f1}} \qquad b_{49,36} = -a_{49,36}(t)$$

$$a_{49,46} = -\frac{\delta A_{S5}\delta t\ h_{cf1S5}(t+\delta t)}{\delta V_{f1}} \qquad b_{49,46} = -a_{49,46}(t)$$

$$a_{49,55} = 2\rho_{f1}(t+\delta t)C_{f1}(t+\delta t) + a_{49,8} + a_{49,20} + a_{49,28} + a_{49,36} + a_{49,46}$$
$$+ \frac{\delta t\ v_{f2f1}(t+\delta t)\bar{\rho}_{f2f1}(t+\delta t)\bar{C}_{f2f1}(t+\delta t)}{\delta V_{f1}} + \frac{\delta t\ v_{of1}(t+\delta t)\bar{\rho}_{of1}(t+\delta t)\bar{C}_{of1}(t+\delta t)}{\delta V_{f1}}$$

$$b_{49,55} = -a_{49,55}(t) + 4\rho_{f1}(t)C_{f1}(t)$$

$$a_{49,56} = \frac{\delta t\ v_{f2f1}(t+\delta t)\bar{\rho}_{f2f1}(t+\delta t)\bar{C}_{f2f1}(t+\delta t)}{\delta V_{f1}} \qquad b_{49,56} = a_{49,56}(t)$$

$$c_{49} = \frac{\delta t}{\delta V_{f1}}\ [q_{cI}(t) + q_{cI}(t+\delta t) + v_{of1}(t)\bar{\rho}_{of1}(t)\bar{C}_{of1}(t)\theta_o(t)$$
$$+ v_{of1}(t+\delta t)\bar{\rho}_{of1}(t+\delta t)\bar{C}_{of1}(t+\delta t)\theta_o(t+\delta t)]$$

and similarly for F2:

$$a_{50,8}\theta_{S1}(t+\delta t) + a_{50,20}\theta_{S2}(t+\delta t) + a_{50,28}\theta_{S3}(t+\delta t) + a_{50,36}\theta_{S4}(t+\delta t)$$

$$a_{50,54}\theta_{S6}(t+\delta t) + a_{50,55}\theta_{f1}(t+\delta t) + a_{50,56}\theta_{f2}(t+\delta t) = b_{50,8}\theta_{S1}(t) + b_{50,20}\theta_{S2}(t)$$

$$+ b_{50,28}\theta_{S3}(t) + b_{50,36}\theta_{S4}(t) + b_{50,54}\theta_{S6}(t) + b_{50,55}\theta_{f1}(t) + b_{50,56}\theta_{f2}(t) + c_{50} = z_{50}$$

with similar expressions for the equation coefficients.

In these expressions, h_{cf1Si} $(i = 1-5)$ is the convection coefficient at the fluid/surface interface (W m^{-2}°C^{-1}), δV_{f1} the volume represented by F1 (m^3), v_{f2f1} the volume flow rate from F2 to F1 (m^3s^{-1}), $\bar{\rho}_{f2f1}$ the mean† density determined on the basis of F1 and F2 temperatures (kg m^{-3}), $\bar{\rho}_{of1}$ the mean density determined on the basis of F1 and 'outside' source temperatures, δA_{Si} the contact area between fluid and surface (m^2), A_s the surface area, v_{of1} the flow rate from outwith the zone—infiltration or zone coupled (m^3s^{-1})—and $\bar{C}_{f2f1}$ the mean specific heat capacity determined on the basis of F1 and F2 temperatures (J kg^{-1}°C^{-1}).

Similar reasoning can be applied to the fluid 2 node. Note that with fluid 1 no contact path exists between node F1 and the floor node S6 since no connection exists in reality. For the same reason no contact exists between F2 and S5 in the fluid 2 equation.

In this example there is an air flow coupling to 'outside'. This may result from infiltration, natural ventilation, mechanical ventilation and/or zone-coupled air flow. If all four flowpaths are present then the v_{of1} terms of coefficients $a_{49,55}$, $b_{49,55}$ and c_{49} should be subdivided into

† An alternative approach is to use the density corresponding to the temperature of the incoming fluid stream. However, since there is no explicit representation of the mixing process the correct approach is debatable.

four distinct quantities (with similar modifications applied to the F2 coefficients).

For the single zone problem considered here, the assumption is made that temperature θ_o is a known boundary condition. When processing multi-zone networks, the θ_o variable, if located in another zone, must be treated as a future time-row quantity and determined by simultaneous solution of the adjacent zone's matrix equation as described later in this chapter. In this case the $\theta_o(t+\delta t)$ term in c_{49}, where θ_o is an active node in the problem, must be removed to the equation's left-hand side to become a coefficient of **A** properly located and linked: the $\theta_o(t)$ coefficient will then become a coefficient of **B**. If the θ_o term relates to infiltration or natural ventilation then θ_o will represent external air temperature—a boundary condition—and so the term can remain as one of the c_{49} collection. If θ_o relates to some known mechanical supply then it can be treated in the same way as infiltration and natural ventilation.

The fluid volume coupling flow rates, v_{f2f1} and v_{f1f2}, which can exist simultaneously, may be determined from an air flow model operating in tandem (chapter 5) or can be prescribed.

4.1.2 Zone contents and plant interaction

Before considering the matrix equation solution method of the next section, it is appropriate to examine possible variations of the single zone model established in the previous section. Such variations are required to represent the capacity effects of zone contents and the interactions with heating, cooling and ventilation systems.

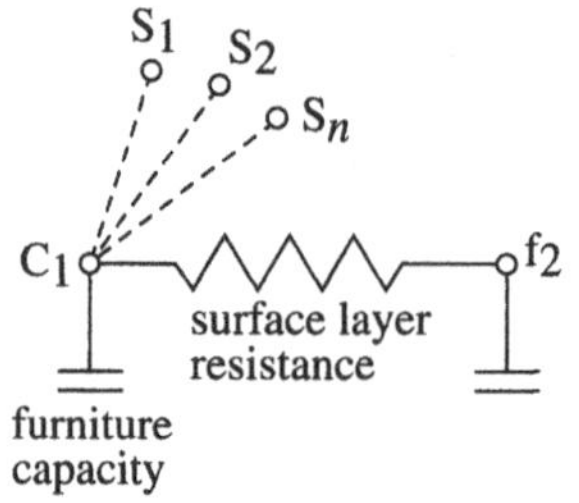

Figure 4.2: Simplified nodal scheme representing zone contents. A nodal connection between C1 and surrounding surfaces can be introduced if required (shown dotted).

Zone contents

Recall the zone contents of figure 3.14 and assume for simplicity that these can be combined into an equivalent volume of material such that the nodal network of figure 4.2 is obtained. The assumption is made that since the contents are contained by fluid F2 then the material temperature will vary little with spatial position and so can be represented by a single node C1. Eqn (3.9) relates to this node type and can be located in the zone matrix of figure 4.1 such that

$$a_{51,56}\theta_{f2}(t+\delta t) + a_{51,57}\theta_{C1}(t+\delta t) = b_{51,56}\theta_{f2}(t) + b_{51,57}\theta_{C1}(t)$$

where

$$a_{51,56} = -\frac{A_{C1}\delta t\, k_{C1}(t+\delta t)}{V_{C1}\delta_{C1f2}} \qquad b_{51,56} = -a_{51,56}(t)$$

$$a_{51,57} = 2\rho_{C1}(t+\delta t)C_{C1}(t+\delta t) + \frac{A_{C1}\delta t\, k_{C1}(t+\delta t)}{V_{C1}\delta_{C1f2}} \qquad b_{51,57} = -a_{51,57}(t) + 4\rho_{C1}(t)C_{C1}(t)$$

Here k_{C1} is the conductivity of the material comprising the contents (W m^{-1}°C^{-1}), C_{C1} the specific heat capacity (J $kg^{-1}°C^{-1}$), A_{C1} the exposed surface area, V_{C1} the contents volume, δ_{C1f2} a characteristic linear dimension, equal to the radius for a sphere, (1/2)a for a cube (where a is a side length), and (a + b + c)/6 for a cuboid (where a, b and c are length, width and height respectively).

This equation can now be added, as the row 51 equation, to give the additions of figure 4.3. Since this equation has a cross-coupling coefficient ($a_{51,56}$) linking to the F2 node, a coefficient entry must also be made for the F2 equation since nodes communicating by non-flow processes must be linked in both directions. Thus coefficients $a_{50,57}$ and $b_{50,57}$ are added, where

$$a_{50,57} = \frac{A_{C1}\,\delta t\, h_{cf2C1}(t+\delta t)}{\delta V_{f2}} \qquad b_{50,57} = a_{50,57}(t)\,.$$

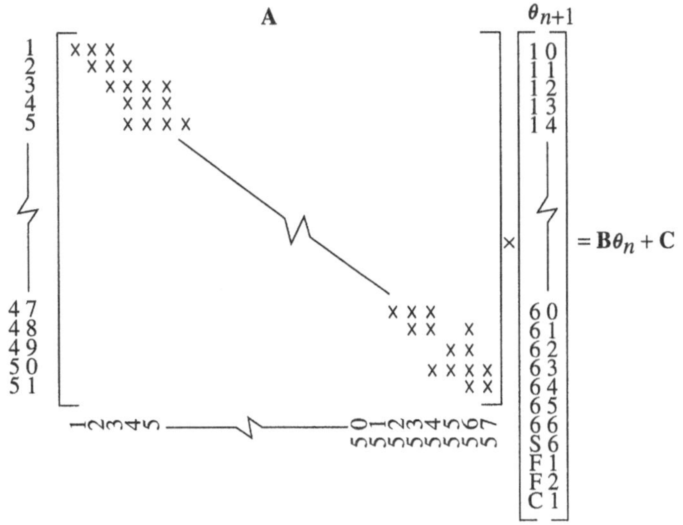

Figure 4.3: Addition of nodal coefficients (to the matrix of figure 4.1) to include zone contents.

Contents capacity is now an integral part of the system and zone performance will be modified accordingly.

Plant interaction

If the modelling objective is to predict the zone's energy requirements to provide stated environmental conditions then it is necessary to allow plant interaction with one or more nodes. An additional coefficient is added to the affected nodal equation(s) and the corresponding entry in the θ_{n+1} vector is a nodal flux term rather than nodal temperature. Locating the plant interaction point at different system nodes has important physical interpretations as follows.

Location at a fluid node implies that the plant input or extract is by convective means since the energy is added to or extracted from the zone air. Such a location is appropriate for the conceptual modelling of air conditioning systems in that heating/cooling flux can be added convectively with any time or mechanical constraint imposed—e.g. system off at 5p.m. or CO_2 levels rising, increase volume flow rate. Chapter 6 discusses approaches to explicit plant modelling.

Location at a surface node implies that the plant input or extract is by convective and radiant means since subsequent energy redistribution will take place by longwave radiation to the surroundings and by convection to the adjacent fluid volume (and, to a lesser extent, by conduction to the next-to-surface material). Some of the energy will flow in or out of the capacity associated with the surface node. This location corresponds to the conceptual modelling of radiant panels and radiators for which an independent surface description has been created in the data of the defining model.

Location at a capacity/insulation node allows the plant input or extract to be directed to some intra-material node. This device is useful for the modelling of systems in which the input energy interacts directly with some material region as with electrical storage units or underfloor heating systems.

It is possible to introduce a number of plant interaction points to handle complex plant input schemes such as found in multiple radiator installations, distributed air conditioning systems and multi-unit electrical storage configurations. In the current example, convective input to the F1 fluid volume is assumed and so coefficients $a_{49,57}$ and $b_{49,57}$ are added to the matrix of figure 4.1, where

$$a_{49,57} = -\ \delta t / V_{f1} \qquad\qquad b_{49,57} = a_{49,57}(t)$$

and the plant terms $q_p(t + \delta t)$ and $q_p(t)$ are appended to the θ_{n+1} and θ_n vectors.

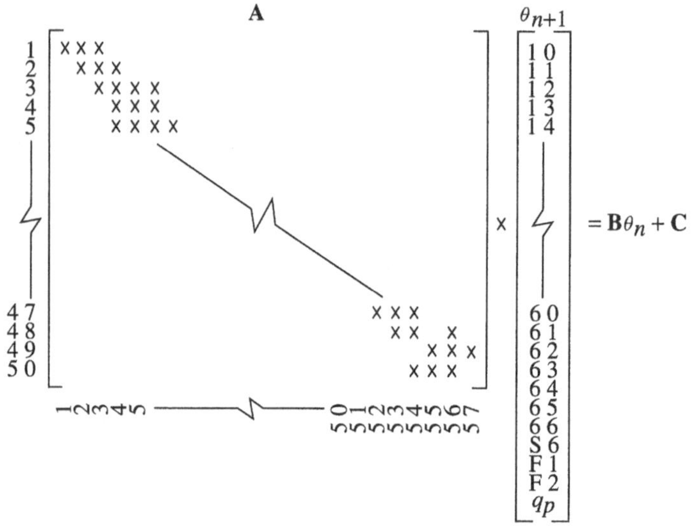

Figure 4.4: Addition of a nodal coefficient (to the matrix of figure 4.1) to include plant interaction potential.

If plant interaction potential is to be allowed with more than one zone node then similar coefficients must be added to the affected nodal equation(s) (as given by a row in the **A** and **B** matrices) with plant variable terms appended as above. This process is shown in figure 4.4.

4.1.3 Multi-zone systems

A collection of single zones can be combined by matrix interlocking to represent some multi-zone system. Figure 4.5 illustrates the interlocking of 4 single zone matrices. Several points should be noted as follows.

Each component matrix is uniquely located and will be of varying size depending on the associated zone shape (that is, the number of bounding surfaces and hence multi-layered

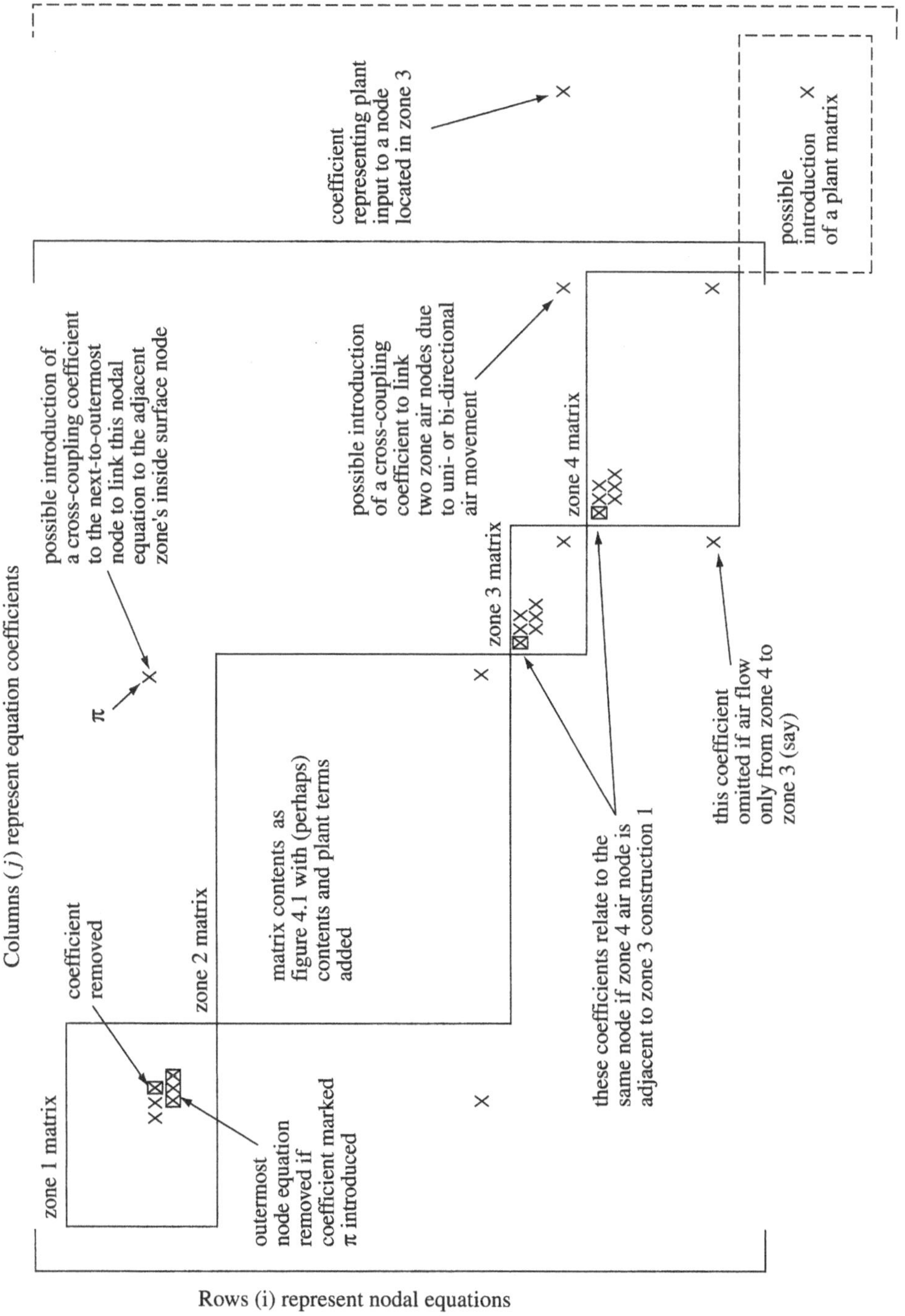

Figure 4.5: A multi-zone matrix formed by interlocking four single-zone matrices as exemplified by figure 4.1.

constructions) and nodal discretisation (perhaps finer subdivision is introduced to improve modelling accuracy in some critical region).

The nodal equations relating to the outermost node(s) in each multi-layered construction will already contain a cross-coupling coefficient linking the node to some air node in an adjacent zone, if the adjacent zone is included in the multi-zone system. This is done to facilitate fast simultaneous solution by matrix partitioning as described in §4.2.

While intra-zone longwave radiation exchange is included within each component matrix, the corresponding exchange between the outermost node and the internal surface nodes pertaining to the equation-set for some adjacent zone must now be introduced. There are two possibilities.

1. The heat generation term of the outermost node can be used to introduce an exchange with the internal surfaces of the adjacent zone by operating one time-step in arrears. This will effectively link the outermost node at the future time-row with the adjacent zone internal surface nodes at the present time-row. This scheme is computationally attractive since it preserves the compact matrix form without introducing, in many cases, a detectable error.
2. If an exact solution is sought then the outermost nodal equation (communicating with some adjacent zone included in the matrix system) can be removed from the component matrix equation-set. The cross-coupling coefficient linking the next-to-outermost surface node to the outermost surface node is relocated to link, instead, to the appropriate inside surface node in the adjacent zone which is already linked, at the future time-row, to the other zone surfaces.

At this point additional cross-coupling coefficients must be introduced to link zone air nodes if inter-zone air movement is to be modelled. In the case of uni-directional air flow between two zone air nodes—as in the case of flow through a small connecting orifice—only one coefficient is added to the receiving node equation. With large openings, such as doorways, where bi-directional air movement can occur, an additional coefficient must be added to both air node equations.

The annotation applied to figure 4.5 summarises the multi-zone matrix formulation procedure, which can be extended to include any number of zones (as a function of the available computing power) to represent whole-building energy balance whilst preserving the important spatial and temporal relationships.

It is at this stage that a further equation-set, representing the energy flows within some distributed plant system, can be introduced to allow the simultaneous processing of an integrated building/plant system. This process is demonstrated in chapter 6.

4.1.4 Treatment of time-dependent properties

The coefficient entries of the **A** matrix relate to the future time-row for which the nodal temperatures are unknown until simultaneous solution has been achieved. It is therefore impossible to evaluate these entries on the basis of the, as yet, uncomputed future time-row state variables. Two processing strategies are possible.

First, coefficient evaluation can proceed one time-step in arrears so that the future time-row **A** matrix is established on the basis of present time-row nodal information, with the present time-row **B** matrix utilising immediate past information. For most practical applications this option will have acceptable consequences; indeed most contemporary energy models hold thermophysical properties constant so that the **A** matrix is, in effect, time invariant. And, of course, as the time-step is reduced, so the temperature variations between time-steps will diminish and the exact formulation be approached.

Second, based on this 'one time-step in arrears' principle it is possible to establish and solve the system of equations to give future time-row nodal conditions. These data can then be used to re-estimate the problem parameters and iteration employed to achieve coefficient convergence. There are few practical applications where such rigour will be required.

4.1.5 Adiabatic boundaries

In some cases it is convenient to introduce an adiabatic plane to reduce the number of zones for modelling. This would be an appropriate solution in the case of a problem exhibiting a high degree of symmetry. While the imposition of an adiabatic plane is straightforward within an analytical method—the heat flux is simply made identically zero at the appropriate location—it is a less trivial task within a numerical method because the space dimension is discretised. Any attempt to impose a zero temperature difference by simply assigning values to the state variables will interfere with the underlying numerical method and lead to an energy imbalance. The required state variable values must instead be attained by the addition or removal of heat at the node(s) in question.

A more straightforward approach is to manipulate the heat transfer parameters. Suppose that the adiabatic plane is located at the outside surface of a wall. This can be represented by assigning a zero emissivity, absorptivity and convection coefficient to the surface. While the heat transfer across the plane will be identically zero, there will still be flux exchanges within the wall and therefore this condition might be markedly different from that implied by setting $q = 0$ in an analytical solution. Where the adiabatic plane is to be located within the wall, this can be approximated by a thin layer of zero conductivity material centred on the plane. The essential point is that the representation of an adiabatic plane within a numerical model requires careful thought!

4.2 Matrix partitioning for fast simultaneous solution

Having established the single zone matrix of figure 4.1, it is now necessary to arrange for simultaneous equation solution to determine the zone's response to a given weather stimulus. If the system matrix is re-established at each successive time-step on the basis of the results from the previous time-step then the system response will evolve over time.

Consider the single zone matrix of figure 4.1, with plant interaction potential at the F1 node. The time dependency of the coefficients of the unknown nodal temperatures (**A**) dictate that the matrix be inverted ($\mathbf{A}^{-1}$) at each time-step in order to achieve the necessary repetitive simultaneous solution. The multiple inversion of such a sparse matrix is computationally inefficient—a heating season simulation at a one hour time-step requiring approximately 5,500† inversions—and for this reason a direct and rapid solution scheme is required. The symmetry of the **A** matrix suggests a linear, step-wise scheme in which the overall matrix is partitioned into a series of elemental sub-matrices, each of which can be processed independently, with the output information from each being brought together to achieve the final solution for a given time-step. The entire zone equation-set can therefore be viewed as a number of interlocking partitioned matrices. The advantages of this approach are threefold. First, only the actual physical schema is addressed; that is, any unfilled elements within the system matrix are not processed

† This number will increase dramatically where intra-zone air and intra-construction moisture flow modelling is included.

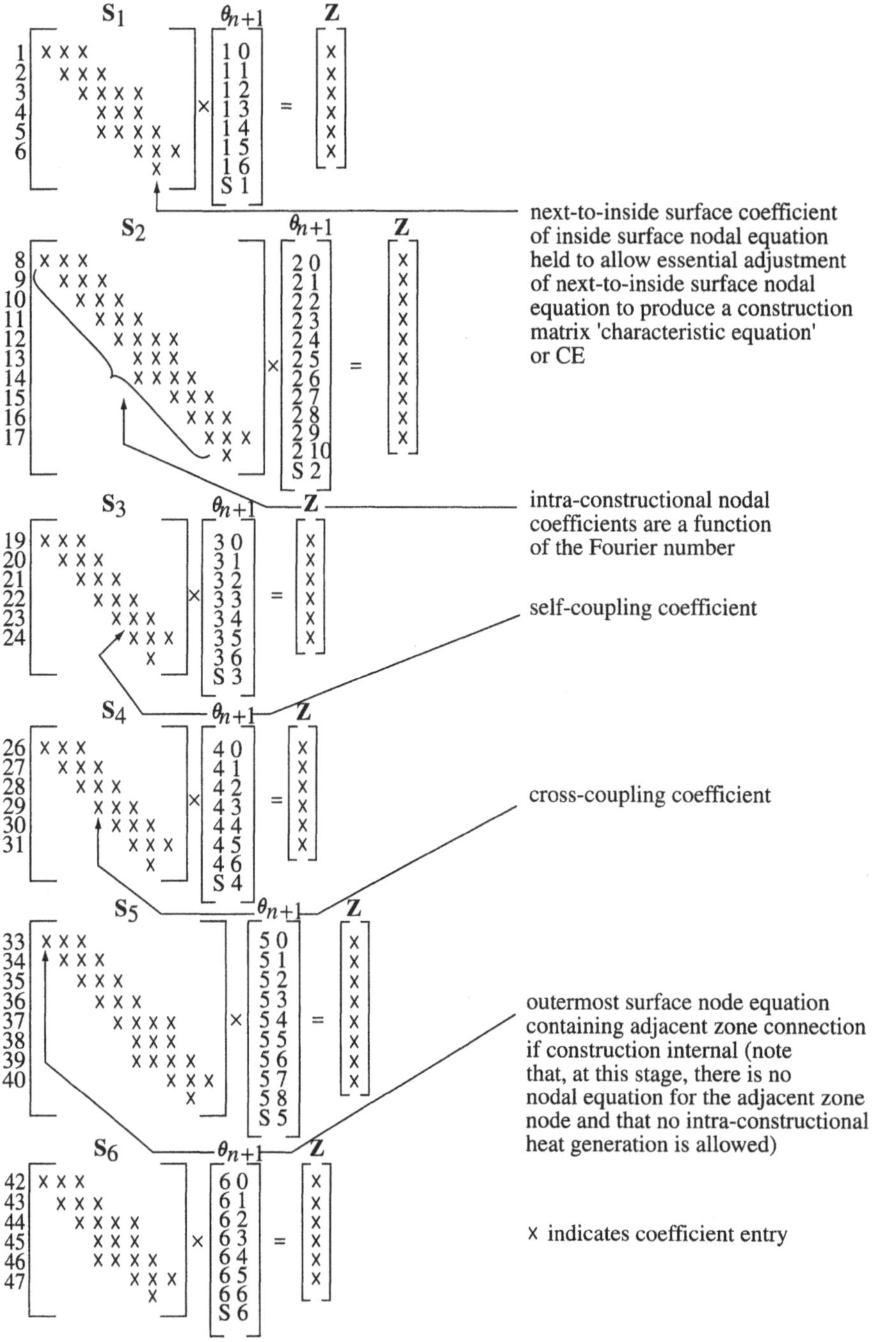

Figure 4.6: Partitioned construction matrix equations.

at any point in the inversion process (in fact, the matrices of figures 4.1 and 4.5 need never be established in computer memory). Second, any partitioned matrix, once inverted, need not be re-inverted at subsequent time-steps unless its contents have changed by a significant amount. Last, because each sub-matrix is processed independently, it is possible to arrange for different matrices to be processed at different frequencies depending on the time constant of the physical system they represent.

The sections that follow cover single- and multi-zone system processing, describe techniques for matrix solution on the basis of comfort criteria, and discuss the treatment of non-linear processes and complex matrix systems.

4.2.1 Single zone solution

For a single zone equation-set, without nodal plant interaction, partitioning is performed according to the following procedure.

A number of partitioned matrices are extracted, each one representing a multi-layered construction within the single zone problem. These matrices contain the coefficients relating to the intra-constructional nodal equations addressing material conduction and heat storage. These sub-matrices will also contain the coefficient of the future time-row temperature of the next-to-inside surface node relating to the surface energy balance equation, to enable adjustment of the next-to-inside surface nodal equation in order to facilitate later modification of the zone energy balance matrix (see below). Partitioned construction matrices are designated by $\mathbf{S}_n$; there are 6 in the problem of figure 4.1 as shown in figure 4.6. It is important to note that only the filled elements of the **S** matrix should be held in memory for processing during matrix reduction. The two-dimensional representations given here are adopted only for reasons of clarity.

Further partitioning of the **A** matrix gives rise to a sub-matrix (designated by **R**) containing the unknown future temperature coefficients of the heat balance equations relating to each internal surface node and each fluid node within the zone. These coefficients represent the zone inter-surface radiation exchange, surface convection and fluid flow processes. At this stage the coefficients of the next-to-inside surface node equations are also held to define the linkage between the **R** and **S** type matrices. The contents of the **R** matrix are shown in figure 4.7.

Since any node can, in addition, have prescribed heat generation due to plant interaction—for example, convective input to air nodes, direct capacity input in the case of under-floor heating systems or electrical storage units, and mixed node injection with radiant systems—the partitioned matrices of figures 4.6 and 4.7 can be modified to incorporate nodal plant injection/extraction terms. Figure 4.8 gives four example matrices for the case of air point injection, surface injection, intra-construction injection and a mixed scheme that incorporates all three injection possibilities.

The solution of the entire zone equation-set—now held in partitioned form—is a 6 stage process which addresses only the 'filled' matrix elements and which accommodates any spatial position of controller location and plant input/extract. The objective is to process each partitioned matrix as far as possible to allow the extraction of one or more 'characteristic equations' (hereafter referred to as a CE) that embody the dynamics of the related component. These CEs are then gathered together to produce a set of whole-system CEs, which relate control nodes (whose temperature or heat flux status will control the simulation path) to the required nodal plant injection/extraction. This CE set is then solved in terms of user imposed control statements before the back substitution phase of the matrix inversion procedure is recommenced. The primary task is to ensure that all nodal terms, representing declared control regions, are 'carried through' to that point in the solution process where control theory is introduced. The

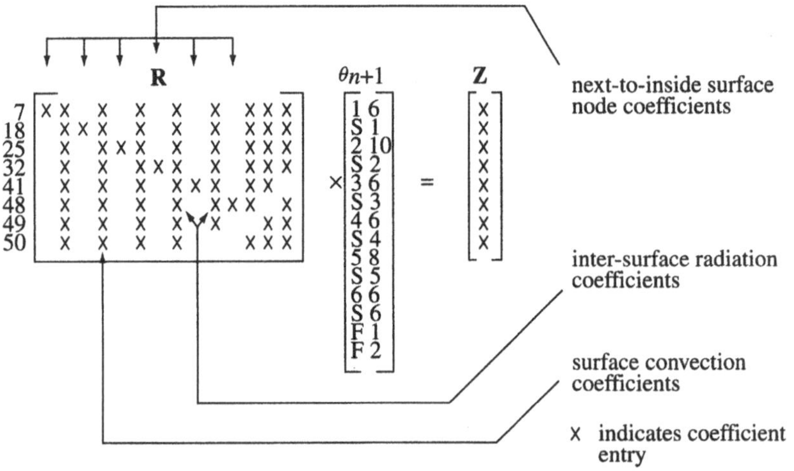

Figure 4.7: Partitioned zone energy balance matrix equation.

solution technique comprises 6 stages as follows.

Stage 1: Each constructional sub-matrix is processed towards the end of the forward reduction stage of a direct Gaussian elimination as detailed in §2.4.3. The reduction continues until the equation relating to the next-to-inside surface node is reached. At this point sub-matrix processing is terminated and the CE containing adjusted matrix coefficients relating to the next-to-inside surface nodal equation extracted. Two possible reduction schemes exist, depending on whether or not intra-constructional node control is active. Figure 4.9 demonstrates the case where a control point is not located within the multi-layered construction and there is no plant interaction potential at an intra-fabric node. In this case the extracted CE will have terms relating to the adjacent zone air node, the next-to-inside surface node, the inside surface node and a composite term relating to the matrix equation right-hand side (a **Z** element representing the entire known time-row).

Figure 4.10 demonstrates the reduction process when a control node is located within the construction. Note that the control point coefficient is carried through and introduced to each subsequent nodal equation (matrix row) so that two CEs emerge, both of which contain control point coefficients.

Figure 4.11 demonstrates matrix forward reduction and CE extraction for the cases of intra-constructional plant interaction when the control point is, and is not, located within a capacity/insulation system.

Stage 2: The CEs, when collected from all constructional sub-matrices, can now be used to eliminate one term from each of the internal surface node equations of the **R** matrix representing radiative, convective, advective, conductive and heat generation processes within a single zone. This is done by algebraic addition of the CE and the **R** matrix row equation which relates to the same construction. Figure 4.12 demonstrates this **R** matrix modification for the cases where a control point is, and is not, located within a multi-layered construction and when no plant interaction nodes are present.

Plant interaction nodes may already exist in the **R** matrix, as shown in figure 4.8, or may be introduced to the $\mathbf{R}'$ matrix by addition of the construction CEs containing one or more plant coefficients. Figure 4.13 demonstrates the $\mathbf{R}'$ matrix to result from a mixed scheme in which

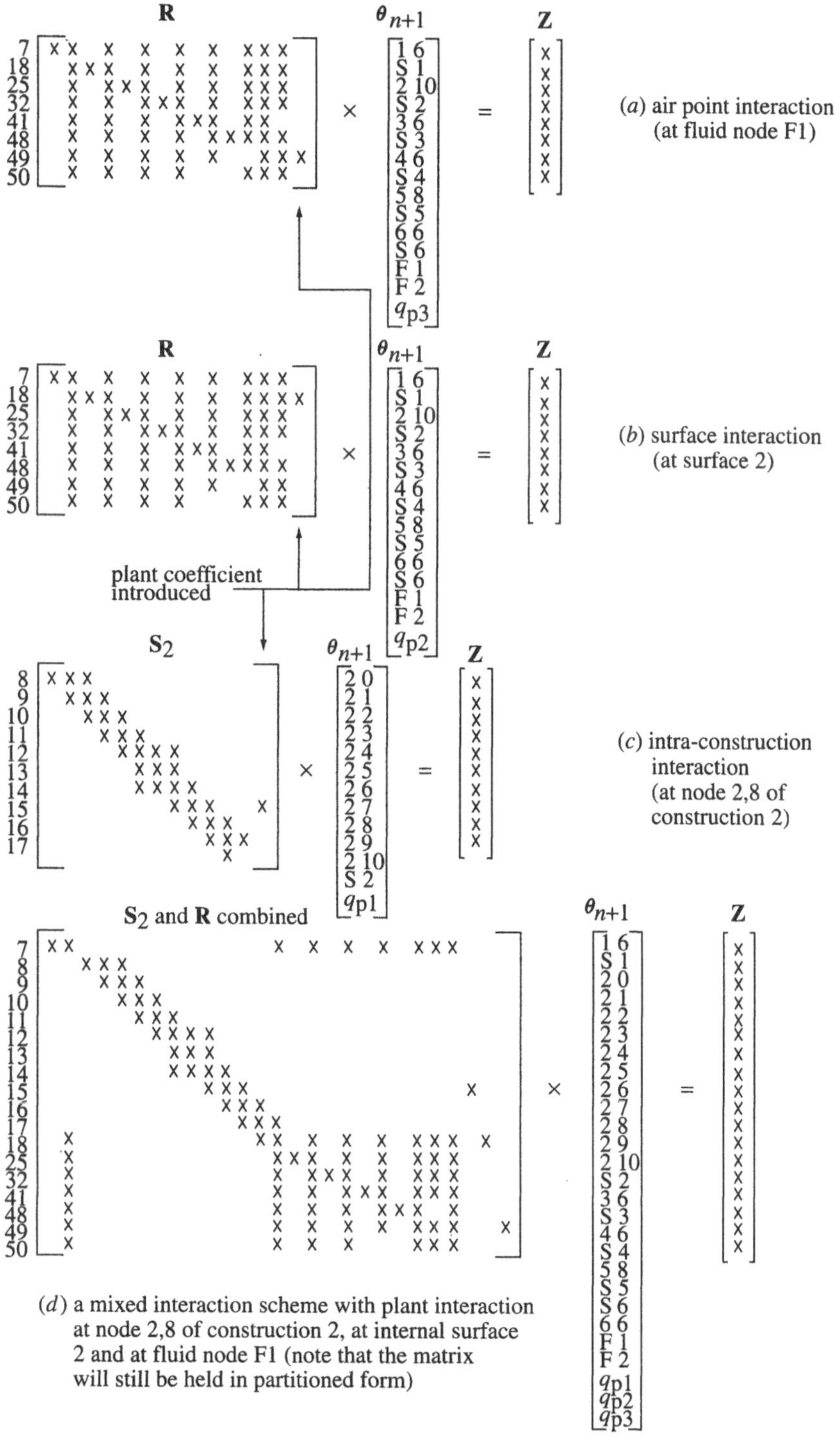

Figure 4.8: Example plant interaction schemes.

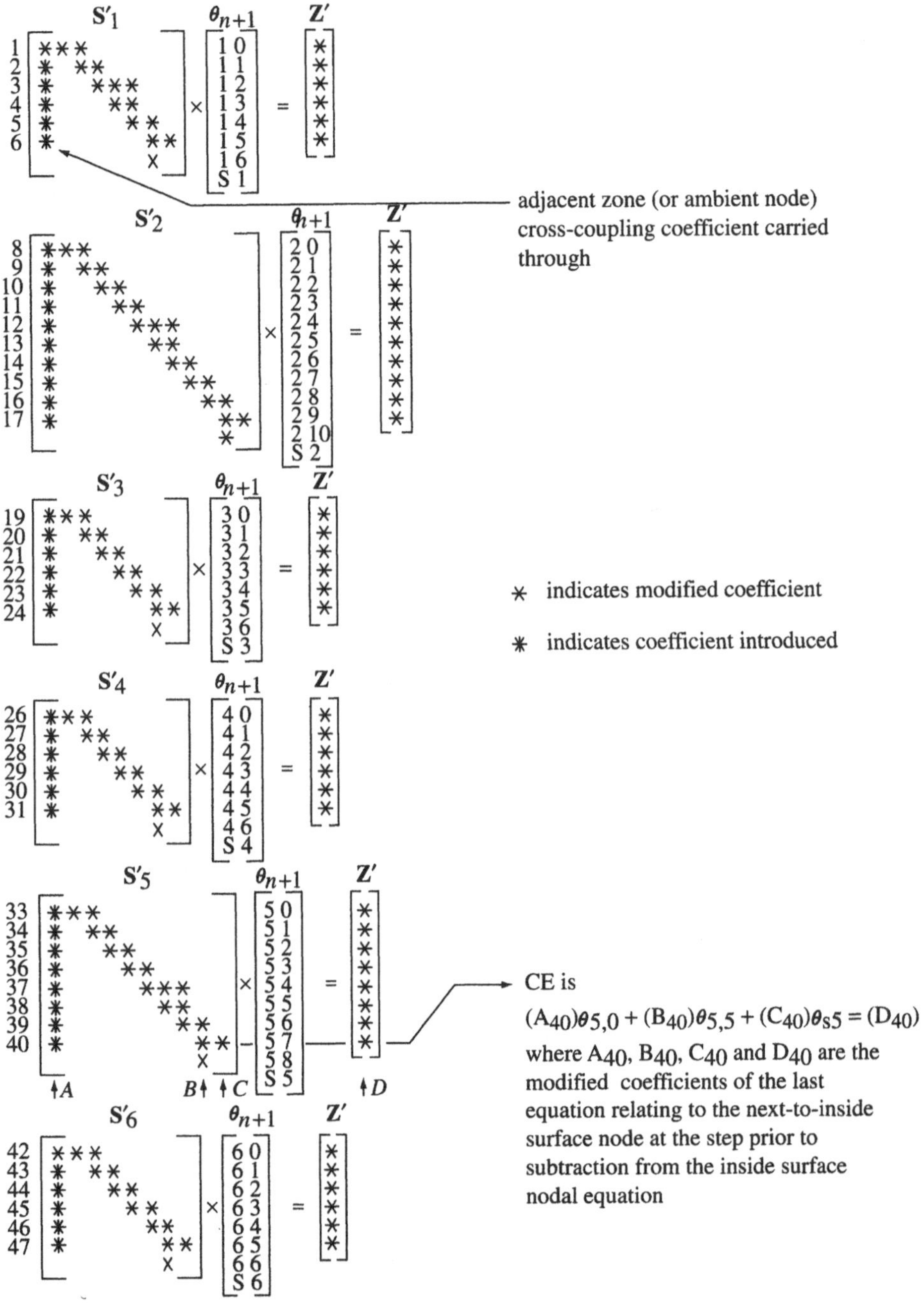

Figure 4.9: Construction matrix reduction: no intra-constructional node control or plant interaction.

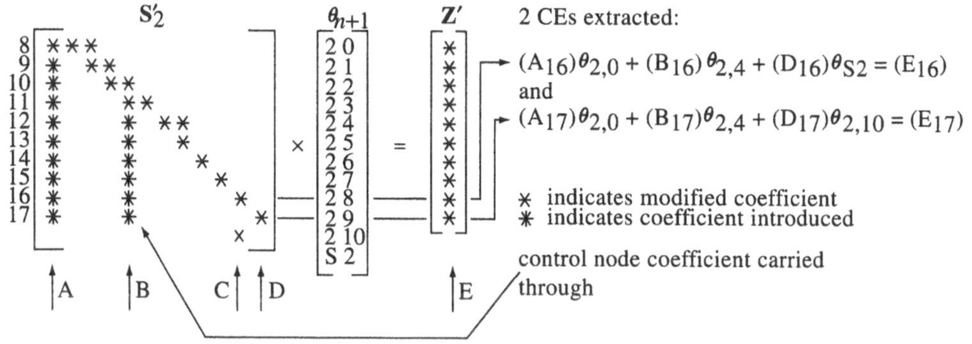

Figure 4.10: Construction matrix reduction: intra-constructional node control; no plant interaction. (Control is imposed on node 2,4.)

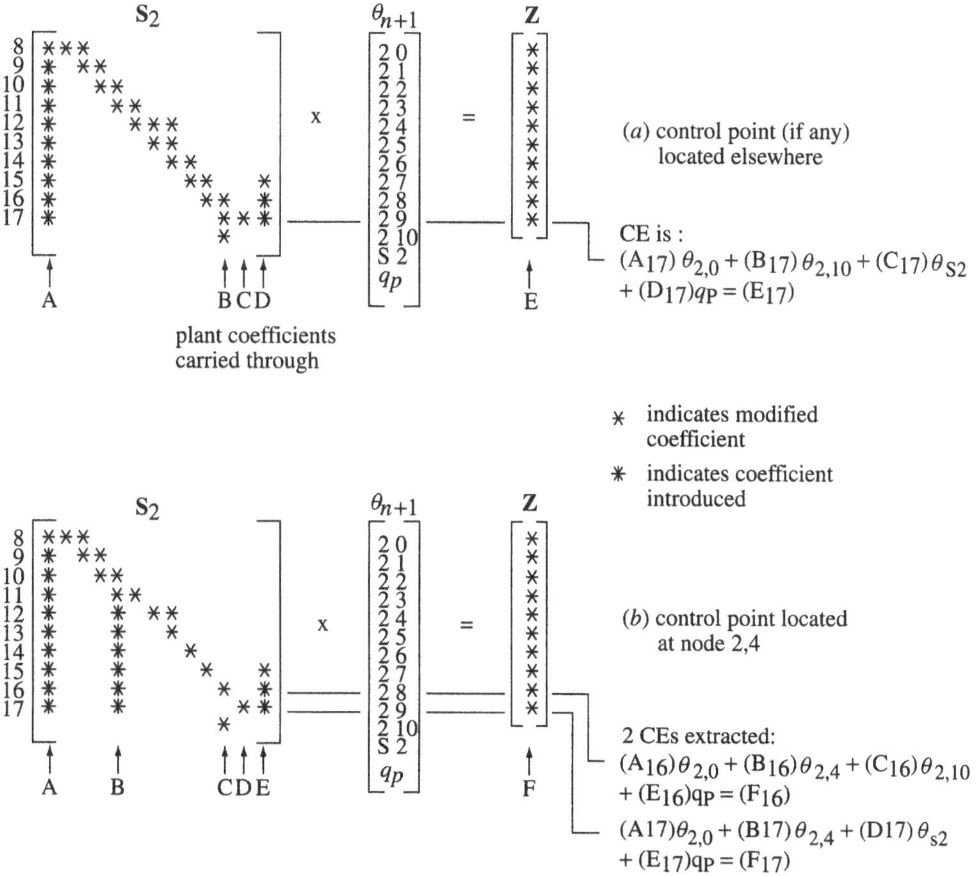

Figure 4.11: Matrix forward reduction and CE extraction for construction with plant interaction.

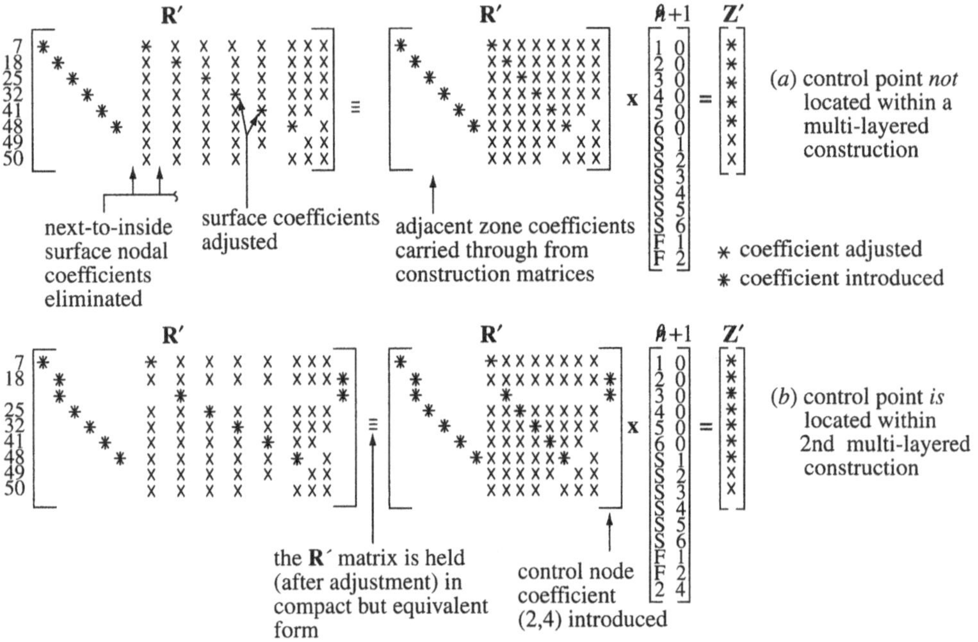

Figure 4.12: Partitioned zone energy balance matrix adjustment by construction CEs.

constructions 2 and 4 each have internal nodes with assigned plant interaction potential.

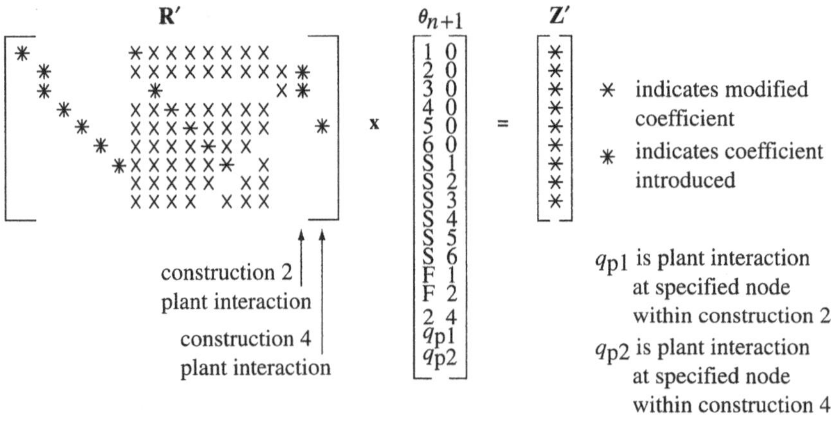

Figure 4.13: The $\mathbf{R}'$ matrix with plant coefficients introduced due to plant terms carried through from construction matrices $\mathbf{S}_2$ and $\mathbf{S}_4$. The matrix indicates that a control point has been located within construction 2 and so two construction CEs are introduced.

Stage 3: After the preceding stages have been achieved for each construction sub-matrix in turn, the modified zone matrix $\mathbf{R}'$ can be processed to the end of its forward reduction, carrying through all control node coefficients relating to surface control node terms or intra-constructional node control coefficients introduced in stage 2. Any plant coefficient must also be carried through. Figure 4.14 demonstrates the range of possible reduction options, in each case assuming only one control node type. Note, however, that any multi-node control scheme is possible although matrix processing will become proportionally more complex.

The CE to emerge at this stage will relate the zone control node(s) to the required plant input or extract, at any user defined plant interaction node(s), in terms of modified **Z** matrix entries and all adjacency conditions.

Stage 4: The CE extracted at stage 3 is now solved for free floating conditions (no plant) or in terms of any user-imposed control statement concerning capacity, time or temperature constraints.

Stage 5: The zone surface temperatures are now obtained by back substitution operations performed on the modified zone matrix.

Stage 6: The intra-constructional node temperatures are obtained by similar back substitution in the appropriate constructional sub-matrices preserved from stage 1.

It is important to note that this solution procedure only addresses existing coefficients—as dictated by the physical schema of the system being simulated—and so the single zone matrix of figure 4.1 is never fully established in computer memory as a complete two-dimensional array. This gives the seed for fast and efficient matrix processing, permitting rapid matrix reconstruction at each time-step and, thereby, model flexibility in use.

In a single zone system, incorporating one control node and one plant interaction node, the CE to emerge from stage 4 will have the form

$$\sum_i \mathrm{A_i}\theta_\mathrm{A} + \mathrm{B}\theta_\mathrm{c} + \mathrm{C}q_\mathrm{p} = \mathrm{D} \tag{4.2}$$

where A, B, C and D are the modified coefficients to emerge from the matrix reduction as shown in figure 4.14, θ_A the adjacent zone temperatures at the future time-row (°C), i the number of adjacent zones, θ_c the temperature of zone node(s) designated a control point at simulation commencement (°C) and q_p the plant interaction with nodes designated interaction points (W or W m^{-2} for air and construction nodes respectively).

To solve this equation, additional information must be harnessed to describe control expectations. In single-zone modelling, adjacent zone temperature information (at the future time-row) cannot be determined since no adjacent zone equation structure is present. With multi-zone simulation no such problem exists (see §4.2.2). For the single-zone case, the adjacent zone temperature can be estimated, set identical to the simulated zone, or established by some independent means (perhaps as a function of the conditions prevailing in the modelled zone or derived from data obtained from a separate simulation of the adjacent zone).

In any event eqn (4.2) will reduce to

$$\mathrm{B}\theta_\mathrm{c} + \mathrm{C}q_\mathrm{p} = \mathrm{D}'$$

where $\mathrm{D}' = \mathrm{D} - \sum \mathrm{A_i}\theta_\mathrm{A}$. This equation can be directly solved if the desired control temperature is specified or if the current plant input or extract is known. It is at this point that controller characteristics are introduced—set-point, control action (proportional, integral, derivative), offset, dead band, hysteresis etc—to allow determination of the actual flux input as a function of the sensed condition. Control system simulation is described in chapter 6 where selected plant systems are considered.

Multiple control nodes and/or plant interaction nodes can be accommodated by simply extracting additional CEs at stage 4 as appropriate. Iterative solution techniques, of the kind suggested in §4.2.2 for multi-zone matrix processing, will then be required.

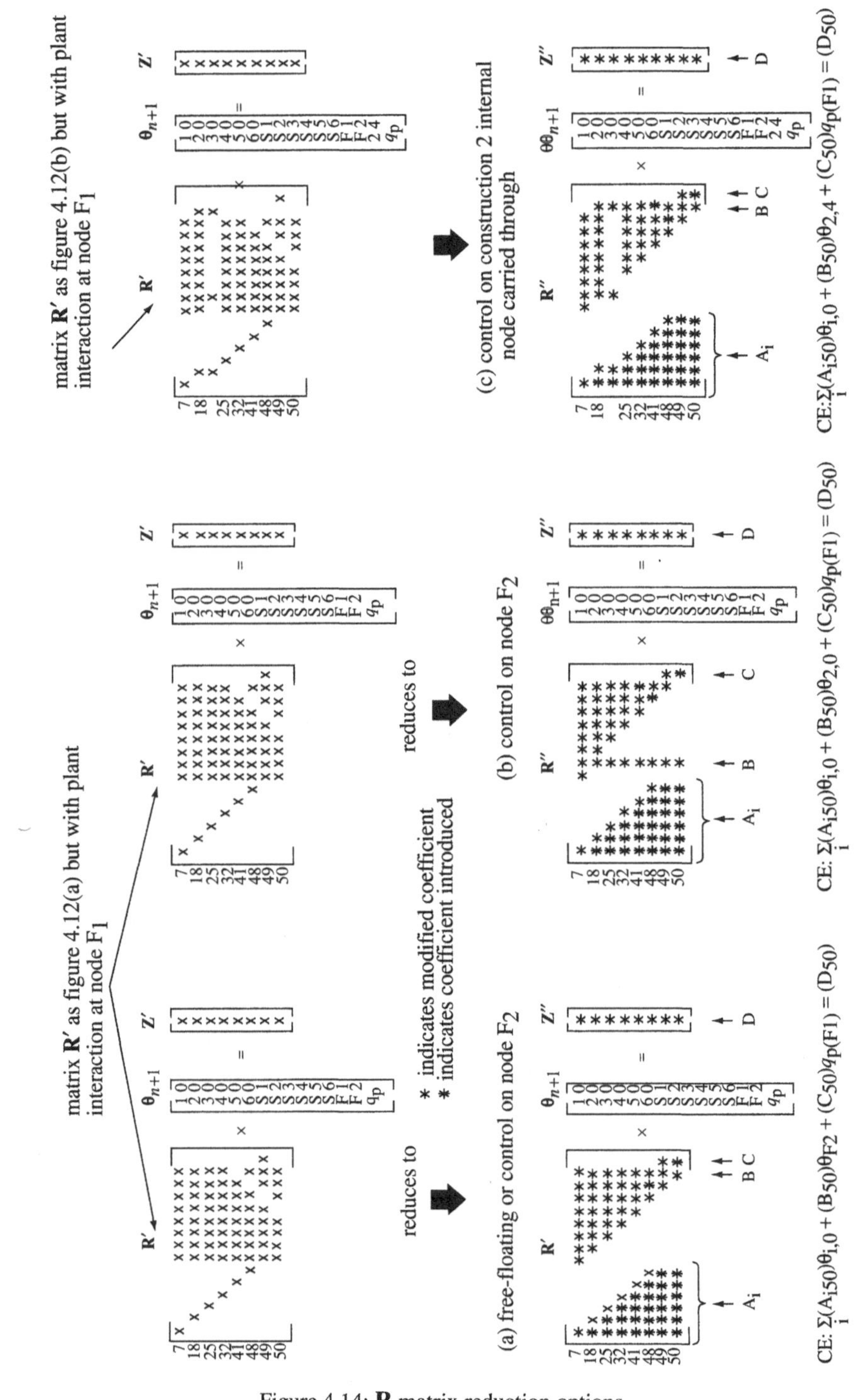

Figure 4.14: **R** matrix reduction options.

4.2.2 Multi-zone solution

With multi-zone configurations, stages 1-4 of the previous section are implemented separately for each zone and the stage 4 CEs extracted to give a set of control/plant interaction equations of the form given by eqn (4.2). Figure 4.15 gives the matrix representation of these CEs.

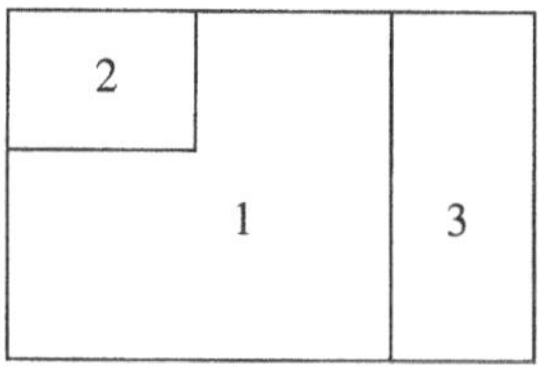

Control condition:
zone 1 – air, zone 1
zone 2 – air, zone 2
zone 3 – air, zone 2

Control CE matrix:

$$\begin{array}{l} \\ \text{zone 1} \\ \text{zone 2} \\ \text{zone 3} \end{array} \begin{array}{c} \mathbf{A}\text{ and }\mathbf{B} \\ \begin{bmatrix} X & X & X & X \\ X & X & & X \\ X & X & & X \end{bmatrix} \end{array} \times \begin{array}{c} \theta\text{ and }\mathbf{q} \\ \begin{bmatrix} \theta_{A1} \\ \theta_{A2} \\ \theta_{A3} \\ q_p \end{bmatrix} \end{array} = \begin{array}{c} \mathbf{D} \\ \begin{bmatrix} X \\ X \\ X \end{bmatrix} \end{array}$$

Figure 4.15: Matrix representation of zone CEs for a 3 zone problem.

If an approximate solution is sought, each zone equation can be solved independently by setting all adjacent conditions at the future time-row to their present time-row values. This allows the future time-row control node temperature and plant interaction node flux to be evaluated for each zone. While this approximation may be adequate for early stage design appraisal it will prove inadequate in many other cases such as the critical examination of control system response.

Where an exact solution is required, a mixed iterative/direct solution scheme is necessary. For this case the adjacent temperatures in each CE are assigned an estimated value (perhaps the present time-row value) to allow the evaluation of the θ_c and q_p terms for each zone. The backward substitution operation (stage 5) is then initiated and continued until each zone air temperature has been evaluated at the future time-row (or zone surface temperatures if the adjacency linkages are defined as proposed in section §4.1.2). The new zone air (or surface) temperatures are then introduced as updated estimates and the solution process iterates, at the same time-step, until the difference between successive predictions are minimised for all zones. To ensure convergence, the error-derived corrector technique introduced in §5.1 may be employed.

4.2.3 Solution on the basis of complex criteria

In real applications zone temperature control is effected at the dictate of a thermostat which is influenced by local air temperature and radiant exchanges with surrounding surfaces. In such cases the sensed temperature is given by

$$\theta_x = \alpha\theta_A + (1-\alpha)\left(\frac{\sum_{i=1}^{N}\omega_{si}\theta_{si}}{\sum_{i=1}^{N}\omega_{si}}\right) \tag{4.3}$$

where θ_x is the sensed temperature, θ_A the local node air temperature, θ_{si} the surface temperature (all in °C), N the number of surrounding surfaces in visual contact, ω_{si} the surface solid angle subtended at the controller, and α the controller convective weighting factor ($\alpha \to 1$ for an aspirated sensor).

The CE to emerge from stage 3 relates the control node temperature to the required plant capacity injected at some interaction node(s). Multi-node control may be enacted by initially assuming that the controlled node is the air point. Then, in the zone CE given by eqn (4.2), it may be assumed that the adjacency conditions are evaluated at the present time-row as described in §4.2.2. The plant capacity term, q_p, is guessed to allow evaluation of the zone air node temperature θ_A ($\theta_c = \theta_A$ in eqn (4.2)). Stage 5 is implemented to establish the zone surface temperatures and the sensed temperature θ_x determined from eqn (4.3). The guessed plant capacity, q_p, is then increased or reduced—within the constraints imposed by the plant capabilities—and the stage 5 procedure reiterated until the sensed temperature is adjusted to the desired value. For exact multi-zone processing it is necessary to 'nest' this procedure within the overall iterative procedure as outlined in §4.2.2.

If simulation control is to be achieved on the basis of comfort considerations, it is possible to arrange that the sensed temperature, θ_x, is evaluated as a function of zone air and mean radiant temperatures and other factors, such as air velocity, humidity level and occupant activity level determined from an appropriate model.

Consider, as an example, the use of dry resultant temperature, θ_{res}, as a comfort index. This is defined as the temperature recorded by a thermometer situated at the centre of a blackened sphere 100mm in diameter and is approximated by

$$\theta_{res} = \frac{\theta_{mrt} + 3.17\theta_A V^{\frac{1}{2}}}{1 + 3.17V^{\frac{1}{2}}}$$

where θ_{mrt} is the zone mean radiant temperature (°C) and V the air speed (m s^{-1}). For sedentary occupation an air speed of 0.1 m s^{-1} is not uncommon and, for this value, the preceding equation reduces to

$$\theta_{res} = 0.5(\theta_{mrt} + \theta_A)$$

where, for comfort, two conditions require to be satisfied (CIBSE 1986): $19 \leq \theta_{res} \leq 23$ and $-5 \leq (\theta_{mrt} - \theta_A) \leq 8$.

Note that it would be incorrect to enforce such conditions rigorously since dry resultant temperature is merely an indicative quantity. More sophisticated indicators of comfort have been devised (Fanger 1972) and it is relatively simple to encapsulate these within a simulation program.

4.2.4 Treatment of non-linear systems

In the present context, a non-linear system is one in which the parameters that underpins the coefficients of the governing equations depend on the equation's state variables. This means that these coefficients cannot be precisely formed since their formulation is the prerequisite of determining the dependent state variables in the first place.

A non-linear system may be processed by applying iteration to the entire matrix equation. Consider eqn (3.11), which relates to building surface layers and includes the inter-surface longwave radiation exchanges. The h_r terms (see §7.5) are determined, at the future time-row, as a function of surface temperatures that are as yet uncomputed. To permit future time-row coefficient matrix formulation (the **A** matrix of figure 4.1) it is usually acceptable, in terms of accuracy, to operate one time-step in arrears. Alternatively, the future time-row surface temperatures may be estimated to allow the estimation of h_r. Steps 1 through 5 (of §4.2.1) are then implemented and the predicted surface temperatures to emerge used to re-establish the **A** matrix if the difference between the guessed and predicted values is significant. In this way, iterative convergence is pursued.

4.3 Mixed frequency inversion

When building and plant matrices are combined—as described in chapter 6—the overall matrix to result will exhibit a temporal mismatch between the components comprising the building and plant. In the former case, time constants will typically be measured in hours; in the latter case in minutes. It is therefore computationally attractive to process different partitioned matrices, representing regions with different time constants, at different frequencies depending on the rate at which region conditions change. This means that the entire matrix system need not be processed at some lowest common denominator time-step.

In the approach, each partitioned matrix is assigned an inversion frequency arrived at from consideration of its related time constant. As the solution scheme of §4.2.1 is implemented, each partitioned matrix is forward reduced according to its assigned frequency. This means that the CE to emerge is held constant, until the next inversion point is reached, independent of sub-interval inversions applied to other partitioned matrices possessing greater inversion frequencies. The technique is obviously an approximation, but has the virtue of improving computational efficiency for marginal penalty, in most practical applications, in accuracy.

This chapter has detailed a mechanism by which large and sparse matrix structures, containing the discretised differential equations representing energy flow within a building, can be efficiently solved. The next chapter considers the extension of the technique to include the explicit modelling of fluid flow—a parameter of the conservation equations derived in §3.2.3—before chapter 6 introduces models of HVAC, renewable energy and control systems.

4.4 References and further reading

Churchill R V (1958) *Operational Mathematics* (2nd edn) (New York: McGraw-Hill)

CIBSE 1986 *CIBSE Guide* (London: Chartered Institute of Building Services Engineers)

Collatz L 1960 *The Numerical Treatment of Differential Equations* (Berlin: Springer-Verlag)

Fanger P O 1972 *Thermal Comfort* (New York: McGraw-Hill)

George A and Liu J W 1981 *Computer Solution of Large Sparse Positive Definite Systems* (New Jersey: Prentice-Hall)

Hartree D R 1958 *Numerical Analysis (2nd edn.)* (Oxford: Oxford Press)

Miranker W L 1981 *Numerical Methods for Stiff Equations and Singular Perturbation Problems* (Boston: Reidel)

Press W H, Flannery B P, Teukolsky S A and Vettlering W T 1986 *Numerical Recipes: the Art of Scientific Computing* (Cambridge University Press)

Stewart G W 1973 *Introduction to Matrix Computations* (New York: Academic Press)

Willoughby R A and Rose D J 1972 *Sparse Matrices and Their Applications* (New York: Plenum)

5

Fluid flow

Buildings and their HVAC systems can be thought of as comprising four interacting fluid flow domains as summarised in figure 5.1.

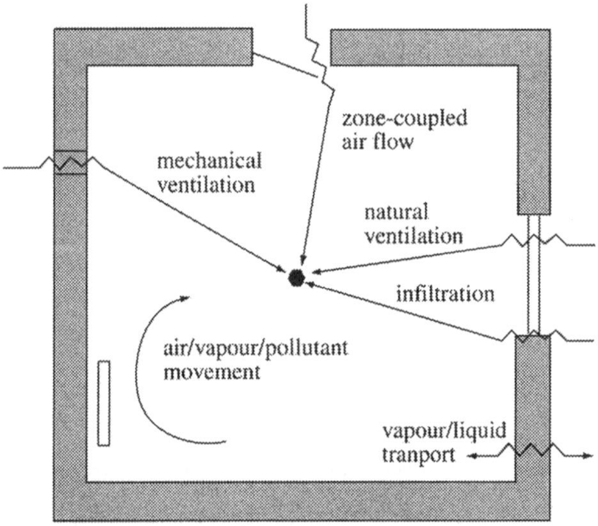

Figure 5.1: Building fluid flow domains.

1. Air/vapour flow through cracks and openings in the building envelope allowing infiltration and natural ventilation.

2. The flow of air/vapour through the leakage paths connecting internal spaces and the distribution networks that exist to service the building's heating, cooling and ventilation demands.

3. The movement of air/vapour/pollutants within the interior spaces of the building or the movement of working fluids within HVAC plant components.

4. The movement of water (in the liquid and vapour states) within the porous materials

comprising the building structure and contents.

In relation to the first three domains, there are essentially three modelling approaches, of progressive detail and applicability: simplified expressions, nodal networks and computational fluid dynamics.

Simplified expressions

These are suited to domain 1 and typically comprise correlations between air flow magnitude and parameters such as temperature difference and wind speed. For a construction of average permeability, for example, the following equation (ASHRAE 1981) has been proposed for use to estimate domestic infiltration rates

$$\mathrm{ACR} = 0.15 + 0.0087\Delta\theta + 0.02\mathrm{v}$$

where ACR is the volume changes per hour, $\Delta\theta$ the inside-to-outside temperature difference (°F) and v the wind speed (mph). The corresponding equation for a 'tight' construction is

$$\mathrm{ACR} = 0.11 + 0.0066\Delta\theta + 0.012\mathrm{v} \ .$$

At best, such expressions can be used to characterise infiltration rates at an early design stage. They have no part to play within a building simulation program.

The nodal network method

This method is applicable to domains 1 and 2. The building and its air handling systems are treated as a collection of nodes representing rooms (or parts of rooms), equipment connection points, ambient conditions etc. Inter-nodal connections are then defined in terms of components such as cracks, doors, fans, ducts etc, each represented by a model that gives the mass flow rate as a function of the pressure difference across the component. Consideration of the conservation of mass at each node leads to a set of non-linear equations that can be solved at successive time steps to characterise the flow throughout the defined network.

Computational fluid dynamics (CFD)

While this method is generally applicable, when applied to buildings it is usually restricted to domain 3. The method is based on the solution of the conservation equations for mass, momentum and energy at discrete points within a room or plant component. For a given boundary condition, numerical methods are employed to solve for the temperature, pressure and velocity fields. It is also possible to determine the distribution of water vapour or pollutants, and to assess the mean age (freshness) of air at different locations within a room. Such information is the prerequisite of an appraisal of indoor air quality and discomfort.

The first approach is generally not used within simulation because it does not provide sufficient resolution while the last two approaches are elaborated in §§5.1 and 5.2 respectively. In relation to the fourth domain, §5.3 introduces a modelling technique based on mass and energy conservation considerations applied to porous building materials.

5.1 The nodal network method

This method is constrained to the steady flow of an incompressible fluid within a network of connected pressure points (nodes) when subjected to successive sets of boundary conditions. In other words, the problem reduces to the calculation of the mass flows through each connection when the nodes represent internal (unknown) and external (known) pressures. Solution for a particular boundary condition is achieved by an iterative approach in which the unknown nodal pressures are repeatedly adjusted until the nodal mass imbalances (residuals) are reduced to insignificance. The flow network may comprise sub-networks, each relating to a different fluid type. There are several implementations of the method in relation to building air flow modelling and the representation of the incompressible working fluids (e.g. water) found within HVAC systems (Cockroft 1979, Clarke 1985, Walton 1989, Feustel and Raynor-Hoosen 1990, Clarke and Hensen 1991, Fürbringer *et al* 1996).

5.1.1 Boundary conditions

In the case of buildings, the surface pressure distribution is wind induced. Its prediction requires information on the prevailing wind—its speed, direction and vertical velocity profile—and, more problematic, the influence of local obstructions and terrain features. Two approaches to the determination of surface pressure distribution are extant: wind tunnel tests applied to scale models, and the use of mathematical models (Launder and Spalding 1974, Hanson *et al* 1982, Häggkvist *et al* 1989, Grosso *et al* 1992, Knoll *et al* 1995). Whatever the approach, it is usual to express the outcome for a given surface in the form of a dimensionless pressure coefficient set as shown in figure 5.2. For any given wind direction, d:

$$C_{id} = \frac{P_{id}}{\frac{1}{2}\rho v_r^2}$$

where C_{id} is the pressure coefficient for surface i and corresponding to a wind direction d, P_{id} the surface pressure (N m^{-2}), ρ the air density (kg m^{-3}) and v_r some reference wind speed corresponding to direction d (m s^{-1}).

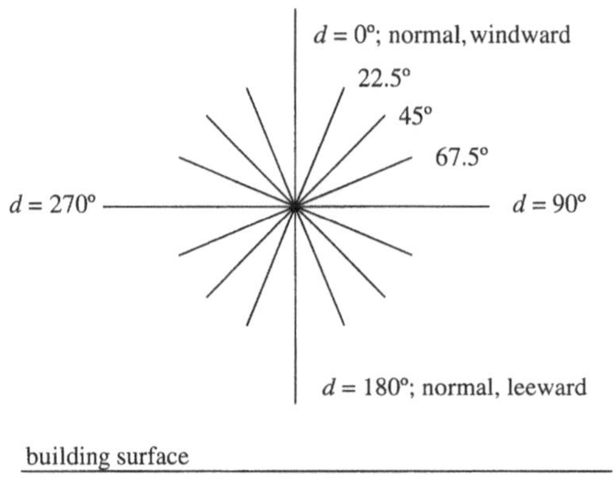

Figure 5.2: A surface pressure coefficient set.

A pressure coefficient set typically comprises 16 compass values at 22.5° intervals so that the coefficient for any particular wind direction may include the influence of an obstruction feature. Note that coefficients can be negative to reflect leeward exposures.

Table 5.1 gives some example pressure coefficient sets for some typical exposures and building length-to-width ratios. De Wit (2001) has stated that the effect of the uncertainty inherent in such pressure coefficients is potentially a major part of the total effect of all possible modelling uncertainties. On the basis of expert opinion applied to published pressure coefficient data, he quantified this uncertainty and concluded that "point estimates on the basis of existing data do not give useful information". Instead, the use of probability distributions applied to existing pressure coefficients data† is proposed.

Where the reference wind speed, v_r, is a local wind speed, it is necessary to modify the free stream wind speed as a function of any height difference and the effect of local terrain roughness. This requires the use of an assumed wind profile. Based on an analysis of the underlying physics, Simiu and Scanlan (1986) suggested a logarithmic form for the wind profile:

$$\frac{v_l}{v_m} = \frac{v_l^*}{v_m^*}\left(\ln\frac{z_l - d_l}{z_{0,l}} \Big/ \ln\frac{z_m - d_m}{z_{0,m}}\right)$$

$$\frac{v_l^*}{v_m^*} \approx \left(\frac{z_{0,l}}{z_{0,m}}\right)^{0.1}$$

where v_l is the local wind speed at a height z_l above the ground, v_m the free stream wind speed measured at some reference height z_m (both m s^{-1}), v_i^* (i = l, m) the atmospheric friction speed (m s^{-1}), $z_{0,i}$ a terrain dependent roughness length (m) and d_i a terrain dependent displacement length (m). Table 5.2 gives typical values for the last two parameters.

Table 5.1: Pressure coefficient sets (from AIVC 2001)

Pressure coefficients at 22.5° intervals	*Context*
0.70/0.53/0.35/-0.08/-0.50/-0.45/-0.40/-0.30/-0.20/-0.30/-0.40/-0.45/-0.50/-0.08/0.35/0.53	1:1 exposed wall
0.20/0.13/0.05/-0.10/-0.25/-0.23/-0.30/-0.28/-0.25/-0.28/-0.30/-0.28/-0.25/-0.10/0.05/0.13	1:1 sheltered wall
0.50/0.38/0.25/-0.13/-0.50/-0.65/-0.80/-0.75/-0.70/-0.75/-0.80/-0.65/-0.50/-0.13/0.25/0.38	2:1 exposed long wall
0.06/-0.03/-0.12/-0.16/-0.20/-0.29/-0.38/-0.34/-0.30/-0.34/-0.38/-0.29/-0.20/-0.16/-0.12/-0.03	2:1 sheltered long wall
0.60/0.40/0.20/-0.35/-0.90/-0.75/-0.60/-0.48/-0.35/-0.48/-0.60/-0.75/-0.90/-0.35/0.20/0.40	1:2 exposed short wall
0.18/0.17/0.15/-0.08/-0.30/-0.31/-0.32/-0.26/-0.20/-0.26/-0.32/-0.31/-0.30/-0.08/0.15/0.16	2:1 sheltered short wall
-0.80/-0.75/-0.70/-0.65/-0.60/-0.55/-0.50/-0.45/-0.40/-0.45/-0.50/-0.55/-0.60/-0.65/-0.70/-0.75	1:1 exposed roof <10°
-0.40/-0.45/-0.50/-0.55/-0.60/-0.55/-0.50/-0.45/-0.40/-0.45/-0.50/-0.55/-0.60/-0.55/-0.50/-0.45	1:1 exposed roof 10-30°
0.30/-0.05/-0.40/-0.50/-0.60/-0.50/-0.40/-0.45/-0.50/-0.45/-0.40/-0.50/-0.60/-0.50/-0.40/-0.05	1:1 exposed roof >30°
-0.70/-0.70/-0.70/-0.75/-0.80/-0.75/-0.70/-0.70/-0.70/-0.70/-0.70/-0.75/-0.80/-0.75/-0.70/-0.70	2:1 exposed roof <10°
-0.70/-0.70/-0.70/-0.70/-0.70/-0.65/-0.60/-0.55/-0.50/-0.55/-0.60/-0.65/-0.70/-0.70/-0.70/-0.70	2:1 exposed roof 10-30°
0.25/0.13/0.00/-0.30/-0.60/-0.75/-0.90/-0.85/-0.80/-0.85/-0.90/-0.75/-0.60/-0.30/0.00/0.13	2:1 exposed roof >30°

Other models have been proposed based on empirical observations (Liddament 1986):

$$\frac{v_l}{v_{10}} = K\, z_l^{\,a}$$

where v_{10} is the free stream wind speed measured at 10m and K, a are terrain dependent constants (table 5.2).

$$\frac{v_l}{v_m} = \frac{\alpha_l (z_l/10)^{\gamma_l}}{\alpha_m (z_m/10)^{\gamma_m}}$$

where α and γ are terrain dependent constants (table 5.2).

These wind profiles are valid for heights between ($20 \times z_0 + d$) and 60-100m. The boundary layer below this band is referred to as the urban canopy for which the wind vector is mostly dependent on obstacles and will require wind tunnel tests or detailed CFD simulations for its accurate resolution. Some care must therefore be taken when using published wind data to appraise buildings located in an urban context.

For cases where data uncertainty is high, it is possible to determine the sensitivity of the

† The actual work was executed in terms of the difference between the pressure coefficients at corresponding points on windward and leeward facades.

Table 5.2: Values of the terrain parameters.

Terrain	K	a	z_0	d	α	γ
Open flat country	0.68	0.17	0.03	0.0	1.00	0.15
Country, scattered wind breaks	0.52	0.20	0.1	0.0	-	-
Rural	-	-	0.5	0.7 h	0.85	0.20
Urban	0.35	0.25	1.0	0.8 h	0.67	0.25
City	0.21	0.33	> 2.0	0.8 h	0.47	0.35
h = building height (m)						

performance predictions to assumed pressure coefficients and wind velocities that characterise the exposure. Where the sensitivity is high, additional analysis is warranted. Paradoxically, simulation is often of greatest use when the defining data is uncertain.

5.1.2 Node definition

It is assumed that nodes (representing discrete, homogeneous fluid volumes) can be characterised by a single temperature, a single static pressure and a height relative to some arbitrary datum. Table 5.3 lists some possible node types and their typical defining data.

Table 5.3: Node types and defining parameters.

Internal, unknown condition	height (m)
Internal, known condition	height (m), total pressure (Pa) and temperature (°C)
Boundary, known pressure	height (m), total pressure (Pa) and temperature (°C)
Boundary, wind pressure	height (m), pressure coefficient set, surface azimuth (° from N)

Note that internal node temperatures are not required where the flow network is to be combined with a building/plant equation-set as described in §6.5. Note also that while the pressure at an internal node is usually unknown, it can be assigned a value. This would be an appropriate action to represent an expansion vessel within a heating system or to mimic a pressurisation test.

5.1.3 Buoyancy effects

Consider figure 5.3, which shows two zones connected by a duct. The zone height is typically given as the average height of all openings when expressed relative to some convenient datum. For the case shown, the ends of the duct are at different heights relative to each other and relative to the nodes representing the zones.

The pressure drop across the component may be determined from Bernoulli's equation for the one-dimensional steady flow of an incompressible fluid:

$$\Delta P = (p_1 + \rho V_1{}^2/2) - (p_2 + \rho V_2{}^2/2) + \rho_i g(z_1 - z_2) \; ; \; i = n, m \tag{5.1}$$

where ΔP is the sum of all friction and dynamic losses (N m^{-2}), p_1, p_2 the entry and exit static pressures (Pa), V_1, V_2 the entry and exit velocities (m s^{-1}), ρ_i the density of the air flowing through the component (kg m^{-3}; i = n or m depending on the direction of the flow), g the acceleration of gravity (m s^{-2}) and z_1, z_2 the entry and exit elevations (m). This equation defines a sign convention for the flow direction: positive from point 1 to point 2 (n to m).

Eqn (5.1) can be simplified by combining related terms. Dynamic pressures are the $\rho V^2/2$ terms, and total pressure is defined to be the sum of static pressure and dynamic pressure—that is $P = p + \rho V^2/2$. If nodes n and m represent large volumes, the dynamic pressures are

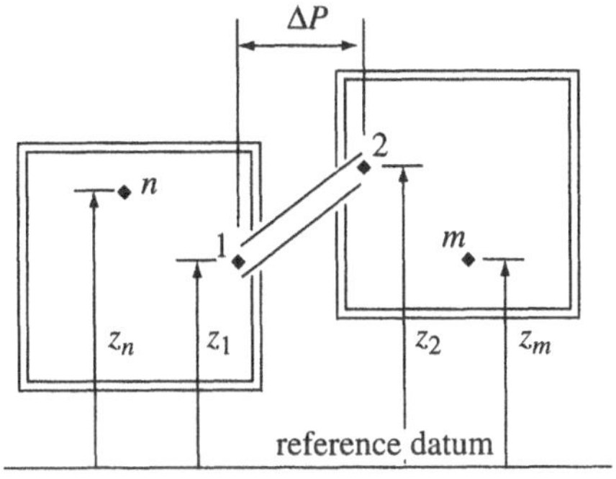

Figure 5.3: Two connected zones.

effectively zero. If the nodes represent some point in a duct or pipe network, there will be a positive dynamic pressure. The pressures at the inlet and outlet of the flow component can be related to the node pressures by the hydrostatic law:

$$P_1 = P_n + \rho_n g(z_n - z_1) = P_n - \rho_n g h_1$$

where $h_1 = z_1 - z_n$, and

$$P_2 = P_m + \rho_m g(z_m - z_2) = P_m - \rho_m g h_2$$

where $h_2 = z_2 - z_m$.

Ignoring dynamic pressures, eqn (5.1) therefore reduces to

$$\Delta P = P_n - P_m + \rho_i g(z_n + h_1 - z_m - h_2) - \rho_n g h_1 + \rho_m g h_2 \; ; i = n, m.$$

The terms $[\rho g(z_n + h_1 - z_m - h_2)]$, $[-\rho_n g h_1]$ and $[\rho_m g h_2]$ are collectively termed the stack pressure, PS, acting on the component:

$$PS = \rho_n g(z_n - z_m) + h_2 g(\rho_m - \rho_n) \text{ ; for the + ve direction}$$

$$PS = \rho_m g(z_n - z_m) + h_1 g(\rho_m - \rho_n) \text{ ; for the – ve direction.}$$

Hensen (1991) has proposed an alternative equation for the calculation of PS in order to overcome any numerical instabilities that might arise where the flow network is complex and sensitive to small pressure variations:

$$PS = (g/2)\left[(\rho_n + \rho_m)(z_n - z_m) + (\rho_m - \rho_n)(h_1 + h_2)\right].$$

By basing the calculation on the average density of the connected nodes, a complex network comprising large and small openings was successfully solved (Pernot and Hensen 1990) in a few iterations.

5.1.4 Component flow models

A flow component is characterised by a type (duct, pipe, pump, crack, doorway etc) and a number of defining parameters. Hensen (1991) has reported on the theoretical basis of a range of models as installed within the ESP-r system and summarised in table 5.4.

Each component has a corresponding model for the evaluation of the flow rate as a function of the pressure drop. For example, a valve may be represented as a conduit with an empirically

Table 5.4: Flow component types and models.

Component	*Model*
Power law volume flow resistance	$\dot{m} = \rho a \Delta P^n$
Power law mass flow resistance	$\dot{m} = a \Delta P^b$
Power law mass flow resistance	$\dot{m} = a\sqrt{\rho}\Delta P^b$
Quadratic law volume flow resistance	$\Delta P = a\dot{m}/\rho + b(\dot{m}/\rho)^2$
Quadratic law mass flow resistance	$\Delta P = a\dot{m} + b\dot{m}^2$
Constant volume flow rate component	$\dot{m} = \rho r_v$
Constant mass flow rate component	$\dot{m} = r_m$
Common orifice flow component	$\dot{m} = C_d A\sqrt{2\rho\Delta P}$
Laminar pipe flow component	$\dot{m} = \dfrac{\rho \Delta P \pi R^4}{8\mu L_p}$
Specific air flow opening	$\dot{m} = 0.65A\sqrt{2\rho\Delta P}$
Specific air flow crack	$\dot{m} = f(\rho, k, \Delta P)$ (see text)
Specific air flow door	$\dot{m} = f(W_d, H, H_r, C_d, \Delta P)$ (see text)
General flow conduit (duct or pipe)	$\dot{m} = A_c\sqrt{\dfrac{2\rho\Delta P}{f\,L_{p/D_h} + \sum C_i}}$;
	$f = 1/2\log(5.74/Re^{0.901} + 0.27k_r/D_h)^2$
General flow inducer (pump or fan)	$\Delta P = \sum_{i=0}^{3} a_i(\dot{m}/\rho)^i$;
	$\dot{q}_{mn} \leq \dot{m}/\rho \leq \dot{q}_{mx}$
General flow corrector	$\dot{m} = \rho\, k_v\left[\dfrac{\Delta P \rho_0}{\Delta P_0 \rho}\right]^{\frac{1}{2}}$
Flow corrector with polynomial local loss	$\dot{m} = A_c\left[\dfrac{2\rho\Delta P}{C}\right]^{\frac{1}{2}}$;
	$C = \sum_{i=0}^{3} a_i(H/H_{100})^i$
Ideal (frictionless) open/shut flow controller	$\dot{m} = 0$ or $\dot{m} = \rho\,\dot{q}$

where $\dot{m}$ is the mass flow rate (kg s^{-1}), a and b are empirical coefficients, n is an empirical exponent, r_v the volume flow rate (m^3s^{-1}), r_m the mass flow rate (kg s^{-1}), A the opening area (m^2), C_d the discharge factor (-), μ the dynamic viscosity (kg $m^{-1}s^{-1}$), Re the Reynolds Number, L_p the pipe length, R the pipe radius, W_c the crack width, L_c the crack length, W_d the door width, H the door height and H_r the door reference height; D_h is the hydraulic diameter, A_c the cross sectional area (m^2), k_r the pipe wall roughness (-), $\sum C_i$ the sum of the local dynamic loss factors (-), $\dot{q}_{mn}$ & $\dot{q}_{mx}$ the minimum and maximum volume flow rates respectively (m^3s^{-1}), ρ & ρ_0 the density and standard density respectively (kg m^{-3}), ΔP_0 the standard pressure (N m^{-2}), k_v the volume flow rate at ρ_0 (m^3s^{-1}) and H & H_{100} are the valve position and fully open position respectively; (all linear dimensions in m).

derived dynamic loss factor, C (-), which is dependent on the valve stem displacement. The mass flow rate, $\dot{m}$ (kg s^{-1}), may then be calculated from

$$\dot{m} = A\left(\frac{2\rho\Delta P}{C}\right)^{\frac{1}{2}}$$

$$C = a_0 + a_1 H/H_{100} + a_2(H/H_{100})^2 + a_3(H/H_{100})^3$$

where A is the cross-sectional area containing the corrector (m^2), H/H_{100} is the relative valve position (-), and a_i are fit coefficients (-).

For the reasons given in §5.1.5, the network solver requires the derivative of the component flow (kg $s^{-1}Pa^{-1}$) at each iteration step. For the above valve model this is given by

$$\frac{d\dot{m}}{d\Delta P} = \frac{0.5\dot{m}}{\Delta P} .$$

If ΔP becomes small, or for cases where an analytical expression for the derivative does not exist, a numerical approximation may be used:

$$\frac{d\dot{m}}{d\Delta P} \approx \frac{\dot{m} - \dot{m}^*}{\Delta P - \Delta P^*}$$

where * denotes the value at the previous iteration step.

In general, the equations that represent the air mass flow rate through simple restrictions may be expressed by an equation of the form

$$\dot{m} = ka(\Delta P)^x \qquad (5.2)$$

where ΔP is the pressure difference across the restriction (N m^{-2}), k an empirical constant that depends on the nature of the flow restriction, 'a' a characteristic dimension such as length or area, and x an empirical exponent.

For a simple orifice at high Reynolds Number (such as a partially open window), x is close to 0.5 (ASHRAE 1981). For cracks and similar restrictions with a large aspect ratio, x is close to 0.65, rising to unity for completely laminar flow. For this class of flow restriction empirical relationships have been established to determine x and k as a function of crack width:

$$x = 0.5 + 0.5 \exp(-W/2)$$

$$k = 9.7(0.0092)^x$$

where W is the crack width (mm) and therefore 'a' of eqn (5.2) becomes the crack length.

With open windows the flow rate (m^3s^{-1}) may be determined from

$$q = C_d A \sqrt{\frac{2\Delta P}{\rho_i}}$$

where C_d is the discharge coefficient (-), A the opening area (m^2), ΔP the pressure difference across the opening (Pa) and ρ_i the density of the incoming air (kg m^{-3}).

Heiselberg *et al* (2001) compared such a model with experimental data and concluded that the discharge coefficient cannot be regarded as a constant in circumstances where the opening area is relatively small. Instead, it will vary as a function of the opening area, the window type (side hung, bottom hung etc) and the prevailing temperature difference. The selection of suitable discharge coefficients is discussed by Andersen (1996).

With large vertical openings, such as doorways, more complex flow patterns occur (Brown 1962a, Cockroft 1979, Allard *et al* 1992, Hensen *et al* 1993).† If a temperature difference exists across such an opening, then air flow can occur in both directions due to the action of small density variations over the door height causing a positive pressure difference at the bottom (or top) of the opening with a corresponding negative pressure difference at the top (or bottom). This situation is illustrated in figure 5.4.

† Similar expressions have been derived for large horizontal openings (Brown 1962b).

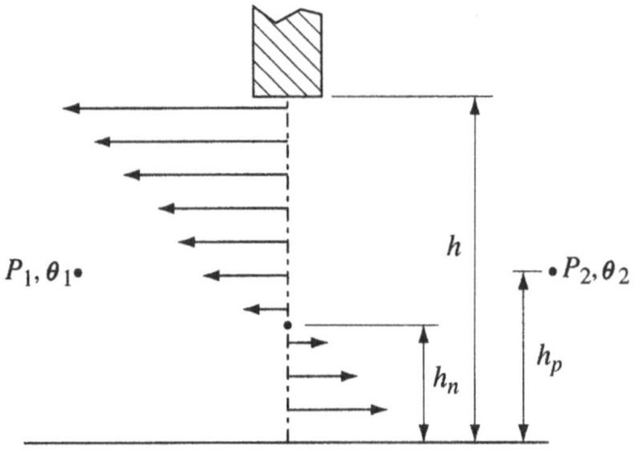

Figure 5.4: Bi-directional air flow across a doorway.

Cockroft (1979) studied this effect and produced an expression for the air flow through such an opening:

$$v = (2/3)[C_D W h (2/\rho)^{\frac{1}{2}} (C_a^{\frac{3}{2}} - C_b^{\frac{3}{2}})/C_t]$$

where C_D is the discharge coefficient (-), W the opening width (m), h the height (m), $C_a = (1 - r_p)C_t + (P_1 - P_2)$, $C_b = (P_1 - P_2) - r_p C_t$, $C_t = gP_a h/R(1/\theta_2 - 1/\theta_1)$, θ_1, θ_2 the absolute temperatures on either side of the opening (K), P_a the atmospheric pressure (N m^{-2}), P_1, P_2 the pressures on either side (N m^{-2}), R the gas constant (J kg^{-1}K^{-1}), $r_p = h_p/h$ and h_p is the height of the reference nodes on either side (m).

On evaluation, this equation yields a sum of real and imaginary parts. Real parts indicate a flow in the positive direction, while imaginary parts indicate a flow in the reverse direction. The equation has a singularity for $C_t = 0$. In this case no buoyancy effect exists and so an expression of the form of eqn (5.2) can be used.

The neutral height h_n—the height at which no net pressure difference can be measured across an opening—is found from

$$h_n = h\left(r_p - \frac{(P_1 - P_2)}{C_t}\right).$$

Temperature differences between internal and external air conditions also give rise to pressure differences according to

$$\Delta P = g\Delta\rho(h - h_n)$$

where $\Delta\rho$ is the density difference (kg m^{-3}). This is the stack effect which predominates in high rise buildings (Tamura and Wilson 1967).

Note that the parameters describing such flow components are uncertain and that the flow field is often strongly influenced by background (or hidden) leakage paths such as behind skirting or through suspended ceilings. Many attempts have been made to quantify the leakage characteristics of building components (Dick 1950, Malinowski 1971, Harris-Bass *et al* 1974) when subjected to a pressure difference or where wind turbulence compresses the air in the internal volume resulting in the induction of fluctuating flows across an opening. CFD models have also been employed to estimate the flow path parameters (Schaelin *et al* 1993). Table 5.5 gives some typical dimensions for building openings and cracks. Similar data are published by the Air Infiltration and Ventilation Centre (AIVC 2001).

Table 5.5: Some example leakage dimensions.

	Typical crack width (mm/m)	*Equivalent area (m^2)*	*Comment*
Window:			
not weather-stripped	1.0†	N/A	Windows of
weather-stripped	0.25†	N/A	average fitting.
ventilation opening	~2.5# (1m window perimeter)	0.02	Based on a 2m × 1m window.
Doors:			
not weather-stripped	1.5†—5.0#	0.0084—0.028	For standard
weather-stripped	0.8†—3.0#	0.0045—0.168	door of
ventilation opening	10# (1m of door perimeter)	0.056	0.8m × 2.0m average fitting.
Walls:			
unplastered	N/A	0.00002†	Based on a
air brick	N/A	0.005—0.033†	10 m^2 wall.
Roof eaves:			
attic ventilation	N/A	0.15#	Based on 20m × 10m (plan area) roof
Floor:			
skirting	1.6†	N/A	
tongue and groove	0.8†	N/A	Floor uncovered.
sub-floor ventilation	1.3—3.2† (1m of external wall length)	N/A	
Flues:			
open fire	N/A	0.033†	
gas fire	N/A	0.013—0.033†	
Ventilators:			
fixed louvre	N/A	0.015†	
constant flow @			
0.9 m s^{-1}	N/A	0.0084†	Based on face
9.0 m s^{-1}	N/A	0.0022†	0.3m × 0.075m.
† data from Dick (1950)	# data from Cockroft (1979)		

It should be noted that it is possible to define a flow network that is insoluble. To avoid this situation two conditions require to be met. First, the pressure of at least one node within the network must be *a priori* known. Second, all unknown nodal pressures must be linkable, by some path, to a node whose pressure is known. Problems can occur where control action closes a flow component and thereby violates the second condition.

5.1.5 Iterative solution procedure

Each non-boundary node is assigned an arbitrary pressure and the connecting components' flow rates determined from the corresponding mass flow model as described in the previous section. The nodal mass flow rate residual (error), R_i (kg s^{-1}), for the current iteration is then determined from

$$R_i = \sum_{k=1}^{K_i} \dot{m}_k$$

where $\dot{m}_k$ is the mass flow rate along the kth connection to node i and K_i is the total number of connections linked to node i.

These residuals are used to determine nodal pressures corrections, $\mathbf{P}^*$, for application to the current pressure field, $\mathbf{P}$:

$$\mathbf{P}^* = \mathbf{P} - \mathbf{C}$$

where $\mathbf{C}$ is a pressure correction vector. The process, which is equivalent to a Newton-Raphson technique (Kreyszig 1979), iterates until convergence is achieved. In the method $\mathbf{C}$ is determined from

$$\mathbf{C} = \mathbf{J}^{-1}\,\mathbf{R} \tag{5.3}$$

where $\mathbf{R}$ is the vector of nodal mass flow residuals and $\mathbf{J}^{-1}$ is the inverse of the square Jacobian matrix whose diagonal elements are given by

$$J_{n,n} = \sum_{i=1}^{L} \left(\frac{\partial \dot{m}}{\partial \Delta P} \right)_i$$

where L is the total number of connections linked to node n. This is equivalent to the rate of change of the node n residual with respect to the node pressure change between each iteration.

The off-diagonal elements of $\mathbf{J}$ are the rate of change of the individual component flows with respect to the change in the pressure difference across the component (at successive iterations):

$$J_{n,m} = \sum_{i=1}^{M} - \left(\frac{\partial \dot{m}}{\partial \Delta P} \right)_i \quad ; n \neq m$$

where M is the number of connections between node n and node m. Note that for internal nodes the summation of the elements comprising each row of the Jacobian matrix are identically zero.

One possibility for the solution of eqn (5.3) is to use LU decomposition with implicit pivoting—known as Crout's method with partial pivoting (Press *et al* 1986). With this method $\mathbf{J}$ is decomposed into a lower triangular matrix, $\mathbf{L}$, and an upper triangular matrix, $\mathbf{U}$, such that $\mathbf{L}\,\mathbf{U} = \mathbf{J}$. This decomposition is then used to solve the linear set

$$\mathbf{J}\,\mathbf{C} = (\mathbf{L}\,\mathbf{U})\,\mathbf{C} = \mathbf{L}\,(\mathbf{U}\,\mathbf{C}) = \mathbf{R}$$

by first solving, by forward substitution, for the vector $\mathbf{Y}$ such that $\mathbf{L}\,\mathbf{Y} = \mathbf{R}$ and then solving (by back substitution) $\mathbf{U}\,\mathbf{C} = \mathbf{Y}$. The advantage of the method is that both substitutions are trivial.

Conservation considerations applied to each node then provide the convergence criterion: $\sum \dot{m}_k \rightarrow 0$ at all internal nodes. As noted by Walton (1982), there may be occasional instances of low convergence with oscillating pressure corrections required at successive iterations. In tests, the observed oscillations followed closely the pattern shown in figure 5.5 where successive pressure corrections are a constant ratio of the previous correction: $C_i = -0.5\,C_i^*$, where * denotes the previous iteration step.

On the basis of this pattern, it is possible to extrapolate to a final solution:

$$P_i = P_i^* - C_i/(1 - r)$$

where r is the ratio of C_i for the current iteration to its value at the previous iteration. The

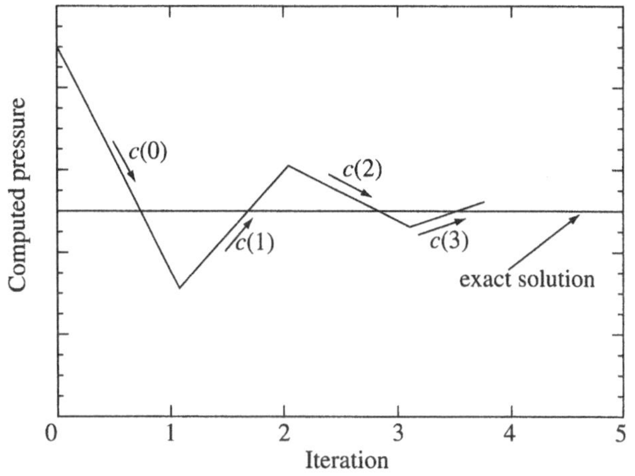

Figure 5.5: Example of successive computed values of pressure and oscillating pressure correction at a single node.

factor $1/(1 - r)$ is a relaxation factor. The extrapolated value of the node pressure can be used in the next iteration. If it is used, then r is not evaluated for that node in the following iteration but only in the one thereafter. In this way r is evaluated only using unrelaxed pressure correction values. This process is similar to Steffensen iteration (Conte and De Boor 1972). The iteration correction method presented above gives a variable and node dependent relaxation factor. When the solution is close to convergence, Newton-Raphson iteration converges quadratically. By limiting the application of the relaxation factor to cases where r is less than some value, such as -0.5, it will not interfere with the rapid convergence.

In practice, the foregoing solution technique may be expected to solve even complex networks in a few iterations. This means that, unlike CFD, the network air flow method will not impose a significant computational burden. This renders the method most suitable for the combined thermal and flow modelling of naturally ventilated buildings (André *et al* 1998, Allard 1998, Hensen 1999).

5.2 Computational fluid dynamics

Although well adapted for building energy application, the nodal network method is limited when it comes to consideration of indoor comfort and air quality. Because momentum effects are neglected, intra-room air movement cannot be studied, while surface convection heat transfer regimes cannot be evaluated because of the low resolution. To overcome these limitations, it is necessary to introduce a computational fluid dynamics (CFD) model† (Patankar and Spalding 1972, Nielsen *et al* 1978) whereby intra-zone air movement may be evaluated and the distribution of the principal parameters determined.

CFD is a complex development field with a rapidly evolving state-of-the-art and general applicability. In recent years its application to buildings—a non-steady, mixed flow (turbulent, laminar and transitional) problem—has grown significantly (Nielsen 1989 & 1994, Jones and

† A method of intermediate detail, termed zonal modelling, may also be employed to determine intra-zone temperature distribution for zones where the momentum effects are small (Inard *et al* 1996, Axley 2001). This method is not considered here.

Whittle 1992, Denev and Stankov 2000a) and attempts have been made to combine CFD and building energy models (Negraõ 1995), to extend CFD to include building features (Schild 1997) and to develop techniques for the realistic representation of HVAC components such as diffusers (Chen and Srebec 1999).

Essentially, a building-integrated CFD model comprises the following elements: room discretisation; a set of equations to represent the conservation of energy, mass, momentum and species; the imposition of boundary conditions; an equation solver; a method to link the CFD, building thermal (chapter 4) and network (§5.1) air flow models; and a means to translate the results to concepts meaningful to a designer. The following sub-sections describes the first 5 elements while §5.4 describes the last.

5.2.1 Domain discretisation

The starting point is to sub-divide the room into a number of finite volumes so that conservation equations for mass, momentum, energy and species concentration may be established and solved for the entire domain (c.f. the mono-volume approach of the nodal network method). While curvilinear co-ordinate systems are commonly employed in CFD analyses, in building applications the geometries are typically orthogonal, facilitating the three dimensional Cartesian gridding technique as illustrated in figure 5.6.

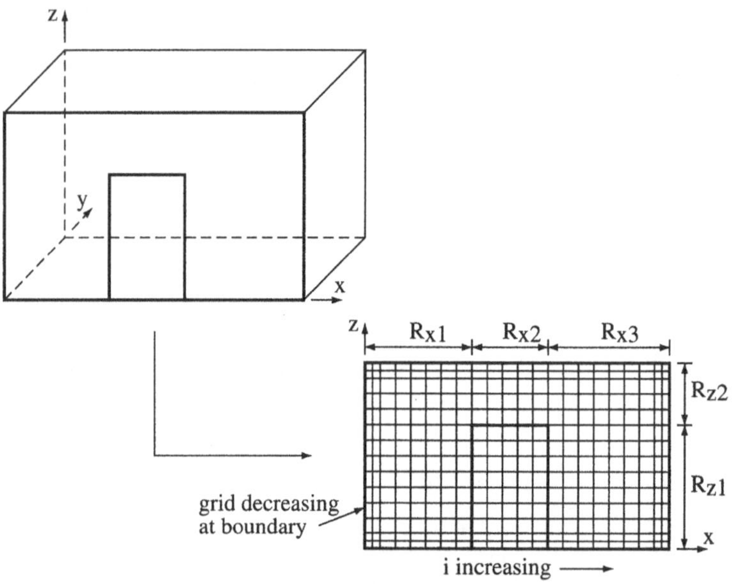

Figure 5.6: CFD domain discretisation.

Each dimension is divided into a number of regions, here 3 in the x-direction and 2 in the z-direction (the y-direction is not shown). The regions are then gridded using a constant or variable spacing evaluated, for example, from

$$x_i = L(i/n)^{c}$$

where x_i is the co-ordinate of grid line i, L the overall dimension of the region, n the number of grid lines and c a power law coefficient. Where $c > 1$, the grid starts fine and becomes coarse as i increases, with $c < 1$ defining the opposite scenario.

Consider the x-direction. The region located to the left of the door has a variable grid which increases with increasing i, the door region has a constant grid, while the region located to the right of the door decreases with increasing i. Negraõ (1995) implemented the scheme of table 5.6 to control the gridding process within the ESP-r system. Other approaches are possible, and some of these are suited to the case of low Reynolds Number models where the near-wall grid is made especially fine.

Table 5.6: Domain gridding parameters.

Number of grid lines in region		*Power law coefficient*	
n > 0	n cells distributed over the region	c = 1	uniform distribution
n < 0	|n| cells distributed symmetrically over the region	c > 1	increasing grid size*
		c < 1	decreasing grid size*

* From the beginning to the end (or middle if n < 0) of the region.

For the case of non-rectangular geometries, or where internal obstructions are present, the above technique may be applied to a rectangular bounding box but with the boundary of the non-participating cells treated as solid surfaces and assigned a boundary condition as shown in figure 5.7 (lower left).

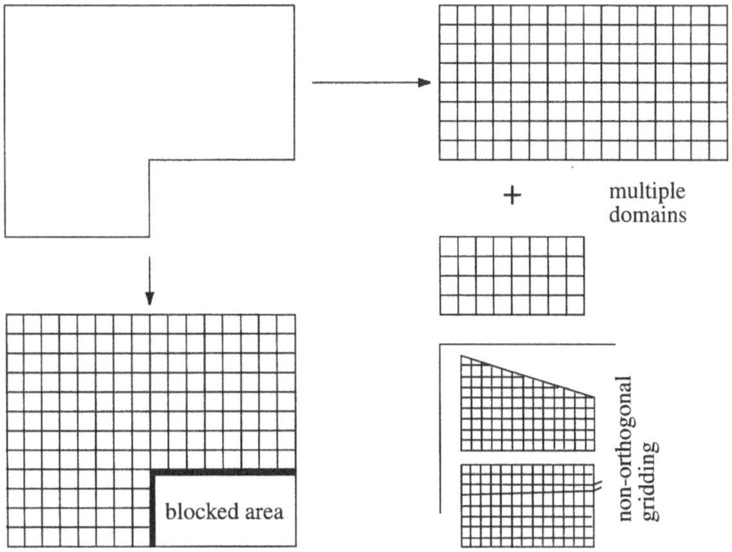

Figure 5.7: Treatment of complex geometries.

Alternatively, a block structured approach may be employed (figure 5.7, upper right) whereby the problem is reduced to separate domains which are then processed independently with the mass, momentum, energy and species exchanges at the interfaces reconciled after each iteration. Such a treatment facilitates the parallel processing of domains but will result in a large number of domains for cases where internal objects are included. In this case the former treatment is probably more applicable.

To accommodate the range of commonly encountered room shapes, Denev and Stankov

(2000b) extended the above gridding scheme to allow one of the three dimensions to be non-orthogonal (figure 5.7, lower right). Other discretisation techniques are described in the literature (Thompson *et al* 1985, Davidson 1990).

5.2.2 Conserving energy, mass, momentum and species concentration

The movement of room air and contaminants may be determined from the solution of discretised mass, momentum, energy and concentration equations when subject to given boundary conditions. In the context of building simulation, the Boussinesq approximation is usually applied whereby the air density is held constant and the effects of buoyancy are included within the momentum equation. In tensor notation, the conservation equations are as follows.

Continuity

$$\frac{\partial \rho}{\partial t} + \frac{\partial}{\partial x_i}(\rho u_i) = 0 \qquad (5.4)$$

Momentum

$$\frac{\partial}{\partial t}(\rho u_i) + \frac{\partial}{\partial x_j}(\rho u_i u_j) = -\frac{\partial p}{\partial x_i} + \frac{\partial}{\partial x_j}\left[\mu\left(\frac{\partial u_i}{\partial x_j} + \frac{\partial u_j}{\partial x_i}\right)\right] - \rho g \beta(\theta_\infty - \theta) \qquad (5.5)$$

Energy

$$\frac{\partial}{\partial t}(\rho H) + \frac{\partial}{\partial x_i}(\rho u_i H) = \frac{\partial}{\partial x_i}\left[\frac{k}{C_p}\frac{\partial H}{\partial x_i}\right] + S_H \qquad (5.6)$$

Concentration

$$\frac{\partial}{\partial t}(\rho C) + \frac{\partial}{\partial x_j}(\rho u_j C) = \frac{\partial}{\partial x_j}\left[\rho D \frac{\partial C}{\partial x_j}\right] + S_C \qquad (5.7)$$

where t is time (s), x_i the co-ordinate axis (= x, y and z), ρ density (kg m^{-3}), u_i the velocity component in the cardinal directions (= u, v, w; m s^{-1}), p the pressure (N m^{-2}), g the gravitational constant (m s^{-2}), μ the viscosity (kg $m^{-1}s^{-1}$), H the specific enthalpy (J $kg^{-1}K^{-1}$), C the species concentration (kg kg^{-1}), k the conductivity (W $m^{-1}K^{-1}$), D the species diffusion coefficient (m^2s^{-1}) for moisture, CO_2 etc, C_p the specific heat (J $kg^{-1}K^{-1}$), β the thermal expansion coefficient of air (K^{-1}), θ_∞ a reference temperature (°C) and S_H, S_C energy and species source terms respectively (W m^{-3} and kg $m^{-3}s^{-1}$).

While the conservation equations can be solved directly or by the technique of Large Eddy Simulation (Deardorff 1970, Davidson and Neilsen 1996), this is a computationally non-trivial task because these techniques require extremely fine meshes to resolve the turbulent fluctuations. The turbulence transport technique (Rodi 1980) is therefore used whereby the instantaneous values of temperature, concentration, velocity, pressure etc may be represented as the sum of their mean and fluctuating components, and the effect of turbulent motion is time-averaged. This gives rise to the following mean conservation equation for an incompressible fluid:

$$\frac{\partial}{\partial t}(\rho \phi) = \frac{\partial}{\partial x_i}\left(\Gamma_\phi \frac{\partial \phi}{\partial x_i} - \rho U_i \phi\right) + S_\phi \qquad (5.8)$$

where ϕ is a transport variable such as continuity ($\phi = 1$), enthalpy, concentration of contaminant or velocity; ρ the density (kg m^{-3}), Γ_ϕ a diffusion coefficient, U_i a mean velocity component (U, V, W), and S_ϕ a mean source term. The transport variables, diffusion coefficients and source terms are then as given in table 5.7 for each conservation equation type. In words, eqn (5.8) can be read as

The rate of increase of ϕ within a fluid element = the rate of increase of ϕ due to diffusion - the net rate of flow of ϕ out of the element + the rate of increase of ϕ due to sources.

Table 5.7: Transport variables (ϕ), diffusion coefficients (Γ_ϕ) and source terms (S_ϕ).

Equation Type	ϕ	Γ_ϕ	S_ϕ
Continuity	1	-	-
Momentum	u_i	μ_{ef}	$-\dfrac{\partial p}{\partial x_i} - \rho g \beta(\theta_\infty - \theta)$
Energy	H	Γ_T	S_H
Species	C	Γ_C	S_C
Turbulent kinetic energy	k	$\dfrac{\mu_{ef}}{\sigma_k}$	$G - C_D \rho \varepsilon - G_b$
Dissipation rate of k	ε	$\dfrac{\mu_{ef}}{\sigma_\varepsilon}$	$C_1 \dfrac{\varepsilon}{k} G - C_2 \rho \dfrac{\varepsilon^2}{k} - C_3 \dfrac{\varepsilon}{k} G_b$

$$\Gamma_T = \frac{\mu}{Pr} + \frac{\mu_t}{\sigma_T} \ ; \ \Gamma_C = \frac{\mu}{Sc} + \frac{\mu_t}{\sigma_C} \ ; \ \mu_{ef} = \mu_t + \mu \ ; \ \rho = \rho(T, C)$$

$$G_b = g\left(\beta_T \frac{\mu_t}{\sigma_T}\frac{\partial T}{\partial x_i} + \beta_C \frac{\mu_t}{\sigma_C}\frac{\partial C}{\partial x_i}\right) \ ; \ G = \mu_t\left(\frac{\partial u_i}{\partial x_j} + \frac{\partial u_j}{\partial x_i}\right)\frac{\partial u_i}{\partial x_j}$$

$$C_D = 1.0 \ ; \ C_1 = 1.44 \ ; \ C_2 = 1.92 \ ; \ \sigma_k = 1.0 \ ; \ \sigma_\varepsilon = 1.3 \ ; \ \sigma_T = 0.9 \ ; \ \sigma_C = 0.9$$

where μ is molecular viscosity (kg $m^{-1}s^{-1}$), μ_t is eddy viscosity, P is pressure (N m^{-2}), g the gravitational acceleration (m s^{-2}), C_p the specific heat (J $kg^{-1}K^{-1}$), q''' is heat generation (W m^{-3}), Pr is the Prandtl Number, Sc is the Schmidt Number, σ_k is the turbulent energy diffusion coefficient, σ_ε is the turbulent energy dissipation diffusion coefficient, σ_T is the turbulent Prandtl Number, σ_C is the turbulent Schmidt Number, β_T is the thermal expansion coefficient (1/K).

The core challenge is to model the effects of turbulent motion on the mean flow and to this end several techniques are available (Wilcox 1993, Launder 1989a, 1989b & 1992, Murakami *et al* 1992, Amano 1984, Nagano and Hisida 1987). To overcome the difficulties of directly modelling turbulent flows, a turbulence transport model is normally used whereby the influence of turbulence on the time averaged motion of air may be determined. A number of modelling techniques have emerged including Reynolds-stress, algebraic-stress and zero-, one- and two-equation eddy-viscosity models. Because such models rely on empirical data, they are not generally applicable to room air flow modelling. Of the possible turbulence transport models, the standard $k - \varepsilon$ model is widely used because of its general applicability and reasonable accuracy (Launder and Spalding 1972 & 1974, Chen 1995). Its function is to determine the eddy viscosity, μ_t, of table 5.7 at each grid point as a function of local values of the turbulent kinetic energy (k) and its rate of dissipation (ε):

$$\mu_t = \rho C_\mu \frac{k^2}{\varepsilon}$$

where C_μ is a dimensionless constant. The distributions of k and ε are then given by the following transport equations:

For k:

$$\frac{\partial}{\partial t}(\rho k) + \frac{\partial}{\partial x_j}(\rho u_j k) = \frac{\partial}{\partial x_j}\left(\frac{\mu_t}{\sigma_k}\frac{\partial k}{\partial x_j}\right) + \mu_t\left(\frac{\partial u_i}{\partial x_j} + \frac{\partial u_j}{\partial x_i}\right)\frac{\partial u_i}{\partial x_j} - \rho\varepsilon - g\beta\frac{\mu_t}{\sigma_t}\frac{\partial \theta}{\partial z} \tag{5.9}$$

For ε:

$$\frac{\partial}{\partial}(\rho\varepsilon) + \frac{\partial}{\partial x_j}(\rho u_j \varepsilon) = \frac{\partial}{\partial x_j}\left(\frac{\mu_t}{\sigma_\varepsilon}\frac{\partial \varepsilon}{\partial x_j}\right) + C_1\frac{\varepsilon}{k}\mu_t\left(\frac{\partial u_i}{\partial x_j} + \frac{\partial u_j}{\partial x_i}\right)\frac{\partial u_i}{\partial x_j} - C_2\rho\frac{\varepsilon^2}{k}$$

$$- C_1\frac{\varepsilon}{k}g\beta\frac{\mu_t}{\sigma_t}\frac{\partial \theta}{\partial z} \tag{5.10}$$

where σ_k, σ_ε, C_1, C_2 are empirical constants. Values for these constants, and C_μ, are established by data fitting techniques applied to a wide range of turbulent flows.

Eqns (5.9) and (5.10) may also be cast in the form of eqn (5.8), with the diffusion and source terms as given in table 5.7. Because these equations contain experimentally derived parameters, the standard $k - \varepsilon$ model is fully valid only for cases of fully turbulent shear layer flow and free jets. Alternative turbulence models have therefore been proposed for use in room air flow modelling, such as the zero-equation (mixing length) model of Chen and Xu (1998), which has shown good agreement with experimental data for three test cases: natural convection with infiltration, forced convection, and mixed convection with displacement ventilation (Srebric *et al* 1999). The attractiveness of this model is that it does not require the solution of the k and ε equations in order to calculate the eddy viscosity. Instead, μ_t is directly related to the local mean velocity through an algebraic expression:

$$\mu_t = 0.3874\rho\bar{V}l$$

where $\bar{V}$ is the local mean velocity and l is the length scale (equal to the distance from the solution point to the nearest solid surface).

This treatment results in a substantial reduction in the computational burden but with the retention of reasonable accuracy in some applications (Beausoleil-Morrison 2000).

Because the standard $k - \varepsilon$ model is valid only for turbulent flow regions, it cannot be used to represent the near-wall condition where viscous effects predominate and the flow is laminar. Instead, logarithmic wall functions are usually employed whereby the form of the velocity and temperature profile within the boundary layer is assumed in order to determine the surface shear stress and convective heat transfer (Launder and Spalding 1974).

For shear stress:

$$\tau = \frac{C_\mu^{1/4} k_p^{1/2} \rho u_p}{\frac{1}{K}\ln\left(\frac{E\rho\Delta x_p C_\mu^{1/4} k_p^{1/2}}{\mu}\right)} \tag{5.11}$$

For convection:

$$\dot{q} = \frac{C_\mu^{1/4} k_p^{1/2} \rho C_p(\theta_p - \theta_w)}{\sigma_t\left[\frac{1}{K}\ln\left(\frac{E\rho\Delta x_p C_\mu^{1/4} k_p^{1/2}}{\mu}\right) + \frac{\pi/4}{\sin(\pi/4)}\left(\frac{A}{k}\right)^{1/2}\left(\frac{Pr}{\sigma_t} - 1\right)\left(\frac{\sigma_t}{Pr}\right)^{1/4}\right]} \tag{5.12}$$

where τ is the wall shear stress (N m^{-2}), $\dot{q}$ the convective heat transfer (W m^{-2}), K the von Kármán constant, Δx_p the distance from the wall to the next-to-wall grid point, P, and A, E are constants representing the wall roughness.

It has been observed that the standard $k - \varepsilon$ model will tend to overpredict the eddy viscosity in regions that are subtly turbulent (Baker *et al* 1994). While this will not impact greatly on the mean flow rate, due to the low diffusion of momentum, it will impact on the calculated heat transfer due to the dominance of thermal diffusion in regions of low flow rate. This can give rise to substantial errors in the prediction of surface convection. Since the convective heat transfer constitutes the pivot point between the air flow and thermal domains, any inaccuracy in its treatment will affect the entire simulation (Beausoleil-Morrison and Clarke 1998). Improving this aspect of CFD is the aim of Low Reynolds Number models (Launder and Sharma 1974, Lam and Bremhorst 1981, Chien 1982, Patel *et al* 1985, Henkes 1990, Stankov and Denev 1996) with enhanced treatment of buoyancy effects (Ince and Launder 1989, Hanjalic and Vasic 1993). Because of stability problems when applied to 3D problems, such models are usually restricted to 2D flows. Even then, they are computationally demanding. Chen (1995) has compared several modified $k - \varepsilon$ models when applied to some typical 2D problems.

An alternative approach is to utilise wall functions derived for the specific case in hand. For example, Yuan *et al* (1993) derived wall functions for the case of vertical walls when subjected to buoyancy driven flow, while others have worked on formulations for horizontal surfaces (Bartzke 1998). Yet another approach is described in §5.4: the replacement of the wall functions with empirical convective heat transfer correlations as elaborated in §7.6.

Whatever the approach, the problem reduces to a set of time-averaged nodal conservation equations for U, V, W, H (or θ) and C and, where the $k - \varepsilon$ model is active, k and ε. These conservation equations may be discretised by the finite volume method of chapter 3 (Versteeg and Malalasekera 1995) to obtain a set of linear equations of the form

$$a_p \phi_p = \sum_i a_i \phi_i + b$$

where ϕ is the relevant variable of state, p designates a domain cell of interest, i designates the neighbouring cells, b relates to the source terms applied at p, and a_p, a_i are the self- and cross-coupling coefficients respectively.

Because these equations are strongly coupled and highly non-linear—that is the equation coefficients and source terms are dependent on the state variables—they must be solved iteratively for a given set of boundary conditions. Moreover, when conflated with the building thermal and network air flow models, the CFD domain equations must be solved in tandem with the other domain equations and repeatedly at each time row throughout a simulation.

5.2.3 Initial and boundary conditions

Initial values of ρ, u_i and H are required at time $t = 0$ for all domain cells. For solid surfaces, the required boundary conditions include the temperature (or flux) at points adjacent to the domain cells. For cells subjected to an in-flow from ventilation openings and doors/windows, the mass/momentum/energy/species exchange must be given in terms of the distribution of relevant variables of state—U, V, W, H, k, ε and C. At outlets, the normal practice is to impose a constant pressure and the conditions $\partial u_n/\partial n = 0$, $\partial H/\partial n = 0$, $\partial k/\partial n = 0$, $\partial \varepsilon/\partial n = 0$, where n indicates the direction normal to the boundary.

Where the CFD model is conflated with the building and network flow models, these boundary conditions will be time dependent. Further, where a method exists to consider the applicability of different near-wall turbulence models (as introduced in §5.4), then parameters such as surface convection coefficients may additionally be assigned as boundary conditions.

5.2.4 Iterative solution procedure

The SIMPLE (Semi-Implicit Method for Pressure-Linked Equations) method is commonly used to solve the set of elliptic flow equations. Essentially, the pressure of each domain cell is linked to the velocities connecting with surrounding cells in a manner that conserves continuity. The method (Patankar 1980, Versteeg and Malalasekera 1995) accounts for the absence of an equation for pressure by establishing a modified form of the continuity equation to represent the pressure correction that would be required to ensure that the velocity components determined from the momentum equations move the solution towards continuity. This is done by using a guessed pressure field to solve the momentum equations for intermediate velocity components U, V and W. These velocities are then used to estimate the required pressure field correction from the modified continuity equation. The nature of these modifications, and the simplifications applied in the process, are detailed in a number of texts (e.g. Chen 1988, Versteeg and Malalasekera 1995). The energy equation, and any other scalar equations (e.g. for concentration), are then solved and the process iterates until convergence is attained. To avoid numerical divergence, it is usual to apply under relaxation to the pressure correction terms. The concentration distribution data may then be post-processed to obtain the local mean age of air using the method of Sandberg (1981). This is a useful index because it directly relates to the distribution of air freshness, a prerequisite of any indoor air quality assessment. Comparisons of measurements with estimates derived from simulation have shown good agreement (e.g. Bartak *et al* 2001).

Variants of the SIMPLE algorithm have been developed in order to reduce the computational burden and assist convergence. These include SIMPLE-Revised (SIMPLER; Patankar 1980), in which the pressure field is obtained directly (i.e. without the need for correction) from a pressure equation derived from the continuity equation, and SIMPLE-Consistent (SIMPLEC; Van Doormal and Raithby 1984), in which the simplifications applied to the momentum/continuity equations to obtain the pressure field correction are less onerous.

The solution of the discretised flow equations may be achieved using the tri-diagonal matrix algorithm favoured because of its modest storage requirements and computational speed. To assist convergence, the grid sweep direction may be alternated between iterations in order to more effectively introduce the boundary conditions into the solution. Where the CFD domain is connected to a flow network, both solvers will be required to operate in tandem (see §5.4).

5.2.5 Results interpretation

The information inherent within the CFD results is usually communicated graphically. For example, figure 5.8† shows the distribution of the mean age of air for a representative 2D room slice. This, along with similar outputs for the other principal parameters (humidity, contaminants, radiant temperature etc), may then be used to assess the indoor air quality and thermal discomfort according to some appropriate standard (such as prENV 1752, ISO EN 7730 and ANSI/ASHRAE 55-1992).

The conflation of CFD and building simulation gives rise to a significant number of relevant indicators, including but not limited to:

1. the variation in vertical air temperature between floor and head height;
2. the absolute temperature of the floor;
3. radiant temperature asymmetry;
4. unsatisfactory ventilation rate;

---

† Within the ESP-r system images such as this are constructed using an algorithm based on the transformation equations derived in §7.3 for shading prediction.

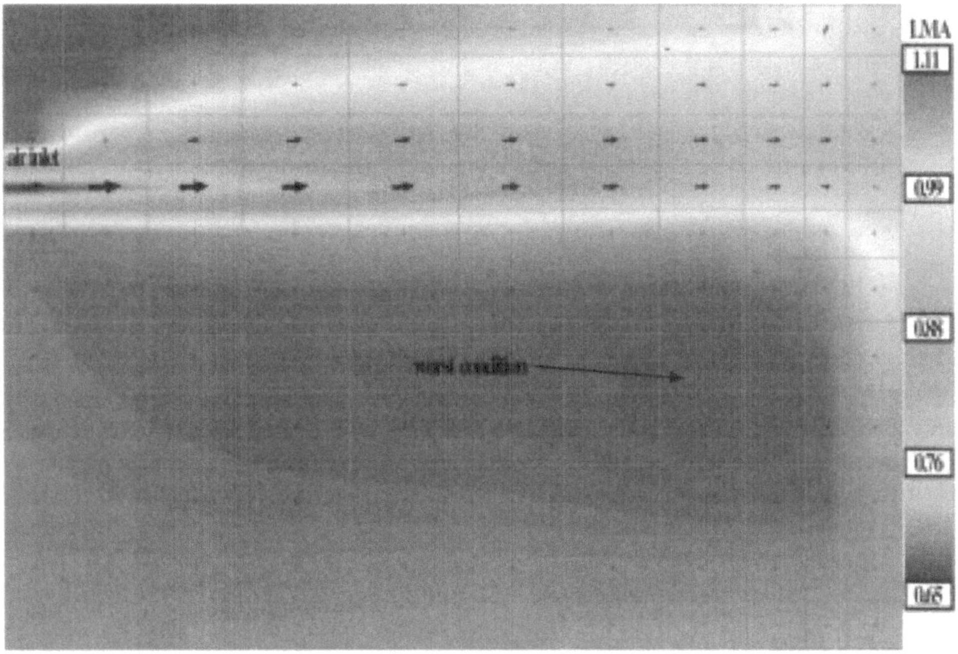

Figure 5.8: Distribution of the local mean age of air.

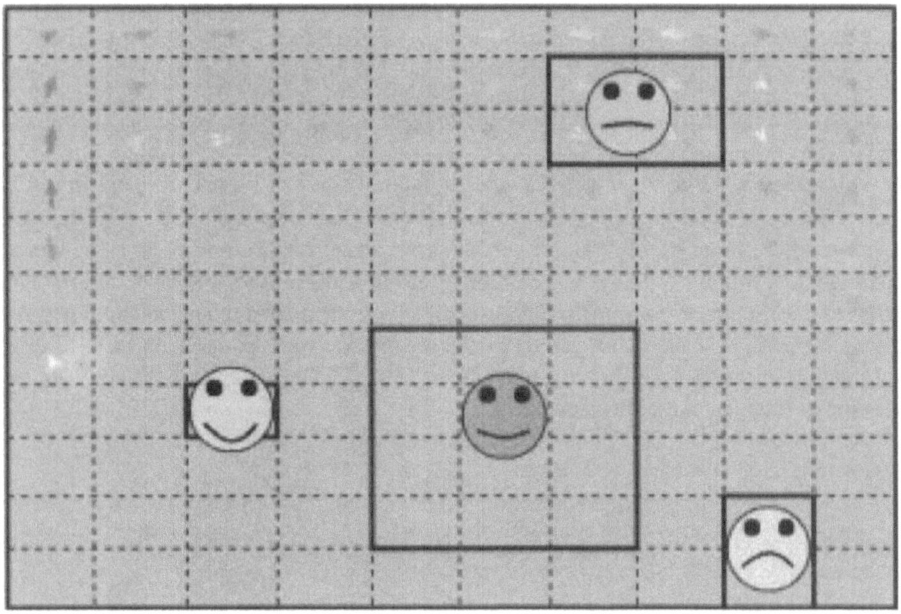

Figure 5.9: Integrated modelling supports the assessment of variations in air quality and thermal comfort.

5. unsatisfactory CO_2 level;
6. local draught assessed on the basis of the turbulence intensity distribution;
7. additional air speed required to off-set an elevated temperature;
8. comfort check based on effective temperature;
9. mean age of air.

As humorously illustrated in figure 5.9, such an approach can differentiate indoor regions in terms of a comprehensive set of performance indicators and standards of performance (Allard 1998).

5.3 Moisture flow within porous media

A significant proportion of houses world-wide are affected by condensation (Hens and Sneave 1991), giving rise to material degradation and mould growth on internal surfaces (Scottish Homes 1991, Dales et al 1991a & 1991b). As the use of biocidal compounds on affected surfaces is only acceptable in special circumstances, there is a general consensus that the preferred strategy is the elimination of the conditions that can lead to mould initiation and proliferation. A key element in such an approach would be the ability to predict where, when and under what conditions mould growth will occur. This requires a model of the moisture flow within building constructions for use in conjunction with the previously derived building and air flow models.

With temperature and partial vapour pressure as the transport potentials for moisture, it is possible to establish a coupled heat and moisture transport model. Nakhi (1995), for example, has applied mass and energy conservation considerations to a homogeneous, isotropic, construction control volume to establish the following moisture flow model within the ESP-r system.

For the moisture term (in one dimension):

$$\rho_o \zeta \frac{\partial (P/P_s)}{\partial t} + \frac{d\rho_l}{dt} = \frac{\partial}{\partial x}\left(\delta_P^\theta \frac{\partial P}{\partial x} + D_\theta^P \frac{\partial \theta}{\partial x} \right) + S \tag{5.13}$$

where ρ is the density (kg m^{-3}); o and l denote porous medium and liquid respectively, ζ the moisture storage capacity (kg kg^{-1}), P the partial water vapour pressure (N m^{-2}), P_s the saturated vapour pressure (N m^{-2}), δ the water vapour permeability (kg Pa^{-1}m^{-1}s^{-1}), D the thermal diffusion coefficient (kg m^{-2}K^{-1}s^{-1}) and S a moisture source term (kg m^{-3}s^{-1}). θ and P denote temperature and pressure driving potentials respectively, with the principal potential given as the subscript. The significance and characteristics of the transport parameters have been the subject of much research (e.g. Galbraith and McLean 1993, McLean and Galbraith 1996).

For the energy term (in one dimension):

$$[\rho_o(c_o + c_v g_v) + c_l \rho_l] \frac{\partial \theta}{\partial t} + h_v \frac{\partial \rho_v}{\partial t} + h_l \frac{\partial \rho_l}{\partial t}$$

$$= \frac{\partial}{\partial x}\left(\lambda \frac{\partial \theta}{\partial x} \right) - \frac{\partial h_s}{\partial x} J_v + q \tag{5.14}$$

where c is the specific heat capacity (J kg^{-1}K^{-1}), subscript v denotes vapour, g the moisture content (kg kg^{-1}), λ the heat conductivity (W m^{-1}K^{-1}), J_v the vapour mass flux (kg m^{-2}s^{-1}), q a source of heat (W m^{-3}) and h_v, h_l, h_s are the enthalpies (J kg^{-1}) of vapour, liquid and moisture flux sources respectively.

For the condensation and evaporation processes, a control equation is implemented as a one-way liquid valve connected to the control volume. When the relative humidity reaches its maximum value, the valve opens to deliver the condensation to an imaginary tank. Conversely, when the relative humidity falls below its maximum value, liquid is returned to the control volume where it re-evaporates. This process has been implemented as a function of the saturation pressure. Special treatment would be required for the case of capillary condensation where, at elevated relative humidities, the moisture begins to fill the entire capillaries' cross section (Galbraith and McLean 1993).

Use of eqns (5.13) and (5.14) allows for the solution of the three dependent variables, P, θ and ρ_l, for each control volume within a multi-layered construction when evolving under the influence of the boundary heat and mass transfers. To achieve this, a finite difference approximation is applied to the equations giving, for moisture:

$$a_i(t+\delta t)P_i(t+\delta t) + \sum_{j=1}^{2} a_j(t+\delta t)P_j(t+\delta t) + m_{li}(t+\delta t) - \gamma\delta t S_i(t+\delta t)$$

$$= a_i(t)P_i(t) + \sum_{j=1}^{2} a_j(t)P_j(t) + m_{li}(t) + (1-\gamma)\delta t S_i(t)$$

$$+ \sum_{j=1}^{2} b_j(t)[\theta_j(t) - \theta_i(t)] + \sum_{j=1}^{2} b_j(t+\delta t)[\theta_j(t+\delta t) - \theta_i(t+\delta t)]$$

where γ is the degree of implicitness, t and $t+\delta t$ refer to the present and future time-rows respectively, m the specific mass (kg m^{-3}) and

$$a_j(t+\delta t) = -\gamma(\delta_P^\theta)_{j\to i}(t+\delta t)\ A\delta t/\delta X_{j\to i}$$

$$a_j(t) = (1-\gamma)(\delta_P^\theta)_{j\to i}(t)\ A\delta t/\delta X_{j\to i}$$

$$a_i(\xi) = \rho\zeta_i V/P_{si}(\xi) - a_{i-1}(\xi) - a_{i+1}(\xi)$$

$$b_j(t+\delta t) = -\gamma(D_\theta^P)_{j\to i}(t+\delta t)\ A\delta t/\delta X_{j\to i}$$

$$b_j(t) = (1-\gamma)(D_T^{P)_{j\to i}}(t)\ A\delta t/\delta X_{j\to i}$$

$$m_l(\xi) = V\rho_l(\xi) \quad (kg_{moisture})$$

$$S_i(\xi) = VS_i(\xi) \quad (kg_{moisture}s^{-1})$$

where A is area, δX the flowpath length, V the node volume and $\xi = t$ or $t+\delta t$. For the energy term:

$$a_i(t+\delta t)\theta_i(t+\delta t) + \sum_{j=1}^{2} a_j(t+\delta t) - \gamma G_i(t+\delta t)\delta t - \gamma h_s\dot{m}_v(t+\delta t)\delta t$$

$$= a_i(t)\theta_i(t) + \sum_{j=1}^{2} a_j(t)\theta_j(t) + (1-\gamma)G_i(t)\delta t + (1-\gamma)h_s\dot{m}_v(t)\delta t$$

$$- h_{li}(t+\delta t)[m_{li}(t+\delta t) - m_{li}(t)] - h_{vi}(t+\delta t)[m_{vi}(t+\delta t) - m_{vi}(t)]$$

where

$$a_j(t+\delta t) = -\gamma A\lambda_{j\to i}(t+\delta t)\delta t/\delta X_{j\to i}$$

$$a_j(t) = (1-\gamma)A\lambda_{j\to i}(t)\delta t/\delta X_{j\to i}$$

$$a_i(\xi) = \rho_o V(c_o + c_v u_v) + c_l m_l - \sum_{j=1}^{2} a_j(\xi).$$

For a homogeneous control volume, the moisture storage capacity, ζ, may be found from an expression by Hansen (1986):

$$\zeta = \frac{u_h}{n} A\phi \left(1.0 - \frac{\ln \phi}{A}\right)^{\frac{-n+1}{n}} ; 0 \leq \phi \leq 1$$

where ϕ is the relative humidity and a mass weighted value of ζ is used for the heterogeneous case. The vapour permeability, δ, is evaluated from

$$\delta = \frac{\delta_a}{\mu} = \frac{1.89923e^{-10}}{\mu} \quad @\ 25^{\circ}C$$

where the vapour resistance factor, μ, is given by $\mu = 1/(a + be^{c\phi})$ for $0 \leq \phi \leq 1$ and a, b and c are constants for a given material.

The coupled moisture/energy equations are then given by

$$\begin{bmatrix} \mathbf{E} \\ \mathbf{M} \end{bmatrix} \times \begin{bmatrix} \theta \\ \mathbf{P} \end{bmatrix} = \begin{bmatrix} \mathbf{B}_e \\ \mathbf{B}_m \end{bmatrix} \tag{5.15}$$

where **E** and **M** are the energy and moisture coefficient matrices respectively, θ, **P** the temperature and vapour pressure vectors, and $\mathbf{B}_e$, $\mathbf{B}_m$ the energy and moisture boundary conditions.

5.4 Linking the building and flow domains

The building thermal, network air flow and CFD models can now be arranged to act co-operatively. This requires a conflation controller to ensure that the CFD model is appropriately configured at each time-step. Beausoleil-Morrison (2001) has implemented such a controller within the ESP-r system.

At the start of a time-step, the zero-equation turbulence model of Chen and Xu (1998) is employed in investigative mode to determine the likely flow regimes at each surface (forced, buoyant, fixed, fully turbulent or weakly turbulent). This information is then used to select appropriate surface boundary conditions while the estimated eddy viscosity distribution is used to initialise the k and ε fields. Utilising this well configured $k - \varepsilon$ model, a second CFD simulation is then initiated for the same time-step.

After the investigative simulation, the nature of the flow at each surface is evaluated from the local Grashof (Gr) and Reynolds (Re) Numbers (Incropera and Dewitt 1985) as determined from the investigative simulation. The Grashof Number is defined as the ratio of the buoyancy and viscous forces and indicates "how buoyant" the flow is adjacent to the surface, while the Reynolds Number is defined as the ratio of the inertial and viscous forces and indicates "how forced" is the flow. The following conditions are relevant:

$Gr/Re^2 \ll 1$	forced convection effects overwhelm free convection;
$Gr/Re^2 \gg 1$	free convection effects dominate;
$Gr \approx Re^2$	both forced and free convection effects are significant.

Based on the outcome, the following procedure is invoked.

Where buoyancy forces are insignificant, the buoyancy term in the z-momentum equation is discarded to improve solution convergence.

Where free convection predominates, the log-law wall functions are replaced by the Yuan *et al* (1993) wall functions and a Dirichlet† boundary condition imposed where the surface is vertical; otherwise a convection coefficient correlation is prescribed and a Neumann†† boundary condition imposed (note that this means that the thermal domain will influence the flow domain but not the reverse).

Where convection is mixed, the log-law wall functions are replaced by a prescribed convection coefficient and a Robin††† boundary condition is imposed.

Where forced convection predominates, the ratio of the eddy viscosity to the molecular viscosity (μ_t/μ), as determined from the investigative simulation, is examined to determine how turbulent the flow is locally:

> $\mu_t/\mu \leq 30$:- the flow is weakly turbulent; the log-law wall functions are replaced by a prescribed convection coefficient and a Neumann boundary condition is imposed;
> $\mu_t/\mu > 30$:- the log-law wall functions are retained and a Dirichlet boundary condition is imposed.

The iterative solution of the flow equations is then re-initiated for the current time-step. For surfaces where h_c correlations are active, these are shared with the building model so that the surface heat flux is effectively imposed on the CFD solution. Where such correlations are not active, the CFD-derived convection coefficients are inserted into the building model's surface energy balance equations.

Where an air flow network is active, the network node representing the room is removed and new network connections are added to effect a coupling with the appropriate domain cell(s) (Negraõ 1995, Clarke *et al* 1995) as shown in figure 5.10.

The flow network may now be solved using the technique of §5.1 when this is placed within an iterative scheme that includes the building and plant models (as shown in figure 6.17) in order to effect the required couplings. Kafetzopoulos and Suen (1995) report on the possible approaches to this iterative coupling.

Denev (1995) has developed a technique to ensure the accurate representation of both mass and momentum exchange in the situation where domain cells and network flow components are of dissimilar size (as illustrated in figure 5.10). The approach increases or reduces the network connection's area to achieve a match with the corresponding domain cell(s) and then adjusts the associated velocity to maintain the correct flow rate. Within the solution process, the adjusted velocity is imposed as a boundary condition to satisfy the flow rate and then the velocity is readjusted in the momentum equation to give the correct momentum. From the viewpoint of the flow network, the air exchanges with the CFD domain are treated as sources or sinks of mass at appropriate points within the flow network solution.

The foregoing procedure is embedded within a higher level controller, which acts to synchronise the customised solvers for the building, network flow and CFD equation-sets. Note that the frequency of invocation of these solvers may differ. For example, in order to reduce the computational burden, the building-side solver might be invoked more frequently than the two flow solvers.

---

† Dirichlet condition: fixed surface temperature $\theta = \theta_S$.

†† Neumann condition: fixed surface heat flux $k \dfrac{\partial \theta}{\partial n} = q$.

††† Robin condition: heat flux proportional to the local heat transfer $k \dfrac{\partial \theta}{\partial n} = h_c(\theta - \theta_S)$.

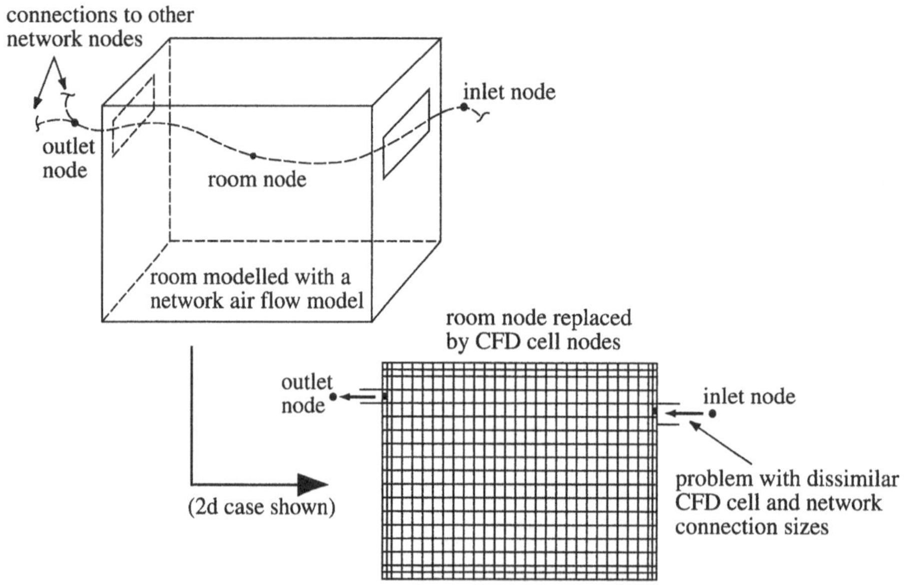

Figure 5.10: Coupling network flow and CFD models.

Note also that it is possible to operate on the basis of partially matched schemes. For example, the building model might comprise several zones, with only a subset addressed by CFD. To further enhance application flexibility, a flow network may be linked to one or several CFD domains, have nodes in common with some, but not all, of the other zones comprising the building model and have extra nodes to represent zones and/or plant components that are outwith the modelled building portion. Each part of such a model would then operate on the basis of best available information (e.g. a zone with no matched air flow model would utilise its user-specified infiltration/ventilation rates).

Where moisture flow is active, the solution of eqn (5.15) proceeds as follows. Because the energy equations can often be linearised, while the moisture equations typically cannot, the two equation systems are processed separately but under global iteration control to handle the coupling effects. This allows each equation-set to be integrated at different frequencies depending on the characteristics of the system they represent.

For the energy equations, the matrix partitioning technique of chapter 4 may be employed. This allows variable time-stepping with iteration incorporated for non-linear cases.

Because of their highly non-linear nature, the moisture flow equations are solved by a Gauss-Seidel method, with linear under-relaxation employed to prevent convergence instabilities in the case of strong non-linearity or where discontinuities occur in the moisture transfer rate at the maximum relative humidity due to condensation. A false time step relaxation factor may also be used. This acts to magnify the vapour storage term at the future time-row and so lessen the difference between the present and future values of the dependent variable. Because some of the terms within the moisture equations are dependent on temperature, the moisture solution is usually constrained to proceed at a time-step equal to or larger than that imposed on the energy equations.

The global iteration control is invoked whenever the liquid mass variations exceed some specified limit. When this occurs, the energy matrix equation is resolved on the basis of the

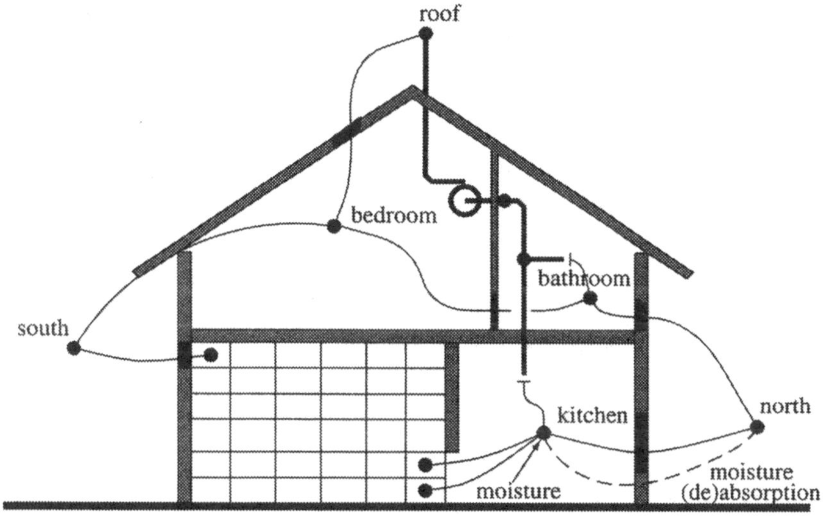

Figure 5.11: Coupling plant and flow models.

recently computed moisture-side variables but with no recalculation of the energy-side parameters. For highly coupled cases, both equation systems are solved at matched and small time-steps.

As illustrated in figure 5.11, a further model can now be added to represent HVAC equipment. The form of this additional model is the subject of the next chapter.

5.5 References and further reading

AIVC 2001 Air Infiltration and Ventilation centre http://www.aivc.org/

Allard F, Bienfait D, Haghighat F, Liebecq G, van der Maas J, Pelletret R, Vandaele L and Walker R 1992 Air Flow Through Large Openings in Buildings *IEA Annex 20 Technical Report* (Lausanne: LESO-PB, EPFL)

Allard F (Ed) 1998 *Natural Ventilation in Buildings; A Design Handbook* (London: James and James)‘

Amano R S 1984 Development of a Turbulence Near-Wall Model and its Application to Separated and Reattached Flows *Numerical Heat Transfer* **7**(1) 59-75

Andersen K T 1996 Inlet and Outlet Coefficients—A Theoretical Analysis *Proc. ROOMVENT '96* 379-90

André P, Kummert M and Nicolas J 1998 Coupling Thermal Simulation and Airflow Calculation for a Better Evaluation of Natural Ventilation Strategies *Proc. Systems Simulation Conf.* (University of Liège)

ASHRAE 1981 *Handbook of Fundamentals* (ASHRAE)

Axley J W 2001 Surface-drag Flow Relations for Zonal Modeling *Building and Environment* **36** 843-50

Baker A J, Williams P T and Kelso R M 1994 Development of a Robust Finite Element CFD Procedure for Predicting Indoor Room Air Motion *Building and Environment* **26** 261-73

Bartak M, Cermak M, Clarke J A, Denev J, Drkal F, Lain M, Macdonald I, Majer M and Stankov P 2001 *Proc. Building Simulation '01* (Rio de Janeiro)

Beausoleil-Morrison I 2000 The Adaptive Coupling of Heat and Air Flow Modelling within Dynamic Whole-Building Simulation *PhD Thesis* (Glasgow: University of Strathclyde)

Beausoleil-Morrison I 2001 The Adaptive Coupling of Computational Fluid Dynamics with Whole-Building Simulation *Proc. Building Simulation 2001* (Rio de Janeiro)

Beausoleil-Morrison I and Clarke J 1998 The Implications of Using the Standard $k-\varepsilon$ Turbulence Model to Simulate Room Air Flows which are not Fully Turbulent *Proc. ROOMVENT '98* 99-106 (Stockholm)

Brown 1962a Natural Convection Through Rectangular Openings in Partitions—1: Vertical Partitions *J. Heat and Mass Transf.* **5**

— 1962b Natural Convection Through Rectangular Openings—2: Horizontal Partitions *J. Heat and Mass Transf.* **5** 869-81

Chen Q and Srebric J 1999 Simplified Diffuser Boundary Conditions for Numerical Room Airflow Models *Report for ASHRAE Project 1009-TRP*

Chen Q and Xu W 1998 A Zero-Equation Turbulence Model for Indoor Airflow Simulation *Energy and Buildings* **28**(2) 137-44

Chen Q 1995 Comparison of Different $k-\varepsilon$ Models for Indoor Air Flow Computations *Numerical Heat Transfer* **28** Part B 353-69

Chien K-Y 1982 Predictions of Channel and Boundary-Layer Flows with a Low-Reynolds-Number Turbulence Model *AIAA Journal* **20**(1) 33-8

Clarke J A 1985 *Energy simulation in building design (1st Edn)* (Bristol: Adam Hilger)

Clarke J A, Dempster W M and Negraõ C 1995 The Implementation of a Computational Fluid Dynamics Algorithm within the ESP-r System *Proc. Building Simulation '95* 166-75 (Madison)

Clarke J A and Hensen J L M 1991 An Approach to the Simulation of Coupled Heat and Mass Flows in Buildings *Indoor Air* **3** 283-96

Cockroft J P 1979 Heat Transfer and Air Flow in Buildings *PhD Thesis* (Glasgow: University of Glasgow)

Conte S D and de Boor C 1972 *Elementary Numerical Analysis: an Algorithmic Approach* (New York: McGraw-Hill)

Dales R E, Burnett R and Zwanenburg H 1991a Adverse Health Effects Among Adults Exposed to House Dampness and Moulds *American Reviews of Respiratory Disease* **143** 505-9

Dales R E, Zwanenburg H, Burnett R and Flannigan C A 1991b Respiratory Health Effects of Home Dampness and Moulds Among Canadian Children *American Journal of Epidemiology* **134** 196-206

Davidson L 1990 Calculation of the Turbulent Buoyancy-Driven Flow in a Rectangular Cavity Using an Efficient Solver and Two Different Low Reynolds Number $k-\varepsilon$ Turbulence Models *Numerical Heat Transfer* **18** 129-47

Davidson L and Neilsen P V 1996 Large Eddy Simulations of the Flow in a Three-Dimensional Ventilated Room *Proc. ROOMVENT '96* **2** 161-8 (Yokohama)

De Wit S 2001 Uncertainty in Predictions of Thermal Comfort in Buildings *PhD Thesis* (Delft: Technische Universiteit)

Deardorff J W 1970 A Numerical Study of Three-Dimensional Turbulent Channel Flow at Large Reynolds Numbers *J. Fluid Mech.* **42** 453-80

Denev J A 1995 Boundary Conditions Related to Near-Inlet Regions and Furniture in Ventilated Rooms *Proc. Application of Mathematics in Engineering and Business* 243-8 (Sofia: Institute of Applied Mathematics and Informatics, Technical University)

Denev J A and Stankov P 2000a Indoor Climate Assessment Using Computer Simulation of Air Flow in a Ventilated Room *Journal of Theoretical and Applied Mechanics* **31**(1)

Denev J A and Stankov P 2000b Non-Orthogonal Grid Generation Techniques Applied to Room Airflow Modeling *Proc. 26th Summer School on Applications of Mathematics in Engineering and Economics (Sozopol)*

Dick J B 1950 The Fundamentals of Natural Ventilation of Houses *J. IHVE* **18** 123-34

Feustel H-E and Raynor-Hoosen A (Editors) 1990 Fundamentals of the Multizone Air Flow Model COMIS *Technical note 29 (Coventry: Air Infiltration and Ventilation Centre)*

Fürbringer J-M, Roulet C-A and Borchiellini R 1996 Evaluation of COMIS *Final Report for IEA Annex 21, Task 12* (Lausanne: Swiss Federal Institute of Technology)

Galbraith G H and McLean R C 1993 The Determination of Vapour and Liquid Transport Coefficients as Input to Combined Heat and Mass Transport Models *Proc. Building Simulation '93* (Adelaide)

Grosso M, Marino D and Parisi E 1992 Wind Pressure Distribution Calculation Program for Multizone Airflow Models *Proc. Building Simulation '95* (Madison)

Hanjalic K and Vasic S 1993 Computation of Turbulent Natural Convection in Rectangular Enclosures with an Algebraic Flux Model *Int. J. Heat Mass Transfer* **36**(14) 3603-24

Hansen K K 1986 Sorption Isotherms: A Catalogue *Technical Report 162/86* (Copenhagen: Technical University of Denmark)

Hanson T, Smith F, Summers D and Wilson C B 1982 Computer Simulation of Wind Flow Around Buildings *Computer-Aided Design* **14**(1) 27-31

Häggkvist K, Svensson U and Taesler R 1989 Numerical Simulations of Pressure Fields Around Buildings *Building and Environment* **24**(1) 65-72

Harris-Bass J, Lawrence P and Kavarana B 1974 Adventitious Ventilation of Houses *Proc. British Gas Symp. on Ventilation of Housing* (London)

Heiselberg P, Svidt K and Nielsen P 2001 Characteristics of Airflow from Open Windows *Building and Environment* **36** 859-69

Henkes R A W M 1990 Natural Convection Boundary Layers *PhD Thesis* (Delft University of Technology)

Hens H and Sneave E (eds) 1991 Annex 14: Condensation and Energy *Final Report* (International Energy Agency)

Hensen J L M 1991 On the Thermal Interaction of Building Structure and Heating and Ventilating System *PhD Thesis* (Eindhoven: University of Eindhoven) ISBN 90-386-0081-X

Hensen J L M 1999 A Comparison of Coupled and De-coupled Solutions for Temperature and Air Flow in a Building *ASHRAE Trans.* **105**(2) 962-9

Hensen J L M, van der Maas J and Roos A 1993 Air and Heat Flow Through Large Vertical Openings *Proc Building Simulation '93* (Adelaide)

Inard C, Bouia H and Dalicieux P 1996 Prediction of Air Temperature Distribution in Buildings with a Zonal Model *Energy and Buildings* **24** 125-32

Ince N Z and B E Launder 1989 On the Computation of Buoyancy-Driven Turbulent Flows in Rectangular Enclosures *Int. J. Heat and Fluid Flow* **10**(2) 110-7

Incropera F P and Dewitt D P 1985 *Fundamentals of Heat and Mass Transfer* (New York: Wiley)

Jones P J and Whittle G E 1992 Computational Fluid Dynamics for Building Air Flow Prediction - Current Status and Capabilities *Building and Environment* **21**(3) 321-38

Kafetzopoulos M G and Suen K O 1995 Coupling of Thermal and Airflow Calculation Programs Offering Simultaneous Thermal and Airflow Analysis *Building Services Eng. Res. and Tech.* **16**(1) 33-6

Knoll B, Phaff J C and de Gids W F 1995 Pressure Simulation Program *Proc. 16th AIVC Conf.* (Palm Springs)

Kreyszig E 1979 Advanced Engineering Mathematics (4th Edn) (New York: Wiley)

Lam C K G and Bremhorst K A 1981 Modified Form of the $k-\varepsilon$ Model for Predicting Wall Turbulence *J. of Fluids Eng.* **103** 456-60

Launder B E 1989a Second-Moment Closure: Present and Future? *Int. J. Heat and Fluid Flow* **10** 282-300

Launder B E 1989b Second-Moment Closure and its Use in Modeling Turbulent Industrial Flows *Int. J. Num. Methods in Fluids* **9** 963

Launder B E 1992 On the Modeling of Turbulent Industrial Flows *Proc. First European Computational Fluid Dynamics Conf.* Brussels 91-102

Launder B E and Sharma B I 1974 Application of the Energy-Dissipation Model of Turbulence to the Calculation of Flow Near a Spinning Disc Letters in *Heat and Mass Transfer* **1** 131-8

Launder B E and Spalding D B 1972 *Mathematical Models of Turbulence* (New York: Academic)

Launder B E and Spalding D B 1974 The Numerical Computation of Turbulent Flows *Computer Methods in Applied Mechanics and Engineering* **3** 269-89

Liddament M W 1986 Air Infiltration Calculation Techniques - An Applications Guide *Report AIC-AG-1-86* (Bracknell: Air Infiltration and Ventilation Centre)

Malinowski H K 1971 Wind Effect on the Air Movement Inside Buildings *Proc. 3rd Conf. on Wind Effects on Buildings and Structures (Tokyo) 125-34*

McLean R C and Galbraith G H 1996 Investigation of Moisture Transmission in Building Materials Under Non-Isothermal Conditions *Final Report for Research Contract GR/K69124* (Swindon: Engineering and Physical Sciences Research Council)

Murakami S, Kato S and Kondo Y 1992 Numerical Prediction of Horizontal Nonisothermal 3-D Jet in Room Based on Algebraic Second-Moment Closure Model *ASHRAE Trans.* **98**(1) 951-61

Nagano Y and Hisida M 1987 Improved Form of the $k-\varepsilon$ Model for Wall Turbulent Shear Flows *J. of Fluids Eng.* **109** 156-60

Nakhi A E 1995 Adaptive Construction Modelling Within Whole Building Dynamic Simulation *PhD Thesis* (Glasgow: Energy Simulation Research Unit, University of Strathclyde)

Negraõ C O R 1995 Conflation of Computational Fluid Dynamics and Building Thermal Simulation *PhD Thesis* (Glasgow: University of Strathclyde)

Nielsen P V, Restivo A and Whitelaw J H 1978 The Velocity Characteristics of Ventilated Rooms *Journal of Fluids Engineering* **100**(3) 291-8

Nielsen P V 1989 Airflow Simulation Techniques Progress and Trends *Proc. 10th AIVC Conf.* 203-23

Nielsen P V 1994 Air Distribution in Rooms-Research and Design Methods *Proc. ROOMVENT '94* Krakow, Poland, V1, 15-35.

Patankar S V and Spalding D B 1972 *A Calculation Procedure for Heat, Mass and Momentum Transfer in Three-Dimensional Parabolic Flows Int. J. Heat Mass Transf.* **15** p.1787

Patankar S V 1980 *Numerical Heat Transfer and Fluid Flow* (New York: Hemisphere)

Patel V C, Rodi W and Scheuerer G 1985 Turbulence Models for Near-Wall and Low Reynolds Number Flows: A Review *AIAA Journal* **23** 1308-19

Pernot C E E and Hensen J L M 1990 Adviezen Inzake de Kwaliteit van het Binnenmilieu van de HeuvelGalerie te Eindhoven *Report TPD-FAGO-RPT-90-42* (University of Eindhoven)

Press W H, Flannery B P, Teukolsky S A and Vettlering W T 1986 *Numerical Recipes: the Art of Scientific Computing* (Cambridge University Press)

Chen Q 1988 Indoor Airflow, Air Quality and Energy Consumption of Buildings *PhD Thesis* (Technical University of Delft) ISBN 90-9002435-2

Rodi W 1980 Turbulence Models and their Applications in Hydraulics—A State of the Art Review (Delft: Int. Association for Hydraulic Research)

Sandberg M 1981 What is Ventilation Efficiency? *Building and Environment* **16** 123-35

Schaelin A, Dorer V, Van der Maas J and Moser A 1993 Improvement of Multizone Model Predictions by Detailed Flow Path Values from CFD Calculations *ASHRAE Trans* **99**(2) 709-20

Schild P 1997 Accurate Prediction of Indoor Climate in Glazed Enclosures *PhD Thesis* (Trondheim: Norwegian University of Science and Technology)

Scottish Homes 1991 *Scottish Housing Condition Survey* (Edinburgh: Scottish Homes)

Simiu E and Scanlan R H 1986 *Wind effects on Structures: An Introduction to Wind Engineering* (Wiley)

Srebric J, Chen Q and Glicksman L R 1999 Validation of a Zero-Equation Turbulence Model for Complex Indoor Airflow Simulation *ASHRAE Trans.* **105**(2)

Stankov P and Denev J 1996 Turbulence Modelling of Low Reynolds Number Effects in 3D Ventilated Rooms *Proc. HERMIS'96 695-702 (Athens) ISBN 960-85176-5-6*

Tamura G T and Wilson A G 1967 Pressure Difference Caused by Chimney Effect in Three High Rise Buildings *ASHRAE Trans.* **73**

Thompson J F, Warsi Z U A and Mastin C W 1985 *Numerical Grid Generation - Foundations and Applications* (Elsevier)

Van Doormal J P and Raithby G D 1984 Enhancements of the SIMPLE Method for Predicting Incompressible Fluid Flows *Numerical Heat Transfer* **7** 147-63

Versteeg H K and Malalasekera W 1995 An introduction to Computational Fluid Dynamics: The Finite Volume Method (Harlow, England: Longman)

Walton G N 1982 Airflow and Multiroom Thermal Analysis *ASHRAE Trans.* **88**(2)

Walton G N 1989 Airflow Network Models for Element-Based Building Airflow Modelling *ASHRAE Trans.* **95**(2) 613-20

Wilcox D C 1993 *Turbulence Modeling for CFD* (California: DCW Industries)

Yuan X, Moser A and Suter P 1993 Wall Functions for Numerical Simulation of Turbulent Natural Convection Along Vertical Plates *Int. J. Heat Mass Transfer* **36**(18) 4477-85

6

HVAC, renewable energy conversion and control systems

In practice, HVAC and related systems are rarely simulated. Instead, components are sized using traditional (mostly steady-state) procedures that employ the heating, cooling and ventilating loads determined from the simulation of the building when ideal plant operation is assumed. There are two main reasons for this situation. First, the number of component types and possible arrangements is large requiring the development of a vast array of models. Second, manufacturers are more likely to produce data on a component's performance than to describe its geometry and material/fluid properties to the level of detail required by simulation. For these reasons, systems simulation has mostly been employed as a research tool. However, given the significance of the dynamic interactions between a building and its plant, there can be little doubt that practitioners would benefit from a performance assessment tool that preserved the dynamic characteristics of both domains.

The ASHRAE Task Group on Energy Requirements for Heating and Cooling of Buildings (Stoecker 1995) defined systems simulation as "predicting the operating quantities within a system (pressures, temperatures, energy- and fluid-flow rates) at the condition where all energy and material balances, all equations of state of working substances, and all performance characteristics of individual components are satisfied".

This chapter demonstrates the formulation of such a dynamic model of a building's environmental control systems, including components for renewable energy conversion and electrical power distribution. The aim is to ensure that these models are compatible with the previously derived models for construction related processes and fluid flow when represented at different levels of resolution. This is achieved by applying the modelling approach of chapters 3 and 4 to HVAC and renewable energy components to establish equations that represent the intra- and inter-component energy and multi-phase mass flows.

To demonstrate the process, models are established for four representative system types: a packaged air handler providing heating, cooling and (de)humidification; an active solar system comprising a solar collector and thermal store; a wet central heating system comprising a boiler, pump, radiators and a hot water cylinder; and a renewable energy conversion system comprising facade-integrated photovoltaic components and ducted wind turbines co-operating with the public electricity supply to service distributed loads. In each case, the integration of the conservation equations with the previously derived building and fluid flow models is described.

Finally, alternative approaches to the modelling of control are described, as required at an early design stage where issues of form and fabric are being addressed and, later, when control itself is the main issue.

6.1 Approaches to systems simulation

Essentially, there are two approaches to explicit systems simulation—sequential and simultaneous—and each have been widely employed.

In the sequential approach, plant components are replaced by an equivalent input/output relationship so that when connected to form a system, the calculated output from one component becomes the input to the next. An iterative solution method is then used to achieve solution convergence throughout the network. The algorithms that represent the individual components may be simplified (e.g. based on manufacturers' data) or detailed (e.g. based on a fundamental mathematical model). The technique has three principal advantages: different modelling methods can be applied to different plant components allowing simplified and fundamental models to coexist; a rapid prototyping approach is fostered because component models can initially be rudimentary; and the discreteness of the approach prevents new models from negatively impacting on the overall solution. Difficulties will arise, however, when control dynamics are included or where a component model requires downstream information that is, at the time of need, undetermined (e.g. where recirculation loops are present). Successful implementations of the sequential approach are described elsewhere (Quick 1982, Hanby 1985, Klein 1990, Gough 1986).

In the simultaneous approach, each plant component is represented by discrete finite volumes (FV), with each one assigned a set of conservation equations depending on the number of phases present and the properties to be conserved (energy, mass, electrical power etc). The matrix equation to emerge for the network of plant components may then be combined with the building and fluid flow matrix equations, control statements superimposed, and the entire equation system solved using appropriate numerical techniques. To reduce the complexity of the overall plant network, it is possible to restrict the number of component FVs, typically to 1 or 2, and then employ an independent algorithm to represent the component's internal processes. The network matrix equation represents the linked resistance/capacitance network, with the component algorithms used to impose behaviour. In either case—low or high resolution discretisation—the problems associated with the sequential approach are overcome. Successful implementations of the simultaneous approach are described elsewhere (Benton 1982, McLean 1982, Kelly *et al* 1984, Clarke and Mac Randal 1984, Tang 1984, Hanby and Clarke 1987, Hensen 1991, Bonin *et al* 1991, Aasem 1993, Buhl *et al* 1993, Chow 1995a). The combinatorial possibilities for systems representation are effectively without limit.

Recall figure 4.5 of §4.1.3. This shows a building equation-set comprising 4 zone matrices and plant interaction terms, q_p, as detailed in figure 4.4. These terms are the heating or cooling power requirements to maintain a given temperature at some specified location(s) (i.e. the nodal temperature(s) being controlled during the simulation). Within such a model, only the most rudimentary account is taken of plant response times and operational inefficiencies (via the governing control law as elaborated in §6.4). When these q_p terms are replaced by an explicit plant model, this limitation is removed.

6.2 HVAC systems

This section applies the simultaneous approach to three example systems and demonstrates how elements of the sequential approach can be used to reduce the overall complexity of the model. While the approach is demonstrated in relation to example systems, it is universally applicable regardless of system type, complexity and scale.

6.2.1 Air conditioning

The function of an air conditioning (AC) system is to deliver appropriately clean air at a given temperature and moisture content as required to offset the sensible and latent loads imposed on the conditioned space. Specific systems may be associated with one of three general types: all air, air and water and packaged. Complete descriptions of these types and their sub-categories are given elsewhere (Jones 1985, ASHRAE 1992).

Irrespective of its categorisation, an AC system can be built from a limited number of components: duct, mixing box, fan, heating coil, cooling coil, boiler, pump, pipe, chiller, heat pump, supply diffuser, damper and n-way diverging/converging junction. Mathematical models of these components are required at different levels of detail in order to support the range of possible design tasks—from a rapid assessment of overall performance, to an analysis of the impact of extended surface geometry on coil efficiency.

The simulation of an AC system is complicated by the fact that the working fluid comprises two phases, dry air and water vapour. Consider the rudimentary FV scheme applied to a packaged air handling unit as shown in figure 6.1. Outside air at temperature θ_o, humidity ratio g_o and enthalpy h_o is mixed with zone return air at temperature θ_r, humidity ratio g_r and enthalpy h_r, and passed to a chilled water cooler, a humidifier and a re-heater to achieve the required supply condition to offset the zone's sensible and latent loads. Using the technique of chapter 3, an energy balance may be formulated at some arbitrary time-row, ξ:

For component 1:

$$m_o h_o + m_r h_r - m_1 h_1 + q_{e1} = \left.\frac{d(\bar{\rho}_1 V_1 h_1)}{dt}\right|_{t=\xi} \tag{6.1}$$

For component 2:

$$m_1 h_1 - m_2 h_2 - m_c h_c + q_{e2} - q_{x2} = \left.\frac{d(\bar{\rho}_2 V_2 h_2)}{dt}\right|_{t=\xi} \tag{6.2}$$

For component 3:

$$m_2 h_2 + m_h h_h - m_3 h_3 + q_{e3} = \left.\frac{d(\bar{\rho}_3 V_3 h_3)}{dt}\right|_{t=\xi} \tag{6.3}$$

For component 4:

$$m_3 h_3 - m_4 h_4 + q_{e4} + q_{x4} = \left.\frac{d(\bar{\rho}_4 V_4 h_4)}{dt}\right|_{t=\xi} \tag{6.4}$$

For component 5:

$$m_4 h_4 - m_5 h_5 + q_{e5} = \left.\frac{d(\bar{\rho}_5 V_5 h_5)}{dt}\right|_{t=\xi} \tag{6.5}$$

where m is the mass flow rate of the air/vapour mixture (kg s^{-1}), h the mixture specific enthalpy (J kg^{-1}), q_{ei} the ith component's heat exchange with the surroundings (W), q_{x2} the cooling coil total heat transfer (W), q_{x4} the re-heater coil total heat transfer (W), $\bar{\rho}_i$ the volume weighted density of component i (kg m^{-3}), V_i the total volume of component i (m^3). The subscripts o and r relate to ambient and return air states respectively, c relates to the cooler moisture extract, and h relates to the humidifier moisture addition.

Since each component is represented by a single node, its thermal inertia is a function of the

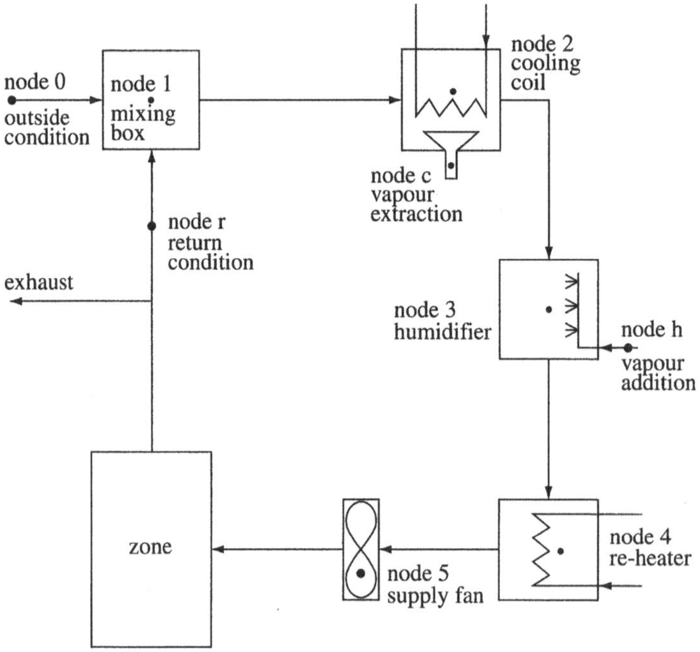

Figure 6.1: A simple model of a packaged air handling unit.

average thermodynamic state as represented by its single density value. This requires the use of an average density:

$$\bar{\rho}_i = \sum_{j=1}^{N}(\rho_j V_j) / \sum_{j=1}^{N}(V_j)$$

where N is the number of distinct intra-component regions. (A more refined approach will result from the introduction of a multi-FV representation whereby the thermal inertia of these intra-component regions is explicitly represented. Such a refinement is demonstrated later in this section.)

The component mass balances, for the dry air and water vapour separately, may be formulated as follows.

For component 1:

$$m_{o(d)} + m_{r(d)} - m_{1(d)} = 0 \Big|_{t=\xi} \tag{6.6}$$

$$m_{o(d)} g_o + m_{r(d)} g_r - m_{1(d)} g_1 = 0 \Big|_{t=\xi} \tag{6.7}$$

For component 2:

$$m_{1(d)} - m_{2(d)} = 0 \Big|_{t=\xi} \tag{6.8}$$

$$m_{1(d)} g_1 - m_{2(d)} g_2 - m_c = \frac{d(\rho_L V_c)}{dt} \Big|_{t=\xi} \tag{6.9}$$

For component 3:

$$m_{2(d)} - m_{3(d)} = 0 \Big|_{t=\xi} \tag{6.10}$$

$$m_{2(d)}g_2 - m_{3(d)}g_3 + m_h = \left.\frac{d(\rho_L V_h)}{dt}\right|_{t=\xi} \tag{6.11}$$

For component 4:

$$m_{3(d)} - m_{4(d)} = 0\Big|_{t=\xi} \tag{6.12}$$

$$m_{3(d)}g_3 - m_{4(d)}g_4 = 0\Big|_{t=\xi} \tag{6.13}$$

For component 5:

$$m_{4(d)} - m_{5(d)} = 0\Big|_{t=\xi} \tag{6.14}$$

$$m_{4(d)}g_4 + m_{5(d)}g_5 = 0\Big|_{t=\xi} \tag{6.15}$$

where $m_{i(d)}$ is the mass flow rate of dry air (kg s^{-1}) associated with component i, g the humidity ratio (kg kg^{-1}), ρ_L the density of the water remaining in the cooler or humidifier (kg m^{-3}), V_c the volume of this water, V_h the humidifier residual water volume, m_c the cooler vapour extraction rate (kg s^{-1}) and m_h the humidifier vapour addition rate (kg s^{-1}).

As in chapter 3, the energy conservation equations are obtained by performing an equal weighting of the explicit and implicit forms of eqns (6.1) through (6.5):

For component 1:

$$[2\bar{\rho}_1(t+\delta t)V_1 + \delta t\ m_1(t+\delta t)]h_1(t+\delta t) - \delta t\ q_{e1}(t+\delta t)$$
$$= [2\bar{\rho}_1(t)V_1 - \delta t\ m_1(t)]h_1(t) + \delta t\ m_o(t+\delta t)h_o(t+\delta t)$$
$$+ \delta t\ m_r(t+\delta t)h_r(t+\delta t) + \delta t\ m_o(t)h_o(t) + \delta t\ m_r(t)h_r(t) + \delta t\ q_{e1}(t).$$

In the absence of a building model, the $q_e(t+\delta t)$, $h_o(t+\delta t)$ and $h_r(t+\delta t)$ terms are removed to the equation right-hand side since they relate to system boundary conditions. Where a building model is active, the $h_r(t+\delta t)$ term will be a building-side state variable, while the $q_{ei}(t+\delta t)$ term can be replaced by introducing an exchange resistance between the component and one or more building FV.

With reference to figure 6.2, which shows the overall system matrix equation for the system of figure 6.1, the equation for component 1 becomes

$$a_{11}h_1(t+\delta t) = b_{11}h_1(t) + c_1 \tag{6.16}$$

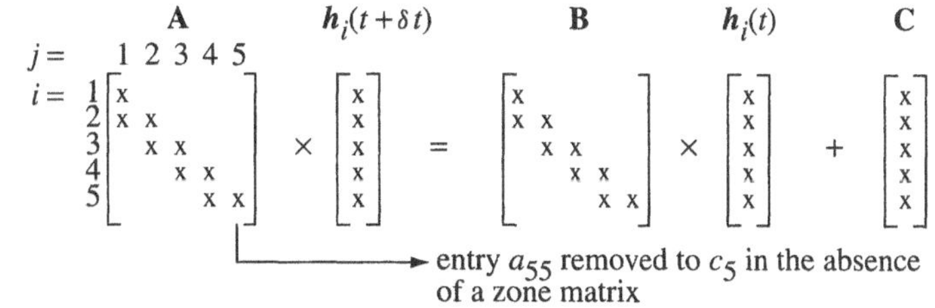

Figure 6.2: AC system energy balance matrix equation, $\mathbf{Ah}(t+\delta t) = \mathbf{Bh}(t) + \mathbf{C}$.

where the subscripts of the a and b coefficients refer to the ith row and jth column position:

$$a_{11} = 2\bar{\rho}_1(t+\delta t)V_1 + \delta t\ m_1(t+\delta t) \qquad b_{11} = 2\bar{\rho}_1(t)V_1 - \delta t\ m_1(t)$$
$$c_1 = \delta t\ m_o(t+\delta t)h_o(t+\delta t) + \delta t\ m_r(t+\delta t)h_r(t+\delta t) + \delta t\ m_o(t)h_o(t)$$
$$+ \delta t\ m_r(t)h_r(t) + \delta t[q_{e1}(t+\delta t) + q_{e1}(t)]\ .$$

For component 2:

$$[2\bar{\rho}_2(t+\delta t)V_2 + \delta t\ m_2(t+\delta t)]h_2(t+\delta t) - \delta t\ m_1(t+\delta t)h_1(t+\delta t) + \delta t\ m_c(t+\delta t)h_c(t+\delta t) - \delta t\ q_{e2}(t+\delta t) + \delta t\ q_{x2}(t+\delta t) = [2\bar{\rho}_2(t)V_2 - \delta t\ m_2(t)]h_2(t) + \delta t\ m_1(t)h_1(t) - \delta t\ m_c(t)h_c(t) + \delta t\ q_{e2}(t) - \delta t\ q_{x2}(t)\ .$$

Because the component representation is rudimentary, this model will require an algorithm for the estimation of the coil total heat transfer and condensate exit condition. The component 2 equation therefore becomes

$$a_{21}h_1(t+\delta t) + a_{22}h_2(t+\delta t) = b_{21}h_1(t) + b_{22}h_2(t) + c_2 \tag{6.17}$$

where

$$a_{21} = -\ \delta t\ m_1(t+\delta t) \qquad b_{21} = \delta t\ m_1(t)$$
$$a_{22} = 2\bar{\rho}_2(t+\delta t)V_2 + \delta t\ m_2(t+\delta t) \qquad b_{22} = 2\bar{\rho}_2(t)V_2 - \delta t\ m_2(t)$$
$$c_2 = -\ \delta t\ m_c(t+\delta t)h_c(t+\delta t) - \delta t\ m_c(t)h_c(t) + \delta t[q_{e2}(t+\delta t) + q_{e2}(t) - q_{x2}(t+\delta t) - q_{x2}(t)]\ .$$

For component 3:

$$[2\bar{\rho}_3(t+\delta t)V_3 + \delta t\ m_3(t+\delta t)]h_3(t+\delta t) - \delta t\ m_2(t+\delta t)h_2(t+\delta t) - \delta t\ m_h(t+\delta t)h_h(t+\delta t) - \delta t\ q_{e3}(t+\delta t) = [2\bar{\rho}_3(t)V_3 - \delta t\ m_3(t)]h_3(t) + \delta t\ m_2(t)h_2(t) + \delta t\ m_h(t)h_h(t) + \delta t\ q_{e3}(t)\ .$$

Again, in the absence of a multi-FV humidifier model, the vapour supply term, $m_h h_h(t+\delta t)$, must be independently assessed and so this equation becomes

$$a_{32}h_2(t+\delta t) + a_{33}h_3(t+\delta t) = b_{32}h_2(t) + b_{33}h_3(t) + c_3 \tag{6.18}$$

where

$$a_{32} = -\ \delta t\ m_2(t+\delta t) \qquad b_{32} = \delta t\ m_2(t)$$
$$a_{33} = 2\bar{\rho}_3(t+\delta t)V_3 + \delta t\ m_3(t+\delta t) \qquad b_{33} = 2\bar{\rho}_3(t)V_3 - \delta t\ m_3(t)$$
$$c_3 = -\ \delta t\ m_h(t+\delta t)h_h(t+\delta t) - \delta t\ m_h(t)h_h(t) + \delta t[q_{e3}(t+\delta t) + q_{e3}(t)]\ .$$

For component 4:

$$[2\bar{\rho}_4(t+\delta t)V_4 + \delta t\ m_4(t+\delta t)]h_4(t+\delta t) - \delta t\ m_3(t+\delta t)h_3(t+\delta t) - \delta t\ q_{e4}(t+\delta t) - \delta t\ q_{x4}(t+\delta t) = [2\bar{\rho}_4(t)V_4 - \delta t\ m_4(t)]h_4(t) + \delta t\ m_3(t)h_3(t) + \delta t\ q_{e4}(t) + \delta t\ q_{x4}(t)$$

which gives

$$a_{43}h_3(t+\delta t) + a_{44}h_4(t+\delta t) = b_{43}h_3(t) + b_{44}h_4(t) + c_4 \tag{6.19}$$

where

$$a_{43} = -\ \delta t\ m_3(t+\delta t) \qquad b_{43} = \delta t\ m_3(t)$$
$$a_{44} = 2\bar{\rho}_4(t+\delta t)V_4 + \delta t\ m_4(t+\delta t) \qquad b_{44} = 2\bar{\rho}_4(t)V_4 - \delta t\ m_4(t)$$
$$c_4 = \delta t[q_{e4}(t+\delta t) + q_{e4}(t) + q_{x4}(t+\delta t) + q_{x4}(t)]\ .$$

For component 5:

$$[2\bar{\rho}_5(t+\delta t)V_5 + \delta t\ m_5(t+\delta t)]h_5(t+\delta t) - \delta t\ m_4(t+\delta t)h_4(t+\delta t) - \delta t\ q_{e5}(t+\delta t) = [2\bar{\rho}_5(t)V_5 - \delta t\ m_5(t)]h_5(t) + \delta t\ m_4(t)h_4(t) + \delta t\ q_{e5}(t)$$

and this becomes

$$a_{54}h_4(t+\delta t) + a_{55}h_5(t+\delta t) = b_{54}h_4(t) + b_{55}h_5(t) + c_5 \quad (6.20)$$

where

$a_{54} = -\,\delta t\; m_4(t+\delta t)$ $\quad$ $b_{54} = \delta t\; m_4(t)$

$a_{55} = 2\bar{\rho}_5(t+\delta t)V_5 + \delta t\; m_5(t+\delta t)$ $\quad$ $b_{55} = 2\bar{\rho}_5(t)V_5 - \delta t\; m_5(t)$

$c_5 = \delta t[q_{e5}(t+\delta t) + q_{e5}(t)]$.

The plant matrix equation of figure 6.2 will normally be combined with its building counterpart (e.g. as represented by figure 4.1). For the present purpose, the two matrix equations are decoupled so that the required zone supply condition is assumed known and the component 5 equation may be re-written as

$$a_{54}h_4(t+\delta t) = b_{54}h_4(t) + b_{55}h_5(t) + c_5'$$

where

$$c_5' = c_5 - [2\bar{\rho}_5(t+\delta t)V_5 + \delta t\; m_5(t+\delta t)]h_5(t+\delta t) .$$

For matrix equations of the form of figure 6.2, but of arbitrary complexity, two solution possibilities exist as follows.

Terms such as q_{x2} and q_{x4}, which were removed to the **C** vector, may be assessed by independent algorithms when operating with latest state variable values within an iterative scheme.

Alternatively, such terms may be removed to the future time-row coefficients matrix, **A**, and replaced by an expanded multi-FV equation-set based on fundamental thermodynamic considerations.

The component mass balance equations are now formulated by taking an equal weighting of the explicit and implicit forms of eqns (6.6) through (6.15):

For component 1:

$$m_{1(d)}(t+\delta t) = m_{o(d)}(t+\delta t) + m_{r(d)}(t+\delta t) + m_{o(d)}(t) + m_{r(d)}(t) - m_{1(d)}(t)$$

and

$$m_{1(d)}(t+\delta t)g_1(t+\delta t) = m_{o(d)}(t+\delta t)g_o(t+\delta t) + m_{r(d)}(t+\delta t)g_r(t+\delta t) + m_{o(d)}(t)g_o(t) + m_{r(d)}(t)g_r(t) - m_{1(d)}(t)g_1(t)$$

which, with reference to figure 6.3, becomes

$$d_{11}m_{1(d)}(t+\delta t) = e_{11}m_{1(d)}(t) + f_1$$

and

$$d_{22}[m_{1(d)}(t+\delta t)g_1(t+\delta t)] = e_{22}[m_{1(d)}(t)g_1(t)] + f_2 .$$

For component 2:

$$m_{1(d)}(t+\delta t) - m_{2(d)}(t+\delta t) = -\,m_{1(d)}(t) + m_{2(d)}(t)$$

and

$$m_{1(d)}(t+\delta t)g_1(t+\delta t) - m_{2(d)}(t+\delta t)g_2(t+\delta t) - m_c(t+\delta t) - 2\rho_L(t+\delta t)V_c/\delta t$$

$$
\underbrace{\begin{bmatrix}
1 &&&&&&&&& \\
& 1 &&&&&&&& \\
1 && -1 &&&&&&& \\
& 1 && -1 &&&&&& \\
&& 1 && -1 &&&&& \\
&&& 1 && -1 &&&& \\
&&&& 1 && -1 &&& \\
&&&&& 1 && -1 && \\
&&&&&& 1 && -1 & \\
&&&&&&& 1 && -1
\end{bmatrix}}_{\mathbf{D},\; j = 1\ 2\ 3\ 4\ 5\ 6\ 7\ 8\ 9\ 10;\; i = 1 \ldots 10}
\times
\underbrace{\begin{bmatrix}
m_{1(d)} \\ m_{1(d)}g_1 \\ m_{2(d)} \\ m_{2(d)}g_2 \\ m_{3(d)} \\ m_{3(d)}g_3 \\ m_{4(d)} \\ m_{4(d)}g_4 \\ m_{5(d)} \\ m_{5(d)}g_5
\end{bmatrix}}_{\phi_i(t+\delta t)}
=
\underbrace{\begin{bmatrix}
-1 &&&&&&&&& \\
& -1 &&&&&&&& \\
-1 && 1 &&&&&&& \\
& -1 && 1 &&&&&& \\
&& -1 && 1 &&&&& \\
&&& -1 && 1 &&&& \\
&&&& -1 && 1 &&& \\
&&&&& -1 && 1 && \\
&&&&&& -1 && 1 & \\
&&&&&&& -1 && 1
\end{bmatrix}}_{\mathbf{E}}
\times
\underbrace{\begin{bmatrix}
m_{1(d)} \\ m_{1(d)}g_1 \\ m_{2(d)} \\ m_{2(d)}g_2 \\ m_{3(d)} \\ m_{3(d)}g_3 \\ m_{4(d)} \\ m_{4(d)}g_4 \\ m_{5(d)} \\ m_{5(d)}g_5
\end{bmatrix}}_{\phi_i(t)}
+
\underbrace{\begin{bmatrix}
X \\ X \\ 0 \\ X \\ 0 \\ X \\ 0 \\ 0 \\ 0 \\ 0
\end{bmatrix}}_{\mathbf{F}}
$$

Figure 6.3: AC system mass balance matrix equation, $\mathbf{D}\phi(t + \delta t) = \mathbf{E}\phi(t) + \mathbf{F}$.

$$= - m_{1(d)}g_1(t) + m_{2(d)}(t)g_2(t) + m_c(t) - 2\rho_L(t)V_c/\delta t\,.$$

or, relative to the figure 6.3 matrix equation:

$$d_{31}m_{1(d)}(t + \delta t) + d_{33}m_{2(d)}(t + \delta t) = e_{31}m_{1(d)}(t) + e_{33}m_{2(d)}(t)$$

and

$$d_{42}[m_{1(d)}(t + \delta t)g_1(t + \delta t)] + d_{44}[m_{2(d)}(t + \delta t)g_2(t + \delta t)]$$

$$= e_{42}[m_{1(d)}(t)g_1(t)] + e_{44}[m_{2(d)}(t)g_2(t)] + f_4$$

assuming that some independent algorithm exists to determine the vapour extraction rate from the air/vapour mixture passing through the cooling coil, and that f_4 is given by

$$f_4 = m_c(t) + m_c(t + \delta t) + 2V_c[\rho_L(t + \delta t) - \rho_L(t)]/\delta t\,.$$

For component 3: $\quad m_{2(d)}(t + \delta t) - m_{3(d)}(t + \delta t) = - m_{2(d)}(t) + m_{3(d)}(t)$

and

$$m_{2(d)}(t + \delta t)g_2(t + \delta t) - m_{3(d)}(t + \delta t)g_3(t + \delta t) + m_h(t + \delta t)$$
$$- 2\rho_L(t + \delta t)V_h/\delta t = - m_{2(d)}g_2(t) + m_{3(d)}(t)g_3(t)$$
$$- m_h(t) - 2\rho_L(t)V_h/\delta t$$

or, with reference to figure 6.3:

$$d_{53}m_{2(d)}(t + \delta t) + d_{55}m_{3(d)}(t + \delta t) = e_{53}m_{2(d)}(t) + e_{55}m_{3(d)}(t)$$

and

$$d_{64}[m_{2(d)}(t + \delta t)g_2(t + \delta t)] + d_{66}[m_{3(d)}(t + \delta t)g_3(t + \delta t)]$$

$$= e_{64}[m_{2(d)}(t)g_2(t)] + e_{66}[m_{3(d)}(t)g_3(t)] + f_6$$

where

$$f_6 = - m_h(t) - m_h(t + \delta t) + 2V_h[\rho_L(t + \delta t) - \rho_L(t)]/\delta t\,.$$

For component 4: $\quad m_{3(d)}(t + \delta t) - m_{4(d)}(t + \delta t) = - m_{3(d)}(t) + m_{4(d)}(t)$

and

$$m_{3(d)}(t + \delta t)g_3(t + \delta t) - m_{4(d)}(t + \delta t)g_4(t + \delta t) \\ = - m_{3(d)}g_3(t) + m_{4(d)}(t)g_4(t)$$

or, with reference to figure 6.3:

$$d_{75}m_{3(d)}(t + \delta t) + d_{77}m_{4(d)}(t + \delta t) = e_{75}m_{3(d)}(t) + e_{77}m_{4(d)}(t)$$

and

$$d_{86}[m_{3(d)}(t + \delta t)g_3(t + \delta t)] + d_{88}[m_{4(d)}(t + \delta t)g_4(t + \delta t)]$$

$$= e_{86}[m_{3(d)}(t)g_3(t)] + e_{88}[m_{4(d)}(t)g_4(t)] \; .$$

For component 5: $m_{4(d)}(t + \delta t) - m_{5(d)}(t + \delta t) = - m_{4(d)}(t) + m_{5(d)}(t)$

and

$$m_{4(d)}(t + \delta t)g_4(t + \delta t) - m_{5(d)}(t + \delta t)g_5(t + \delta t) \\ = - m_{4(d)}g_4(t) + m_{5(d)}(t)g_5(t) \; .$$

Again, with reference to figure 6.3:

$$d_{97}m_{4(d)}(t + \delta t) + d_{99}m_{5(d)}(t + \delta t) = e_{97}m_{4(d)}(t) + e_{99}m_{5(d)}(t)$$

and

$$d_{108}[m_{4(d)}(t + \delta t)g_4(t + \delta t)] + d_{1010}[m_{5(d)}(t + \delta t)g_5(t + \delta t)]$$

$$= e_{108}[m_{4(d)}(t)g_4(t)] + e_{1010}[m_{5(d)}(t)g_5(t)]$$

since $m_{5(d)}$ is (here) a known zone mass flow rate.

To handle the case where a component is connected to more than one other component, the distribution system must be explicitly represented. This introduces additional FVs to represent the branching points. It is then the job of the network flow model of chapter 5 to apportion the flows to each branch based on the prevailing pressure differences (see Hensen 1991 for further discussion on the conflation of plant component and network flow models). In the absence of a flow model, the branching point diversion ratios are required as inputs, in much the same way that recirculation ratios are specified for use in design calculations. For example, the component 1 diversion ratio for the present system might be given as $m_o/m_r = 0.25$. Aasem (1993) proposed the use of a fictitious, mass-less component for use in the representation of complex networks in cases where a network flow model is not present.

Where nodes located downstream from a control valve do not experience a flow rate adjustment until some time after valve operation due to a network transport delay, this can be modelled by delaying the introduction of the modified mass flow rate to the matrix coefficient entry until some later matrix formulation depending on the node location, fluid velocity and simulation time-step.

The matrix equations of figures 6.2 and 6.3 may now be solved, at any time-step, given the existence of component algorithms to establish the q and m terms as present within the **C** and **F** vectors, and as a function of any user-specified control constraints. One of several possible methods is to proceed as follows.

Step 1: at each time-step, establish

$$\mathbf{A}\mathbf{h}(t + \delta t) = \mathbf{B}\mathbf{h}(t) + \mathbf{C}$$

$$\mathbf{D}\phi(t + \delta t) = \mathbf{E}\phi + \mathbf{F}$$

and initialise $q_{ei}(t + \delta t)$, $q_{x2}(t + \delta t)$, $q_{x4}(t + \delta t)$, $m_c(t + \delta t)$ and $m_h(t + \delta t)$ to some suitable value (e.g the values established at the previous time step).

Step 2: assume no humidification or dehumidification and determine the network humidity ratios, $g_i'(t + \delta t)$, from

$$\phi(t + \delta t) = \mathbf{D}^{-1}[\mathbf{E}\phi(t) + \mathbf{F}] . \tag{6.21}$$

Set $\Delta g = (g_5' - g_5)$, where g_5 is the desired humidity ratio (kg kg^{-1}).

$$\text{If } \Delta g \begin{cases} < 0 & \text{go step 3} \\ > 0 & \text{go step 4} \\ = 0 & \text{go step 5 .} \end{cases}$$

Step 3: humidification required—determine $m_h(t + \delta t)$ to give required $g_5(t + \delta t)$ from iterative application of eqn (6.21). Then go to step 5.

Step 4: dehumidification required—determine $m_c(t + \delta t)$ to give required $g_5(t + \delta t)$ from iterative application of eqn (6.21). From cooler algorithm determine minimum $q_{x2}(t + \delta t)$ to give required $g_2(t + \delta t)$.

Step 5: estimate $q_{ei}(t + \delta t)$ from known component/boundary conditions. (Note that this step is not required where a building model is present to allow the explicit representation of the coupling between the component and its surroundings).

Step 6: with $q_{x4}(t + \delta t)$ remaining at its initial value and $q_{x2}(t + \delta t)$ set at the value determined at step 4 (or its initial value if step 4 was omitted), determine circuit enthalpies, $h_i'(t + \delta t)$, from

$$\mathbf{h}(t + \delta t) = \mathbf{A}^{-1}[\mathbf{B}\mathbf{h}(t) + \mathbf{C}] . \tag{6.22}$$

Set $\Delta h = (h_5' - h_5)$, where h_5 is the desired supply enthalpy (J kg^{-1}).

$$\text{If } \Delta h \begin{cases} < 0 & \text{go step 7} \\ > 0 & \text{go step 8} \\ = 0 & \text{go step 9 .} \end{cases}$$

Step 7: re-heat required—determine $q_{x4}(t + \delta t)$ from iterative application of eqn (6.22), then go to step 9.

Step 8: further cooling then humidification required—determine $q_{x2}(t + \delta t)$ from iterative application of eqn (6.22). From cooler algorithm assess new $m_c(t + \delta t)$ and determine new $m_h(t + \delta t)$ to give required $g_5(t + \delta t)$ from iterative application of eqn (6.21).

Step 9: the desired supply conditions are now achieved corresponding to a minimum cooler and re-heater energy input: $|q_{x2}(t + \delta t)| + q_{x4}(t + \delta t)$.

Since the duties and entering/ leaving air/vapour states are now known, any active component algorithms can be used to assess the internal operational requirements. Should the component be unable to perform as required then its duty can be set to its limit value and the effect on supply conditions established from eqns (6.21) and (6.22). In the presence of a building model, any deviation from the required supply condition will result in a departure from the desired environmental condition. If component operational constraints are imposed prior to simulation then the limit condition may be reached when the cooler algorithm is invoked at steps 4 or 8. In this case the supply conditions will not be met and the simulation will proceed with a correspondingly greater demand at subsequent time-steps until the building load diminishes.

The important point to note is that the above procedure is independent of component location since the matrix topology defines the network configuration.

The next two sub-sections describe the formulation of an internal component process model—firstly, in algorithmic form suitable for use as described in the foregoing solution procedure and, secondly, in a form that allows the removal of the $q_{x2}(t + \delta t)$ and $q_{x4}(t + \delta t)$ terms to the matrix equation left-hand side (future time-row).

6.2.1.1 Component process models: algorithmic

As indicated previously, it is possible to establish an algorithm to represent a component's internal operation, and use this in conjunction with the system matrix equation, which represents component inertia and the inter-component connections, to attain a solution. Many algorithmic formulations are possible (e.g. Myers *et al* 1967, James and Marshall 1973, Hanby and Clarke 1987, Yuill and Wray 1990) and one is elaborated here to illustrate the technique.

Consider the annotated cooling coil schematic of figure 6.4. The following procedure, based on the sensible heat ratio method, may be used to calculate cooling coil performance from known inlet conditions.

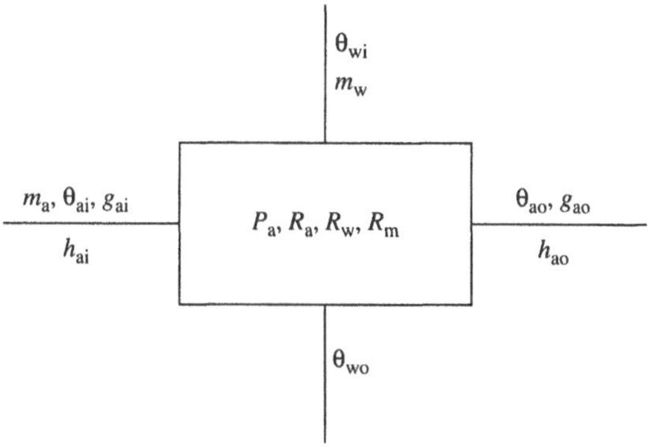

Figure 6.4: Quantities defining the state of a cooling coil.

Step 1: at the commencement of a time-step, the following quantities are known—inlet water temperature and mass flow rate, θ_{wi} and m_w; inlet air dry bulb temperature, mass flow rate, humidity ratio and enthalpy, θ_{ai}, m_a, g_{ai} and h_{ai}; air, water and metal thermal resistances, R_a, R_w and R_m; coil surface area, A; atmospheric pressure, P_a; and air and water specific heat capacities, C_{pa} and C_{pw}.

Step 2: calculate the coil bypass factor, β, from $\beta = \exp[-A/(C_{pa}m_aR_a)]$.

Step 3: set coil effectiveness, E, to the value calculated at previous time-step.

Step 4: guess the sensible heat ratio, SHR.

Step 5: calculate the coil overall thermal transmittance, U, the number of heat transfer units, NTU, and the capacity-rate ratio, CRR, from

$$\phi_1 = m_aC_{pa}/SHR \qquad \phi_2 = m_wC_{pw}$$
$$C_{min} = \min(\phi_1,\phi_2) \qquad C_{max} = \max(\phi_1,\phi_2)$$
$$CRR = C_{min}/C_{max} \qquad U = 1/[(R_aSHR) + R_m + R_w]$$
$$NTU = AU/C_{min}$$

Step 6: establish if guessed E and SHR correspond by applying

$$E = \frac{1 - e^{[-NTU(1 - CRR)]}}{1 - CRR\ e^{[-NTU(1 - CRR)]}}$$

for $CRR \rightarrow 1$, $E = NTU/(1 + NTU)$.

Step 7: if E and SHR do not correspond, iterate from step 4.

Step 8: evaluate coil heat transfer from

$$q = C_{min}E(\theta_{ai} - \theta_{wi})\ .$$

Step 9: using Q, calculate the outlet air enthalpy, h_{ao}.

Step 10: calculate the saturation enthalpy, h_s, at the coil surface temperature, from

$$h_s = (h_{ao} - \beta h_{ai})/(1 - \beta)\ .$$

Step 11: determine the coil surface temperature, θ_s, and the saturation humidity ratio, g_s, from the saturation enthalpy and atmospheric pressure.

Step 12: calculate the outlet air temperature, θ_{ao}, and humidity ratio, g_{ao}, from

$$\theta_{ao} = \beta(\theta_{ai} - \theta_s) + \theta_s \quad \text{and} \quad g_{ao} = \beta(g_{ai} - g_s) + g_s\ .$$

Step 13: calculate the corresponding sensible heat ratio, SHR', from

$$SHR' = (\theta_{ai} - \theta_{ao})C_{pa}/(h_{ai} - h_{ao})\ .$$

Step 14: compare SHR' to SHR and, if significantly different, iterate from step 3.

Step 15: eventually, perhaps after changing coil parameters to attain the desired coil performance, terminate algorithm and insert q from step 8 in system matrix equation to give final circuit enthalpies and humidity ratios.

Aasem (1993) introduced a special component type into the ESP-r system which serves as a numerical wrapper for algorithmic component models. This was done to enable the reuse of existing algorithms, such as those found within TRNSYS (Klein 1990), within the simultaneous approach.

6.2.1.2 Component process models: numerical

A more fundamental approach is to introduce additional FVs to explicitly represent the intra-component states and processes. Consider, for example, the coil model suggested by Holmes (1982):

$$C_w \frac{d\theta_1}{dt} = \frac{\theta_o - \theta_1}{R_1} - \frac{\theta_1 - \theta}{R_{mw}} \tag{6.23}$$

$$C_m \frac{d\theta}{dt} = \frac{\theta_1 - \theta}{R_{mw}} - \frac{\theta}{R_a + R_4} \tag{6.24}$$

$$\theta = \frac{\theta_2(R_a + R_4)}{R_4}$$

where $\theta_o = \theta_{wi} - \theta_{ai}$, $\theta_1 = \theta_{wo} - \theta_{ai}$, $\theta_2 = \theta_{ao} - \theta_{ai}$, $R_1 = 1/(m_w C_{pw})$, R_{mw} is the metal plus water film thermal resistance, $R_4 = 1/(m_a C_{pa})$, C_w the water thermal capacity (J °C^{-1}), C_m the metal thermal capacity and C_{pw}, C_{pa} the specific heat capacities of water and air respectively (J kg^{-1}°C^{-1}).

Eqns (6.23) and (6.24) define a two FV component model. As shown in figure 6.5, a finite difference approximation of these equations, using the technique of chapter 3, gives rise to a two equation extension to the system of equations shown in figure 6.2.

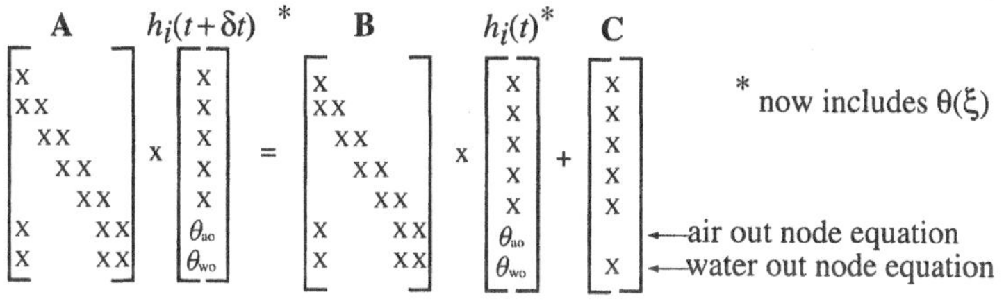

air equation

$$-\left(\frac{2C_m(R_a+R_4)}{\delta t R_4}+\frac{(R_a+R_4)}{R_{mw}R_4}\right)\theta_{ai}(t+\delta t)$$

$$+\left(\frac{2C_m(R_a+R_4)}{\delta t R_4}+\frac{(R_a+R_4)}{R_{mw}R_4}+\frac{1}{R_{mw}}\right)\theta_{ao}(t+\delta t)-\frac{1}{R_{mw}}\theta_{wo}(t+t)$$

$$=-\left(\frac{2C_m(R_a+R_4)}{\delta t R_4}+\frac{(R_a+R_4)}{R_{mw}R_4}\right)\theta_{ai}(t)$$

$$+\left(\frac{2C_m(R_a+R_4)}{\delta t R_4}+\frac{(R_a+R_4)}{R_{mw}R_4}-\frac{1}{R_{mw}}\right)\theta_{ao}(t)+\frac{1}{R_{mw}}\theta_{wo}(t)$$

water equation

$$\left(\frac{2C_w}{\delta t}-\frac{(R_a+R_4)}{R_{mw}^2}+\frac{1}{R_{mw}}\right)\theta_{ai}(t+\delta t)-\frac{(R_a+R_4)}{R_{mw}^2}\theta_{ao}(t+\delta t)$$

$$+\left(\frac{2C_w}{\delta t}+\frac{1}{R_1}+\frac{1}{R_{mw}}\right)\theta_{wo}(t+\delta t)$$

$$=-\left(\frac{2C_w}{\delta t}-\frac{1}{R_{mw}}+\frac{(R_a+R_4)}{R_{mw}^2}\right)\theta_{ai}(t)-\frac{(R_a+R_4)}{R_{mw}^2}\theta_{ao}(t)$$

$$+\left(\frac{2C_w}{\delta t}-\frac{1}{R_1}-\frac{1}{R_{mw}}\right)\theta_{wo}(t)+\frac{1}{R_1}\left[\theta_{wi}(t)+\theta_{wi}(t+\delta t)\right]$$

Figure 6.5: Addition of air- and water side equations.

Several models of this type have been developed for use in the simulation of air conditioning systems (e.g. Tang 1984, Aasem 1993, Chow 1995b). The effort required to develop such models, combined with the potentially large number of component variants, has given rise to the need for a model construction capability based on the synthesis of primitive parts. Such an approach is elaborated in the next section.

6.2.1.3 Modelling by 'primitive parts'

While the component-based approach is widely used in contemporary programs, it has limited applicability because, typically, component models are established to serve a specific purpose or are based on range-restricted empirical data.

In practice, the types of AC systems vary dramatically, requiring that flexible component models be available to facilitate system representation at various levels of abstraction. Several programs have addressed this issue—e.g. SPARK (Buhl *et al* 1993), ZOOM (Bonin *et al* 1991), IDA (Sahlin 1993) and CLIM2000 (Bonneau *et al* 1993). In an attempt to eliminate the need for pre-formed component models, Chow (1995b, *et al* 1997, 1998) proposed the concept of representation by primitive parts (PP). Each PP represents a fundamental heat or mass transfer process and may be combined with other PPs in order to synthesise a model on the basis of extended component descriptions—in the same way that a building model might be synthesised on the basis of geometrical, constructional and operational inputs using the PPs of chapter 3. The essential finding of Chow's work was that a model of any AC system can be constructed from a collection of 27 PPs.

The approach utilises decomposition: an air handling unit, for example, comprises air damper, mixing chamber, cooling coil, humidifier, reheater and fan sub-models; the fan comprises impeller, motor and casing sub-models; the casing comprises metal and insulation sub-models; and the insulation comprises conduction, convection and moisture flow sub-models. Within the PP approach, these last heat and mass transfer sub-models are the PPs. Table 6.1 lists the 27 PPs when organised into 10 categories.

Table 6.1: The 27 PPs for air conditioning systems modelling (from Chow 1995b).

Category	*PP*	*Category*	*PP*
1	Thermal conduction	6	Flow converger
	1.1 solid to solid		6.1 moist air
	1.2 with ambient solid		6.2 two phase fluid
			6.3 one phase fluid
			6.4 leak-in moist air
2	Surface convection	7	Flow upon water spray
	2.1 with moist air		7.1 moist air
	2.2 with two phase fluid		
	2.3 with one phase fluid		
	2.4 with ambient		
3	Surface radiation	8	Fluid injection
	3.1 with local surface		8.1 water/steam to moist air
	3.2 with ambient surface		
4	Flow upon surface	9	Fluid accumulator
	4.1 moist air; 3 nodes		9.1 moist air
	4.2 two phase fluid; 3 nodes		9.2 liquid
	4.3 one phase fluid; 3 nodes		
	4.4 moist air; 2 nodes		
	4.5 one phase fluid; 2 nodes		
5	Flow divider and inducer	10	Heat injection
	5.1 flow diverger		10.1 to solid
	5.3 flow multiplier		10.2 to vapour-generating fluid
	5.3 flow inducer		10.3 to moist air

Categories 1 through 3 describe the three modes of heat transfer at the surface of a solid: conduction, convection and radiation. Some 8 PPs are available, depending on whether the state variables of the three possible contact nodes—ambient, neighbouring surface and internal solid—are known boundary conditions of the problem or are unknown and must be solved throughout the simulation.

Category 4 comprises 5 PPs that describe the thermal behaviour of a fluid when flowing over a solid surface. These cover liquids, gases and air/vapour mixtures, with no restriction placed on the physical shape or composition of the surface (whether it be the internal surface of a ventilating fan or the exposed surface of a duct heater).

Categories 5 and 6 describe the processes associated with the mixing, diverging or induction of fluid streams. These PPs are used to represent T-junctions, mixing boxes and fans. The flow multiplier PP exists to adjust the fluid flow rate to simplify the component modelling task or to act as a filter (e.g. for use with a steam trap model to allow only liquid water to pass).

Category 7 refers to the direct interaction of air and water streams, as in the case of air washers and cooling towers.

Category 8 describes the injection of water or steam into a moist air stream as in the case of a humidifier.

Category 9 covers fluid accumulators such as an expansion tank within a hydronic circuit or the air in a conditioned space.

Category 10 covers the heat injection processes as associated, for example, with air or water electric heating elements.

To illustrate the construction of a component model using PPs, consider the counter-flow, chilled water cooling coil as shown in figure 6.6. The straight tube section of the coil may be modelled by 4 nodes (denoted S, A_1, W_M and W_1) and 2 connections (denoted A_0 and W_0). Application of the approach described in §6.2.1 would give rise to the following matrix equation.

$$\begin{bmatrix} C(1) & C(2) & C(3) & 0 & 0 & 0 \\ C(4) & C(5) & 0 & 0 & C(11) & 0 \\ C(6) & 0 & C(7) & C(8) & 0 & C(12) \\ 0 & 0 & C(9) & C(10) & 0 & 0 \end{bmatrix} \times \begin{bmatrix} S \\ A_1 \\ W_M \\ W_1 \\ A_0 \\ W_0 \end{bmatrix} = \begin{bmatrix} C(13) \\ C(14) \\ C(15) \\ C(16) \end{bmatrix}.$$

where, for convenience, the coefficients corresponding to the 4 internal nodes are sequentially numbered, followed by the coefficients relating to the 2 connections and 4 present time-row terms.

Such a tube model may be synthesised using PP 4.3 and PP 4.4, where the former PP is given by

$$\begin{bmatrix} A(43,1) & A(43,2) & 0 & 0 \\ A(43,4) & A(43,5) & A(43,6) & A(43,11) \\ 0 & A(43,8) & A(43,9) & 0 \end{bmatrix} \times \begin{bmatrix} S \\ W_M \\ W_1 \\ W_0 \end{bmatrix} = \begin{bmatrix} B(43,1) \\ B(43,2) \\ B(43,3) \end{bmatrix}$$

and the latter PP by

$$\begin{bmatrix} A(44,1 & A(44,2) & 0 \\ A(44,3) & A(44,4) & A(44,6) \end{bmatrix} \times \begin{bmatrix} S \\ A_1 \\ A_0 \end{bmatrix} = \begin{bmatrix} B(44,1) \\ B(44,2) \end{bmatrix}$$

where $A(i,j)$ is a future time-step coefficient, $B(i,j)$ is a present time-step coefficient and i corresponds to the PP number. The key point is that the 16 coefficients of the first equation may be expressed in terms of the coefficients of the last two equations. This is illustrated in table 6.2.

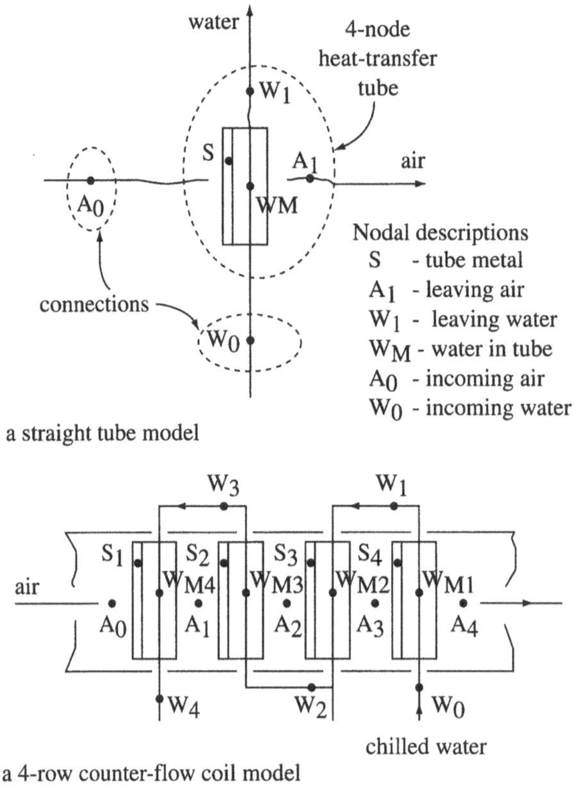

Figure 6.6: Construction of a cooling coil model using PPs.

Within the present model, the thermal resistance of the tube wall is assumed to be negligible. Such an assumption may be removed by including PP 1.1—by simply adding in the new coefficients as required. Finally, the model of the complete cooling coil is realised as the union of the (here) 4 matrix equations, one for each tube.

The PP approach is compatible with the notions of equation-based modelling (Sahlin and Sowell 1989, Buhl *et al* 1990, Sowell and Moshier 1995), hierarchical model decomposition (Laret 1989, Bjork 1989), component model standardisation (Dubois 1990, Augenbroe and Winkelmann 1991, Bring *et al* 1992) and neutral model formats (Bring *et al* 1992, Nataf 1995). A common theme across these topics is the notion of an input-output independent component model representation by which model applicability and program interoperability is enhanced. Using PPs, alternative input-output relationships are represented by different PP groupings. Consider an air duct model in which the temperature of the air surrounding the exposed surface is expressed as a boundary condition. If this model is required for use in a context in which this temperature is an interface variable for numerical solution, then all that is required is to substitute PP 2.4 by PP 2.1.

The premise underlying the PP approach is that the best way to analyse real systems is to provide a mechanism to permit the incorporation of additional process models as the focus of the problem narrows or as the component in question becomes more dominant. Because each PP is a model of a distinct physical phenomenon, component models can call on any number of PPs in order to attain the required modelling resolution. An air-to-water heat exchanger may

Table 6.2: Expression of the tube model in terms of PP coefficients.

Matrix equation coefficient	*PP 4.3*[a]	*PP4.4*[b]
C(1) =	A(43,1) +	A(44,1)
C(2) =		A(44,2)
C(3) =	A(43,2)	
C(4) =		A(44,3)
C(5) =		A(44,4)
C(6) =	A(43,4)	
C(7) =	A(43,5)	
C(8) =	A(43,6)	
C(9) =	A(43,8)	
C(10) =	A(43,9)	
C(11) =		A(44,6)
C(12) =	A(43,11)	
C(13) =	B(43,1) +	B(44,1)
C(14) =		B(44,2)
C(15) =	B(43,2)	
C(16) =	B(43,3)	

[a] Flow upon surface for single phase fluid $S - W_1, W_0$.
[b] Flow upon surface for moist air $S - A_1, A_0$.

be simply represented by PP 4.4 and PP 4.5, or be made arbitrarily detailed by interconnecting finned tubes, each represented by a combination of PP 4.1, PP 1.1 and PP 4.3.

6.2.2 Active solar

Figure 6.7 shows the elements of an active solar system. The flat plate collector (air or water) supplies some heating load directly or, in times of excess, delivers the heat to a thermal store. McLean (1982) formulated a numerical model for the system of figure 6.7.

When applying the method of chapter 3, many of the resulting conservation equations will be identical to those already derived. For example, the collector back plate FVs will adhere to eqn (3.10), glazing FVs to eqn (3.5) and fluid nodes to eqn (3.12). This section considers only those FV types not previously derived.

Heat exchanger

A four FV model may be used to represent the sensible heat exchange between the cold and hot fluid regions. Theoretically, the maximum rate of heat transfer between the fluids is given by

$$q = C_{min}(\theta_{hi} - \theta_{ci})$$

where $C_{min} = \min(\dot{m}_h C_{ph}, \dot{m}_c C_{pc})$, $\dot{m}$ the mass flow rate (kg s^{-1}), C_p the specific heat capacity (J $kg^{-1}{}^{\circ}C^{-1}$), θ_{hi} the hot fluid inlet temperature (°C), θ_{ci} the cold fluid inlet temperature, and h and c indicate the hot and cold fluids respectively.

In reality, the heat transfer rate is less than this theoretical maximum as expressed by the exchanger effectiveness: E = (actual heat transfer)/(theoretical maximum). Table 6.3 gives E as a function of the flow geometry.

As before, conservation equations can be derived by inserting the FV flux quantities corresponding to the present and future time-rows of some arbitrary time-step into eqn (3.1) and averaging the result. For any fluid stream, f, this gives for the future time-row

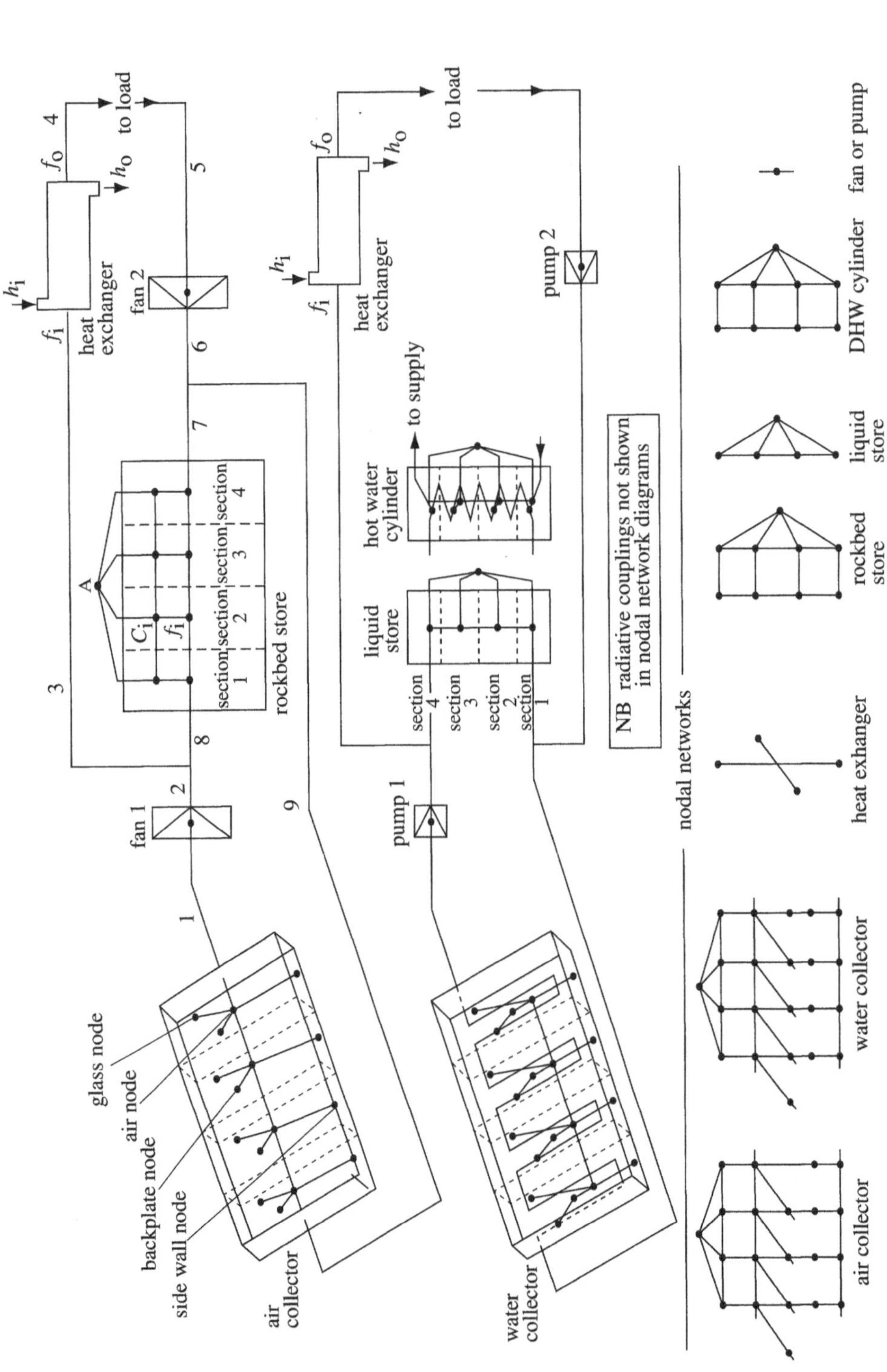

Figure 6.7: Active solar system components and FV scheme.

Table 6.3: Heat exchanger effectiveness (from Kays and London 1964).

Flow geometry	*Effectiveness*
Parallel	$\dfrac{1 - \exp[-N(1 + C)]}{(1 + C)}$
Counterflow	$\dfrac{1 - \exp[-N(1 - C)]}{1 - C\exp[-N(1 - C)]}$
Crossflow	
C_{max}, C_{min} unmixed	$1 - \exp \dfrac{C}{N^{-0.22}} [\exp(-N^{0.78}C) - 1]$
C_{max}, C_{min} mixed	$\left(\dfrac{1}{1 - \exp(-N)} + \dfrac{C}{1 - \exp(-NC)} - \dfrac{1}{N}\right)^{-1}$
C_{max} mixed, C_{min} unmixed	$(1/C)(1 - \exp[C(1 - \exp(-N))])$
C_{max} unmixed, C_{min} mixed	$1 - \exp\{[1 - \exp(-NC)]/C\}$
Shell and tube (1 shell pass; 2, 4, 6 tube passes)	$2\left[1 + C + (1 + C^2)^{0.5}\left(\dfrac{1 + \exp[-N(1 + C^2)^{0.5}]}{1 - \exp[-N(1 + C^2)^{0.5}]}\right)\right]^{-1}$
N = number of transfer units = UA/C_{min}; $C = C_{min}/C_{max}$	

$$\frac{\rho_f(t + \delta t)C_{pf}(t + \delta t)\delta V_f}{\delta t}\theta_f(t + \delta t) - \frac{\rho_f(t)C_{pf}(t)\delta V_f}{\delta t}\theta_f(t)$$
$$= m_f(t + \delta t)C_{pf}(t + \delta t)\theta_{fi}(t + \delta t) - m_f(t + \delta t)C_{pf}(t + \delta t)\theta_{fo}(t + \delta t)$$
$$\pm EC_{min}(t + \delta t)[\theta_{hi}(t + \delta t) - \theta_{ci}(t + \delta t)] + q_e(t + \delta t) \qquad (6.25)$$

where ρ_f is the bulk fluid density (kg m^{-3}), C_{pf} the bulk fluid specific heat (J kg^{-1}°C^{-1}), δV_f the fluid volume (m^3), δt the time-step (s), m_f the fluid mass flow rate (kg s^{-1}), θ_f the bulk fluid temperature (°C), θ_{fi} the inlet fluid temperature, θ_{fo} the outlet fluid temperature and q_e the heat loss to the surroundings (W).

And at the present time-row:

$$\frac{\rho_f(t + \delta t)C_{pf}(t + \delta t)\delta V_f}{\delta t}\theta_f(t + \delta t) - \frac{\rho_f(t)C_{pf}(t)\delta V_f}{\delta t}\theta_f(t)$$
$$= m_f(t)C_{pf}(t)\theta_{fi}(t) - m_f(t)C_{pf}(t)\theta_{fo}(t) \pm EC_{min}(t)[\theta_{hi}(t) - \theta_{ci}(t)] + q_e(t)\,. \qquad (6.26)$$

Assuming that the bulk fluid temperature can be expressed as a weighted average of inlet and outlet conditions such that

$$\theta_f = \alpha\theta_{fi} + (1 - \alpha)\theta_{fo} \qquad (6.27)$$

then combination of eqns (6.25) and (6.26) gives, after rearrangement, for both the hot (f = h) and cold (f = c) fluids

$$[2\alpha\rho_f(t + \delta t)C_{pf}(t + \delta t)\delta V_f - \delta t\, m_f(t + \delta t)C_{pf}(t + \delta t) + \delta t\, EC_{min}(t + \delta t)]\theta_{fi}(t + \delta t)$$
$$+ [2(1 - \alpha)\rho_f(t + \delta t)C_{pf}(t + \delta t)\delta V_f + \delta t\, m_f(t + \delta t)C_{pf}(t + \delta t)]\theta_{fo}(t + \delta t)$$
$$- \delta t\, EC_{min}(t + \delta t)\theta_{xi}(t + \delta t) - \delta t\, q_e(t + \delta t)$$

$$= [2\alpha \rho_f(t)C_{pf}(t)\delta V_f + \delta t\ m_f(t)C_{pf}(t) - \delta t\ EC_{min}(t)]\theta_{fi}(t)$$

$$+ [2(1-\alpha)\rho_f(t)C_{pf}(t)\delta V_f - \delta t\ m_f(t)C_{pf}(t)]\theta_{fo}(t) + \delta t\ EC_{min}(t)\theta_{xi}(t) + \delta t\ q_e(t)\ .$$

If accuracy considerations render eqn (6.27) unacceptable then further fluid volume subdivision will be necessary, with the foregoing procedure repeated for each hot/cold fluid pairing.

Rockbed thermal store

Consider the rockbed thermal store of figure 6.7, shown segmented into four isothermal volumes, with each volume assigned two FVs: a bulk fluid node and a bulk capacity node. Eqn (3.12) defines the nodal simulation equation for the fluid node so that for any rockbed segment, i, receiving fluid from segment i – 1:

$$[2\rho_{fi}(t+\delta t)C_{fi}(t+\delta t)\delta V_{fi} + \delta A_{ci}\delta t\ h_{ci}(t+\delta t)$$

$$+ \delta t\ v_{i-1,i}(t+\delta t)\rho'_{i-1,i}(t+\delta t)C'_{i-1,i}(t+\delta t)]\theta_{fi}(t+\delta t)$$

$$- \delta A_{ci}\delta t\ h_{ci}(t+\delta t)\theta_{ci}(t+\delta t) - \delta t\ v_{i-1,i}(t+\delta t)\rho'_{i-1,i}(t+\delta t)C'_{i-1,i}(t+\delta t)\theta_{fi-1}(t+\delta t)$$

$$- \delta t\ q_{fi}(t+\delta t) = [2\rho_{fi}(t)C_{fi}(t)\delta V_{fi} - \delta A_{ci}\delta t\ h_{ci}(t) - \delta t\ v_{i-1,i}(t)\rho'_{i-1,i}(t)C'_{i-1,i}(t)]\theta_{fi}(t)$$

$$+ \delta A_{ci}\delta t\ h_{ci}(t)\theta_{ci}(t) + \delta t\ v_{i-1,i}(t)\rho'_{i-1,i}(t)C'_{i-1,i}(t)\theta_{fi-1}(t) + \delta t\ q_{fi}(t) \qquad (6.28)$$

where h_{ci} is the convective heat transfer coefficient connecting the fluid and capacity segments (W m^{-2}°C^{-1}), δA_{ci} the exposed surface area of the segment capacity, δV_{fi} the segment fluid volume, $v_{i-1,i}$ the fluid volume flow rate from segment i – 1 to i (m^3s^{-1}), θ_{fi} the temperature of the fluid in segment i (°C), θ_{ci} the temperature of the capacity in segment i, θ_{fi-1} the temperature of fluid in segment i – 1, and q_{fi} the heat exchange with the environment surrounding segment i (W); ρ_{fi}, C_{fi} are the density and specific heat capacity of the segment i fluid (kg m^{-3} and J kg^{-1}°C^{-1}) and $\rho'_{i-1,i}$, $C'_{i-1,i}$ the density and specific heat capacity of the fluid evaluated at the mean temperature of the two connected segments.

Occasionally a preheated liquid is obtained by passing conduits through the rockbed. The q_{fi} terms facilitate the modelling of such a device by allowing the removal of heat as a function of any thermostatic or time-based schedule.

For real systems, it is difficult to quantify the h_{ci}, δA_{ci} and δV_{fi} terms and so it is desirable to operate with a volumetric convective heat transfer coefficient:

$$h_{vi} = h_{ci}\delta A_{ci}/\delta V_{fi}\ .$$

Lof and Hawley (1948) derived empirical relationships that give h_{vi} as a function of rockbed parameters:

$$h_{vi} = 650(\rho v/A_b d)^{0.7}$$

where A_b is the rockbed cross-sectional area (m^2), v the air flow rate (m^3s^{-1}), ρ the density of the particles (kg m^{-3}) and d the equivalent spherical diameter of the particles (m) given by

$$d = (6V_p/\pi N)^{1/3}$$

where V_p is the net particle volume (m^3) and N is an estimate of the number of particles.

The rockbed bulk capacity FV is represented by a modified form of eqn (3.11):

$$[2\rho_{ci}(t+\delta t)C_{ci}(t+\delta t) + \delta t\ h_{vi}(t+\delta t)]\theta_{ci}(t+\delta t)$$

$$- \delta t\, h_{vi}(t+\delta t)\theta_{fi}(t+\delta t) - \delta t\, q_{ci}(t+\delta t)/\delta V_{ci}$$

$$= [2\rho_{ci}(t)C_{ci}(t) - \delta t\, h_{vi}(t)]\theta_{ci}(t) + \delta t\, h_{vi}(t)\theta_{fi}(t) + \delta t\, q_{ci}(t)/\delta V_{ci} \tag{6.29}$$

where ρ_{ci}, C_{ci} are the density and specific heat capacity of the segment i capacity and δV_{ci} the net volume of the material in the segment.

Liquid thermal store

As with the rockbed, a liquid thermal store will require subdivision although in a vertical direction (figure 6.7). Each FV is then represented by eqn (3.12):

$$[2\rho_{fi}(t+\delta t)C_{fi}(t+\delta t)\delta V_i + \delta A_{si}\delta t\, U_{ci}(t+\delta t) + \delta t\, v_s(t+\delta t)\rho'_{s,i}(t+\delta t)C'_{s,i}(t+\delta t)$$

$$+ \delta t\, v_{i-1,i}(t+\delta t)\rho'_{i-1,i}(t+\delta t)C'_{i-1,i}(t+\delta t)]\theta_i(t+\delta t)$$

$$- \delta A_{si}\delta t\, U_{ci}(t+\delta t)\theta_c(t+\delta t) - \delta t\, v_s(t+\delta t)\rho'_{s,i}(t+\delta t)C'_{s,i}(t+\delta t)\theta_s(t+\delta t)$$

$$- \delta t\, v_{i-1,i}(t+\delta t)\rho'_{i-1,i}(t+\delta t)C'_{i-1,i}(t+\delta t)\theta_{i-1}(t+\delta t) - \delta t[q_{el}(t+\delta t) + q_k(t+\delta t)]$$

$$= [2\rho_{fi}(t)C_{fi}(t)\delta V_i - \delta A_{si}\delta t\, U_{ci}(t) - \delta t\, v_s(t)\rho'_{s,i}(t)C'_{s,i}(t)$$

$$- \delta t\, v_{i-1,i}(t)\rho'_{i-1,i}(t)C'_{i-1,i}(t)]\theta_i(t) + \delta A_{si}\delta t\, U_{ci}(t)\theta_c(t) + \delta t\, v_s(t)\rho'_{s,i}(t)C'_{s,i}(t)\theta_s(t)$$

$$+ \delta t\, v_{i-1,i}(t)\rho'_{i-1,i}(t)C'_{i-1,i}(t)\theta_{i-1}(t) - \delta t[q_{el}(t) + q_k(t)]$$

where U_{ci} is the tank wall thermal transmittance (W m^{-2}°C^{-1}), δA_{si} the total tank surface area associated with segment i (m^2), v_s the supply flow rate (direct cylinder case) from collector to store (m^3s^{-1}), $v_{i-1,i}$ the volume flow rate coupling segments i – 1 and i, q_{el} the electrical resistance heat input (W), q_k the segment heat injection (W) from the collector in the indirect cylinder case:

$$q_k = A_k U_k(\theta_s - \theta_i)$$

where A_k is the conduit surface area associated with segment i and U_k the conduit thermal transmittance (W m^{-2}°C^{-1}).

For the case of an indirect cylinder, an additional node per segment will be required to represent the change in collector fluid condition as it passes through the associated conduit. Eqn (6.28) is the representative equation for this FV but with $v_{i-1,i}$ representing the conduit flow rate, h_{ci} replaced by the conduit thermal transmittance and ci, fi representing the segment and conduit fluids respectively.

Latent thermal store

Eqn (6.28) is again the appropriate conduit fluid equation, with the terms changed as for the liquid thermal store case and ci representing the phase change material.

The phase change material is represented by eqn (6.29) but with the heat generation term expanded to account for the latent energy absorbed or released during the phase change:

$$[2\rho_{ci}(t+\delta t)C_{ci}(t+\delta t) + \delta t\, h_{vi}(t+\delta t)]\theta_{ci}(t+\delta t) - \delta t\, h_{vi}(t+\delta t)\theta_{fi}(t+\delta t)$$

$$- \delta t[q_{ci}(t+\delta t) + q_x(t+\delta t)] = [2\rho_{ci}(t)C_{ci}(t) - \delta t\, h_{vi}(t)]\theta_{ci}(t)$$

$$+ h_{vi}(t)\delta t\theta_{fi}(t) + \delta t[q_{ci}(t) + q_x(t)] \tag{6.30}$$

where q_{ci} is the heat loss to the surroundings (W) and q_x the stored or released latent flux (W).

During sensible cooling or heating $q_x(\xi) = 0$. When the phase change temperature is reached, $q_x(t + \delta t)$ is used to ensure that isothermal conditions prevail by arranging that $\theta_{ci}(t + \delta t) = \theta_{ci}(t)$. By this device, the total latent energy in 'storage' is known at any time from the summation history:

$$\sum^{n} q_x(t + \delta t)\delta t$$

where n is the number of time steps since the phase change began, q_x is +ve or -ve at any given time-row depending on the direction of the phase change. If this summation at any time exceeds the latent heat associated with the change in phase, or the summation reduces to zero, then $q_x(t + \delta t)$ is set to zero in eqn (6.30) and sensible cooling/heating recommences.

The active solar system of figure 6.7 can now be made discrete by distributing nodes in a manner that reflects the design issues in hand or the importance of particular energy exchanges: one possible scheme is shown superimposed.

Table 6.4 lists the final conservation equations for the FV types comprising this system and figure 6.8 shows the whole system matrix equation to result for the example problem considered here. As with the air conditioning system, a corresponding mass balance matrix equation can be established for use in conjunction with a corresponding flow network to determine the component pressure drops and working fluid mass flow rates.

6.2.3 Wet central heating

Figure 6.9 shows the main components of a wet central heating system: boiler, pump, radiators and hot water cylinder, all linked by distribution pipes and subjected to control action.

Tang (1984) has modelled this system type using the method of chapter 3. Again, many of the FV conservation equations to result will be identical to those derived in chapter 3. For example, the fluid in a pipe may be represented by eqn (3.15), a radiator surface by eqn (3.11), while some boiler and radiator water FVs will conform to the formulations of eqns (3.12) or (3.16). The following derivations relate only to those FV types not previously considered.

Boilers

The FV equation-set for a boiler will depend on the type of boiler being modelled. For the present purpose a tubeless, steel shell, domestic boiler is assumed for which the products of combustion are well mixed (allowing a single FV combustion chamber representation), and the heat source may be considered as a radiating plane, at the average combustion chamber temperature, parallel to the heat transfer surface. One possible discretisation scheme is then as shown in figure 6.9: 4 vertical subdivisions with FVs placed to represent the outer shell, inner shell and water volumes at each level, in addition to the combustion chamber and ambient air.

The conservation equation pertaining to the ambient air (A) and water (w1—w4) FVs are given by eqns (3.12) and (3.15) respectively, but with modified surface heat transfer coefficients in the water equations as described later. For the combustion chamber node (G), the right-hand side of eqn (3.1) is re-expressed as

$$\sum_{j=1}^{N} h_{Tij}(\theta_{ij} - \theta_G)|_{t=\xi} + \rho_G C_{pG} v_G(\theta_A - \theta_G)|_{t=\xi} + \bar{W} q_G|_{t=\xi} \qquad (6.31)$$

where h_{Tij} is the total heat transfer coefficient at the inner surface of the inner shell (denoted by i) associated with the jth FV (W m^{-2}°C^{-1}), θ_{ij} the inner shell temperature at the jth section

Table 6.4: Characteristic equations and coefficient formulae, active solar system.

FV type	*Characteristic equation*
Collector wall (homogeneous elements and element interfaces).	$-A_1(t+\delta t)\theta_{I-1}(t+\delta t)+A_2(t+\delta t)\theta_I(t+\delta t)$ $- A_3(t+\delta t)\theta_{I+1}(t+\delta t) - A_4(t+\delta t)q_I(t+\delta t)$ $= A_1(t)\theta_{I-1}(t)+A_5(t)\theta_I(t)+A_3(t)\theta_{I+1}(t)+A_4(t)q_I(t)$
Internal surface, rockbed bulk storage and latent storage.	$-B_1(t+\delta t)\theta_{I-1}(t+\delta t)+B_2(t+\delta t)\theta_I(t+\delta t)$ $- B_3(t+\delta t)\theta_{I+1}(t+\delta t) - \sum_{j=1}^{N} B_{4,j}(t+\delta t)\theta_j(t+\delta t)$ $- B_5(t+\delta t)q_I(t+\delta t) = B_1(t)\theta_{I-1}(t)+B_6(t)\theta_I(t)$ $+ B_3(t)\theta_{I+1}(t)+\sum_{j=1}^{N} B_{4,j}(t)\theta_j(t)+B_5(t)q_I(t)$
Collector, heat exchanger, rockbed and liquid store fluid.	$-\sum_{f=1}^{N} C_f(t+\delta t)\theta_f(t+\delta t)+C_1(t+\delta t)\theta_b(t+\delta t)$ $- \sum_{o=1}^{M} C_o(t+\delta t)\theta_o(t+\delta t) - C_2(t+\delta t)q_I(t+\delta t)$ $= \sum_{f=1}^{N} C_f(t)\theta_f(t)+C_3(t)\theta_b(t)+\sum_{o=1}^{M} C_o(t)\theta_o(t)+C_2(t)q_I(t)$

θ_f = bulk fluid temperature; θ_b = fluid flow stream temperature, θ_o = boundary or outer surface temperature.

Coefficient formulae

for boundary between homogeneous elements:	for a homogeneous element:
$A_1(\xi)=(k_A(\xi)R_c(\xi)+2\delta x_{I-1,I}) \times \delta t/\delta x_{I-1,I}R_c(\xi)\delta_{I-1,I+1}$	$A_1(\xi)=k(\xi)\delta t/\delta x_I^2$
$A_2(\xi)=2\rho_I(\xi)C_I(\xi)+A_1(\xi)+A_3(\xi)$	$A_2(\xi)=2\rho_I(\xi)C_I(\xi)+A_1(\xi)+A_2(\xi)$
$A_3(\xi)=(k_B(\xi)R_c(\xi)+2\delta x_{I+1,I}) \times \delta t/\delta x_{I+1,I}R_C(\xi)\delta_{I-1,I+1}$	$A_3(\xi)=k(\xi)\delta t/\delta x_I^2$
$A_4(\xi)=\delta t/(\delta_{I-1,I+1}\delta_{J-1,J+1}\delta_{K-1,K+1})$	$A_4(\xi)=\delta t/\delta x_I\delta x_J\delta x_K$
$A_5(\xi)=2\rho_I(\xi)C_I(\xi)-A_1(\xi)-A_3(\xi)$	$A_5(\xi)=2\rho_I(\xi)C_I(\xi)-A_1(\xi)-A_3(\xi)$
$q_I(\xi)=q_P(\xi)$	$q_I(\xi)=q_P(\xi)$
for internal surfaces:	for rockbed and latent stores:
$B_1(\xi)=k_A(\xi)\delta t/\delta x_{I-1,I}\delta_{I-1,I+1}$	$B_1(\xi)=0$
$B_2(\xi)=2\rho_I(\xi)C_I(\xi)+B_1(\xi)+B_3(\xi) + \sum_{j=1}^{N} B_{4,j}(\xi)$	$B_2(\xi)=2\rho_I(\xi)C_I(\xi)+B_3(\xi)$
$B_3(\xi)=h_c(\xi)\delta t/\delta_{I,I-1}$	$B_3(\xi)=h_v(\xi)\delta t$
$B_{4,j}(\xi)=h_{cj,I}(\xi)\delta t/\delta_{I,I-1}$	$B_{4,j}(\xi)=0$
$B_5(\xi)=\delta t/(\delta_{I,I-1}\delta_{J-1,J+1}\delta_{K-1,K+1})$	$B_5(\xi)=\delta t/\delta V$
$B_6(\xi)=2\rho_I(\xi)C_I(\xi)-B_1(\xi)-B_3(\xi) - \sum_{j=1}^{N} B_{4,j}(\xi)$	$B_6(\xi)=2\rho_I(\xi)C_I(\xi)-B_3(\xi)$
$q_I(t)=q_P(t)+q_{SI}(t)+q_{RI}(t)+q_{SI}(t+\delta t) + q_{RI}(t+\delta t)$	$q_I(\xi)=q_P(\xi)+q_x(\xi)$
$q_I(t+\delta t)=q_P(t+\delta t)$	$q_x(\xi)$ used only in latent store applications

(continued)

Table 6.4 (continued)

for collector, rockbed and liquid store fluids (for which $q_I = q_c + q_k$, $h_{c,o} = U_o$, M = 2, o = 1 refers to the inlet fluid and o = 2 the stratified store contents):	for heat exchanger:
$C_f(\xi) = v_f(\xi)\rho_f(\xi)c_f(\xi)\delta t/\delta V_I$	$C_{f=1}(t) = \dfrac{2\alpha\rho_f(t)c_f(t)\delta V}{\delta t} + m(t)C(t)$ $C_{f=2}(t) = \dfrac{2(1-\alpha)\rho_I(t)C_I(t)\delta V}{\delta t} - m(t)C(t)$ $C_{f=1}(t+\delta t) = \dfrac{2\alpha\rho_I(t+\delta t)C_I(t+\delta t)\delta V}{\delta t} - m(t+\delta t)C(t+\delta t) + C_{min}(t+\delta t)$ $C_{f=2}(t+\delta t) = \dfrac{2(1-\alpha)\rho_I(t+\delta t)C_I(t+\delta t)\delta V}{\delta t} + m(t+\delta t)C(t+\delta t)$
$C_o(\xi) = h_{c,o}(\xi)A_o\delta t/\delta V_I$	$C_o(\xi) = 0$
$C_1(\xi) = 2\rho_I(\xi)C_I(\xi) + \sum_{f=1}^{N} C_f(\xi) + \sum_{o=1}^{M} C_o(\xi)$	$C_1(\xi) = C_{min}(\xi)$
$C_2(\xi) = \delta t/\delta V_I$	$C_2(\xi) = 1$
$C_3(\xi) = 2\rho_I(\xi)C_I(\xi) - \sum_{f=1}^{N} C_f(\xi) - \sum_{o=1}^{M} C_o(\xi)$	$C_3(\xi) = -C_{min}(\xi)$
$q_I(\xi) = q_P(\xi)$	$q_I(\xi) = q_e(\xi)$

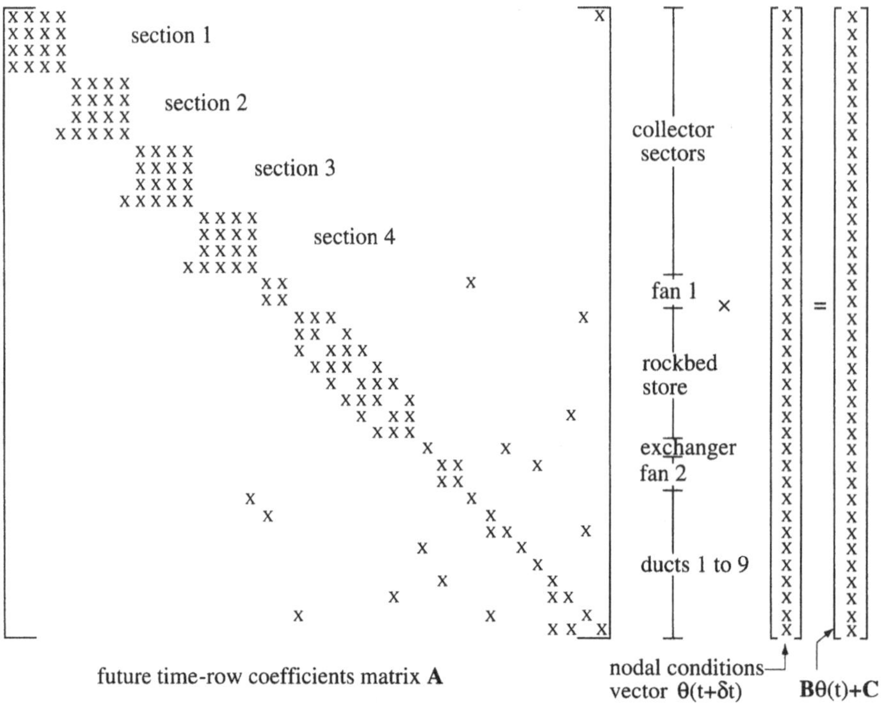

Figure 6.8: Active solar system energy balance matrix equation.

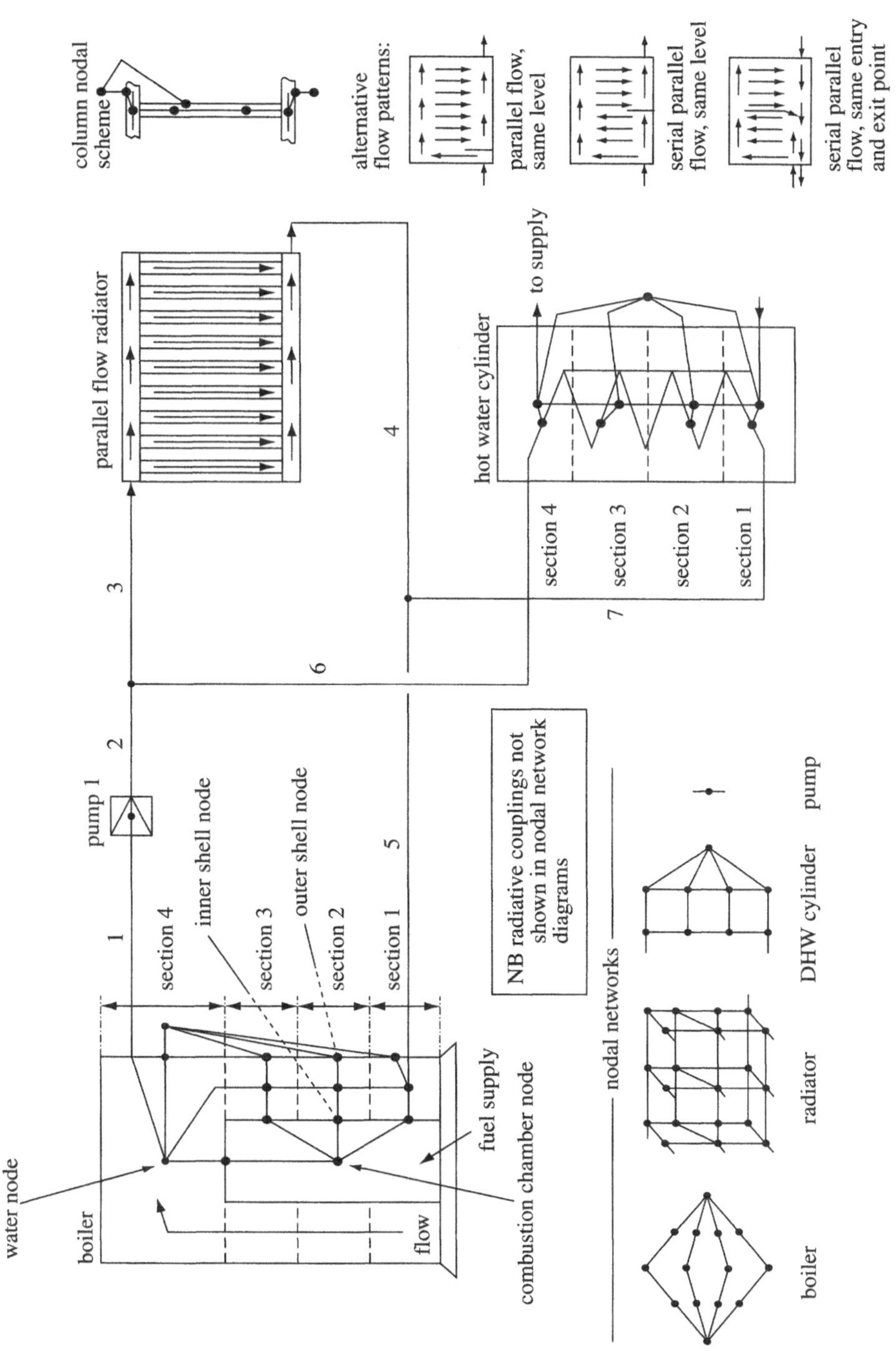

Figure 6.9: Wet central heating system components and FV scheme.

(°C), θ_G the average combustion chamber temperature, ρ_G the density of the combustion products (kg m^{-3}), C_{pG} the specific heat capacity of the combustion products (J kg^{-1}°C^{-1}), v_G the volume flow rate of supply air (m^{3}s^{-1}), θ_A the supply air temperature, $\bar{W}$ the fuel supply rate (kg s^{-1}), q_G the fuel heat content (J kg^{-1}) and N the number of vertical water FVs (here 4).

Equating expression (6.31) to the capacity term of eqn (3.1) ($\rho_G C_{pG} \delta V \partial\theta_G/\partial t$) and averaging the result over some arbitrary time-step ($t \rightarrow t + \delta t$) gives, after rearrangement

$$\Big[2\rho_G(t+\delta t)C_{pG}(t+\delta t)\delta V_G + \delta t \sum_{j=1}^{N} h_{Tij}(t+\delta t)$$

$$+\,\delta t\, v_G(t+\delta t)\rho_G(t+\delta t)C_{pG}(t+\delta t)\Big]\,\theta_G(t+\delta t) - \delta t \sum_{j=1}^{N} h_{Tij}(t+\delta t)\theta_{ij}(t+\delta t)$$

$$-\,\delta t\, v_G(t+\delta t)\rho_G(t+\delta t)C_{pG}(t+\delta t)\theta_A(t+\delta t) - \delta t\, \bar{W}(t+\delta t)q_G(t+\delta t)$$

$$= \left[2\rho_G(t)C_{pG}(t)\delta V_G - \delta t \sum_{j=1}^{N} h_{Tij}(t) - \delta t\, v_G(t)\rho_G(t)C_{pG}(t)\right]\theta_G(t)$$

$$+\,\delta t \sum_{j=1}^{N} h_{Tij}(t)\theta_{ij}(t) + \delta t\, v_G(t)\rho_G(t)C_{pG}(t)\theta_A(t) + \delta t\, \bar{W}(t)q_G(t) \tag{6.32}$$

where δV_G is the combustion chamber volume (m^3).

For inner shell FVs, adjacent to the 3 vertical water segments, the convective heat transfer at the water-side interface will be enhanced due to the occurrence of nucleate pool boiling because the surface temperature will be greater than the water saturation temperature. Assuming that the bulk water temperature is subcooled, the water-side convective heat transfer is given by

$$(q/A)_{total} = (q/A)_{boiling} + (q/A)_{forced\ convection}$$

where q is the heat flux (W) and A the heat transfer area (m^2). This equation may be re-expressed in terms of a forced convection coefficient, h_c, and a subcooled nucleate boiling coefficient, h_b (both W m^{-2}°C^{-1}):

$$(q/A)_{total} = (h_c + h_b)(\theta_{wj} - \theta_{ij})\ ;\ j = 1, 2, 3.$$

Holman (1981) suggested that h_c may be obtained from an empirical formulation:

$$Nu = 0.019\, Re^{0.8}\, Pr^{0.4}$$

where the Nusselt (Nu), Reynolds (Re) and Prandtl (Pr) Numbers are as defined in §7.6.

For subcooled, nucleate pool boiling, Rohsenow (1952) correlated experimental data to obtain

$$\frac{C(\theta_{wj} - \theta_{SAT})}{h_{fg}Pr} = C_{sf}\left[\frac{(q/A)_{boiling}}{\mu h_{fg}}\left(\frac{g\sigma}{g(\rho_w - \rho_v)}\right)^{0.5}\right]^{0.33}$$

where C is the specific heat capacity of the saturated water (J kg^{-1}°C^{-1}), θ_{SAT} the saturation temperature (°C), θ_{wj} the bulk water temperature of section j, h_{fg} the enthalpy of vaporisation (J kg^{-1}) and C_{sf} a constant 0.013 for water to copper (Holman 1981). Further, μ is the water viscosity (kg m^{-1}s^{-1}), g the gravitational constant (m s^{-2}) and σ the surface tension at the liquid/vapour interface ($N\ m^{-2}$); ρ_w is the density of saturated water ($kg\ m^{-3}$) and ρ_v the density of the saturated water vapour.

Table 6.5 gives simplified empirical relationships for h_b in relation to submerged surface boiling at atmospheric pressure. For non-atmospheric pressures an empirical modification is required:

$$h_b' = h_b(p/p_{AT})^{0.4}$$

where h_b' is the boiling coefficient at pressure p and h_b the boiling coefficient at atmospheric pressure, p_{AT}.

Table 6.5: Simplified relationships for h_b for water at atmospheric pressure (from Holman 1981).

Surface	h_b	*Condition*
Horizontal	$1024(\theta_w - \theta_{SAT})^{1/3}$	$(q/A)_b \leq 16\ kW\ m^{-2}$
	$5.56(\theta_w - \theta_{SAT})^3$	$16 < (q/A)_b < 240\ kW\ m^{-2}$
Vertical	$537(\theta_w - \theta_{SAT})^{1/7}$	$(q/A)_b \leq 3\ kW\ m^{-2}$
	$7.96(\theta_w - \theta_{SAT})^3$	$3 < (q/A)_b < 63\ kW\ m^{-2}$

For an inner shell node at the jth section, the net flux is given by

$$h_{Tij}\delta A_{ij}(\theta_G - \theta_{ij})|_{t=\xi} + (h_{cij} + h_{bij})\delta A_{ij}(\theta_{wj} - \theta_{ij})|_{t=\xi} \qquad (6.33)$$

where δA_{ij} is the surface area associated with the jth section FV (m^2).

Substitution of this flux within eqn (3.1) for the two time-row poles of some arbitrary time-step gives

$$\{2\rho_{ij}(t+\delta t)C_{pij}(t+\delta t)\delta V_{ij} + \delta A_{ij}\ \delta t\ h_{Tij}(t+\delta t) + \delta A_{ij}\delta t[h_{cij}(t+\delta t) + h_{bij}(t+\delta t)]\}\theta_{ij}(t+\delta t)$$
$$- \delta A_{ij}\ \delta t\ h_{Tij}(t+\delta t)\theta_G(t+\delta t) - \delta A_{ij}\delta t[h_{cij}(t+\delta t) + h_{bij}(t+\delta t)]\theta_{wj}(t+\delta t)$$
$$= \{2\rho_{ij}(t)C_{pij}(t)\delta V_{ij} - \delta A_{ij}\ \delta t\ h_{Tij}(t) - \delta A_{ij}\delta t[h_{cij}(t) + h_{bij}(t)]\}\theta_{ij}(t)$$
$$+ \delta t\ h_{Tij}(t)\delta A_{ij}\theta_G(t) + \delta A_{ij}\ \delta t\ [h_{cij}(t) + h_{bij}(t)]\theta_{wj}(t)\ . \qquad (6.34)$$

For the top section inner shell FV (i4), the water-side surface velocity can be assumed to approach zero so that in eqn (6.34) $h_{ci4}(\xi) \to 0$. With the outer shell FVs (o1 $\to$ o4), only forced convection occurs at the water-side interface so that eqn (6.34) is appropriate but with subscript ij replaced by oj ($j = 1, 2, 3, 4$). Setting $h_{boj}(\xi) = 0$ and $\theta_G(\xi) = \theta_A(\xi)$ gives, for the ojth FV

$$[2\rho_{oj}(t+\delta t)C_{poj}(t+\delta t)\delta V_{oj} + \delta A_{oj}\ \delta t\ h_{Toj}(t+\delta t) + \delta A_{oj}\delta t\ h_{coj}(t+\delta t)]\theta_{oj}(t+\delta t)$$
$$- \delta A_{oj}\ \delta t\ h_{Toj}(t+\delta t)\theta_A(t+\delta t) - \delta A_{oj}\delta t\ h_{coj}(t+\delta t)\theta_{wj}(t+\delta t)$$
$$= [2\rho_{oj}(t)C_{poj}(t)\delta V_{oj} - \delta A_{oj}\ \delta t\ h_{Toj}(t) - \delta A_{oj}\delta t\ h_{coj}(t)]\theta_{oj}(t)$$
$$+ \delta t\ h_{Toj}(t)\delta A_{oj}\theta_A(t) + \delta A_{oj}\delta t\ h_{coj}(t)\theta_{wj}(t)\ . \qquad (6.35)$$

In this case, the total heat transfer coefficient, $h_{Toj}(\xi)$, may be separated into its convective and radiant parts. This would be required, for example, in cases where large inter-surface temperature differences are expected and the analysis purpose is to evaluate the impact of different levels of casing insulation.

Radiators

Figure 6.9 also shows a parallel flow, flat plate radiator with high level inlet and low level outlet connections. The top horizontal tube is connected to an identical bottom tube by a number of finned columns. For each column segment a discrete FV scheme may be established (here 7 FVs are employed). The inter-connection of these segments then gives the complete radiator model. Alternatively, to minimise the total number of FVs, the central and peripheral finned columns can be grouped and considered as single regions so that the complete radiator model will now comprise 21 FVs. In either case only three FV types are present: water FVs conforming to eqn (3.15), horizontal tube wall FVs conforming to eqn (6.35), and finned column tube wall FVs requiring a modification to eqn (6.35) to account for the presence of fins. For this last FV type, the total heat transfer coefficient, $h^*_{Toj}(\xi)$, is given by

$$h^*_{Toj}(\xi) = \frac{1}{\delta A_T}\left[h_{Toj}(\xi)\delta A_{oj} + 2(h_{Toj}(\xi)p\,k\,\delta A_f)^{\frac{1}{2}} \tanh ml\right]$$

$$+ \frac{h_{roj}(\xi)\left[\theta_A(\xi) - 0.5\left(\sum_{i=1}^{M} f_{oj\to i}\theta_{si}^4\right)^{0.25}\right]\left[h_{Toj}(\xi)\delta A_{oj} + 2(h_{Toj}(\xi)p\,k\,\delta A_f)^{\frac{1}{2}} \tanh ml + 2k\,\delta A_f/\cosh ml\right]}{\delta A_T h_{Toj}(\xi)[\theta_{oj}(\xi) - \theta_A(\xi)]}$$

where, with reference to figure 6.10, δA_{oj} is the surface area of the finned tube base (m^2), p the fin perimeter (m), k the conductivity of the fin material (W m^{-1}°C^{-1}), δA_f the fin cross-sectional area, $m = (h_{Toj}(\xi)p/k\,\delta A_f)^{\frac{1}{2}}$, l the fin length (m), h_{roj} the radiative heat transfer coefficient at the radiator surface (W m^{-2}°C^{-1}), $f_{oj\to i}$ the view factor between oj and room surface i, M the number of room surfaces in radiative communication with oj, θ_{si} the temperature of surface j (°C), $\delta A_T = \delta A_{oj} + l\,p/2$ and ξ refers to the time-row.

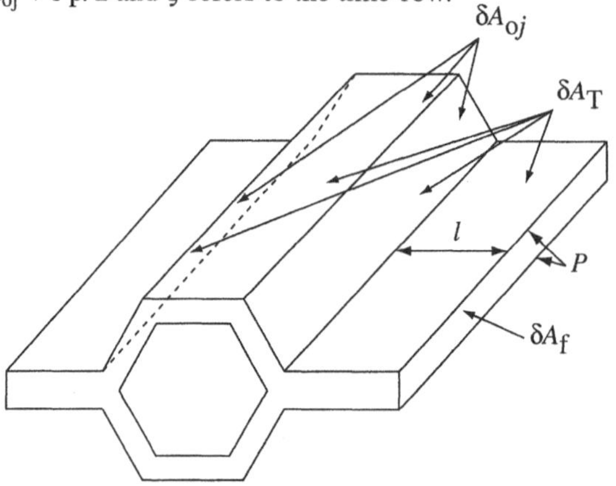

Figure 6.10: A finned radiator column segment.

Hot water cylinder

This is treated in the same manner as the liquid thermal store of the active solar system, requiring the use of eqns (3.12) and (6.28).

Pump

The heat gain characteristic of a pump may be included by considering the pump as an equivalent length of tube with an internal energy source derived from considerations of electrical power conversion to heat.

Table 6.6 lists the conservation equations for the FV types comprising the wet central heating system and figure 6.11 shows the system matrix equation to result for the example problem shown in figure 6.9. As with previous systems, a corresponding mass balance matrix equation may be established to enable the imposition of flow regulation in co-operation with a flow model as established in chapter 5.

6.3 New and renewable energy conversion systems

Renewable energy (RE) systems have typically been pursued at the strategic level, with the grid connection of medium-to-large scale hydro stations, bio-gas plant and wind farms. In order to avoid problems with network balancing and power quality, it has been estimated that the deployment of RE systems with limited control possibilities should be restricted to around 25% of total installed capacity (EA 1999). This limitation is due to the intermittent nature of RE sources, requiring controllable, fast responding reserve capacity to compensate for fluctuations in output; and energy storage to compensate for non-availability. To achieve a greater penetration of RE, alternative deployment approaches will be required. For an equivalent installed capacity of 2000 MW (i.e. a conventional power station), table 6.7 gives the number of systems required to maintain the same installed capacity as the technology scale reduces. Not surprisingly, the replication extent is inversely proportional to the deployment scale. Less obvious, perhaps, is the increased opportunity for energy demand reduction and economic stimulation at the reduced scale where the 'greening' of society can more readily be enacted (Born *et al* 2001).

Table 6.7: The spectrum of RE deployment opportunities.

Scale	*Replication extent*†
Power station	1 @ 2000 MW
Wind farm	100 @ 20 MW
Wave power	4000 @ 0.5 MW
Micro gas turbine	40,000 @ 50 kW
Building-integrated RE system	200,000 @ 10 kW

† The number of RE replications would be 3-5 times greater if the requirement is to match energy production.

Recent developments in small scale new and renewable energy systems (NRE)—in the form of photovoltaic modules, micro gas turbines, fuel cells and ducted wind turbines—have given rise to the concept of a micro power approach whereby NRE technologies are embedded within the built environment. In conjunction with demand reduction measures, the approach requires the matching of local supply potentials to optimised demand: passive solar and other technologies are used to reduce energy requirements, and active NRE systems are used to meet a significant proportion of the residual demand. Any energy deficit is met from the public electricity supply operating co-operatively with the NRE systems. For the approach to be successful, NRE systems deployment and energy efficiency measures must be pursued together. Consider the following scenario.

For a northern European climate characterised by a mean annual global horizontal solar irradiance of 150 W m^{-2}, the mean power production from a photovoltaic component of 13% conversion efficiency is approximately 20 W m^{-2}. For a mean wind speed of 5 m s^{-1}, the power produced by a micro wind turbine will be of a similar order of magnitude (but with a different profile shape). The point is that such power densities are significantly lower than a typical

Table 6.6: Characteristic equations and coefficient formulae, wet central heating system.

FV location	*Characteristic equation*
Radiator shell and pump/pipe casing.	$-A_1(t+\delta t)\theta_b(t+\delta t) + A_2(t+\delta t)\theta_m(t+\delta t)$ $- A_3(t+\delta t)\theta_o(t+\delta t) - A_4(t+\delta t)q_I(t+\delta t)$ $= A_1(t)\theta_b(t) + A_5(t)\theta_m(t) + A_3(t)\theta_o(t) + A_4(t)q_I(t)$
Boiler combustion chamber; boiler, pipe and pump water.	$-\sum_{f=1}^{N} B_f(t+\delta t)\theta_f(t+\delta t) + B_1(t+\delta t)\theta_b(t+\delta t)$ $-\sum_{o=1}^{M} B_o(t+\delta t)\theta_o(t+\delta t) - B_2(t+\delta t)q_I(t+\delta t)$ $= \sum_{f=1}^{N} B_f(t)\theta_f(t) + B_3(t)\theta_b(t) + \sum_{o=1}^{M} B_o(t)\theta_o(t) + B_2(t)q_I(t)$
Water cylinder fluid and surfaces.	As table 6.4.

θ_b = bulk fluid temperature, θ_f = fluid temperature, θ_o = boundary or outer surface temperature, θ_m = membrane or shell temperature.

Coefficient formulae

$A_1(\xi) = h_{T,i}(\xi)\delta A_I \delta t/\delta V$	$B_o(\xi) = h_{T,o}(\xi)\delta A_I \delta t/\delta V$
$A_2(\xi) = 2\rho_I(\xi)C_I(\xi) + A_1(\xi) + A_3(\xi)$	$B_1(\xi) = 2\rho_I(\xi)C_I(\xi) + \sum_{f=1}^{N} B_f(\xi) + \sum_{o=1}^{M} B_o(\xi)$
$A_3(\xi) = h_{T,o}(\xi)\delta A_o \delta t/\delta V$	$B_f(\xi) = m_f(\xi)C_f(\xi)\delta t/\delta V$
$A_4(\xi) = \delta t/\delta V$	$B_2(\xi) = \delta t/\delta V$
$A_5(\xi) = 2\rho_I(\xi)C_I(\xi) - A_1(\xi) - A_3(\xi)$	$B_3(\xi) = 2\rho_I(\xi)C_I(\xi) - \sum_{f=1}^{N} B_f(\xi) - \sum_{o=1}^{M} B_o(\xi)$

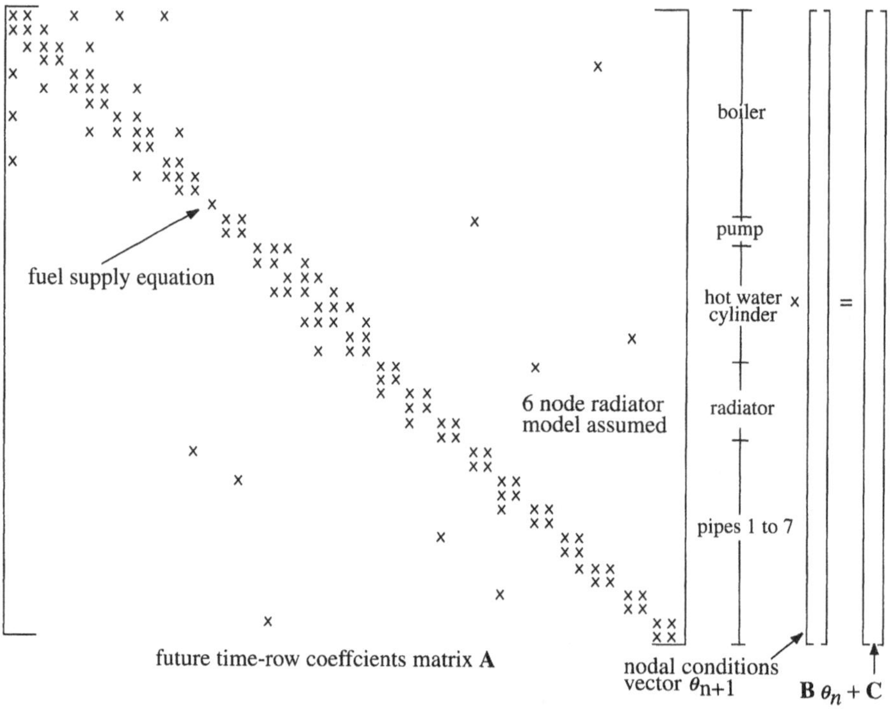

Figure 6.11: Wet central heating system energy balance matrix equation.

building's energy demand. In the UK, for example, an office building will have a demand of the order of 300 kWh $m^{-2}y^{-1}$ (DETR 1994). Assuming 250 working days in the year, this translates to approximately 50 W m^{-2} of floor area or 100 W m^{-2} of facade area, i.e 5 times the available renewable resource! This situation changes dramatically when energy efficiency measures are firstly put in place. Consider the case where aggressive measures result in a reduced demand of 70 kWh $m^{-2}y^{-1}$, which translates to approximately 24 W m^{-2} of facade area. Now the supply matches the demand, quantitatively if not temporally: the challenge then is to overcome the temporal mismatch without recourse to importing/exporting from/to the public electricity supply (PES).

The goal within a micro power approach is to utilise energy efficiency and/or passive solar measures to reduce the overall energy demand and adjust the demand profiles to facilitate NRE systems integration. At times where the demand exceeds the NRE supply, the deficit is met either from local storage or from the PES. At times when the demand is less than the available NRE supply, the excess is used to charge the local storage system.

Two possibilities exist to avoid exporting to the PES when the storage system is fully charged. Either the excess can be directed to a local dump load or the RE systems can be forced to operate at lower efficiency (i.e. non-optimally) by adjusting the load conditions experienced by the micro generators. In the case of photovoltaic components, for example, this can be done by increasing the load voltage above that of the maximum power point. Such a non-exporting arrangement minimises the risk of compromising power quality and overcomes the concerns of the network operators with respect to the connection of distributed, stochastic RE systems.

The identification of the optimum mix of demand reduction measures and NRE systems for deployment is a task ideally suited to the integrated simulation approach. For example, simulations can be performed to assess the performance of different energy efficiency measures when deployed jointly and severally. This might involve the assessment of the potential of daylight utilisation to reduce electricity consumption for lighting, transparently insulated facade components to off-set heating loads, advanced glazings to reduce heat loss/gain, and smart control to enforce minimum heating/lighting set-points. In this way the energy demand profile may be reshaped to temporally accommodate the introduction of NRE systems. After the required level of demand reduction has been achieved, simulations can be performed to identify appropriate NRE technology combinations to meet the residual temporal demand, eliminate inherent conflicts (e.g. between a facade-integrated photovoltaic system and daylight utilisation) and minimise the capacity of the required energy storage system. Chapter 8 describes the use of integrated simulation to explore such a scenario.

Because NRE components typically straddle the thermal and electrical domains—that is, they interact with a building's fabric, plant, lighting and control systems—they cannot be modelled in isolation. Instead, a coupled modelling approach is necessary: this requires the establishment of a power flow model of the building's electrical network and power producing/consuming equipment (Stagg and El-Abiad 1968, Gross 1986).

6.3.1 Electrical power flow

The approach to network air flow, as elaborated in Chapter 5, can be applied to form such a power flow model, as demonstrated by Kelly (1998). An electrical circuit may be conceived as a network of nodes representing the junctions between conducting elements and locations where power is extracted to feed loads or added from the PES or local NRE systems.

Application of Kirchhoff's current law to some arbitrary node, i, with n connected nodes, forms the basis for the network power flow solution:

$$\sum_{j=1}^{n} \tilde{I}_{i,j} = 0. \tag{6.36}$$

Here the emphasis is on alternating current, so that $\tilde{I}$ is a complex quantity, i.e $\tilde{I} = I(\cos\theta + j\sin\theta)$, where θ is the phase angle. Separating out the generation (G), transmission (T) and load (L) currents in eqn (6.36) gives

$$\sum_{j=1}^{nG} \tilde{I}_{Gi,j} - \sum_{j=1}^{nT} \tilde{I}_{Li,j} + \sum_{j=1}^{nL} \tilde{I}_{Ti,j} = 0\,. \tag{6.37}$$

In a building simulation context, $\tilde{I}_G$ and $\tilde{I}_L$ are treated as boundary conditions in the power flow solution by arranging that each have counterpart models that connect the thermal and electrical domains. Examples include photovoltaic components and luminaire models respectively.

The current transmitted between nodes is now expressed in terms of the prevailing (complex) voltage differences and connecting component admittances, $\tilde{Y}$ (the inverse of the component's impedance):

$$\tilde{I}_{Ti,j} = (\tilde{V}_i - \tilde{V}_j)\tilde{Y}_{i,j}\,. \tag{6.38}$$

Substituting eqn (6.38) into eqn (6.37) and rearranging terms gives

$$\left[\sum_{j=1}^{nT}(\tilde{V}_i - \tilde{V}_j)\tilde{Y}_{i,j}\right] = \sum_{j=1}^{nL} \tilde{I}_{Li,j} - \sum_{j=1}^{nG} \tilde{I}_{Gi,j}\,. \tag{6.39}$$

While eqn (6.39) offers a suitable means of determining the network voltages, the solution is only achievable where the boundary conditions are given as current flows. Since the electrical loads/sources are usually defined in terms of their power demand/supply, it is necessary to solve the network in terms of power flows. At any network node, i, the (complex) power flow is given by

$$\tilde{S}_i = \tilde{V}_i \tilde{I}_i' \tag{6.40}$$

where $\tilde{I}'$ is the conjugate of $\tilde{I}$. Eqn (3.39) may then be rewritten as

$$\tilde{V}_i \left[\sum_{j=1}^{nT}(\tilde{V}_i - \tilde{V}_j)\tilde{Y}_{i,j}\right]' = \sum_{j=1}^{nL} \tilde{S}_{Li,j} - \sum_{j=1}^{nG} \tilde{S}_{Gi,j} \tag{6.41}$$

which, for solution convenience, may be rearranged to give

$$\tilde{V}_i \left[\sum_{j=1}^{nT} \tilde{V}_j \tilde{\mathbf{Y}}_{i,j}\right]' = \sum_{j=1}^{nL} \tilde{S}_{Li,j} - \sum_{j=1}^{nG} \tilde{S}_{Gi,j} \tag{6.42}$$

where $\tilde{\mathbf{Y}}_{i,i} = \sum_{j=1}^{nT} \tilde{Y}_{i,j}$ and $\tilde{\mathbf{Y}}_{i,j} = -\tilde{Y}_{i,j}$.

Eqn (6.42) can be split into its real (P) and reactive (Q) power components:

$$\left[\sum_{j=1}^{nT} V_i(V_j \mathbf{Y}_{i,j})\right] \cos(\theta_i - \theta_j - \gamma_{i,j}) = \sum_{j=1}^{nL} P_{Li,j} - \sum_{j=1}^{nG} P_{Gi,j},$$

$$\left[\sum_{j=1}^{nT} V_i(V_j Y_{i,j})\right] \sin(\theta_i - \theta_j - \gamma_{i,j}) = \sum_{j=1}^{nL} Q_{Li,j} - \sum_{j=1}^{nG} Q_{Gi,j} \,. \tag{6.43}$$

At each node in the network there are 2n non-linear equations of the form given by eqn (6.43). There are also 2n unknowns: the voltage magnitude at each node and the voltage phase angle, θ_i. At a given time, assumed values of V and θ can be inserted to determine the nodal power residuals. In a manner analogous to that used to obtain the iterative solution of a fluid flow network (§5.1.5), these residuals may then be used to obtain nodal voltage/phase angle corrections from

$$\begin{bmatrix} \Delta P_1 \\ \Delta P_2 \\ \cdot \\ \cdot \\ \cdot \\ \Delta P_n \\ \cdots \\ \Delta Q_1 \\ \Delta Q_2 \\ \cdot \\ \cdot \\ \cdot \\ \Delta Q_n \end{bmatrix} = - \begin{bmatrix} \frac{\partial P_1}{\partial V_1} \cdots \frac{\partial P_1}{\partial V_n} & \frac{\partial P_1}{\partial \theta_1} \cdots \frac{\partial P_1}{\partial \theta_n} \\ \cdot \quad \cdot & \cdot \quad \cdot \\ \cdot \quad \cdot & \cdot \quad \cdot \\ \frac{\partial P_n}{\partial V_1} \cdots \frac{\partial P_n}{\partial V_n} & \frac{\partial P_n}{\partial \theta_1} \cdots \frac{\partial P_n}{\partial \theta_n} \\ \cdots \quad \cdots & \cdots \quad \cdots \\ \frac{\partial Q_1}{\partial V_1} \cdots \frac{\partial Q_1}{\partial V_n} & \frac{\partial Q_1}{\partial \theta_1} \cdots \frac{\partial Q_1}{\partial \theta_n} \\ \cdot \quad \cdot & \cdot \quad \cdot \\ \cdot \quad \cdot & \cdot \quad \cdot \\ \frac{\partial Q_n}{\partial V_1} \cdots \frac{\partial Q_n}{\partial V_n} & \frac{\partial Q_n}{\partial \theta_1} \cdots \frac{\partial Q_n}{\partial \theta_n} \end{bmatrix} \times \begin{bmatrix} \Delta V_1 \\ \Delta V_2 \\ \cdot \\ \cdot \\ \cdot \\ \Delta V_n \\ \cdots \\ \Delta \theta_1 \\ \Delta \theta_2 \\ \cdot \\ \cdot \\ \cdot \\ \Delta \theta_n \end{bmatrix} \tag{6.44}$$

where ΔP is the real power residual, ΔQ the reactive power residual, ΔV the nodal voltage correction and $\Delta\theta$ the nodal voltage phase angle correction.

This matrix equation may be iteratively established and solved, using the method described in §5.1.5, until the nodal power imbalance is reduced to insignificance. The nodal voltages may then be used to determine the network currents, branch power flows, required power imports etc. Note, however, that the method does not allow for the effects of transient behaviour as would be required to study the impact of particular schemes on power quality.

6.3.2 Electrical component models

The solution of such a power flow network requires models for the connecting components, sources of power and loads. Kelly (1998) has described models for conductors, transformers, combined heat and power units and RE components while other researchers have developed models of wind turbines (Grant *et al* 1994, Born 2001). Three of these models, corresponding to a conductor, a photovoltaic panel and a ducted wind turbine, are described here to exemplify the approach.

Single phase conductor

Consider the single phase conductor of Figure 6.12. The current flow is given by

$$\begin{bmatrix} \tilde{Z}_{ph} & 0 \\ 0 & \tilde{Z}_n \end{bmatrix} \begin{bmatrix} \tilde{I}_{ph} \\ \tilde{I}_n \end{bmatrix} = \begin{bmatrix} \tilde{V}_{iph} - \tilde{V}_{jph} \\ \tilde{V}_{in} - \tilde{V}_{jn} \end{bmatrix}.$$

where $\tilde{Z}_{ph}$ is impedance ($= R_{ph} + jX_{ph}$), R and X are resistance and inductance respectively (Ω) and ph and n relate to the phase conductor and neutral lines.

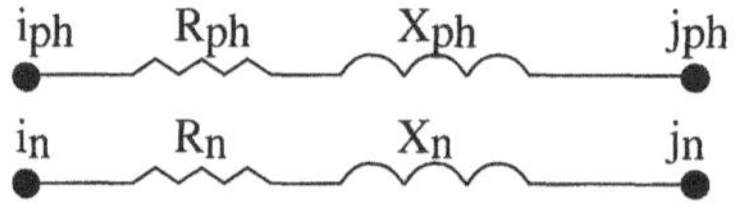

Figure 6.12: Single phase conductor.

The model can be simplified by referencing the voltages to neutral, achieved by subtracting the equation for the current flow through the neutral from the current flow through the phase conductor:

$$\begin{bmatrix} \tilde{Z}_{ph} & -\tilde{Z}_n \\ 0 & 0 \end{bmatrix} \begin{bmatrix} \tilde{I}_{ph} \\ \tilde{I}_n \end{bmatrix} = \begin{bmatrix} (\tilde{V}_{iph} - \tilde{V}_{in}) - (\tilde{V}_{jph} - \tilde{V}_{jn}) \\ 0 \end{bmatrix}. \tag{6.45}$$

Applying Kirchhoff's current law gives $\tilde{I}_n = -\tilde{I}_{ph}$, allowing the neutral current to be eliminated from eqn (6.45):

$$(\tilde{Z}_{ph} + \tilde{Z}_n)\,\tilde{I}_{ph} = (\tilde{V}_{iph-n} - \tilde{V}_{jph-n}) \tag{6.46}$$

where $\tilde{V}_{iph-n} = \tilde{V}_{iph} - \tilde{V}_{in}$ and $\tilde{V}_{jph-n} = \tilde{V}_{jph} - \tilde{V}_{jn}$. Rearranging eqn (6.46) gives the expression for the phase current:

$$\tilde{I}_{ph} = \tilde{Y}_{ph}(\tilde{V}_{iph-n} - \tilde{V}_{jph-n}) \tag{6.47}$$

where $\tilde{Y}_{ph} = [\tilde{Z}_{ph} + \tilde{Z}_n]^{-1}$.

Finally, eqn (6.47) can be transformed to give an expression for the power flow from node j to i:

$$\tilde{V}_{iph-n}[\tilde{I}_{ph}]' = \tilde{V}_{iph-n}[\tilde{Y}_{ph}]'\,[\tilde{V}_{iph-n} - \tilde{V}_{jph-n}]'.$$

The admittance, $\tilde{Y}_{ph}$, is the value used in eqns (6.38)—(6.43). It characterises the single-phase conductor and links nodes i and j.

Facade-integrated photovoltaic components

Consider the facade-integrated photovoltaic component shown in Figure 6.13. Solar radiation is transmitted to the surface of the photovoltaic layer through the glass cover by the processes described in §7.4.4. Before this flux is applied to the corresponding FV conservation equation its magnitude is reduced to reflect the fact that not all the absorbed solar radiation, α_i, will be converted to heat since a proportion is converted to electrical energy:

$$\alpha_i' = \alpha_i - q_{ei}$$

where α_i' is the actual absorption and q_{ei} is the PV power output (W), which may be determined from a model described by Buresch (1983):

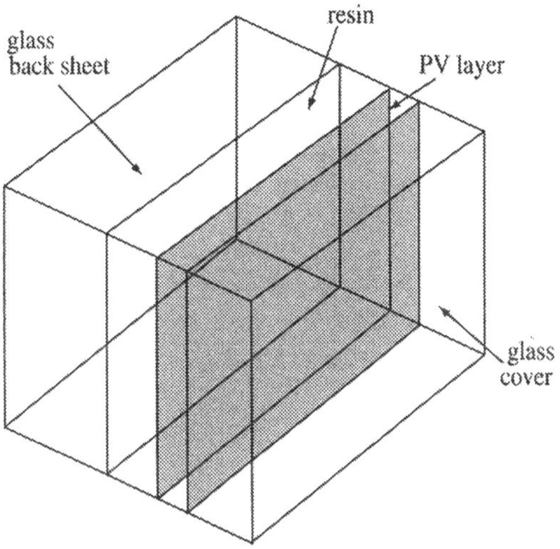

Figure 6.13: A facade-integrated photovoltaic component.

$$q_{ei} = nc\left[V_i I_g\left(1 - \exp\left\{\frac{eV_i}{\ell\sigma\;{}_i}\right\}\right) - \frac{V_i I_{sc}\alpha'_i}{\alpha'_{i(ref)}}\right].$$

where θ_i is the temperature (K) of the PV material node (as determined from the conduction model), V_i the node voltage, I_g the light generated current, I_{sc} the short circuit current, λ the electron charge (1.6×10^{-19}Coulombs), n the number of series connected cells, c the number of parallel connected cells, and σ the Stefan-Boltzmann constant (1.38×10^{-23}W m^{-2}K^{-4}). The light generated current is calculated as a function of the solar energy absorbed in the PV layer, α_i, when referenced to the solar absorption, $\alpha'_{i(ref)}$, corresponding to the standard test condition. Note that the thermal and electrical energy balance equations of the PV model are coupled, with the model translating thermal phenomena and applying the output power to the corresponding node in the electrical network. Note also that the PV cell temperature, θ_i, is given by the temperature of the corresponding intra-constructional node represented by eqn (3.5).

Ducted wind turbines

The power output, P_w, from a conventional wind turbine may be characterised by

$$P_w = 0.5 C_P \rho \pi R^2 v^3 \qquad (6.48)$$

where v is the wind speed (m $^{-1}$), R the rotor radius (m), ρ the air density (kg m^{-3}) and C_P a power coefficient that represents the effectiveness of the turbine rotor. For a given turbine, C_P is a function of the tip speed ratio t_s:

$$t_s = \frac{\omega R}{v} \qquad (6.49)$$

where ω is the angular velocity of the rotor.

Ideas for wind energy conversion in an urban environment have ranged from the notion of placing wind turbines on roofs, to building-integrated ducted wind turbines (Grant *et al* 1994). With reference to figure 6.14, when the wind blows at right angles to the face of a tall building, stagnation will occur at about two-thirds of the total height. Below this level, a rolling vortex is formed; above it, the air rises to pass over the roof. If the roof is flat, it will separate from the upwind edge, possibly reattaching some distance downstream. A ducted wind turbine (DWT) is designed to draw air from the high-pressure region on the upwind face of the building and exhaust it into the low-pressure region above the flat roof. The device will operate efficiently over a 60° range of wind directions, with the ducting serving to dampen any turbulence in the air stream, which can be a serious problem when other buildings are in close proximity.

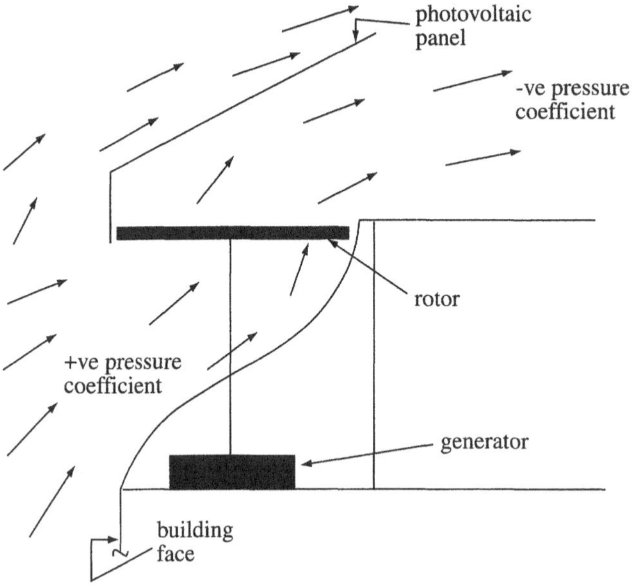

Figure 6.14: A building-integrated ducted wind turbine.

Because most conventional grid-connected wind turbines are constrained to turn at constant angular velocity (ω), as the wind speed varies the value of t_s (and hence C_P) will change. A DWT, being a small unit for autonomous use, is unlikely to be connected to the electrical grid and so can operate in variable-speed mode. Maintaining a constant value for t_s allows the DWT to operate at a constant (optimum) value of C_P. Combining eqns (6.49) and (6.48):

$$P_w = T\omega = \frac{0.5 C_P \rho \pi R^2 \omega R}{t_s^3} = K\omega^3$$

where K is a constant and the torque T at the rotor shaft follows a quadratic characteristic, $T = K\omega^2$. In practice it is a simple matter to control the generator to produce a quadratic torque characteristic at the rotor shaft, and so a near-constant value of C_P may be achieved.

This mode of operation can be abandoned in high wind speeds, where it is necessary to limit the power produced. Here, the electrical load on the generator may be increased, reducing the rotor speed, stalling its blades and reducing C_P to the required value. At low wind speeds (below about 5 m s^{-1}), the power output will fall below the values indicated in eqn (6.49) as a result of disproportionate mechanical losses and low generator efficiency.

In order to increase the overall power density, a prototype DWT (figure 6.14) has been established that incorporates a PV component within its spoiler (Clarke *et al* 2000, Grant and Dannecker 2000). Section 8.3 describes the application of DWTs (and PV) to a real building.

6.4 Control systems

The basic operational principles of building control systems are widely described in the literature (e.g. BS 1967, Wolsey 1975, Letherman 1981, Dexter 1983, Dexter and Graham 1984, Hartman 1988). Traditionally, control system modelling is achieved using response functions whereby the differential equations relating to system processes and control components are converted into their algebraic counterpart using the Laplace transform as described in chapter 2. For a linear, time invariant system, this gives rise to transfer functions (Leigh 1992, Liptak 1995) that relate control action to effect. Such an approach has limited applicability in the non-linear, time varying context of integrated building performance simulation. In response, a numerical approach to control system modelling has emerged (Hitchin 1991) and numerical simulation has been used as the basis of a virtual testbed for algorithm evaluation (Haves *et al* 1996).

Within the approach described here, a control system is conceived as a collection of control loops, with each loop serving to manipulate a related model parameter such as room temperature, boiler fuel supply rate, luminaire voltage, window opening extent or shading device position. In fact, any model parameter may be 'actuated', whether it corresponds to a real entity, such as a dimmer switch, or an abstract entity, such as room flux input/extract. This is a useful feature in that ideal control regimes can be established at an early design stage in order to constrain simulations in support of design exploration. The imposition of the desired temperature profile on a room in order to explore design options is a simple example. Real control system response characteristics can then be included at a later stage.

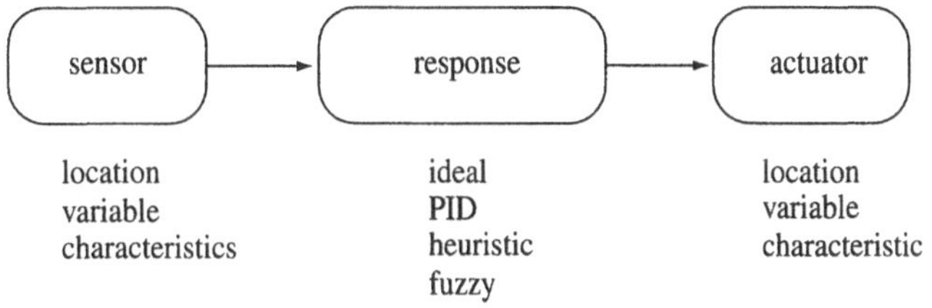

Figure 6.15: Elements of a control loop.

As shown in Figure 6.15, a control loop comprises three basic elements: a sensor to detect a model parameter or group of parameters, an actuator to deliver the control action and a regulation law to represent the characteristics of the sensor and actuator devices (delay, hysteresis, off-set etc) and determine the control action as a function of the sensed condition. Table 6.8 gives some typical examples of each element.

Table 6.8: Examples of control loop elements.

Sensed/actuated parameters	*Regulation law*	*Sensed/actuated parameters*	*Regulation law*
time of day/year	ideal	CO_2 level	duty cycling
climate	PID combinations	ventilation rate	load shedding
various temperatures	optimum start/stop	room air velocity	adaptive
glare and illuminance	weather compensation	hygro-thermal properties	fuzzy logic
luminaire status	cascade	humidity	neural network
occupancy	enthalpy cycle		

Once established, the control system may be imposed on the solution of the coupled equation-sets representing the domains comprising the integrated model. This is best achieved by implementing the control action within the solution procedure in order to influence the future time-row state variables, either directly as described in §4.2 or via an iterative algorithm.

At each time-step, the condition detected by the sensor is fed to some algorithm representing controller response. This algorithm acts to modify one or more model parameters prior to final matrix equation solution or reformulation. In this way, the sensed condition(s) may be controlled as a function of a prescribed state (ideal regulation), a deviation (proportional control), a deviation rate of change (derivative control) or a deviation past history (integral control). Controller types can be combined (proportional + integral action etc) and the effects of response rates or learning algorithms incorporated.

For a proportional controller, regulation will take the form

$$\theta_p = K_p \theta_\delta \tag{6.50}$$

where θ_δ is the deviation of the input signal from the defined set point, θ_p the corresponding change in the output signal and K_p the proportional gain factor, which is ideally a constant and will be dependent on the adjustment of the controller.

To improve controller response or to remove offset, a derivative and/or integral component may be added to the basic proportional control action. Derivative action is defined by

$$\theta_d = K_p T_d (d\theta_\delta/dt)$$

where T_d is the derivative action time; and integral action by

$$\theta_i = \frac{K_p}{T_i} \int \theta_\delta dt$$

where T_i is the integral action time. A three component, or PID, scheme can now be established as

$$\theta_{p+d+i} = K_p \left[\theta_\delta + T_d(d\theta_\delta/dt) + (1/T_i) \int \theta_\delta dt \right].$$

In cases where the sensor or actuator has an inherent lag, it is possible to locate an additional equation-set within the system matrix equation. Consider a thermostatic radiator valve (TRV). The sensor is a wax filled capsule which expands against a spring to cause the valve to throttle water flow. The valve stroke is a continuous function of the sensed temperature deviation and so control action is proportional. Valve manufacturers will normally have data that describes the relationship between the sensed temperature and valve position, and between valve position and flow rate. These data can be re-expressed in the form of figure 6.16 allowing the proportional gain to be determined for any operating pressure.

Applying an energy balance to the wax capsule:

$$\rho CV \frac{d\theta_w}{dt} = \sum_j (h_c + h_r) A(\theta_j - \theta_w)$$

where θ_w is the wax temperature (°C), ρ the wax density (kg m^{-3}), C the wax specific heat capacity (J kg^{-1} $°C^{-1}$), h_c, h_r the convective and radiative heat transfer coefficients at the sensor head surface (W m^{-2} $°C^{-1}$) V the capsule volume (m^3), A the sensor head surface area (m^2), j the number of heat transfer paths (with air, radiator inlet water, surrounding surfaces etc) and θ_j the contact temperatures (air or surface).

The final TRV simulation equation is obtained by approximating the derivative by a finite difference and concatenating the results obtained at the present and future time-row of some arbitrary time step:

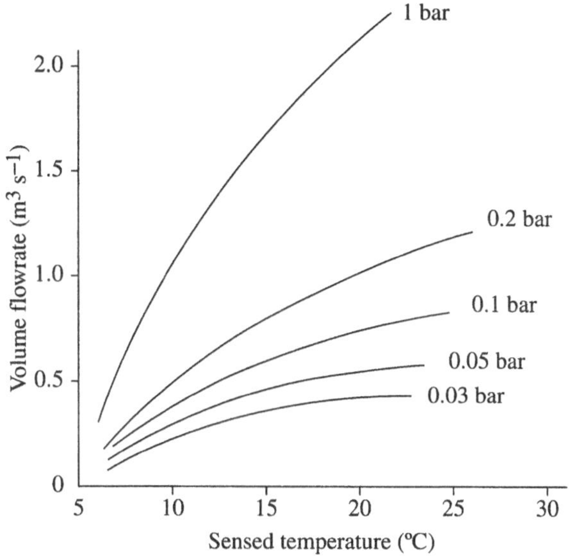

Figure 6.16: Characteristic curves for a thermostatic radiator valve.

$$[2\rho(t+\delta t)C(t+\delta t)V + A\delta t\, h_c(t+\delta t) + A\delta t\, h_r(t+\delta t)]\theta_w(t+\delta t)$$
$$- A\delta t\, h_c(t+\delta t)\theta_A(t+\delta t) - A\delta t\, h_r(t+\delta t)\theta_s(t+\delta t)$$
$$= [2\rho(t)C(t)V - A\delta t\, h_c(t) - A\delta t\, h_r(t)]\theta_w(t)$$
$$- A\delta t\, h_c(t)\theta_A(t) + A\delta t\, h_r(t)\theta_s(t)\,. \qquad (6.51)$$

This equation defines sensor response and can be incorporated within the system matrix equation to give the TRV sensed temperature for use in the determination of the mass flow rate from eqn (6.50).

As a second example, and for the problem of figure 6.11, eqn (6.51) can be added to the matrix equation to represent (say) a room thermostat, with an on/off controller acting to link the thermostat to the boiler fuel supply rate of the combustion node equation.

MacQueen (1997) has established control algorithms in a form suitable for use within a numerical simulation program. Essentially, each algorithm encompasses three elements: spatial, temporal and logical. Spatial elements relate to the distributed parameters that may be controlled (such as temperature and flow rate), temporal elements relate to the scheduling of control actions (including event management and controller response time), and logical elements relate to the control system characteristic and capability (such as hysteresis, and optimum start control). Utilising these concepts, Clarke *et al* (2001) investigated the possibility of placing simulation at the core of the supervisory element of a building energy management system in order to compare the efficacy of alternative control actions before enacting the best one. Of the many possible control strategies (Martin and Banyard 1998), table 6.9 lists a few examples that might be enabled by a simulation assisted control approach.

Recall the building-side solution technique of §4.2 and the concept of a characteristic equation that links the zone's control and actuated points. A building control strategy may be

Table 6.9: Some examples of simulation assisted control.

Control focus	Optimised parameter	Control focus	Optimised parameter
HVAC operation	start/stop time	district heating	match to load
night cooling	hours of operation	under-floor heating	period of operation
night set-back	set-back temperature	mixed ventilation	avoid overheating
boiler sequencing	heating system eff.	ice store charging	hours of operation
load shedding	energy consumption	ground heat pump	thermal storage
CHP	hours of operation		

established as one or more control functions, where a control function is associated with one or more zones to prescribe the time dependent environmental conditions and impose plant capacity restrictions. Each control function will then encapsulate the following information.

The control point location. A zone can be controlled as a function of its nodal state variables. Typical locations include air nodes, surface nodes, intra-capacity nodes (if the zone represents an electrical storage heater for example) or some fictitious node representing the weighting of several other nodal values. This last type allows control on the basis of realistic sensor characteristics or comfort considerations (e.g. the use of resultant temperature).

The controller type. Either the assumption of ideal regulation may be made or realistic response characteristics imposed (e.g. PID and sensor lag).

The interaction point location. This defines the point within the zone where flux interaction will occur depending on the system type. For example, an air conditioning system will interact with the zone's air nodes, a radiant heating panel with air and surface nodes, and an underfloor heating system with intra-constructional nodes.

The plant schedule. By sub-dividing the year into typical day types, and these days into typical periods, it is possible to adapt the control action over time by assigning different control constraints to each period. Examples include free-floating where no control is imposed, preheating (or cooling) where the aim is to raise (or lower) the control point temperature to the set-point required for the following period, flux injection/extraction where the intention is to attain a desired environmental condition, and the creation of an explicit link to an active plant network to test the ability of the plant to provide the required conditions.

Such facilities permit the creation of complex time varying control regimes to allow the examination of plant operation and the determination of optimum start times, night set-back temperatures and so on. In this way idealised control and/or real plant interactions can be intermixed to facilitate the design of an efficient and effective HVAC system.

6.5 Linking the building, flow and systems models

The matrix equation representing a connected plant system can now be interlocked with the building-side matrix equation as shown previously in figure 4.5. Three points are stressed here.

First, because this matrix equation is sparse, it is not established within computer memory as a two-dimensional array. Instead, it is held as a linked list of one-dimensional arrays, where each array corresponds to the component equation-set to emerge from the partitioning process (see chapter 4). These linked arrays define the topology of the whole matrix equation and are used in conjunction with coefficient generator routines to establish the component equation-sets at each computational time-step.

Second, as a general rule, the plant-side matrix equation will be substantially smaller than its building-side counterpart. For example, the total number of equations in the central heating system model of figure 6.9 is approximately 150, while a building-side model for an average-sized house will require approximately 1000 equations. It is therefore possible to process the plant model as a single equation-set rather than applying partitioning. Several sparse matrix solution techniques are available that are suitable for this purpose (e.g. Aasem 1993).

Third, the combined matrix equation will not yet define a soluble system because it will contain more unknowns than equations. This is overcome by the addition of control information to link certain state variables (e.g. room air temperature to the mass flow rate induced by a fan).

The simultaneous solution of the combined matrix equation then gives the evolution of the distributed state variables. Figure 6.17 summarises one possible solution procedure based on the iterative solution of nested domain equations. Such a procedure allows domain specific solvers to be used co-operatively. Given that the building time constants are generally greater than those related to the plant, the approach taken is to process the plant system at the same or greater frequency than those domains associated with the building. In this way, the plant equations may be solved for small time steps to accurately represent the effect of control action, while the more slowly evolving building may be solved less frequently. Where required, the processing frequencies may be matched and/or increased.

At each building-side time step, and for a given climate boundary condition, the air/liquid flow networks corresponding to the building and plant are established, control considerations imposed and the equations solved using the techniques of §5.1. Solution of these networks give the air and working fluid flow rates throughout the building and within the plant system respectively.

Using the techniques of §6.3, the electrical power flow network representing building-side entities (e.g. lighting, small power, photovoltaic facades, the public electricity supply etc) and plant components (e.g. fans, pumps, CHP plant etc) is established, constrained by control action and the equations solved.

The building-side multi-zone matrix equation is then established using the latest estimates of the fluid/power flows and plant induced flux injections/extractions. Equation solution is achieved by the method of chapter 4 whereby the control action is an intrinsic part of the solution process. This gives the building-side temperatures and heat flows.

Using the newly computed intra-construction temperatures, the moisture flow matrix equation is established and solved using the techniques of §6.3. This gives the moisture content distribution within the building fabric.

Using the building temperatures and air flow rates as boundary conditions, the CFD domain equations are established and solved using the technique of §5.2. This gives the intra-zone state in terms of the distribution of temperature, velocity, pressure, contaminants etc.

The building temperatures and air/liquid flow rates are then used, along with relevant control loops, to establish and solve the plant heat and mass flow matrix equations using the technique of §6.2. Solution of these equations gives the plant temperatures and flow rates.

Finally, where the coupling between the building and plant is highly non-linear (i.e. the properties of state depend on the state variables), iteration is employed to improve accuracy as illustrated in figure 6.17.

6.6 References and further reading

ASHRAE 1992 AC Systems and Equipment *ASHRAE Handbook* (ASHRAE)

Aasem E O 1993 Practical Simulation of Buildings and Air Conditioning Systems in the Transient Domain *PhD Thesis* (Glasgow: University of Strathclyde)

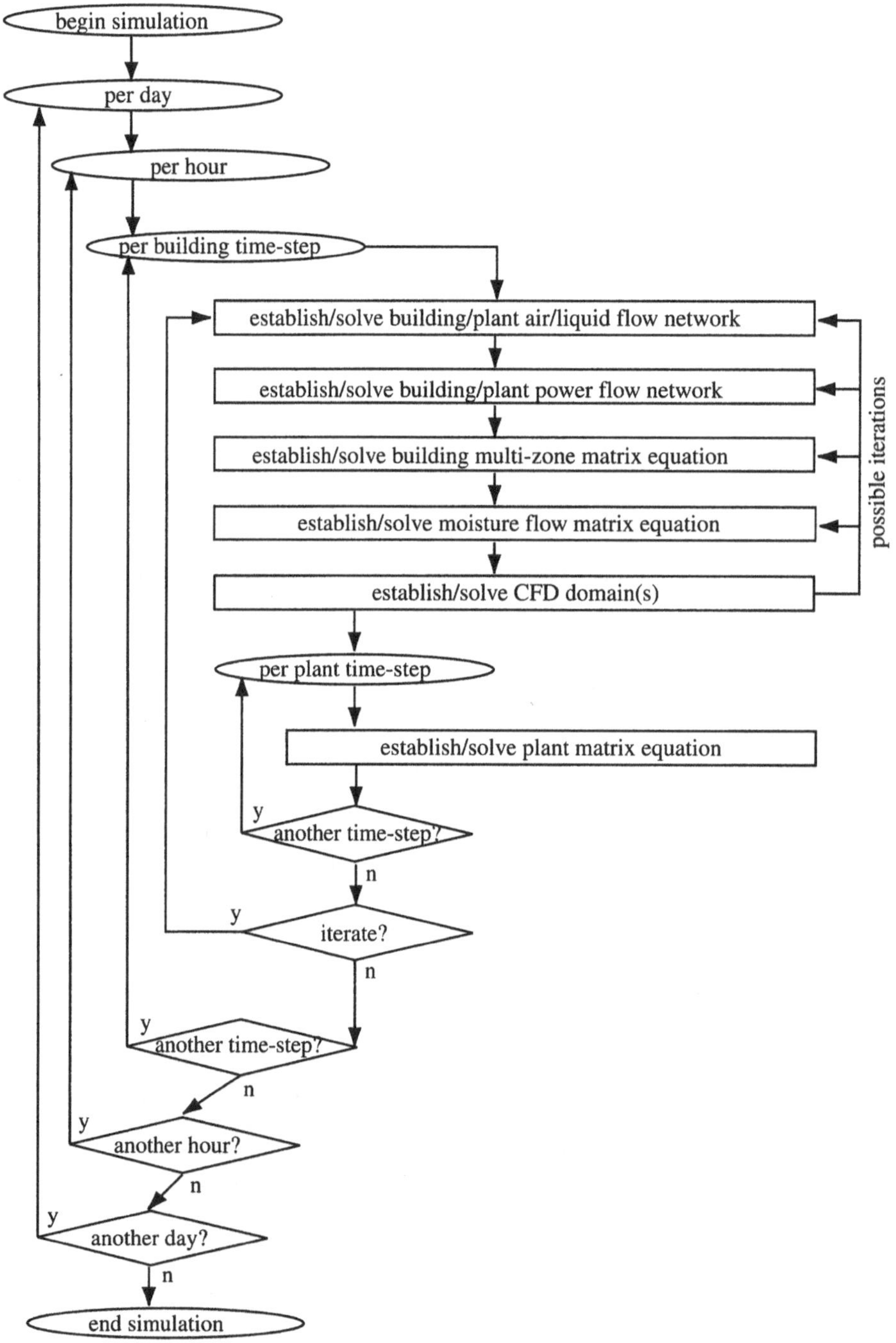

Figure 6.17: Iterative solution of nested domain equations.

Augenbroe G and Winkelmann F 1991 Integration of Simulation into the Building Design Process *Proc. Building Simulation '91* Sophia Antipolis, 367-74

Benton R 1982 Generalised Modelling and Simulation Software Tools for Building Systems *Proc. CIB W67 3rd Int. Symp. on Energy Conservation in the Built Env.* (Dublin) Born F 2000 Aiding Renewable Energy Integration through Complementary Demand-Supply Matching *PhD Thesis* (Glasgow: University of Strathclyde)

Bjork B 1989 Product Models of Buildings and their Relevance to Building Simulation *Proc. Building Simulation '89* Vancouver, 193-8

Bonin J L, Butto C, Dufresne J L, Grandpeix A, Joly J Y, Lahellec A, Platel V and Rigal M 1991 The ALMETH Project ZOOM Code: Results and Perspectives *Proc. Building Simulation '91* 355-63 (Sophia Antipolis: CSTB) .

Bonneau D, Rongere F X, Covalet D and Gautier B 1993 CLIM2000: Modular Software for Energy Simulation in Buildings *Proc. Building Simulation '93* Adelaide, 85-91

Born F J, Clarke J A, Johnstone C M, Kelly N J, Burt G, Dysko A, McDonald J and Hunter I B B 2001 On the Integration of Renewable Energy Systems Within the Built Environment *Building Serv. Eng. Res. & Technol.* **22**(1) 3-13

Bring A, Sahlin P and Sowell E 1992 The Neutral Model Format for Building Simulation *Research Project Report* (Stockholm, Department of Building Services Engineering, Royal Institute of Technology)

BS 1967 Glossary of Terms Used in Automatic Controlling and Regulating Systems (London: British Standards Institution)

Buhl W F, Erdem E, Nataf J M, Winkelmann F C, Moshier M A and Sowell E F 1990 The US EKS: Advances in the SPANK- based Energy Kernel System *Proc. 3rd Int. Conf. on System Simulation in Buildings* University of Liege, 107-50

Buhl W F, Erdem F C, Winkelmann F C and Sowell E F 1993 Recent Improvements in SPARK: Strong Component Decomposition, Multivalued Objects, and Graphical Interface *Proc. Building Simulation '93* Adelaide, 283-9

Buresch M 1983 *Photovoltaic Energy Systems - Design and Installation* (New York: McGraw-Hill)

Chow T T 1995a Generalization in Plant Component Modelling *Proc. Building Simulation '95* 48-55 (Madison: University of Wisconsin)

— 1995b Air-conditioning Plant Component Taxonomy by Primitive Parts *PhD Thesis* University of Strathclyde, Glasgow.

Chow T T, Clarke J A and Dunn A 1997 Primitive Parts: An Approach to Air-Conditioning Component Modelling *Energy and Buildings* **26** 165-73

Chow T T, Clarke J A and Dunn A 1998 Theoretical Basis of Primitive Part Modelling *ASHRAE Trans.* **104**(2)

Clarke J A and Mac Randal D F 1984 Development of a Prototypical Component-Based Energy Modelling System *Final report for Grant GR/D/26610* (Swindon: Science and Engineering Research Council)

Clarke J A, Cockroft J, Conner S, Hand J W, Kelly N, Moore R, O'Brien T and Strachan P A 2001 Control in Building Energy Management Systems: the Role of Simulation *Proc. Building Simulation '01* (Rio de Janeiro)

Clarke J A, Conner S and Johnstone C M 2000 Development of Demand and Supply Management Systems for Network Connected Photovoltaic Systems *Proc. 16th European Photovoltaic Solar Energy Conf.* (Glasgow)

DETR 1994 Best Practice Programme - Introduction to Energy Efficiency in Buildings (London: Department of Transport and the Regions)

Dexter A L 1983 Self-Tuning Control Algorithm for Single-Chip Microcomputer Implementation *Proc. IEE* **130**(5)

Dexter A L and Graham W J 1984 A Simple Self-Tuning Controller for Heating Plant Applications *Proc. Int. Symp. on the Perf. of HVAC Systems and Controls in Buildings (Garston: Building Research Establishment)*

Dubois A M 1990 Component Model Documentation and Management: the Evolution of the Proforma and Modelotheque *Proc. 3rd Int. Conf. on Systems Simulation in Buildings* University of Liege, 17-39.

Gough M C B 1986 B-Tas: A Software Package for Component-Bases HVAC System Modelling *Proc. CIB 5th Int. Symp.* (Bath)

Grant A D, Nasr el_Din S A and Kilpatrick J 1994 Development of a ducted wind energy converter *Wind Engineering* **18**(6) 297-304

Grant A D and Dannecker R 2000 A Hybrid PV/Wind Energy Module for Integration in Buildings *Proc. 16th European Photovoltaic Solar Energy Conf.* (Glasgow)

Gross C A 1986 *Power Systems Analysis (2nd Edn)* (New York: Wiley)

Hanby V 1985 The Application of Component-Based Simulation to HVAC Systems Design *Final Report for Grant GR/C/46376 (Swindon: Engineering and Physical Science Research Council)*

Hanby V I and Clarke J A 1987 Catalogue of UK HVAC component models *Final report for SERC grant GR/D/07459* (Glasgow: University of Strathclyde)

Hartman T B 1988 Dynamic Control: Fundamentals and Considerations *ASHRAE Trans* **2**(1)

Haves P, Norford L K, DeSimone M and Mei L 1996 A Standard Simulation Testbed for the Evaluation of Control Algorithms and Strategies *ASHRAE Trans.*

Hensen J L M 1991 On the Thermal Interaction of Building Structure and Heating and Ventilating System *PhD Thesis* (Eindhoven: University of Technology)

Hitchin E R 1991 Control Options in Building Energy Simulation Programs *Research Report 91/1* (UK Building Environmental Performance Analysis Club)

Holman J P 1981 *Heat Transfer* (New York: McGraw Hill)

Holmes M J 1982 The Simulation of Heating and Cooling Coils for Performance Analysis *Internal Report* (London: Ove Arup & Partners)

EA 1999 Workshop on Connection of Photovoltaic Systems (EA technology)

James R W and Marshall 1973 S A Dynamic Analysis of a Refrigeration System *J. Institute of Refrigeration*

Jones W P 1985 *Air conditioning Engineering* (London: Arnold)

Kelly N J 1998 Towards a Design Environment for Building-Integrated Energy Systems: The Integration of Electrical Power Flow Modelling with Building Simulation *PhD Thesis* (Glasgow: University of Strathclyde)

Kelly G, Park C, Clark D and May W 1984 HVACSIM+, a Dynamic Building/HVAC/Control Systems Simulation Model *Proc. Workshop on HVAC Controls Modeling and Simulation* (Atlanta: Georgia Institute of Technology)

Klein S A 1990 TRNSYS, A Transient System Simulation Program *Report 38-13* (Madison: University of Wisconsin)

Laret L 1989 Building & HVAC Simulation: the Need for Well-suited Models *Proc. Building Simulation '89* Vancouver, 199-204

Leigh J R 1992 *Control Theory: A Guided Tour* (London: Peregrinus)

Letherman K M 1981 Automatic Controls for Heating and Air Conditioning (Oxford: Pergamon)

Liptak B G 1995 *Process Control* (Oxford: Butterworth-Heinemann)

Lof G O G and Hawley R W 1948 Unsteady State Heat Transfer Between Air and Loose Solids *Ind. Eng. Chem.* **40** 1061-66

MacQueen J 1997 The Modelling and Simulation of Energy Management Control Systems *PhD Thesis* (Glasgow: University of Strathclyde)

Martin A J and Banyard C P 1998 Library of System Control Strategies *Application Guide AG7/98 (Bracknell: Building Services Research and Information Association)*

McLean D J 1982 The Simulation of Solar Energy Systems *PhD Thesis* (Glasgow: University of Strathclyde)

Myers G E, Mitchell J W and Norman R F 1967 The Transient Response of Crossflow Heat Exchangers, Evaporators and Condensers *ASME Journal of Heat Transfer* **89**(1) 75-80

Nataf J M 1995 A Translator from NMF to SPARK *Proc. Building Simulation '95* (Madison: University of Wisconsin) 529-36

Quick J P 1982 Simultaneous Solution of Room Response and Plant Performance *Proc. Int. Conf. on Systems Simulation in Buildings* (University of Liege)

Rohsenow W M 1952 A Method of Correlating Heat Transfer Data for Surface Boiling Liquids *ASME Trans.* **94** 969

Sahlin P and Sowell E F 1989 A Neutral Format for Building Simulation Models *Proc. Building Simulation '89* Vancouver, 147-54

Sahlin P 1993 IDA Modeller, a Man-Model Interface for Building Simulation *Proc. Building Simulation '93* Adelaide, 299-305

Sowell E F and Moshier M A 1995 HVAC Component Model Libraries for Equation-Based Solvers *Proc. Building Simulation '95* (Madison: University of Wisconsin) 189-96

Stagg G W and El-Abiad A H 1968 *Computer Methods in Power Systems Analysis* (New York: McGraw-Hill)

Stoecker W (Ed) 1975 *Procedures for Simulating the Performance of Components and Systems for Energy Calculations* (ASHRAE)

Tang D C 1984 Modelling of Heating and Air Conditioning Systems *PhD Thesis (Glasgow: University of Strathclyde)*

Wolsey W H 1975 Basic Principles of Automatic Control (London: Hutchinson Educational)

Yuill G K and Wray C P 1990 Overview of the ASHRAE TC4.7 Annotated Guide to Models and Algorithms for Energy Calculations Relating to HVAC Equipment *ASHRAE Trans* **96**(1)

7

Energy-related subsystems

The theories of chapters 2 through 6 require that geometrical quantities, weather variables and heat transfer parameters be available at the time of equation formation. This chapter describes procedures for the determination of these essential data. These procedures may be invoked prior to a simulation or at each simulation time-step in the case of time dependent parameters.

7.1 Weather

In most simulations the aim is to test alternative design possibilities against relatively short period data that characterise typical or extreme weather conditions for the location in question. In this way undesirable options may be disregarded before the near final scheme is subjected to long term simulations, usually annual, to determine energy consumption trends. The selection of an annual weather collection requires some care. Of particular importance is the applicability of the data in relation to the design problem in hand. Generally, two conditions require to be met:

> Portions of the collection should correspond to the different levels of severity under which the building will operate, e.g extreme and typical conditions in the winter, summer and transition seasons.
>
> The collection overall should support an assessment of cost-in-use.

Typically, the former condition is the easier to satisfy since it is usually possible to locate apposite data sequences within collections that do not necessarily correspond to the location in question and therefore will have an unacceptable overall severity. The latter condition will usually require the existence of site-specific data and a technique to rate severity. That said, Degelman (1997) has presented findings that demonstrate that the results from typical week simulations can be made to correlate well with those derived from annual simulations.

7.1.1 Availability of weather data

Table 7.1 summarises the main weather parameters required for simulation. While many weather observation centres exist throughout the world, most will only collect a subset of these data, with solar radiation often excluded. Table 7.2, for example, lists the data that has historically been measured at some UK locations.

Table 7.1: Required weather parameters.

Dry bulb temperature (°C)
Wet bulb temperature (°C)
Wind speed (m s^{-1})
Wind direction (° from north)
Atmospheric pressure (bar)
Net longwave radiation (W m^{-2})
Precipitation (mm)
Global horizontal (or direct normal) solar radiation (W m^{-2})
Diffuse horizontal solar radiation (W m^{-2})
and, where solar radiation data is not available
Cloud cover and type (%, -)
Sunshine hours (hr)

In the US, the National Climatic Data Center (NCDC) has published weather collections for locations throughout the world (NCDC 1993), while the American Society of Heating, Refrigeration and Air Conditioning (ASHRAE) and the National Renewable Energy Laboratory (NREL) have released weather data for use in building simulation (ASHRAE 1997, NREL 1995). Most notable, TMY2[#] (Typical Meteorological Year) data sets have been produced by NREL for 239 locations in the US and the IWEC (International Weather for Energy Calculations) collection, comprising hourly weather data for 227 non-US locations, has been published by ASHRAE.

In Europe, the Meteonorm system (Meteonorm 2000) is a comprehensive source of weather data covering some 2400 meteorological stations worldwide. In addition to measured data, the system offers models for the calculation of inclined surface irradiance and supports data exporting in a variety of program-ready formats.

The European Solar Radiation Atlas (CEC 1996), published by the European Commission, contains monthly means of global, diffuse and beam solar radiation as well as sunshine hours for representative sites throughout Europe. Work is underway to extend this resource by adding information on the spectral components (UVA, UVB, illuminance etc) of surface irradiation under clear, overcast and intermediate skies (Page 2001). National data-sets have also been produced along with technical information on their use (e.g. CIBSE 2001).

Often a simulation exercise will require local weather data that takes account of micro climate effects. The International Association for Urban Climate (IAUC 2001) is an organisation concerned with all aspects of urban climatology and air quality. Through its members it is possible to source urban climate data for particular locations.

7.1.2 Weather collection classification

Irrespective of the source of weather data, a collection (usually annual) must be classified in terms of its severity. A Test Reference Year (TRY) is a weather collection which, when judged against some relevant criteria, is deemed to be representative. While many organisations/countries have produced TRYs, the procedures used are markedly different and a standard has yet to be established. Consider the following examples.

As distinct from the earlier TMY data sets that comprise different elements and adhere to a different presentation convention.

Table 7.2: UK solar radiation stations and parameters measured.

Station	*Latitude*	*Longitude*	*Elevation (m)*	*Element measured*
Lerwick	60° 08' N	01° 11' W	82	G, D, L, B, SS
Eskdalemuir	55° 19' N	03° 12' W	242	G, D, L, B, SS
Aldergrove	54° 39' N	06° 13' W	68	G, D, L, B, SS
Aberporth	52° 08' N	04° 34' W	133	G, D, SS
Cardington	52° 06' N	00° 25' W	29	G, D, SS
London	51° 31' N	00° 07' W	77	G, D, L, SS
Kew	51° 28' N	00° 19' W	5	G, D, L, B, SS, I, F
Bracknell	51° 23' N	00° 47' W	73	G, D, L, SS, I, F, N, S, E, W
Jersey	49° 13' N	02° 12' W	83	G, D, L, B, SS
Aberdeen	57° 10' N	02° 05' W	35	G
Dunstaffnage	56° 28' N	05° 26' W	3	G
Dundee	56° 27' N	03° 04' W	30	G, B
Hurley	51° 32' N	00° 49' W	43	G

G is global horizontal solar radiation, D diffuse horizontal solar radiation, I direct normal solar radiation, N/S/E/W are total solar radiation on vertical surfaces facing the cardinal points, SS sunshine hours, B radiation balance, L total horizontal illumination and F diffuse horizontal illumination.

North America

Stamper (1977) and NCDC (1976) developed a procedure that eliminates those years that contain months with extreme high or low mean temperatures until only one year, the TRY, remains. This is achieved by marking those months within the period that, in terms of mean monthly temperature, can be described as hottest or coldest as shown in table 7.3. The procedure then continues by marking the next-to-hottest July, the next-to-coldest January and so on until one year remains without any marked months. The resulting TRY is considered useful for comparative studies but not for the estimation of long term energy consumption.

Table 7.3: TRY selection method.

1. hottest	July	13. coolest	July
2. coldest	January	14. mildest	January
3. hottest	August	15. coolest	August
4. coldest	February	16. mildest	February
5. hottest	June	17. coolest	June
6. coldest	December	18. mildest	December
7. hottest	September	19. coolest	September
8. coldest	March	20. mildest	March
9. warmest	May	21. coolest	May
10. coolest	November	22. warmest	November
11. warmest	October	23. coolest	October
12. coolest	April	24. warmest	April

Japan

Saito and Matsuo (1974) based TRY selection not on weather parameters but on the cooling and heating loads to result from the application of the data to a standardised design problem. These loads were computed hour-by-hour over a ten year period, for two different enclosures and four different orientations. This gave eight different ten year profiles. The procedure was then repeated for each individual year in the period of record and the yearly profile considered 'nearest' to the ten-year profile (over all eight cases) declared the TRY.

Denmark

Lund (1976) and Anderson (1974) developed a selection procedure based on a rigorous statistical analysis applied to the daily mean and maximum dry bulb temperatures, and the daily total of solar radiation. Months with abnormal weather conditions are excluded at the outset. Then months with typical mean values of the three weather parameters are selected by comparing mean values for each month with the mean value for the same month but established from the whole period of record. Finally, months with typical variations of the three parameters were selected. This was done by comparing the deviation of the three parameters from the previously selected monthly mean values, with the corresponding deviations for the whole period. Each month was then rank ordered according to these criteria and the TRY selected.

South Africa

This procedure involved the selection of typical hot days on the basis of either the daily maximum sol-air temperature or daily maximum dry bulb temperature occurring on 10%, 5% and 2.5% of the days in the period considered. Typical cold days are selected on the basis of daily minimum temperatures. A study initiated by the UK Building Research Establishment and the Meteorological Office proposed a similar method (IHVE 1973).

United Kingdom

The Chartered Institute of Building Services Engineers produced an 'Example Year' based on a selection method proposed by Holmes and Hitchin (1978). Using global and diffuse solar radiation, wind speed, dry bulb temperature and degree days, the method eliminates any year containing a monthly mean value that varies by more than two standard deviations from the long term mean for the month. The recommended example year is October 1964-September 1965 from the Kew observatory or, if a calendar year is required, the 1967 collection. These data are considered adequate for predicting energy demands but not for peak load estimation.

Typical Meteorological Years (TMY; NCDC 1981) and Weather Years for Energy Calculations (WYEC; ASHRAE 1985) have been produced in an attempt to overcome the two main deficiencies of a TRY: the paucity of solar data and the method's tendency to exclude extreme conditions.

Crawley (1998) has reported the results from a project to examine the applicability of these classification approaches. Two major findings emerged. First, TRY-like weather years should be avoided because they do not represent typical long-term weather patterns. Second, synthesised weather years, such as TMY and WYEC, will usually give a better match between predicted and actual consumption over the long term.

7.1.3 Climate severity assessment

Excluding the Japanese procedure, each of the classification methods suffer from the fact that only simple synoptic data is used to discriminate between different collections and no attempt is made to include the characteristics of the building in the selection procedure.

To address these issues, Markus *et al* (1984) developed a technique to assess the severity of a given weather collection in terms of a Climate Severity Index (CSI). The CSI indicates, by a single number on a dimensionless scale, the thermal stress placed on a building. Of course, a

CSI already exists in the form of the simple degree day concept, but this is inadequate because it excludes the effects of major weather parameters and only takes account of building characteristics in a rudimentary manner via the base temperature. Figure 7.1 summarises the approach used to construct the CSI for the UK.

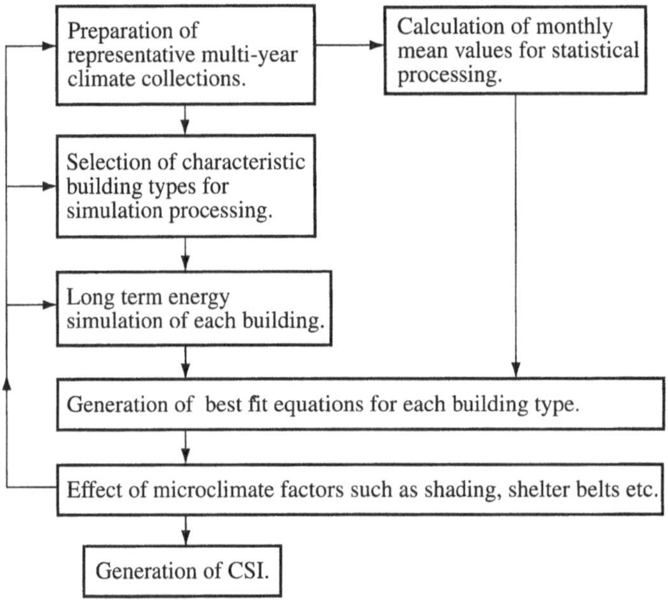

Figure 7.1: Formulation of the CSI.

Step 1: A dynamic simulation program is selected and used to determine the energy behaviour of selected designs when each are subjected to a range of weather collections.

Step 2: Monthly mean values of each of the main weather parameters are extracted from the collections for later use in step 5.

Step 3: To determine the range of designs to be processed, the principal design parameters must firstly be identified. In the case of housing, for example, window size, insulation level, capacity level, capacity position and air permeability might be considered adequate determinants of energy use. If each parameter can exist at one of three levels then there will be 243 (3^5) potential designs. That is, any specific house can be regarded as a unique combination of the possible levels of the five parameters.

Step 4: Long term simulations are now conducted. Since the processing of each design combination against the (potentially) many weather collections is computationally prohibitive, a 1/9th replicate sub-set (randomly selected) was employed.

Step 5: Monthly energy requirements are extracted from the simulation results and subjected, along with the monthly mean weather parameter values, to curve fitting techniques to establish, for each design, a best fit relationship of the form

$$E = a\bar{\theta}^2 + b\bar{R}_d^2 + c\bar{R}_f^2 + d\bar{V}^2 + e\bar{\theta} + f\bar{R}_d + g\bar{R}_f + h\bar{V} + i\bar{\theta}\bar{R}_d$$

$$+ j\bar{\theta}\bar{R}_f + k\bar{\theta}\bar{V} + l\bar{R}_d\bar{R}_f + m\bar{R}_d\bar{V} + n\bar{R}_f\bar{V} + o$$

where E is the monthly energy requirement (kWh), $\bar{\theta}$ the monthly mean temperature (°C), $\bar{R}_d$

the monthly mean direct normal solar radiation (W m^{-2}), $\bar{R}_f$ the monthly mean diffuse horizontal solar radiation (W m^{-2}), $\bar{V}$ the monthly mean wind speed (m s^{-1}) and 'a' through 'o' are the least squares coefficients.

Table 7.4 gives the equations to result for the case of housing when subjected to five yearly weather collections for Kew (South England), Aberporth (Wales), Eskdalemuir (Scottish Borders) and Lerwick (Shetland).

Step 6: The CSIs are now generated by a four stage process as follows.

1: By entering the maximum and minimum climate parameters (to which the designs were subjected during simulation) into each regression equation, an energy difference between maximum and minimum loading can be established for each design and for each weather parameter. An example of this is given in table 7.5 for house types 16 (average) and 26 (passive solar).

Table 7.5: Contribution of the climate parameters to the energy difference between maximum and minimum loadings.

Climate parameter	*Max/min condition*	*Energy difference (kWh)*	
		House 16	*House 26*
$\bar{\theta}$	15.3/-1.9	1186	1596
$\bar{R}_d$	155.2/0.6	120	194
$\bar{R}_f$	94.4/3.3	2	361
$\bar{V}$	10.6/2.7	250	269
All	-	1559	2420

2: By then computing the ratio of the individual energy differences to the overall totals, the contribution of each particular weather variable is obtained.

3: If the best combination of climate parameters (lowest energy case) is made, arbitrarily, to correspond to a CSI of 2 (to allow later extension of the range when colder climates are added) and the worst combination a CSI of (say) 10, the contribution of each individual weather parameter can be determined by multiplying these upper and lower CSI values by the previously determined ratios. Figure 7.2 shows one way of expressing the result: the contribution of each weather parameter to the CSI is given for a particular design. Alternatively, a CSI equation can be formulated for each design to give the contribution of each weather parameter to the total CSI. Table 7.6 lists the final CSI equations for the housing example considered here. Note that although the weather parameter contributions will vary for different designs, the normalisation technique ensures that the total CSI for each design, when subjected to the same weather collection, will be essentially the same (see table 7.7). This means that CSI maps (equivalent to degree day maps) can be constructed for all locations with available monthly mean weather data by the application of any equation from table 7.6, and that each weather collection can be assigned a unique CSI value.

4: A family of curves can now be produced that relate climate severity to energy consumption for the processed designs with the remaining designs assessed by interpolation. Figure 7.3 gives example curves for three house types.

In addition to weather collection rating, the CSI has strategic uses. For any country for which a CSI map exists, the CSI for a given location can be read off and used, in conjunction with the curves of figure 7.3, to determine the relative energy consequences of adopting alternative housing mixes or locations. Alternatively, the curves of figure 7.3 can be redrawn for the case of constant energy as shown in figure 7.4. This is achieved by determining (from the

Table 7.4: Least squares coefficients of the energy regression equations.

Case#	$\bar{\theta}^2$	$\bar{R}_d^2$	$\bar{R}_f^2$	$\bar{V}^2$	$\bar{\theta}$	$\bar{R}_d$	$\bar{R}_f$	$\bar{V}$	$\bar{\theta}\bar{R}_d$	$\bar{\theta}\bar{R}_f$	$\bar{\theta}\bar{V}$	$\bar{R}_d\bar{R}_f$	$\bar{R}_d\bar{V}$	$\bar{R}_f\bar{V}$	constant
1	0.9100	0.0076	0.0276	0.1630	-64.6656	-0.0896	-1.4820	16.5864	0.0472	-0.0096	-1.1807	-0.0272	-0.0492	-0.0295	850.0271
2	0.6844	0.0116	0.0276	0.0934	-50.3507	-1.4727	-3.5503	19.2228	-0.0204	0.2580	-1.3318	-0.0256	0.0130	-0.0659	590.2528
3	0.9803	0.0220	0.0413	0.0513	-93.3554	-4.3369	-3.5748	26.0130	0.0713	0.0780	-1.8874	-0.0418	0.0353	-0.1049	1308.1035
4	1.2703	0.0295	0.0409	-1.0818	-99.8170	-5.2093	-2.5492	40.8774	0.0694	0.1374	-1.3193	-0.0495	0.0449	-0.2570	1253.9896
5	1.0045	0.0067	0.0274	0.1509	-59.9448	-0.5894	-1.7462	25.1951	0.0372	0.0296	-1.5546	-0.0220	-0.0039	-0.0727	724.4446
6	1.2734	0.0137	0.0446	0.1706	-76.3315	-1.9088	-4.6295	26.8021	0.0217	0.1322	-1.7218	-0.0361	0.0152	-0.0678	996.7990
7	0.3256	0.0229	0.0287	-0.2266	-87.8016	-3.8537	-3.1527	30.3013	-0.0423	0.2072	-1.9627	-0.0389	0.0285	-0.1022	1269.7730
8	1.2973	0.0231	0.0517	0.2335	-83.8376	-3.4787	-4.3549	29.2712	0.0570	0.1344	-1.9535	-0.0479	0.0008	-0.1304	1100.1423
9	1.9297	0.0370	0.0664	-0.0897	-130.1218	-9.1346	-5.9252	41.9583	0.2313	0.2819	-2.5699	-0.0502	0.0855	-0.3076	1701.7621
10	1.0291	0.0094	0.0386	0.1051	-69.9616	-0.7531	-3.6782	15.4652	0.0376	0.0810	-0.8655	-0.0280	-0.0148	-0.0607	926.7016
11	1.4308	0.0235	0.0493	-0.0346	-99.1432	-4.6952	-4.7566	22.5083	0.0976	0.1527	-1.3431	-0.0429	0.0291	-0.1259	1324.1257
12	1.2771	0.0173	0.0467	0.0231	-68.5724	-2.2832	-5.0732	30.5679	0.0351	0.2211	-1.8449	-0.0394	0.0213	-0.1063	819.5991
13	1.0605	0.0112	0.0233	-0.5264	-76.8607	-2.2238	-2.3820	38.4960	0.0193	0.1039	-1.7694	-0.0272	0.0651	-0.0897	936.4577
14	0.9605	0.0105	0.0315	0.1391	-54.2414	-1.3047	-3.0575	18.6869	0.0192	0.1033	-1.4864	-0.0271	-0.0037	-0.0384	667.8197
15	0.9630	0.0081	0.0303	0.1799	-69.0387	0.0668	-1.5254	25.5187	0.0433	-0.0093	-1.6541	-0.0307	-0.0562	-0.0265	905.2008
16	1.0952	0.0110	0.0261	0.0491	-83.9506	-2.4921	-2.5211	31.0448	0.0248	0.0553	-1.9939	-0.0253	0.0449	-0.0693	1114.5220
17	0.2538	0.0165	0.0209	0.0221	-51.6724	-1.8704	-2.5772	21.8033	-0.0575	0.2840	-1.3268	-0.0290	-0.0117	-0.0879	621.1768
18	0.9906	0.0055	0.0246	0.1172	-55.2493	-0.7056	-1.7556	16.5954	0.0406	0.0289	-1.1329	-0.0172	-0.0018	-0.0742	665.1710
19	1.0004	0.0092	0.0291	0.1121	-72.1607	-1.6364	-3.1017	28.0738	0.0172	0.0803	-1.8924	-0.0266	0.0426	-0.0266	951.8901
20	1.1712	0.0098	0.0396	0.1618	-58.2987	-0.8520	-3.9856	16.6714	0.0188	0.1146	-1.1084	-0.0270	-0.0047	-0.0823	714.2599
21	0.8543	0.0128	0.0283	0.0179	-61.0690	-1.9737	-2.7151	21.3190	0.0111	0.1359	-1.4775	-0.0274	0.0003	-0.0895	759.9731
22	1.0812	0.0235	0.0420	0.1279	-86.0188	-4.1079	-3.9742	40.7104	0.0380	0.2388	-2.8644	-0.0426	0.0001	-0.1410	1099.4317
23	1.4335	0.0235	0.0497	-0.0359	-98.9667	-4.6606	-4.7857	22.4904	0.0960	0.1532	-1.3453	-0.0432	0.0282	-0.1240	1322.3968
24	0.8388	0.0044	0.0245	0.1201	-51.3552	-0.3429	-2.2132	26.1886	0.0137	0.0930	-1.7062	-0.0176	-0.0127	-0.0655	604.8176
25	1.2439	0.0173	0.0423	-0.0268	-86.5663	-2.9538	-4.9234	21.4916	0.0413	0.1606	-1.3686	-0.0391	0.0445	-0.0659	1163.1544
26	0.2439	0.0576	0.0259	0.0739	-96.4461	-10.2289	-6.4945	33.1380	0.2211	0.9374	-1.9247	-0.0208	-0.0182	-0.0396	1087.7877
27	0.9905	0.0100	0.0386	0.1294	-78.1842	0.0041	-3.1972	16.0948	0.0379	0.0383	-0.9732	-0.0348	-0.0558	-0.0312	1092.4584
28	0.6281	0.0415	0.0515	0.0205	-117.0153	-8.2064	-5.7879	43.1737	0.0303	0.5098	-2.9572	-0.0533	0.0436	-0.2821	1656.9116
29	1.4444	0.0540	0.0628	-0.3205	-111.9623	-10.0009	-7.4048	33.4261	0.2210	0.6378	-1.9180	-0.0529	0.0085	-0.2701	1380.3015
30	1.0526	0.0106	0.0407	0.1167	-82.6358	0.1481	-3.1904	25.5825	0.0354	0.0391	-1.4604	-0.0383	-0.0635	-0.0291	1145.3155
Coeff.	a	b	c	d	e	f	g	h	i	j	k	l	m	n	o

Energy equation: $E = a\bar{\theta}^2 + b\bar{R}_d^2 + c\bar{R}_f^2 + d\bar{V}^2 + e\bar{\theta} + f\bar{R}_d + g\bar{R}_f + h\bar{V} + i\bar{\theta}\bar{R}_d + j\bar{\theta}\bar{R}_f + k\bar{\theta}\bar{V} + l\bar{R}_d\bar{R}_f + m\bar{R}_d\bar{V} + n\bar{R}_f\bar{V} + o$

See table 7.6 for a definition of the design parameter combinations comprising each case.

equations) the altered CSI condition which would ensure energy equity against some standard house type: type 16 in this case. This allows, for any CSI, the determination of an altered CSI that defines a region within which the house type of interest can be placed without incurring an energy penalty relative to the base house type.

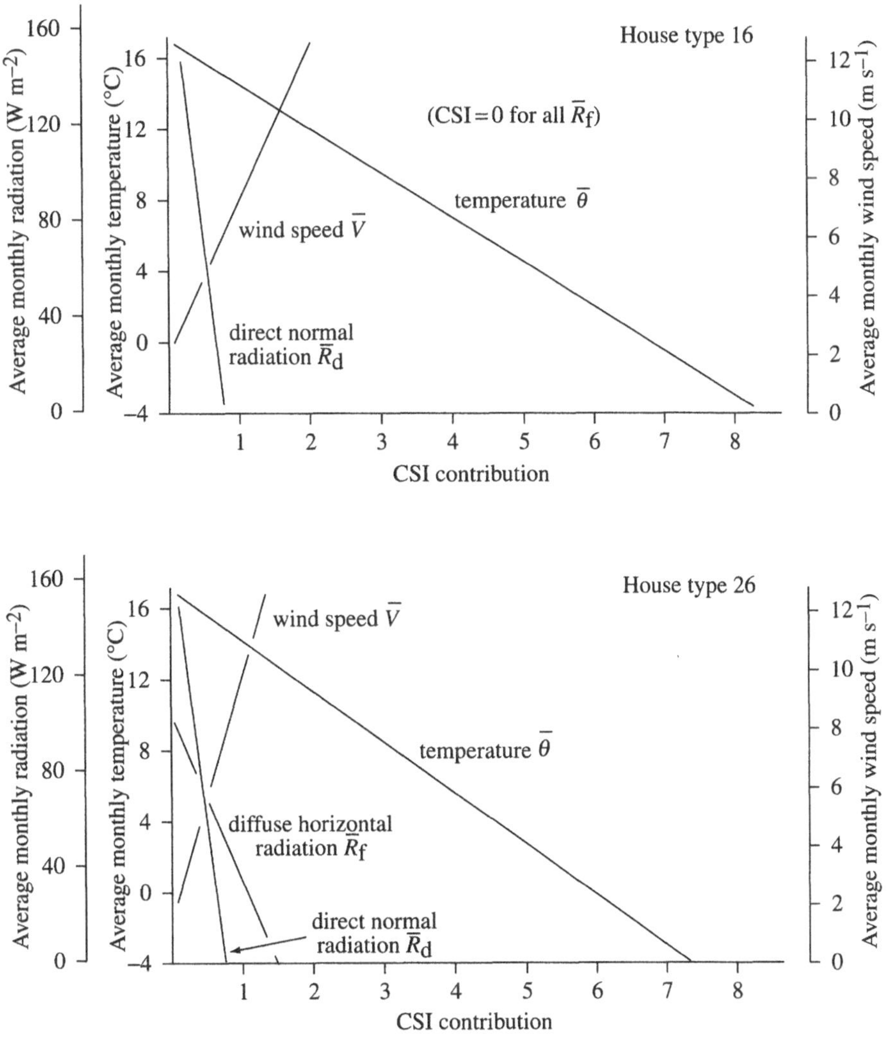

Figure 7.2: Contribution of the climate parameters to the CSI.

Most modelling systems will provide a range of pre-constructed weather files and support the importing/exporting of data in alternative formats. In addition, it is usually possible to graph the weather data and extract relevant statistics such as period degree days, max/min conditions, irradiance totals and the frequency of occurrence of specified conditions. The existence of a feature to support the rank ordering of collections in terms of application related severity considerations is atypical.

Table 7.6: Final CSI equations. $CSI = (a\bar{\theta} + e_a) + (b\bar{R}_d + e_b) + (c\bar{R}_f + e_c) + (d\bar{V} + e_d)$
$= a\bar{\theta} + b\bar{R}_d + c\bar{R}_f + d\bar{V} + e_t$

House type	*Design combination†*	*a*	*b*	*c*	*d*	e_a	e_b	e_c	e_d	e_t
1	0/1/0/1/0	-0.356	-0.007	-0.008	0.127	6.118	1.281	0.860	-0.234	8.024
2	2/2/0/1/1#	-0.372	-0.003	-0.008	0.185	6.391	0.521	0.808	-0.339	7.381
3	1/0/0/1/0	-0.409	-0.005	-0.002	0.136	7.022	0.799	0.243	-0.250	7.814
4	2/1/1/1/0	-0.402	-0.003	-0.007	0.129	6.908	0.513	0.733	-0.236	7.918
5	0/2/0/0/1	-0.359	-0.004	-0.007	0.210	6.167	0.605	0.750	-0.386	7.136
6	1/2/1/0/1	-0.409	-0.002	-0.002	0.201	7.026	0.269	0.199	-0.368	7.125
7	2/0/0/1/0	-0.436	-0.001	-0.002	0.143	7.500	0.258	0.192	-0.262	7.688
8	1/1/1/0/1	-0.405	-0.001	-0.004	0.198	6.966	0.127	0.441	-0.363	7.171
9	1/0/1/0/0	-0.350	-0.011	-0.002	0.137	6.018	1.952	0.196	-0.251	7.914
10	0/2/1/1/1	-0.417	-0.005	-0.001	0.125	7.159	0.910	0.070	-0.230	7.909
11	1/2/1/1/0	-0.421	-0.005	-0.000	0.116	7.238	0.939	0.031	-0.213	7.995
12	2/2/1/0/1#	-0.375	-0.003	-0.004	0.225	6.441	0.518	0.392	-0.413	6.937
13	2/2/0/0/0	-0.401	-0.003	-0.001	0.202	6.894	0.529	0.072	-0.370	7.125
14	1/2/0/1/1	-0.403	-0.003	-0.000	0.200	6.928	0.557	0.018	-0.367	7.136
15	0/0/0/0/0	-0.333	-0.008	-0.008	0.166	5.720	1.359	0.888	-0.304	7.663
16	1/1/0/0/0	-0.400	-0.004	-0.000	0.183	6.870	0.775	0.016	-0.366	7.325
17	2/1/0/1/1#	-0.375	-0.005	-0.004	0.172	6.445	0.938	0.436	-0.315	7.504
18	0/0/0/1/1	-0.400	-0.001	-0.006	0.173	6.873	0.249	0.644	-0.318	7.448
19	1/2/0/0/0	-0.408	-0.001	-0.002	0.206	7.108	0.243	0.190	-0.377	7.047
20	0/1/1/1/1#	-0.386	-0.006	-0.001	0.171	6.631	1.054	0.114	-0.313	7.486
21	1/1/0/1/1	-0.435	-0.000	-0.000	0.189	7.469	0.031	0.044	-0.347	7.197
22	1/0/0/0/1	-0.392	-0.002	-0.001	0.233	6.744	0.423	0.072	-0.427	6.812
23	1/1/1/1/0	-0.422	-0.005	-0.000	0.117	7.254	0.912	0.037	-0.214	7.989
24	0/1/0/0/1#	-0.370	-0.003	-0.002	0.257	6.359	0.545	0.172	-0.471	6.606
25	2/2/1/1/0	-0.426	-0.002	-0.005	0.129	7.322	0.272	0.506	-0.236	7.863
26	2/0/1/1/1#	-0.346	-0.005	-0.015	0.127	5.954	0.804	1.541	-0.233	8.066
27	0/0/1/1/0	-0.157	-0.038	-0.001	0.043	2.703	6.511	0.145	-0.079	9.280
28	2/0/1/0/0	-0.385	-0.006	-0.003	0.154	6.609	1.063	0.281	-0.282	7.761
29	1/0/1/1/1	-0.383	-0.007	-0.005	0.121	6.576	1.131	0.550	-0.221	8.036
30	0/1/1/0/0	-0.355	-0.009	-0.004	0.141	6.100	1.607	0.424	-0.258	7.873

† A/B/C/D/E:

A = capacity level—0 low, 1 medium, 2 high
B = capacity position—1 inside, 2 middle, 3 outside
C = window size—0 standard, 1 large
D = infiltration level—0 standard; 1 tight
E = insulation level—0 standard; 1 super
\# indicates blind control active

Table 7.7: Each equation gives the same total CSI.

Climate condition	*Contribution to the CSI for house type ...* 16	26	30
$\bar{\theta} = 15.3$	0.75	0.67	0.67
$\bar{R}_d = 155.2$	0.16	0.08	0.20
$\bar{R}_f = 94.2$	0.02	0.18	0.05
$\bar{V} = 2.7$	0.12	0.11	0.12
Total	1.05	1.04	1.04
$\bar{\theta} = -1.9$	7.63	6.61	6.78
$\bar{R}_d = 0.6$	0.77	0.80	1.60
$\bar{R}_f = 3.3$	0.02	1.49	0.41
$\bar{V} = 10.6$	1.60	1.11	1.24
Total	10.02	10.01	10.03

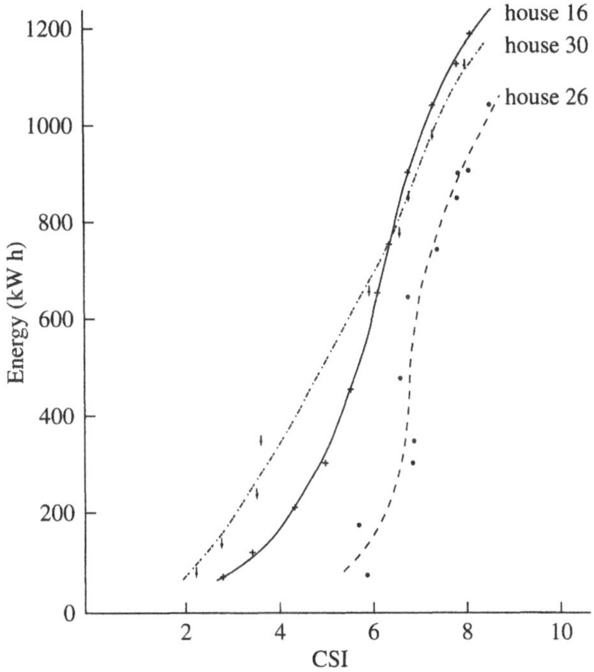

Figure 7.3: CSI/ energy relationship for three house types.

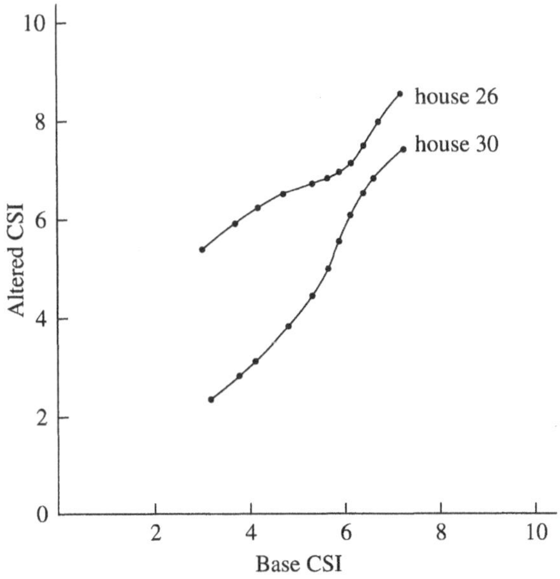

Figure 7.4: Energy equity curves (relative to house type 16).

7.2 Geometrical considerations

This section outlines procedures for the calculation of the geometrical quantities required by the conservation equations derived earlier. These are but a sub-set of the operations required to support computer graphics applications generally (Foley and Van Dam 1982).

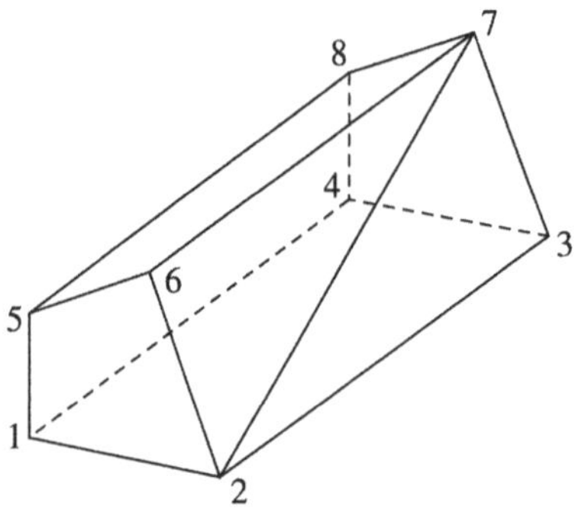

Figure 7.5: A general-shaped bounded volume.

Consider the general-shaped volume of figure 7.5 as formed by 7 planar polygons. Shape aside, the bisection of polygon 2,3,7,6 might typically represent a surface composed from different multi-layered constructions (e.g. an opaque and a transparent portion). One possible geometrical representation technique is to specify the topography as a collection of points relative to some arbitrarily chosen Cartesian co-ordinate system. The topology is then given as an ordered list of these co-ordinates. Thus, for the example of figure 7.5, and with each vertex denoted as the point (x_i y_i z_i), the topology is given by

Polygon	Number of Vertices	Ordered Description
1	4	1, 2, 6, 5
2	3	2, 3, 7
3	4	3, 4, 8, 7
4	4	4, 1, 5, 8
5	3	2, 7, 6
6	4	8, 5, 6, 7
7	4	1, 4, 3, 2

Here the convention is to describe each polygon as a collection of vertices specified anti-clockwise when the polygon is viewed from outwith the zone it bounds. Any intra-polygon hole is then specified by reversing the vertex order relative to the main body convention.

This structure is convenient since scaling and rotation operations will modify the topography but not the topology. This means that the latter need only be held once for any shape classification. Although this scheme is well suited to building simulation, more advanced surface representation techniques exist (Rogers and Adams 1976) which are better suited to other application fields where geometries are primarily non-orthogonal.

This section defines a number of basic operations that can be applied to the above data to determine quantities such as area, volume, perimeter length and selected angles. Other operations, such as three-dimensional rotation, are detailed in §7.3 where shading and insolation is considered.

For any planar polygon, p, consider the following summations applied to the polygon vertices.

$$\left.\begin{aligned} XSUM_p &= \sum_{i=1}^{NV}(y_i z_j - z_i y_j) \\ YSUM_p &= \sum_{i=1}^{NV}(z_i x_j - x_i z_j) \\ ZSUM_p &= \sum_{i=1}^{NV}(x_i y_j - y_i x_j) \end{aligned}\right\} \begin{array}{l} j = I + 1; \\ \text{for } j > NV, j = 1 \end{array}$$

where NV is the total number of vertices in the polygon. The polygon area is then given by

$$AREA_p = 0.5(XSUM_p^2 + YSUM_p^2 + ZSUM_p^2)^{\frac{1}{2}} .$$

Since the clockwise conventions used to define openings produce negative areas, the algebraic summation of the polygons comprising a given surface gives the net area for the case where the polygon contains a window. The perimeter length of any polygon is given by

$$PERIM_p = \sum_{i=1}^{NV}[(x_j - x_i)^2 + (y_j - y_i)^2 + (z_j - z_i)^2]^{\frac{1}{2}} \quad ; j = i + 1; \text{for } j > NV, j = 1 .$$

It is usual to define the orientation of a body face by azimuth and elevation angles as shown in figure 7.6.

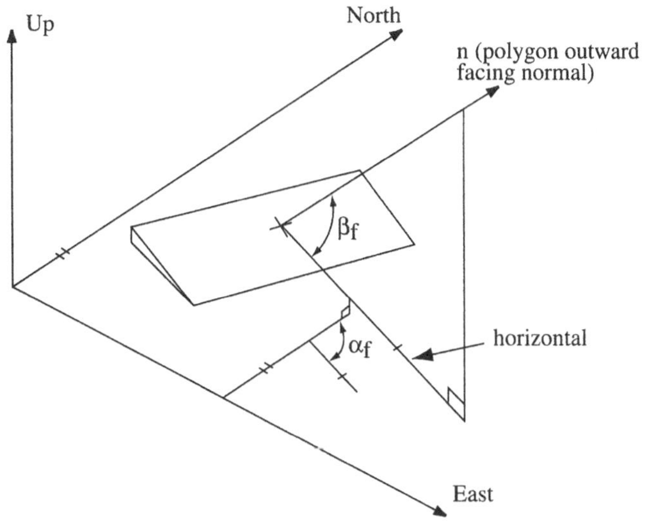

Figure 7.6: Polygon azimuth (α) and elevation (β) angles.

Here the azimuth is defined as the clockwise angle between the co-ordinate system's Y-axis (North) and the projection of the polygon's outward facing normal onto the XY plane (usually made to represent the horizontal plane). The plane elevation is defined as the angle between the outward facing normal and the projection of this normal onto the XY (horizontal) plane. Adopting this convention gives, for the azimuth

$$\alpha_f = \tan^{-1}(XSUM_p/YSUM_p)$$

where, for $YSUM_p = 0$,

$$\alpha_f = -90° \text{ for } XSUM_p < 0$$

$$\alpha_f = 0° \text{ for } XSUM_p = 0$$

$$\alpha_f = 90° \text{ for } XSUM_p > 0$$

and, for the elevation:

$$\beta_f = \tan^{-1}[ZSUM_p/(XSUM_p^2 + YSUM_p^2)^{\frac{1}{2}}]$$

where, for $XSUM_p^2 + YSUM_p^2 = 0$,

$$\beta_f = -90° \text{ for } ZSUM_p < 0$$

$$\beta_f = 0° \text{ for } ZSUM_p = 0$$

$$\beta_f = 90° \text{ for } ZSUM_p > 0.$$

The volume contained by the bounding polygons is given as the algebraic sum of the volumes of the prisms formed by connecting the vertices of each polygon to the origin. Note that since polygons specified as holes (clockwise ordering) will still bound the volume, their vertex ordering must firstly be reversed. The contained volume is given by

$$VOL = \frac{1}{6}\sum_{j=1}^{NP}(x_{j1}XSUM_j + y_{j1}YSUM_j + z_{j1}ZSUM_j)$$

where $(x_{j1}\ y_{j1}\ z_{j1})$ are the co-ordinates of the first mentioned vertex in polygon j and NP is the total number of polygons.

7.3 Shading and insolation

The solar energy incident on a building is influenced by shading as caused by parts of itself, surrounding buildings, facade features and natural obstructions such as trees, Shading and insolation must therefore be determined as a function of solar position and target/obstruction geometry. Indeed, it is an astute strategy to use design parameters such as orientation, shape and obstruction geometry to so modify the shading/insolation patterns that environmental performance is improved without recourse to mechanical intervention. Within a simulation program the requirement is to determine the time variation in external surface insolation and the corresponding variation for internal surfaces determined separately for each possible window/surface combination. These data are summarised in figure 7.7: external surface data are used in the prediction of external opaque surface solar absorption and reflection, and transparent surface absorption, transmission and reflection; internal surface data are required to track the window transmitted shortwave energy to its first internal reflection (§7.4 discusses its treatment thereafter).

One approach to the determination of these data is as follows. Starting with a collection of target and obstruction objects, each defined relative to an arbitrary site co-ordinate system XYZ, the XZ plane is relocated to the plane of the target body face of interest. The obstruction objects can then be projected, parallel to the sun's rays, onto the face and the projected image expressed in 2D relative to the local face co-ordinate system. A grid can then be superimposed on the face and each grid cell tested for overlap with the individual shadow polygons. Prediction accuracy and speed may then be controlled by simply adjusting the number of grid cells.

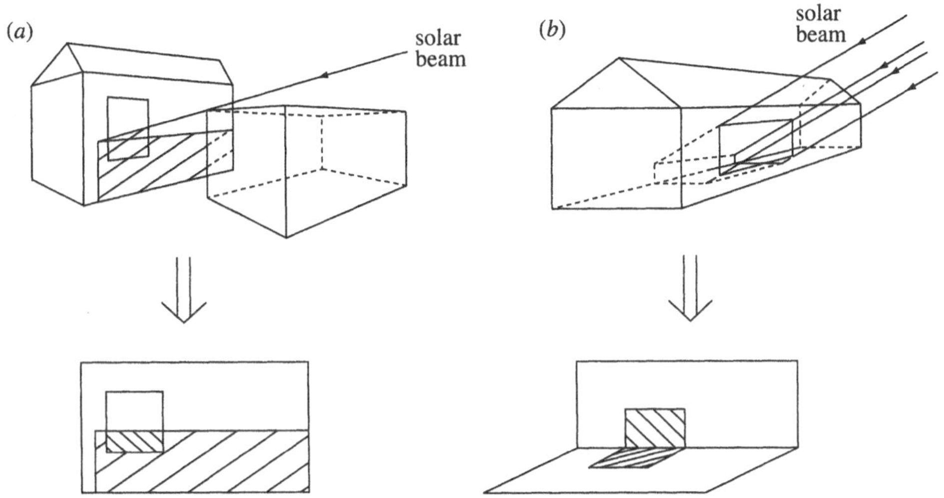

Figure 7.7: Shading/insolation—(a) external surfaces, (b) internal surfaces.

7.3.1 Insolation transformation equations

Figure 7.8 shows a target and an obstruction body located in a right-hand Cartesian co-ordinate system. Body I is a general-shaped object for which insolation data are required. Body J is also a general-shaped object (an adjacent building or some facade feature) which is the cause of shading on body I. Body J has N vertices given by $(x_i\ y_i\ z_i)$, $i = 1, 2, \ldots, N$.

The objective of the following derivation is to generate a set of transformation equations to allow any obstruction body vertex to be projected onto each target body face in turn, with the projected co-ordinates expressed relative to a co-ordinate system relocated to the face in question. This allows the surface shading to be determined by simple two-dimensional operations.

Translation: Move the co-ordinate system origin to the first defined vertex in the first face of the target body. As shown in figure 7.8, this gives a new co-ordinate system in which any original point (x y z) translates to a new point $(x'\ y'\ z')$ according to a translation matrix:

$$(x'\ y'\ z') = (x\ y\ z\ 1)\begin{bmatrix} 1 & 0 & 0 \\ 0 & 1 & 0 \\ 0 & 0 & 1 \\ -x0 & -y0 & -z0 \end{bmatrix} \tag{7.1}$$

where (x0 y0 z0) is the new origin in old co-ordinates, i.e. the components of translation in the X, Y and Z directions.

Rotation: The translated axes are now subjected to a X', Y' and Z' axis rotation to align the $X''Z''$ plane of the new co-ordinate system $X''Y''Z''$ with the plane of the target body face, with the new Y''-axis pointing away from the sun. The rotation angles α, β and γ, as shown in figure 7.9, should be regarded as clockwise Z'-, X'- and Y'-axis rotations when viewed from the co-ordinate system origin. This three-axis rotation will result in a localised co-ordinate system allowing two-dimensional polygon manipulation after the insolation polygons have been established by projection of the obstruction bodies. Any point $(x'y'z')$ transforms to the point $(x''y''z'')$ according to the matrix relationship

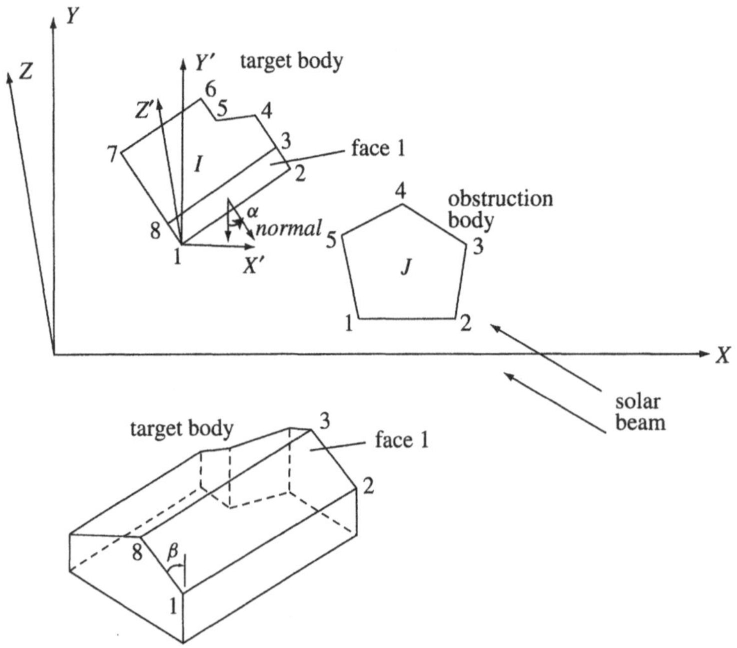

Figure 7.8: Geometry of a target and obstruction body.

$$(x'' \; y'' \; z'') = (x' \; y' \; z') \begin{bmatrix} 1 & 0 & 0 \\ 0 & \cos\beta & -\sin\beta \\ 0 & \sin\beta & \cos\beta \end{bmatrix} \begin{bmatrix} \cos\gamma & 0 & \sin\gamma \\ 0 & 1 & 0 \\ -\sin\gamma & 0 & \cos\gamma \end{bmatrix} \begin{bmatrix} \cos\alpha & -\sin\alpha & 0 \\ \sin\alpha & \cos\alpha & 0 \\ 0 & 0 & 1 \end{bmatrix} \quad (7.2)$$

where clockwise axes rotations (looking to the positive side of the origin) are positive. Substitution of eqn (7.1) in eqn (7.2) gives the final axes transformation matrix:

$$(x'' \; y'' \; z'') = (x' \; y' \; z') \begin{bmatrix} 1 & 0 & 0 \\ 0 & 1 & 0 \\ 0 & 0 & 1 \\ -x0 & -y0 & -z0 \end{bmatrix} \begin{bmatrix} 1 & 0 & 0 \\ 0 & \cos\beta & -\sin\beta \\ 0 & \sin\beta & \cos\beta \end{bmatrix}$$

$$\times \begin{bmatrix} \cos\gamma & 0 & \sin\gamma \\ 0 & 1 & 0 \\ -\sin\gamma & 0 & \cos\gamma \end{bmatrix} \begin{bmatrix} \cos\alpha & -\sin\alpha & 0 \\ \sin\alpha & \cos\alpha & 0 \\ 0 & 0 & 1 \end{bmatrix}. \quad (7.3)$$

The Z' axis rotation, α, is related to the face azimuth, a_f, by $\alpha = 180 - a_f$ $(0 \le \alpha_f \le 360)$, the X' axis rotation, β, is related to the face elevation, β_f, by $\beta = -\beta_f$ $(-90 \le \beta_f \le 90)$ and the Y' axis rotation, γ, is related to the angle, γ_f (made by the line joining the first two vertices of the face with the horizontal), by $\gamma = -\gamma_f$ $(0 \le \gamma_f \le 360)$.

Projection: The vertices of each obstruction object can now be projected onto the $X''Z''$ plane of the $X''Y''Z''$ co-ordinate system to give the vertices of the projected shadow polygons relative to the local face co-ordinate system. Figure 7.10 shows the projection of an arbitrary point $(x'' \; y'' \; z'')$ onto an arbitrary target face with relocated co-ordinate system $X''Y''Z''$ to give

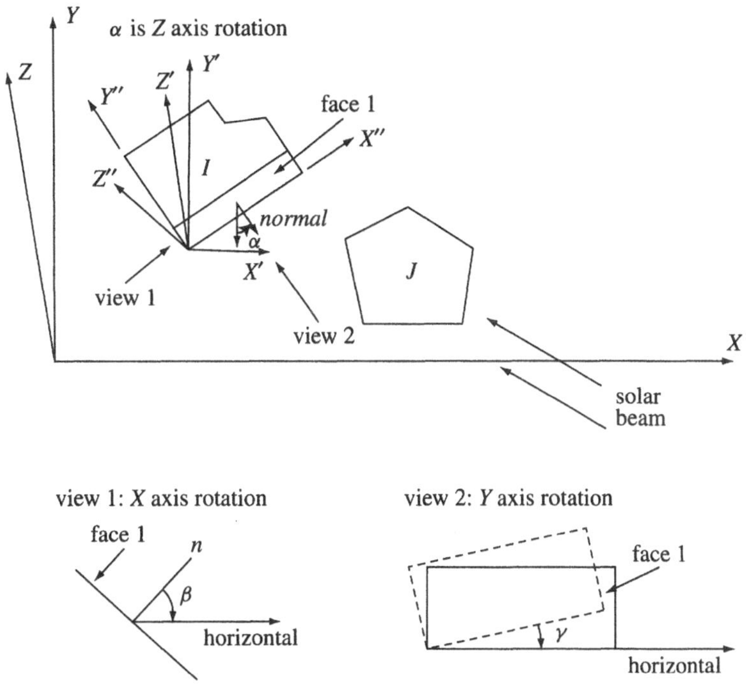

Figure 7.9: Axes transformation: axes rotations shown positive. Note that γ (in view 2) is the angle between the first two numbered vertices in the face and the horizontal, and is zero in this case.

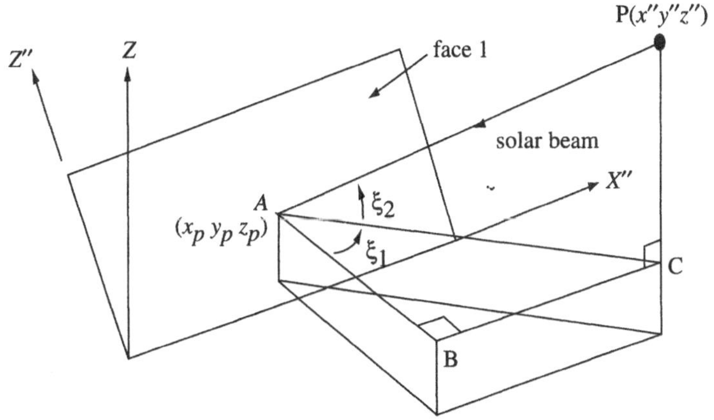

Figure 7.10: Projection of a point P onto the X''Z'' plane.

the projected point $(x_P\ y_P\ z_P)$.

Noting that $y_p = 0$:

$$\tan \xi_1 = BC/AB$$

$$= (x'' - x_P)/y'' \quad ;\ \text{for } \xi_1 \text{ positive as shown}$$

$$= (x_P - x'')/y'' \quad ;\ \text{for } \xi_1 \text{ negative}$$

$$\Rightarrow x_P = x'' \pm y'' \frac{\sin \xi_1}{\cos \xi_1} .$$

Also,

$$\cos \xi_1 = AB/AC = y''/AC$$

$$\Rightarrow AC = y''/\cos \xi_1 \tag{7.4}$$

and

$$\tan \xi_2 = PC/AC$$

$$= (z'' - z_P)/AC \quad ;\ \text{for } \xi_2 \text{ positive as shown}$$

$$= (z_P - z'')/AC \quad ;\ \text{for } \xi_2 \text{ negative}$$

$$\Rightarrow z_P = z'' \pm AC \tan \xi_2 . \tag{7.5}$$

Substituting eqn (7.4) in eqn (7.5) gives

$$z_P = z'' \pm y'' \tan \xi_2 / \cos \xi_1 \quad \text{and} \quad y_P = 0 .$$

Note that if $y'' > 0$ then point P lies 'behind' the face relative to the sun and is therefore omitted from processing. This will have algorithmic implications in the case of bodies partially behind and partially in front of the face in question because only part of the body will then shade the target face. In matrix notation, projection is given by

$$(x_P\ y_P\ z_P) = (x''\ y''\ z'') \begin{bmatrix} 1 & 0 & 0 \\ \pm \dfrac{\sin \xi_1}{\cos \xi_1} & 1 & \pm \dfrac{\tan \xi_2}{\cos \xi_1} \\ 0 & 0 & 1 \end{bmatrix} \tag{7.6}$$

where decisions on the addition and subtraction operators are made on the basis of the sign of ξ_1 and ξ_2. These angles are pseudo solar azimuth and elevation angles formed by direct transformation of the real solar angles in the XYZ co-ordinate system. This transformation is shown in figure 7.11 and the procedure for arriving at ξ_1 and ξ_2 is given in §7.3.3.

7.3.2 The complete translation, rotation and projection equations

Eqns (7.3) and (7.6) can now be combined:

$$(x_P\ y_P\ z_P) = (x\ y\ z\ 1) \begin{bmatrix} 1 & 0 & 0 \\ 0 & 1 & 0 \\ 0 & 0 & 1 \\ -x0 & -y0 & -z0 \end{bmatrix} \begin{bmatrix} 1 & 0 & 0 \\ 0 & \cos \beta & -\sin \beta \\ 0 & \sin \beta & \cos \beta \end{bmatrix}$$

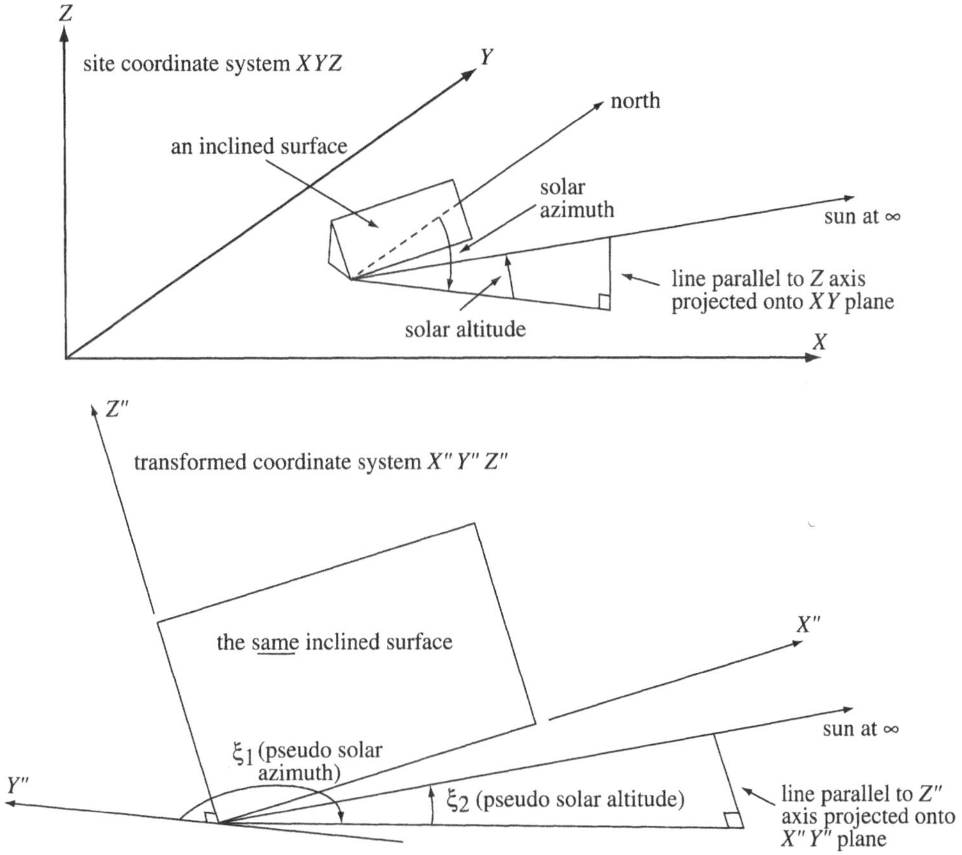

Figure 7.11: Transformation of solar position.

$$\times \begin{bmatrix} \cos\gamma & 0 & \sin\gamma \\ 0 & 1 & 0 \\ -\sin\gamma & 0 & \cos\gamma \end{bmatrix} \begin{bmatrix} \cos\alpha & -\sin\alpha & 0 \\ \sin\alpha & \cos\alpha & 0 \\ 0 & 0 & 1 \end{bmatrix} \begin{bmatrix} 1 & 0 & 0 \\ \pm\dfrac{\sin\xi_1}{\cos\xi_1} & 1 & \pm\dfrac{\tan\xi_2}{\cos\xi_1} \\ 0 & 0 & 1 \end{bmatrix}$$

where (x0 y0 z0) defines the new face origin expressed in old co-ordinates; β is related to the face elevation angle, γ the local face x-axis tilt, α is related to the face azimuth and ξ_1, ξ_2 are related to the solar azimuth and altitude angles respectively when these angles are re-expressed relative to the local face co-ordinate system. Expanding this transformation matrix equation gives

$$(x_P \; y_P \; z_P) = [(x - x0), (y - y0)\cos\beta - (z - z0)\sin\beta, (y - y0)\sin\beta + (z - z0)\cos\beta]$$

$$\times \begin{bmatrix} \cos\gamma & 0 & \sin\gamma \\ 0 & 1 & 0 \\ -\sin\gamma & 0 & \cos\gamma \end{bmatrix} \begin{bmatrix} \cos\alpha & -\sin\alpha & 0 \\ \sin\alpha & \cos\alpha & 0 \\ 0 & 0 & 1 \end{bmatrix} \begin{bmatrix} 1 & 0 & 0 \\ \pm\dfrac{\sin\xi_1}{\cos\xi_1} & 1 & \pm\dfrac{\tan\xi_2}{\cos\xi_1} \\ 0 & 0 & 1 \end{bmatrix}.$$

Assuming $x - x0 = x_T$, $y - y0 = y_T$ and $z - z0 = z_T$, then

$$(x_P\ y_P\ z_P) = (x_T \cos\gamma \cos\alpha + y_T \sin\beta \sin\gamma \cos\alpha - z_T \cos\beta \sin\gamma \cos\alpha$$

$$+ y_T \cos\beta \sin\alpha + z_T \sin\beta \sin\alpha, \ - x_T \cos\gamma \sin\alpha$$

$$- y_T \sin\beta \sin\gamma \sin\alpha + z_T \cos\beta \sin\gamma \sin\alpha + y_T \cos\beta \cos\alpha$$

$$+ z_T \sin\beta \cos\alpha, \ x_T \sin\gamma - y_T \sin\beta \cos\gamma + z_T \cos\beta \cos\gamma)$$

$$\times \begin{bmatrix} 1 & 0 & 0 \\ \pm\dfrac{\sin\xi_1}{\cos\xi_1} & 1 & \pm\dfrac{\tan\xi_2}{\cos\xi_1} \\ 0 & 0 & 1 \end{bmatrix}$$

$$= (x_{TR}\ y_{TR}\ z_{TR}) \begin{bmatrix} 1 & 0 & 0 \\ \pm\dfrac{\sin\xi_1}{\cos\xi_1} & 1 & \pm\dfrac{\tan\xi_2}{\cos\xi_1} \\ 0 & 0 & 1 \end{bmatrix}$$

where the point $(x_{TR}\ y_{TR}\ z_{TR})$ is the original point (x y z) but expressed relative to some new co-ordinate system arrived at by axes translation and rotation only. That is

$$x_{TR} = x_T \cos\gamma \cos\alpha + y_T \sin\beta \sin\gamma \cos\alpha - z_T \cos\beta \sin\gamma \cos\alpha$$

$$+ y_T \cos\beta \sin\alpha + z_T \sin\beta \sin\alpha$$

$$y_{TR} = - x_T \cos\gamma \sin\alpha - y_T \sin\beta \sin\gamma \sin\alpha + z_T \cos\beta \sin\gamma \sin\alpha$$

$$+ y_T \cos\beta \cos\alpha + z_T \sin\beta \cos\alpha$$

$$z_{TR} = x_T \sin\gamma - y_T \sin\beta \cos\gamma + z_T \cos\beta \cos\gamma \ . \tag{7.7}$$

Further matrix multiplication gives, for translation, rotation and projection

$$x_P = x_T \cos\gamma \cos\alpha + y_T \sin\beta \sin\gamma \cos\alpha - z_T \cos\beta \sin\gamma \cos\alpha$$

$$+ y_T \cos\beta \sin\alpha + z_T \sin\beta \sin\alpha - x_T \cos\gamma \sin\alpha(\pm\tan\xi_1)$$

$$- y_T \sin\beta \sin\gamma \sin\alpha(\pm\tan\xi_1) + z_T \cos\beta \sin\gamma \sin\alpha(\pm\tan\xi_1)$$

$$+ y_T \cos\beta \cos\alpha(\pm\tan\xi_1) + z_T \sin\beta \cos\alpha(\pm\tan\xi_1)$$

$$y_P = - x_T \cos\gamma \sin\alpha - y_T \sin\beta \sin\gamma \sin\alpha + z_T \cos\beta \sin\gamma \sin\alpha$$

$$+ y_T \cos\beta \cos\alpha + z_T \sin\beta \cos\alpha$$

$$z_P = - x_T \cos\gamma \sin\alpha(\pm\frac{\tan\xi_2}{\cos\xi_1}) - y_T \sin\beta \sin\gamma \sin\alpha(\pm\frac{\tan\xi_2}{\cos\xi_1})$$

$$+ z_T \cos\beta \sin\gamma \sin\alpha(\pm\frac{\tan\xi_2}{\cos\xi_1}) + y_T \cos\beta \cos\alpha(\pm\frac{\tan\xi_2}{\cos\xi_1})$$

$$+ z_T \sin\beta \cos\alpha(\pm\frac{\tan\xi_2}{\cos\xi_1}) + x_T \sin\gamma - y_T \sin\beta \cos\gamma + z_T \cos\beta \cos\gamma \ . \tag{7.8}$$

7.3.3 An insolation algorithm

The transformation equations can be utilised as the basis of an algorithm to determine external and internal surface shading/insolation as a function of building geometry, obstructions geometry and solar position.

For external surfaces, each face of the target building is processed in turn and the surrounding obstructions, including other obstructing parts of the target building, projected onto the plane of each face. This is achieved as follows.

Step 1: Transform the current solar position, relative to the site co-ordinate system XYZ, to pseudo solar azimuth and elevation angles (ξ_1 and ξ_2) expressed relative to the local face co-ordinate system. This can be done by expressing the sun's position as a distant point (x y z) for insertion, along with the target face angles (α, β, γ) in eqns (7.7). The point to emerge (x_{TR} y_{TR} z_{TR}) is then re-expressed as angles ξ_1 and ξ_2 as shown in figure 7.11.

Step 2: Each vertex of an obstruction body is projected onto the plane of the target face by application of eqns (7.8) to give a shadow 'image'. This will appear as a two-dimensional representation of a three-dimensional object, i.e. with all vertices of the image in the same local plane, $X''Z''$.

Step 3: Remove all internal line segments to leave only the shadow polygon which may or may not intersect the target face.

Step 4: Final shading estimation is determined by simple point containment tests applied to a grid superimposed on the target face. Each grid point is assigned a value of zero to indicate that the point is insolated. Any point contained by a shadow polygon is then reassigned a value of one to indicate shading. This allows both magnitude and point of application estimation, with accuracy controlled by varying the grid size. Point containment can be established by radiating a line away from the point in an arbitrary direction. If the number of intersections with the polygon edge is odd the point is contained and, if even, it is not. Appendix E gives a coded algorithm which is suitable for this purpose.

Step 5: For internal surfaces, insolation patch position and magnitude can be assessed by applying a grid to each window and projecting each grid point onto the internal surfaces—using eqns (7.8)—to establish, again by point containment tests, the grid point/receiving surface pairings.

The time-series data to emerge for each external and internal surface, along with the geometrical quantities, can now be utilised in the solar computations of the next section.

7.4 Shortwave radiation processes

Consider figure 7.12, which details the interactions between a building and the incident direct/diffuse solar radiation. The shortwave flux incident on external opaque surfaces will be partially absorbed and partially reflected, while some portion of the absorbed component may be transmitted to the corresponding interior surface, by conduction, to elevate the inside surface temperature and so enter the building via surface convection and longwave radiation exchange. Likewise, a portion of the absorbed component will cause outside surface temperature elevation and so give rise to a re-release of energy to ambient. If a multi-layered construction is opaque overall but has transparent elements located towards its outermost surface, some portion of the incident direct and diffuse radiation will also be transmitted inward until it strikes the intra-constructional opaque interface. Here, absorption and reflection will again occur, the latter giving rise to further absorptions and interface reflections as the flux travels outwards; the process continuing, essentially instantaneously, until the incident flux has been redistributed.

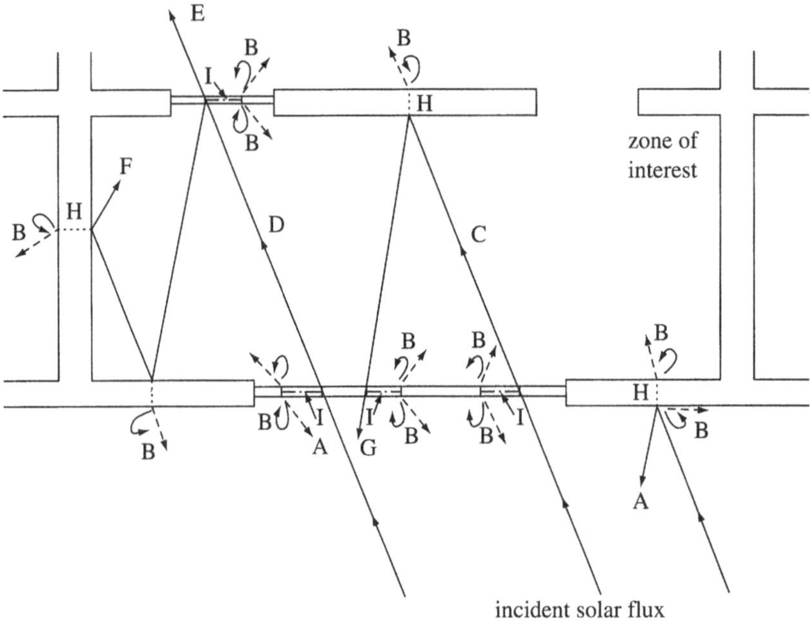

Figure 7.12: Building/solar interaction.
A—reflected shortwave flux; B—flux emission by convection and long-wave radiation; C—shortwave flux transmission to cause opaque surface insolation; D—shortwave transmission to cause transparent surface insolation; E—shortwave transmission to adjacent zone; F—enclosure reflections; G—shortwave loss; H—solar energy penetration by transient conduction; I—solar energy absorption prior to retransmission by the processes of B.

With windows, the direct and diffuse shortwave flux is reflected, absorbed and transmitted at each interface with the internally absorbed component being transmitted inward and outward by the processes of conduction, convection and longwave radiation exchange.

The transmitted direct beam continues onward to cause internal surface insolation as a function of the zone geometry. The subsequent treatment of this incident flux will depend on the nature of the receiving surface(s): absorption and reflection for an opaque surface, or absorption, reflection and transmission (to another zone or back to outside) in the case of a transparent surface. If the internal surface is a specular reflector then the reflected beam's onward path may be tracked by some suitable technique until diminished to insignificance. If the zone surface is a diffuse reflector then the apportioning of the reflected flux to other internal surfaces may be determined by weighting factors derived from the zone view factors (§7.5.2). The same technique may be applied to the transmitted diffuse beam.

The causal effect of these shortwave processes is then represented by the conservation equations of chapter 3, given that the shortwave flux injection at appropriate finite volumes can be established at each computational time-row. The requirement therefore is to establish the time-series of shortwave flux injection for finite volumes representing external opaque and transparent surfaces (§7.4.3), intra-constructional elements, where these are part of a transparent multi-layered construction (§7.4.4), and internal opaque and transparent surfaces (§7.4.5).

7.4.1 Solar position

It is usual to express the position of the sun in terms of altitude and azimuth angles that depend on site latitude, solar declination and local solar time. Figure 7.13 illustrates these angles, which are discussed in detail and established as mathematical expressions in a number of texts (IHVE 1973, Duffie and Beckman 1980, Muneer 1997).

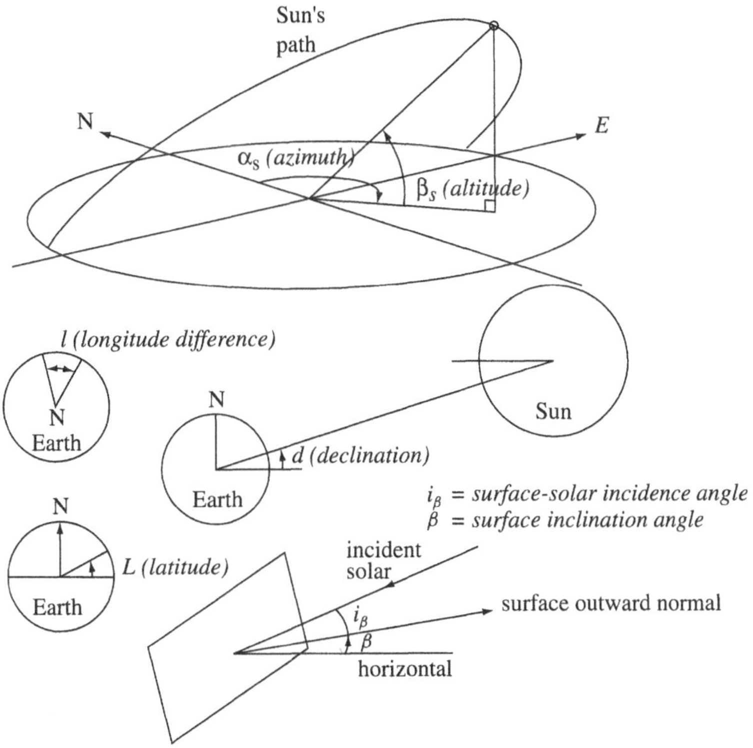

Figure 7.13: Solar angles.

The solar declination may be determined from

$$d = 23.45 \sin(280.1 + 0.9863Y)$$

where d is the solar declination (°) and Y the year day number (January 1 = 1, February 1 = 32 etc). The solar altitude is then obtained from

$$\beta_s = \sin^{-1}[\cos L \cos d \cos \theta_h + \sin L \sin d]$$

where β_s is the solar altitude, L the site latitude (north +ve) and θ_h the hour angle, which is the angular expression of solar time and is positive for times before solar noon and negative for times thereafter:

$$\theta_h = 15(12 - t_s)$$

where t_s is the solar time (or local apparent time). This is a time scale which relates to the apparent angular motion of the sun across the sky vault with solar noon corresponding to the point in time at which the sun traverses the meridian of the observer. Note that solar time does not necessarily coincide with local mean (or clock) time, t_m, with the difference given by

$$t_s - t_m = \pm l/15 + e_t + \delta \tag{7.9}$$

where l is the longitude difference (°), e_t the equation of time (hours) and δ a possible correction for daylight saving (hours). The longitude difference is the difference between an observer's actual longitude and the longitude of the mean or reference meridian for the local time zone. The difference is negative for locations to the west of the reference meridian and positive to the east. For the UK, the reference meridian is at 0° and local mean time is known as Greenwich Mean Time (GMT). For this case eqn (7.9) becomes

$$t_s = \mathrm{GMT} \pm l'/15 + e_t + \delta$$

where l' is the actual longitude of the observer (°). The equation of time makes allowance for the observed disturbances to the earth's rate of rotation:

$$e_t = 9.87\sin(1.978Y - 160.22) - 7.53\cos(0.989Y - 80.11)$$

$$- 1.5\sin(0.989Y - 80.11)\,.$$

The solar azimuth angle is given by

$$\alpha_s = \sin^{-1}[\cos d \sin\theta_h / \cos\beta_s]\,.$$

In applying this equation it is necessary to distinguish between northern and southern latitudes; the azimuthal corrections of table 7.8 are applicable.

Table 7.8: Azimuthal correction factors (from IHVE 1973).

Condition	*Time*	*Northern latitude*	*Southern latitude*
x < y	a.m.	α_s	180 - α_s
	p.m.	360 - α_s	180 + α_s
x = y	a.m.	90	90
	p.m.	270	270
x > y	a.m.	180 - α_s	α_s
	p.m.	180 + α_s	360 - α_s
$x = \cos\theta_h$	$y = \tan d \cot L$		

The angle of incidence of the direct beam, as shown in figure 7.13, may be found from

$$i_\beta = \cos^{-1}[\sin\beta_s \cos(90 - \beta_f) + \cos\alpha_s \cos\omega \sin(90 - \beta_f)]$$

where i_β is the angle between the incident beam and the surface's normal vector, ω the surface-solar azimuth ($= |\alpha_s - \alpha_f|$) and α_f, β_f are the surface azimuth and elevation respectively. Note that negative values of $\cos(i_\beta)$ imply that the surface in question faces away from the sun and is therefore not directly insolated.

While the foregoing expressions represent an accuracy that is commensurate with the requirements of building simulation, more exacting formulations exist (Yallop 1992) and are available in computer-ready format (Muneer 1997).

7.4.2 Solar radiation prediction

The intensity of extraterrestrial solar radiation, when integrated over all wavelengths, is termed the solar constant. Because of the elliptical orbit of the earth around the sun, the value of this 'constant' varies throughout the year:

$$I'_{sc} = I_{sc}\{1 + 0.033 \cos[(360 - Y)/370]\}$$

where I'_{sc} is the corrected solar constant (W m^{-2}), I_{sc} is the solar constant evaluated at the equinox, and Y is the year day number.

The value normally assigned to I_{sc} is 1353 W m^{-2} (Thekaekara 1973), with the transformation to the (extraterrestrial)horizontal plane given by

$$I_{eh} = 1353[1 + 0.033 \cos(0.0172024Y)] \sin \beta_s \; .$$

The spectral composition of solar radiation, that is the wavelength dependent irradiance, is as shown in figure 7.14. Although the entire spectrum spans the range from X-rays (< 0.01 μm) to radio waves (> 100 m), some 99.9% of the total energy is contained within the range 0.22—10.94 μm.

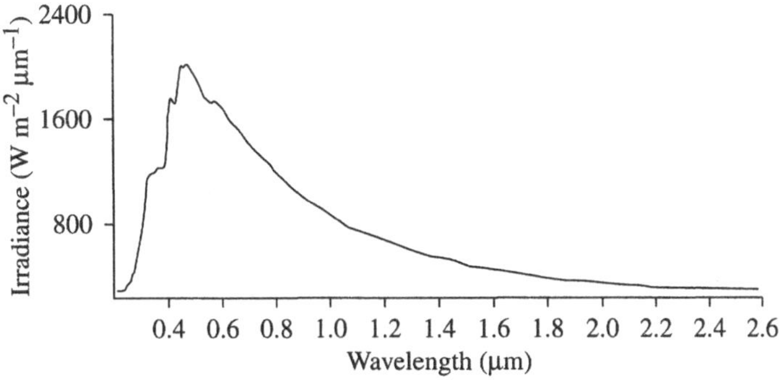

Figure 7.14: The solar spectrum.

To determine the terrestrial irradiance in the absence of measurements, the extraterrestrial intensity, I_{eh}, may be modified to account for the effects of atmospheric transmission. This topic has sustained a significant research effort over time (e.g Ångström 1930, Liu and Jordan 1960, Barbaro *et al* 1979 and Grindley *et al* 1995). As the radiation traverses the atmosphere, scattering and absorption occurs due to the natural and anthropogenic related presence of gases, aerosols and pollutants. The result is that some portion of the solar power is 'lost', while the remaining portion comprises directional and diffuse components. Muneer (1997) describes the following model for the estimation of these components under clear, overcast and intermediate skies.

For clear sky conditions, the diffuse horizontal irradiance, I_{fh}, may be determined from

$$I_{fh} = I_{eh}[1 - 0.1(1 - \tau_\alpha)(1 - m + m^{1.06})]\tau_g \tau_o \tau_w$$

$$\times \left[\frac{0.5(1 - \tau_r)}{1 - m + m^{1.02}} + \frac{0.84(1 - 10^{-0.045^{0.7m}})}{1 - m + m^{1.02}} \right] \qquad (7.10)$$

where m is the air mass corresponding to the prevailing solar altitude and atmospheric pressure, and τ_r, τ_o, τ_w, τ_g, τ_α are the components of atmospheric transmission relating to Rayleigh (Davies *et al* 1975), ozone and water vapour (Lacis and Hansen 1974), mixed gases (Bird and Hulstrom 1979) and aerosol (Ma and Iqbal 1983) scattering respectively.

The global horizontal irradiance, I_{gh}, is then given by

$$I_{gh} = (I_{dh} + I_{fh})\left(\frac{1}{1 - r_s r'_\alpha}\right) \tag{7.11}$$

where I_{dh} is the direct horizontal irradiance given as $I_{eh}\tau_r\tau_\alpha\tau_g\tau_o\tau_w$, r_s the ground reflectance, $r'_\alpha = 0.0685 + 0.17(1 - \tau'_\alpha)$ and τ'_α the Rayleigh scattering transmittance computed at $m = 1.66$.

For an overcast sky the beam radiation is zero and eqns (7.10) and (7.11) remain valid although the accuracy level will fall given the model's inability to accommodate the different possible cloud density distributions. Muneer *et al* (1996) estimated this accuracy to lie between 6% for a clear sky and 30% for an overcast sky. For real skies, comprising clear and overcast portions, various blending mechanisms have been proposed, such as the use of a clearness index, K_B $(= I_{dh}/I_{eh})$, to determine the diffuse to beam ratio, DBR $(= I_{fh}/I_{dh})$, from a generalised correlation (Muneer *et al* 1996):

$$\mathrm{DBR} = 0.285 K_B^{-1.006}.$$

For locations where cloud cover observations are available, Kimura and Stephenson (1969) developed a procedure for the estimation of the direct normal, I_{dnc}, and diffuse horizontal, I_{fhc}, irradiance under real sky conditions:

$$I_{dnc} = I_{dn}F\left(1 + \frac{I_{fh}}{I_{dn}\sin\beta_s}\right)$$

where I_{dn} is the terrestrial direct normal irradiance under clear sky conditions (Wm^{-2}) and F is a function of the cloud type such that

$$F = (1 - CC/10)\left(\frac{\sin\alpha_s}{p + \sin\alpha_s} + \frac{(q-1)}{[0.691 - 0.137\sin\beta_s + 0.394\sin^2\beta_s]}\right) \tag{7.12}$$

where p and q are constants as given in table 7.9 and CC is cloud cover.

Assuming that information on cloud type and amount is available for four reference layers—as specified in the US standard—then the cloud cover can be obtained from

$$CC = T_{ca} - 0.5\left[\left(\sum_{j=1}^{4} C_j\right)_{c1} + \left(\sum_{j=1}^{4} C_j\right)_{c2} + \left(\sum_{j=1}^{4} C_j\right)_{c3}\right]$$

where c1 refers to cirrus cloud, c2 to cirrostratus and c3 to cirrocumulus; T_{ca} is the total cloud amount and C_j is the estimated cloud amount (on a scale 0-10) for the jth layer of the stated cloud type.

The cloudy sky diffuse horizontal intensity is then obtained from

$$I_{fhc} = (I_{dn}\sin\beta_s + I_{fh})(CCF - F) \tag{7.13}$$

where CCF is a cloud cover factor $(= q + rCC + sCC^2)$ and q, r and s are constants as given in table 7.9.

The direct and diffuse irradiance, whether synthesised or measured can now be used in the determination of the insolation of exposed locations throughout the building.

7.4.3 Inclined surface irradiance

The total radiation incident on an exposed opaque or transparent surface of arbitrary inclination β_f and azimuth α_f has three components: direct beam, ground reflected and sky diffuse. Where

Table 7.9: Constants p, q, r and s for use with eqns (7.12) and (7.13).

Season	*p*	*q*	*r*	*s*
Spring		1.06	0.012	-0.0084
Mar 21	0.071			
Apr 21	0.097			
May 21	0.121			
Summer		0.96	0.033	-0.0106
Jun 21	0.134			
Jul 21	0.136			
Aug 21	0.122			
Autumn		0.95	0.030	-0.0108
Sep 21	0.092			
Oct 21	0.073			
Nov 21	0.063			
Winter		1.14	0.003	-0.0082
Dec 21	0.057			
Jan 21	0.058			
Feb 21	0.060			

important, an estimate of the shortwave flux reflections from surrounding buildings may be estimated as a function of the flux incident on the corresponding face of the target building.

The direct component is relatively straightforward to determine since it involves only angular operations on the known direct horizontal irradiance. To determine the ground reflected component, the ground may be considered a diffuse reflector with representative view factors used to associate portions of the reflected radiation with each building surface. Estimation of the sky diffuse component is more problematic because of the anisotropic nature of the sky radiance distribution.

Direct beam component

This is given by

$$I_{d\beta} = I_{dh} \cos i_\beta / \sin \beta_s$$

where $I_{d\beta}$ is the direct intensity on the inclined surface (W m^{-2}) and I_{dh} is the direct horizontal intensity (W m^{-2}).

Ground reflected component

For an unobstructed vertical surface (i.e. $\beta_f = 0$), the view factor between the surface and the ground, and between the surface and the sky, is in each case 0.5 and so the radiation intensity at the surface due to isotropic ground reflection is given by

$$I_{rv} = 0.5(I_{dh} + I_{fh})r_g$$

where I_{rv} is the ground reflected total radiation incident on the vertical surface (W m^{-2}), I_{fh} is the horizontal diffuse radiation (W m^{-2}) and r_g is the ground reflectivity. For a surface of non-vertical inclination a simple view factor modification is introduced so that

$$I_{r\beta} = 0.5[1 - \cos(90 - \beta_f)](I_{dh} + I_{fh})r_g$$

where $I_{r\beta}$ is the ground reflected radiation incident on a surface of inclination β_f.

Sky diffuse component

Three approaches are prominent in the treatment of anisotropic sky conditions.

In the first (Temps and Coulson 1977, Klucher 1979), the sky diffuse component on an inclined surface is determined by an expression that increases the intensity of the diffuse flux due to circumsolar activity and horizon brightening:

$$I_{s\beta} = I_{fh}\{0.5[1 + \cos(90 - \beta_f)]\}\ \{1 + [1 - (I_{fh}^2/I_{Th}^2)]\sin^3 0.5\beta_f\}$$
$$\times\ \{1 + [1 - (I_{fh}^2/I_{Th}^2)]\cos^2(i_\beta)\sin^3(90 - \beta_s)\}$$

where $I_{s\beta}$ is the sky diffuse radiation incident on a surface of inclination β_f (W m^{-2}) and I_{Th} the total horizontal radiation, $I_{dh} + I_{fh}$. When the sky is completely overcast, $I_{fh}/I_{Th} = 1$ and the expression reduces to the isotropic sky case.

In the second approach (Hay 1979), the known horizontal diffuse radiation is assumed to comprise a uniform background diffuse component and a circumsolar component, with a weighting applied according to the degree of sky isotropy. The sky diffuse component is given by

$$I_{s\beta} = I_{fh}[(I_{dh}/I_{eh})\cos i_\beta/\cos(90 - \beta_s) + 0.5(1 + \cos(90 - \beta_f))(1 - I_{dh}/I_{eh})]\ .$$

Again, as overcast sky conditions are approached so $I_{dh} \rightarrow 0$ and the isotropic expression is obtained.

In the third approach (Perez *et al* 1993), the sky diffuse irradiance is considered to be fully anisotropic:

$$I_{s\beta} = I_{fh}[(1 - F_1)\cos^2 0.5\beta_f + F_1(a_0/a_1) + F_2\sin\beta_f]$$

where F_1 and F_2 are circumsolar and horizon brightness coefficients, and a_0, a_1 correct for the angle of incidence of the circumsolar radiation on the inclined and horizontal surfaces respectively:

$$a_0 = \max[0, \cos i_\beta]$$

$$a_1 = \max[\cos 85°, \cos\beta_s]\ .$$

The brightness coefficients are given by

$$F_1 = \max[0, (f_{11} + f_{12} m I_{fh}/I'_{sc} + (\pi/180)\,Z\,f_{13})]$$

$$F_2 = f_{21} + f_{22} + (\pi/180)\,Z\,f_{23}$$

where Z is the zenith angle and the 'f' factors are as given in table 7.10 for a sky clearness, ε_i, evaluated from

$$\varepsilon_i = \frac{([I_{fh} + I'_{sc}/I_{fh}] + 5.535\times10^{-6}\,Z^3)}{(1 + 5.535\times10^{-6}\,Z^3)}\ .$$

The validation sub-group of the EC's PASSYS project concluded that the Perez *et al* model gave the best overall performance when applied to a wide range of locations (Jensen 1994). Other formulations for the sky diffuse component can be found in the literature (Steven and

Unsworth 1980, Muneer 1990).

Table 7.10: Coefficients for use in the Perez *et al* model.

ε_i *(bin)*	*1*	*2*	*3*	*4*	*5*	*6*	*7*	*8*
from	1.000	1.065	1.230	1.500	1.950	2.800	4.500	6.200
to	1.065	1.230	1.500	1.950	2.800	4.500	6.200	-
f_{11}	-0.0083	0.1299	0.3297	0.5682	0.8730	1.1326	1.0602	0.6777
f_{12}	0.5877	0.6826	0.4869	0.1875	-0.3920	-1.2367	-1.5999	-0.3273
f_{13}	-0.0621	-0.1514	-0.2211	-0.2951	-0.3616	-0.4118	-0.3589	-0.2504
f_{21}	-0.0596	-0.0189	0.0554	0.1089	0.2256	0.2878	0.2642	0.1561
f_{22}	0.0721	0.0660	-0.0640	-0.1519	-0.4620	-0.8230	-1.1272	-1.3765
f_{23}	-0.0220	-0.0289	-0.0261	-0.0140	0.0012	0.0559	0.1311	0.2506

The foregoing equations allow the computation of direct and diffuse shortwave radiation impinging upon exposed external surfaces. These flux quantities, when multiplied by surface absorptivity, are the shortwave nodal heat generation terms, q_{SI}, of the conservation equations derived for exposed surface layers in chapter 3.

7.4.4 Reflection, absorption and transmission within transparent media

The chapter 3 conservation equations for multi-layered constructions included the possibility of internal nodal heat generation. This section describes an approach to the computation of intra-construction nodal shortwave flux absorption for the case where a construction is partially or wholly transparent.

The solar spectrum occupies that part of the electromagnetic spectrum extending from 0.22—10.9 μm and therefore the reflectivity, r, of a window system is given by

$$r = \int_{\lambda=0.22}^{\lambda=10.9} I_\lambda r_\lambda d\lambda \Big/ \int_{\lambda=0.22}^{\lambda=10.9} I_\lambda d\lambda$$

where I_λ is the spectral irradiance (W m^{-2}), r_λ the spectral reflectivity (-) and λ the wavelength (μm). The objective of this section is to consider techniques for the determination of r (and transmittance/absorptance) for various window systems.

Consider figure 7.15, which shows a transparent layer of thickness L subjected to an ambient beam (or shortwave flux transmitted from an adjacent layer). It is necessary to treat this radiation as two polarised vectors, one parallel (||) and one perpendicular ($\perp$) to the substrate. Assuming solar radiation to be unpolarised, then the intensity of the incident radiation will consist of equal quantities of each component of polarisation so that

$$I_1 = 0.5(I_{||} + I_{\perp}) .$$

For each electric vector, the interface reflectivity is given by Fresnel's specular reflection equations (Vasicek 1960):

$$r_{||} = \frac{\tan^2(\theta_i - \theta_r)}{\tan^2(\theta_i + \theta_r)} \qquad r_{\perp} = \frac{\sin^2(\theta_i - \theta_r)}{\sin^2(\theta_i + \theta_r)}$$

where $r_{||}$ is the reflectivity for the parallel vector, $r_{\perp}$ the reflectivity for the perpendicular vector, θ_i the beam angle of incidence, and θ_r the corresponding angle of refraction. The interface transmissivity follows from

$$\tau_\xi = 1 - r_\xi$$

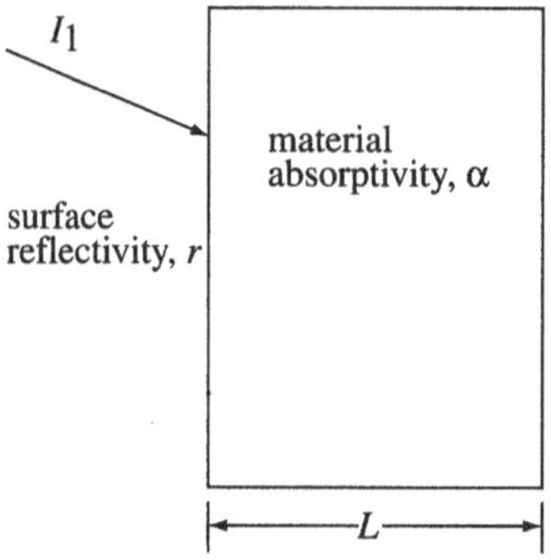

Figure 7.15: A transparent layer subjected to shortwave flux.

where ξ is $\parallel$ or $\perp$ as required.

As the separate radiation vectors pass through the transparent medium absorption occurs and the beam intensity can be assumed to diminish according to Bouguer's law (Duffie and Beckman 1980), which is based on the assumption that the absorbed radiation is proportional to the local intensity within the medium and the path length:

$$dI/dx = -KI$$

where dx is the penetration path length, dI the intensity reduction and K the absorption extinction coefficient. Table 7.11 gives some typical values for K.

Table 7.11: Typical extinction coefficient values.

Material	*K* (cm^{-1})
Polyvinyl fluoride (Tedlar)	1.4
Teflon	0.59
Polyethylene (Mylar)	2.05
Ordinary window glass	0.3
White glass	0.04
Heat absorbing glass	1.3 (to 2.7)

Integrating over the total path length $0 \rightarrow L/\cos\theta_r$, where L is the medium thickness normal to the surface of the substrate, gives

$$\int_{I_1}^{I_2} dI/I = \int_0^{L/\cos\theta_r} -K dx$$

$$\Rightarrow \ln(I_2/I_1) = -KL/\cos\theta_r$$

$$\Rightarrow I_2 = I_1 \exp(-KL/\cos\theta_r)$$

and so the overall absorption fraction is given by

$$\alpha = (I_1 - I_2)/I_1 = 1 - \exp(-KL/\cos\theta_r)\ . \quad (7.14)$$

The index of refraction, μ, is defined by Snell's Law:

$$\mu = \sin\theta_i/\sin\theta_r\ . \quad (7.15)$$

This index is subject to variation depending on the particular wavelength being considered: a phenomenon termed dispersion. In general, the refractive index decreases as the wavelength increases and the rate of decrease is greatest at the shorter wavelengths. Equations describing the dispersion behaviour of different materials can be found in the literature (Optical Society of America 1978). Now

$$\cos\theta_r = (1 - \sin^2\theta_r)^{\frac{1}{2}}$$

and so, from eqn (7.15):

$$\cos\theta_r = \left(1 - \frac{\sin^2\theta_i}{\mu^2}\right)^{\frac{1}{2}}.$$

Substitution in eqn (7.14) gives the absorptivity as a function of the prevailing incidence angle:

$$\alpha = 1 - \exp\left[-KL/\left(1 - \frac{\sin^2\theta_i}{\mu^2}\right)^{\frac{1}{2}}\right].$$

Consider now the superimposition (on figure 7.15) of the flowpath of one radiation vector as it undergoes multiple reflections at each interface and between-interface absorption as shown in figure 7.16. The total absorption, A, within the transparent element is given by

$$A = \left(\sum^{\infty} F_{A2} - \sum^{\infty} F_{R1}\right) + \left(\sum^{\infty} F_{A1} - \sum^{\infty} F_{R2}\right) \tag{7.16}$$

where F_{A1} is the flux arriving at interface 1 from interface 2, F_{A2} the flux arriving at interface 2 from interface 1, F_{R1} the flux reflected at interface 1 towards interface 2, and F_{R2} the flux reflected at interface 2 towards interface 1. Now

$$\sum^{\infty} F_{A2} = I_1(1 - r_{1\xi})\tau_a + I_1(1 - r_{1\xi})\tau_a^3 r_{2\xi} r_{1\xi} + I_1(1 - r_{1\xi})\tau_a^5 r_{2\xi}^2 r_{1\xi}^2 + \cdots$$

$$= I_1(1 - r_{1\xi})\tau_a(1 + \tau_a^2 r_{2\xi} r_{1\xi} + \tau_a^4 r_{2\xi}^2 r_{1\xi}^2 + \cdots)$$

where τ_a is $(1 - \alpha)$ and relates to the incidence angle at interface 1, and ξ is $\parallel$ or $\perp$, the parallel and perpendicular components corresponding to the incidence angle at interface 1. Noting that the bracketed terms constitute an infinite summation with common ratio $\tau_a^2 r_{2\xi} r_{1\xi}$, it follows that

$$\sum^{\infty} F_{A2} = \frac{I_1(1 - r_{1\xi})\tau_a}{1 - \tau_a^2 r_{2\xi} r_{1\xi}}.$$

Similarly,

$$\sum^{\infty} F_{A1} = \frac{I_1(1 - r_{1\xi})\tau_a^2 r_{2\xi}}{1 - \tau_a^2 r_{2\xi} r_{1\xi}} \; ; \; \sum^{\infty} F_{R1} = \frac{I_1(1 - r_{1\xi})}{1 - \tau_a^2 r_{2\xi} r_{1\xi}} \; ; \; \sum^{\infty} F_{R2} = \frac{I_1(1 - r_{1\xi})\tau_a r_{2\xi}}{1 - \tau_a^2 r_{2\xi} r_{1\xi}}.$$

Substitution in expression (7.16) gives

$$A = \frac{I_1(1 - r_{1\xi})[\tau_a - 1 + \tau_a^2 r_{2\xi} - \tau_a r_{2\xi}]}{1 - \tau_a^2 r_{2\xi} r_{1\xi}}. \tag{7.17}$$

If the transparent element exists in isolation, with radiation incident only at interface 1, then this expression gives the absorption of the parallel and perpendicular components, which when

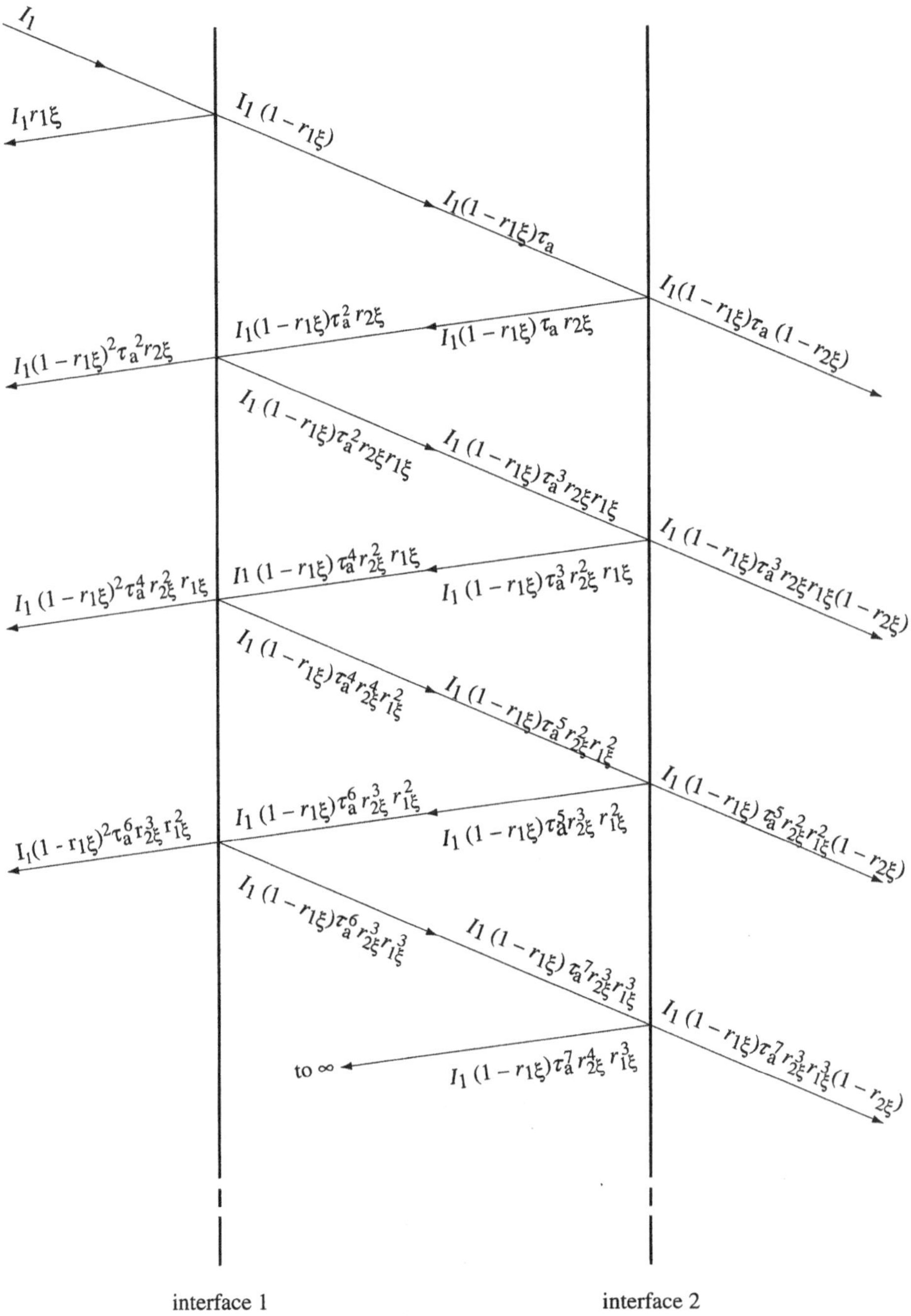

Figure 7.16: Multiple reflections and material absorption for a component of polarisation ξ.

summed gives the total absorption within the element. If shortwave radiation is simultaneously impinging on interface 2 (as with single glazing externally insolated and receiving internal reflections from surrounding surfaces) then further absorption of the second beam, I_2, will occur. Eqn (7.17) is then applied with $I_1 = I_2$ and the total absorption obtained by summing the component absorptions of each individual beam. If the transparent element is a member of a multi-layered group, then two possibilities exist: some portion of the radiation impinging on interface 1 is transmitted on to an adjacent transparent element, with some portion being subsequently retransmitted back across the common interface; or the second element is opaque. In the latter case, the total absorption at the opaque surface is given by

$$I_1(1 - r_{1\xi})\tau_a(1 - r_{2\xi}) + I_1(1 - r_{1\xi})\tau_a^3 r_{2\xi} r_{1\xi}(1 - r_{2\xi}) + I_1(1 - r_{1\xi})\tau_a^5 r_{2\xi}^2 r_{1\xi}^2(1 - r_{2\xi}) + \cdots$$

$$= \frac{I_1(1 - r_{1\xi})\tau_a(1 - r_{2\xi})}{1 - \tau_a^2 r_{2\xi} r_{1\xi}} . \qquad (7.18)$$

In the former case, eqn (7.18) (after both vector components have been added) defines the total onward flux transmitted to the adjacent element. If the interface reflectivities are known then it is possible to process the second element independently by treating the onward transmitted flux from element 1 as the initial flux incident on element 2. In this way the process continues for one forward pass through the multi-layered construction until all elements have been considered, or until an opaque element is encountered. The flux transmitted from one element to the next will be partly polarised even if the original beam was not. It is therefore necessary to process each component of polarisation separately before combining the overall value for the entire system.

This is not the end of the process, however, since each element will reflect flux back to the element from which it initially received the shortwave energy. This reflection is given by

$$I_i r_{i1\xi} + I_i(1 - r_{i1\xi})^2 \tau_{ai}^2 r_{i2\xi} + I_i(1 - r_{i1\xi})^2 \tau_{ai}^4 r_{i2\xi}^2 r_{i1\xi} + I_i(1 - r_{i1\xi})^2 \tau_{ai}^6 r_{i2\xi}^3 r_{i1\xi}^2 + \cdots$$

$$= I_i r_{i1\xi} + \frac{I_i(1 - r_{i1\xi})^2 \tau_{ai}^2 r_{i2\xi}}{1 - \tau_{ai}^2 r_{i2\xi} r_{i1\xi}}$$

where i is the element number.

It is possible to establish an algorithm to transmit a single flux package from one transparent element to the other as calculation 'sweeps' are made in alternating directions until the flux quantities diminish to insignificance. In this way the absorption of each element can be determined as well as the overall transmission of the combined system.

Let A_i be the absorption for element i and τ be the combined system transmission. Consider the volumetric subdivision of chapter 3 applied to the transparent element of figure 7.15, with nodes situated at the element boundaries and at the centre plane. The fraction of the total absorption associated with each node is determined by a volumetric weighting. Given that the total path length of radiation passing from one boundary to the other is given by $L/\cos\theta_r$, then the fraction received by a boundary node is

$$q_{SI} = A_i L / 4 \cos\theta_r$$

or

$$q_{SI} = A_i L / \{4[1 - (\sin^2\theta_i)/\mu^2]^{\frac{1}{2}}\} . \qquad (7.19)$$

And for a centre plane node:

$$q_{SI} = A_i L / \{2[1 - (\sin^2\theta_i)/\mu^2]^{\frac{1}{2}}\} .$$

If the boundary node separates two transparent elements then the nodal shortwave heat generation terms will have contributions from both elements. If the node separates a transparent element and an opaque element, then the absorptions of eqns (7.19) and (7.18) must be added:

$$q_{SI} = A_i L / \{4[1 - (\sin^2\theta_i)/\mu^2]^{\frac{1}{2}}\} + \frac{I_i(1 - r_{i1\xi})(1 - r_{i2\xi})\tau_{ai}}{1 - \tau_{ai}^2 r_{i2\xi} r_{i1\xi}} .$$

As a low energy feature, some window systems incorporate thin films evaporated onto the glass surface to alter the reflectance. Figure 7.17 shows the measured spectra of various oxide films that aim to reduce reflectance in the shortwave region and increase reflectance in the longwave region.

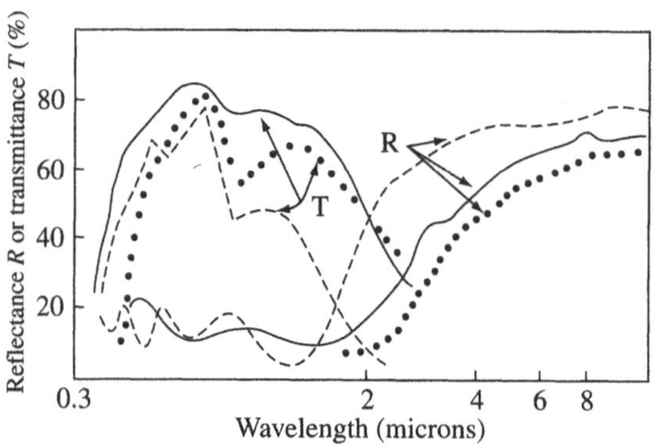

Figure 7.17: Measured spectra of oxide films (from Howson *et al* 1984). Full curve, cadmium 2:1 tin oxide; broken curve, indium 10% tin oxide; dotted curve, indium oxide.

Vasicek (1960) suggested that a thin film is one for which

$$\mu_f \delta t < 2.5\lambda_d$$

where μ_f is the refractive index of the film, δt the film thickness (μm) and λ_d the spectral wavelength (μm) for which the reflectance modification is required. Films that violate this condition can be treated in the same manner as glazing elements as detailed above. With thin films, the change of phase of the radiation vector as it traverses the film must be included in the analysis. Vasicek has studied the effects of the multiple reflections occurring within a thin film bounded by air and a transparent substrate and produced an expression for the interface reflectance:

$$r_\xi = \left(\frac{am_1 + bm_2 - cm_3 - m_4}{am_1 + bm_2 + cm_3 + m_4} \right)^2$$

where a, b, c and m are as given in table 7.12.

Several researchers (Vasicek 1960, Heavens 1955, Born and Wolf 1965, Rousseau and Mathieu 1973) have produced formulations for multiple thin film systems analysis based on matrix combination of individual films.

7.4.5 Intra-zone shortwave distribution

The formulations of §7.4.3 permit the calculation of the direct and diffuse irradiance of exposed building surfaces. For opaque surfaces, irradiance modification by surface

Table 7.12: Values of a, b, c and m (from Rousseau and Mathieu 1973).

Coefficient	*Normal incidence*	*Non-normal incidence*	
		// component	*⊥ component*
a	μ_0/μ_s	$\frac{\mu_0 \cos\theta_f}{\mu_s \cos\theta_i}$	$\frac{\mu_0 \cos\theta_i}{\mu_s \cos\theta_f}$
b	μ_0	$\frac{\mu_0}{\cos\theta_i}$	$\mu_0 \cos\theta_i$
c	$\frac{1}{\mu_s}$	$\frac{\cos\theta_f}{\mu_s}$	$\frac{1}{\mu_s \cos\theta_f}$
m_{11}	$\cos\delta_f$	$\cos\delta_f$	$\cos\delta_f$
m_{12}	$\frac{i \sin\delta_f}{\mu_f}$	$\frac{i \cos\theta_f \sin\delta_t}{\mu_f}$	$\frac{i \sin\delta_t}{\mu_f \cos\theta_f}$
m_{21}	$i\mu_f \sin\delta_t$	$\frac{i\mu_f \sin\delta_t}{\cos\theta_f}$	$i\mu_f \cos\theta_f \sin\delta_t$
m_{22}	$\cos\delta_f$	$\cos\delta_f$	$\cos\delta_f$

μ_0 = refractive index of air
μ_f = refractive index of film
μ_s = refractive index of substrate
θ_f = film angle of refraction
θ_i = beam angle of incidence
δ_t = film thickness (μm)
δ_f = change of phase = $2\pi\mu_f t \cos\theta_f/\lambda_d$

absorptivity and shading factors will give the shortwave heat injection to be applied to surface nodes via the excitation matrix (**C**) of chapter 4. For a transparent system, the formulations of the previous section allow the assessment of the internal shortwave absorption for injection at the intra-construction nodes. The eventual heat exchange between surface nodes and the surrounding air (by convection) and other surfaces (by longwave radiation exchange) will then follow from the conservation matrix equation solution.

The formulations of §7.4.4 allow the assessment of overall system properties such as transmissivity, absorptivity and reflectivity. This section addresses the use of these properties, in conjunction with the shading and insolation time-series information discussed in §7.3, to estimate the apportioning of shortwave energy between internal surfaces.

For a window system, the transmitted portion of the direct beam can be evaluated from

$$Q_{dt} = \frac{I_{dh}}{\sin\alpha_s} \tau_{i\beta}(1 - P_g)A_g \cos i_\beta \qquad (7.20)$$

where Q_{dt} is the transmitted direct beam flux (W), $\tau_{i\beta}$ the overall transmissivity for a given flux incidence angle, P_g the window shading factor (proportion of 1) and $A_g \cos i_\beta$ the apparent window area (m^2). The τ value can be determined by the techniques of §7.4.4 or, alternatively, by reference to published data for different window arrangements and glass types (Pilkington 1973) or special purpose software applications for the evaluation of window thermal/solar properties such as Window 4.1 (http://windows.lbl.gov/software/window/window.html) and

WIS (http://erg.ucd.ie/software_pub.html).

If, as is often the case, more than one internal surface will share this transmitted radiation then the flux defined by eqn (7.20) can be applied to those internal surfaces defined by the insolation data determined by the technique of §7.3. Any internal surface will then receive a heat injection given by

$$q_{Si} = Q_{dt} P_i \Omega_i / A_i$$

where Ω_i is the surface absorptivity, A_i the surface area (m^2) and P_i the proportion of the window direct beam transmission that strikes the surface in question (proportion of 1). The first reflected flux is given by

$$q_{Ri} = q_{Si}(1 - \Omega_i)/\Omega_i \ .$$

The accumulated flux reflections from each surface can now be further processed to give the final apportioning between all internal surfaces. If the usual assumption of diffuse reflections is made then apportionment can be decided on the basis of enclosure view factor information as described in the following section. For the case of specular reflections a recursive ray tracing technique can be employed, perhaps based on the radiosity technique also described in the next section.

Where the internal surface is composed of opaque and transparent portions, there will be onward transmission of incident shortwave flux to a connected zone or back to the outside. This can have a significant impact within buildings incorporating passive solar features. Application of eqn (7.20), with the ($I_{dh}/\sin\alpha_s$) term set to the incident flux value and appropriate adjustments made to the $\cos i_\beta$ and P_g terms, will then give the re-transmitted flux.

The diffuse beam transmission can be determined from

$$Q_{ft} = (I_{s\beta} + I_{r\beta}) A_g \cos 51^\circ \tau_{51}$$

where τ_{51} is the overall transmissivity corresponding to a 51° incidence angle, representing the average approach angle for anisotropic sky conditions.

This flux quantity can now be processed by the technique described for the direct beam: internal surface distribution on the basis of specular or diffuse reflections as described in the following section.

7.5 Longwave radiation processes

Heat transfer by longwave radiation† exchange between two communicating surfaces is an important issue in building energy modelling but one which introduces mathematical complexity due to non-linear behaviour and the spatial problems caused by complex geometries and inter-surface obstructions. In the following derivations internal and external exposed surfaces are treated separately because, in the latter case, detailed knowledge of surrounding surfaces (sky, buildings, ground) is usually unavailable.

The radiation flux emitted by a perfect 'black' body is given by

$$q_b = \sigma A \theta^4 \tag{7.21}$$

where q_b is the black body radiation flux (W), σ the Stefan-Boltzmann constant (1.3×10^{-23} W $m^{-2}K^{-4}$) and θ the body's absolute temperature (K).

† That is radiation of wavelength between 0.28 and 2.8 μm.

Real building materials do not behave as black bodies and deviate in their ability to absorb the incoming longwave energy completely. The radiant flux emitted by such a 'grey' body is given by a temperature dependent modification to eqn (7.21):

$$q = \varepsilon \sigma A \theta^4$$

where q is the grey body radiation flux (W) and ε the temperature dependent surface emissivity.

Kirchhoff's law states that the emissivity of a surface is equal to its absorptivity. Although true for construction materials under normal temperature conditions, the law is violated by some coatings, which aim to promote heat emission whilst minimising absorption, and by materials at high temperatures where emission and absorption can take place at substantially different wavelengths. However, in the derivations that follow, Kirchhoff's law is accepted and so emissivity replaces absorptivity throughout.

7.5.1 Exchange between internal surfaces

Within an enclosure the radiation emitted by all surfaces will, after multiple reflections, be totally re-absorbed and, in the process, redistributed. Assuming that no energy is lost by longwave transmission directly to an adjacent enclosure, and that the surfaces are diffuse reflectors, then a recursive solution is possible as follows.

The initial fluxes emitted by each surface are tracked to first reflection and the surface absorptions determined. For example, if four grey surfaces form an enclosure as shown in figure 7.18, then the flux emitted by each surface is given by

$$q_1 = \varepsilon_1 \sigma A_1 \theta_1^4 \quad q_2 = \varepsilon_2 \sigma A_2 \theta_2^4$$
$$q_3 = \varepsilon_3 \sigma A_3 \theta_3^4 \quad q_4 = \varepsilon_4 \sigma A_4 \theta_4^4$$

and, at first reflection, the absorption at each surface will have contributions as follows

$$\begin{array}{lllll}
a_1' = & & +q_2 f_{2\to1}\varepsilon_1 & +q_3 f_{3\to1}\varepsilon_1 & +q_4 f_{4\to1}\varepsilon_1 \\
a_2' = & +q_1 f_{1\to2}\varepsilon_2 & & +q_3 f_{3\to2}\varepsilon_2 & +q_4 f_{4\to2}\varepsilon_2 \\
a_3' = & +q_1 f_{1\to3}\varepsilon_3 & +q_2 f_{2\to3}\varepsilon_3 & & +q_4 f_{4\to3}\varepsilon_3 \\
a_4' = & +q_1 f_{1\to4}\varepsilon_4 & +q_2 f_{2\to4}\varepsilon_4 & +q_3 f_{3\to4}\varepsilon_4 &
\end{array}$$

where a_i' is the total flux absorption at surface i from all surfaces after the first reflection (W), and $f_{j\to i}$ the geometric view factor between surface j and i.

A single flux quantity can now be determined for each surface that represents the total apparent flux emission for processing to the next reflection:

$$r_i' = a_i'(1 - \varepsilon_i)/\varepsilon_i \;\; ; i = 1, 2, 3, 4$$

where r_i' is the flux reflected at surface i after first reflection (W). After the second reflection, the total absorption at each surface is given by

$$\begin{array}{llllll}
a_1'' = & a_1' & & +r_2' f_{2\to1}\varepsilon_1 & +r_3' f_{3\to1}\varepsilon_1 & +r_4' f_{4\to1}\varepsilon_1 \\
a_2'' = & a_2' & +r_1' f_{1\to2}\varepsilon_2 & & +r_3' f_{3\to2}\varepsilon_2 & +r_4' f_{4\to2}\varepsilon_2 \\
a_3'' = & a_3' & +r_1' f_{1\to3}\varepsilon_3 & +r_2' f_{2\to3}\varepsilon_3 & & +r_4' f_{4\to3}\varepsilon_3 \\
a_4'' = & a_4' & +r_1' f_{1\to4}\varepsilon_4 & +r_2' f_{2\to4}\varepsilon_4 & +r_3' f_{3\to4}\varepsilon_4 &
\end{array}$$

where a_i'' is the total absorption of flux at surface i from all surfaces after the second reflection (W). The flux reflections are then given by

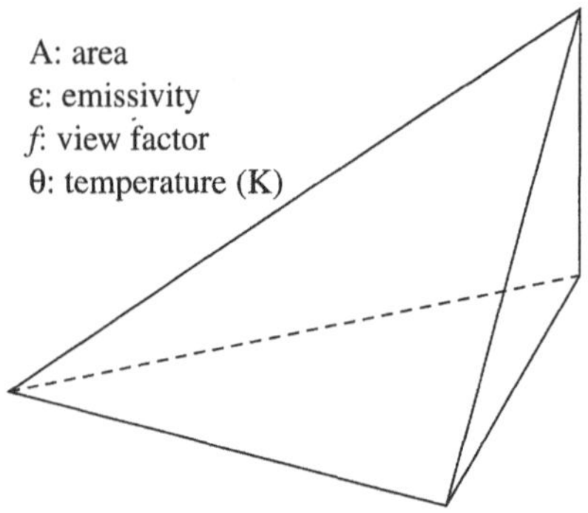

Figure 7.18: Four grey surfaces bounding an enclosure.

$$r_1'' = (a_1'' - a_1')(1 - \varepsilon_1)/\varepsilon_1 \qquad r_2'' = (a_2'' - a_2')(1 - \varepsilon_2)/\varepsilon_2$$
$$r_3'' = (a_3'' - a_3')(1 - \varepsilon_3)/\varepsilon_3 \qquad r_4'' = (a_4'' - a_4')(1 - \varepsilon_4)/\varepsilon_4$$

where the absorptions and reflections at each recursive step may be determined from

$$\left.\begin{aligned} a_i^n &= a_i^{n-1} + \sum_{j=1}^{N} r_j^{n-1} f_{j\to i}\varepsilon_i \\ r_i^n &= (a_i^n - a_i^{n-1})(1 - \varepsilon_i)/\varepsilon_i \end{aligned}\right\} \begin{aligned} &1 \le n \le \infty \\ &a_i^0 = 0 \\ &r_i^0 = q_i \\ &f_{i\to i} = 0 \end{aligned}$$

and in practice the recursive process continues until the reflected flux is reduced to insignificance. In most applications incorporating conventional building materials (high emissivity) this condition will be reached after about three recursive steps. Significantly more recursions will be required where low emissivity surfaces are present.

In relation to the formulation of the surface energy conservation equations of chapter 3, if the objective is to determine a linearised h_r value for each surface pair—to improve accuracy by allowing the future time-row flux exchange to be evaluated, in part, on the basis of the future time-row surface temperatures—then the relevant inter-surface energy exchange at each recursive step must be extracted and separately summed.

On the other hand, if it is deemed acceptable to determine the flux exchange only on the basis of present time-row temperatures, then it is possible to establish a set of surface radiosity balance equations. Note however that this will not give the net exchange between each surface pairing from which an effective h_r values can be computed. Consider the surface radiosity, which is equal to the sum of the emitted and subsequently reflected intensities:

$$R_i = \varepsilon_i \sigma \theta_i^4 + \rho_i \sum_{j=1}^{N} R_j f_{j\to i} A_j / A_i$$

$$= \varepsilon_i \sigma \theta_i^4 + \rho_i \sum_{j=1}^{N} R_j f_{i\to j} \qquad (7.22)$$

where R_i is the radiosity of surface i (W m^{-2}), ρ_i the reflectivity of surface i and N the total number of participating surfaces. Eqn (7.22) can be re-expressed as

$$\sum_{j=1}^{N} \frac{\delta_{ij} - \rho_i f_{i \to j}}{\varepsilon_i} R_j = \sigma \theta_i^4 \tag{7.23}$$

where

$$\delta_{ij} = \begin{cases} 1 \text{ for } i = j \\ 0 \text{ for } i \neq j \end{cases} .$$

For all participating surfaces, eqn (7.23) can be written in matrix form as

$$\mathbf{AR} = \mathbf{\Phi}$$

where

$$\mathbf{A} = \begin{bmatrix} m_{11} & m_{12} & . & . & . & m_{1N} \\ m_{21} & . & . & . & . & . \\ . & . & . & . & . & . \\ . & . & . & . & . & . \\ . & . & . & . & . & . \\ m_{N1} & . & . & . & . & m_{NN} \end{bmatrix} \quad \mathbf{R} = \begin{bmatrix} R_1 \\ R_2 \\ . \\ . \\ . \\ R_N \end{bmatrix} \quad \mathbf{\Phi} = \begin{bmatrix} \sigma\theta_1^4 \\ \sigma\theta_2^4 \\ . \\ . \\ . \\ \sigma\theta_N^4 \end{bmatrix}$$

$$m = (\delta_{ij} - \rho_i f_{i \to j}) / \varepsilon_i$$

and the solution for all surface radiosities is given by

$$\mathbf{R} = \mathbf{A}^{-1}\mathbf{\Phi}$$

which is equivalent to

$$R_i = \sum_{j=1}^{N} m'_{ij} \sigma \theta_j^4 \quad ; i = 1, 2, 3, \ldots, N$$

where m'_{ij} are the elements of $\mathbf{A}^{-1}$.

Given the radiosity of each surface, the net surface flux due to all surface interactions is given by

$$q_i = [\varepsilon_i \sigma \theta_i^4 - (1 - \rho_i) R_i] / \rho_i \ .$$

While the implementation of the recursive or matrix method is relatively trivial, run-times may be excessive for large multi-zone systems involving long-term simulations. An alternative approach is to employ analytical methods, which are often of equivalent accuracy.

Consider the worked example of table 7.13 arrived at by applying the recursive method to the geometry of figure 7.18. Note that only the flux emitted by surface 1 is detailed, with all subsequent reflections shown. The three flux quantities arriving at any surface are not combined to give a single reflective quantity for further onward processing but, instead, are processed as distinct quantities. This is done to allow the identification of principal components as the original flux decays: combination will greatly improve computational efficiency.

By inspection, it can be seen that by the third reflection some 98.7% of the original flux emission from surface 1 has been re-absorbed and that this rises to approximately 99.7% at the fourth reflection: these values are typical for most building enclosures at or near comfort conditions. Recognising that the absorptions at surfaces 2, 3 and 4, at first reflection, are equivalent to the first term of all two-surface interactions

Table 7.13: Application of the recursive techniques to the problem of figure 7.17.

- 1:343.5
  - 2:91.6$_2$
    - 1:61$_3$
      - 2:0.4$_2$
        - 1:0.03$_3$
        - 3:0.02^3
        - 4:0.03
      - 3:0.2$_4$
        - 1:0.07
        - 2:0.07
        - 4:0.08
      - 4:0.5$_4$
        - 1:0.01
        - 2:0.01
        - 3:0.01
    - 3:3.8$_3$
      - 1:1.0$_4$
        - 2:0.07
        - 3:0.04
        - 4:0.08
      - 2:1.0$_4$
        - 1:0.07
        - 3:0.04
        - 4:0.08
      - 4:1.2$_4$
        - 1:0.03
        - 2:0.03
        - 3:0.02
    - 4:6.9$_3$
      - 1:0.2$_4$
        - 2:0.01
        - 3:0.01
        - 4:0.02
      - 2:0.2$_4$
        - 1:0.01
        - 3:0.01
        - 4:0.02
      - 3:0.1$_4$
        - 1:0.03
        - 2:0.03
        - 4:0.04
  - 3:57.3$_2$
    - 1:15.3$_3$
      - 2:1.0$_4$
        - 1:0.07
        - 3:0.04
        - 4:0.08
      - 3:0.6$_2$
        - 1:0.17$_3$
        - 2:0.17^3
        - 4:0.19
      - 4:1.1$_4$
        - 1:0.03
        - 2:0.03
        - 3:0.02
    - 2:15.3$_3$
      - 1:1.0$_4$
        - 2:0.07
        - 3:0.04
        - 4:0.08
      - 3:0.6$_4$
        - 1:0.17
        - 2:0.17
        - 4:0.19
      - 4:1.1$_4$
        - 1:0.03
        - 2:0.03
        - 3:0.02
    - 4:17.2$_3$
      - 1:0.5$_4$
        - 2:0.03
        - 3:0.02
        - 4:0.04
      - 2:0.5$_4$
        - 1:0.03
        - 3:0.02
        - 4:0.04
      - 3:0.3$_4$
        - 1:0.08
        - 2:0.08
        - 4:0.10
  - 4:103.1$_2$
    - 1:3.1$_3$
      - 2:0.2$_4$
        - 1:0.01
        - 3:0.01
        - 4:0.02
      - 3:0.1$_4$
        - 1:0.03
        - 2:0.03
        - 4:0.04
      - 4:0.2$_2$
        - 1:0.01$_3$
        - 2:0.01^3
        - 3:0.004
    - 2:3.1$_3$
      - 1:0.2$_4$
        - 2:0.01
        - 3:0.01
        - 4:0.02
      - 3:0.1$_4$
        - 1:0.03
        - 2:0.03
        - 4:0.04
      - 4:0.2$_4$
        - 1:0.01
        - 2:0.01
        - 3:0.004
    - 3:1.9$_3$
      - 1:0.5$_4$
        - 2:0.03
        - 3:0.02
        - 4:0.04
      - 2:0.5$_4$
        - 1:0.03
        - 3:0.02
        - 4:0.04
      - 4:0.6$_4$
        - 1:0.02
        - 2:0.02
        - 3:0.01

Surface i	θ (K)	Area (m^2)	ε	$f_{i\text{->}j}$ for j = 1	2	3	4
1	295	1	0.8	0	0.33	0.33	0.33
2	297	1	0.8	0.33	0	0.33	0.33
3	293	1	0.5	0.33	0.33	0	0.33
4	295	1	0.9	0.33	0.33	0.33	0

$$1 \rightarrow 2 \rightarrow 1 \rightarrow 2 \rightarrow 1 \rightarrow 2 \cdots \rightarrow \infty$$
$$1 \rightarrow 3 \rightarrow 1 \rightarrow 3 \rightarrow 1 \rightarrow 3 \cdots \rightarrow \infty$$
$$1 \rightarrow 4 \rightarrow 1 \rightarrow 4 \rightarrow 1 \rightarrow 4 \cdots \rightarrow \infty$$
etc

then it is possible to devise an expression for these infinite series to account for the terms at first reflection and selected terms at all odd numbered reflections thereafter. The terms accounted for by all two-surface interactions are shown by subscript 2 in table 7.13. Further inspection reveals that three-surface interactions

$$1 \rightarrow 2 \rightarrow 3 \rightarrow 1 \rightarrow 2 \rightarrow 3 \rightarrow 1 \rightarrow 2 \rightarrow 3 \cdots \rightarrow \infty$$
$$1 \rightarrow 2 \rightarrow 4 \rightarrow 1 \rightarrow 2 \rightarrow 4 \rightarrow 1 \rightarrow 2 \rightarrow 4 \cdots \rightarrow \infty$$
etc

give rise to expressions that account for all terms identified by subscript 3. Similar reasoning for four surface interactions can be applied to account for terms shown by subscript 4 and so on depending on the accuracy required. Of course, the greater the number of interacting surfaces, the nearer will this technique approach the correct solution.

The following derivation applies to this analytical method and gives a simple expression for h_r such that when all possible pairings are treated the near total flux field has been processed.

Consider any two grey surfaces oriented arbitrarily in 3-dimensional space and consider the flux emitted by surface 1 as it is reflected between each surface to infinity as shown in figure 7.19. The radiant flux emitted by surface 1 is given by $\varepsilon_i \sigma A_1 \theta_1^4$ and, of this, some portion $f_{1\rightarrow 2}$ will arrive at surface 2. Of this amount, $\varepsilon_i \sigma A_1 \theta_1^4 f_{1\rightarrow 2}$, a portion ε_2 will be absorbed and $(1 - \varepsilon_2)$ will be diffusely reflected. Given $f_{2\rightarrow 1}$ as the geometric view factor between surface 2 and 1, the flux to arrive back at surface 1 is $(1 - \varepsilon_2)\varepsilon_1 \sigma A_1 \theta_1^4 f_{1\rightarrow 2} f_{2\rightarrow 1}$. Some portion of this, $(1 - \varepsilon_2)(1 - \varepsilon_1)\varepsilon_1 \sigma A_1 \theta_1^4 f_{1\rightarrow 2} f_{2\rightarrow 1}$, is reflected from surface 1 and the inter-reflection process continues to infinity.

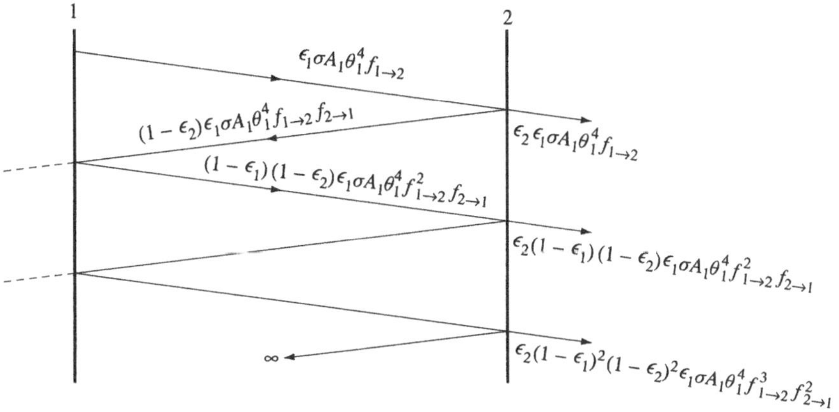

Figure 7.19: Flux absorbed at surface 2 due to two-surface interaction.

Focusing on surface 2, the radiant flux intercepted and absorbed due to infinite two-surface interactions (between surfaces 1 and 2 in this case) is given by

$$q_{(2)1\rightarrow 2} = \varepsilon_2 \varepsilon_1 \sigma A_1 \theta_1^4 f_{1\rightarrow 2} + \varepsilon_2(1 - \varepsilon_1)(1 - \varepsilon_2)\varepsilon_1 \sigma A_1 \theta_1^4 f_{1\rightarrow 2}^2 f_{2\rightarrow 1}$$

$$+\ \varepsilon_2(1-\varepsilon_1)^2(1-\varepsilon_2)^2\varepsilon_1\sigma A_1\theta_1^4 f_{1\to2}^3 f_{2\to1}^2 + \cdots\infty$$

$$= \varepsilon_2\varepsilon_1\sigma A_1\theta_1^4 f_{1\to2}[1 + (1-\varepsilon_1)(1-\varepsilon_2)f_{1\to2}f_{2\to1} + (1-\varepsilon_1)^2(1-\varepsilon_2)^2 f_{1\to2}^2 f_{2\to1}^2 + \cdots\infty]$$

where subscript (2) indicates two-surface interaction.

Since the terms contained in the square bracket comprise a geometrical progression, with common ratio $CR = (1-\varepsilon_1)(1-\varepsilon_2)f_{1\to2}f_{2\to1}$, the series, when summed to infinity, gives $1/(1-CR)$ and so

$$q_{(2)1\to2} = \frac{\varepsilon_2\varepsilon_1\sigma A_1\theta_1^4 f_{1\to2}}{1-(1-\varepsilon_1)(1-\varepsilon_2)f_{1\to2}f_{2\to1}}. \tag{7.24}$$

In a similar manner, three-surface interactions can be considered as shown in figure 7.20. For this case the flux absorbed at surface 2 is given by

$$q_{(3)1\to2} = \varepsilon_2(1-\varepsilon_n)\varepsilon_1\sigma A_1\theta_1^4 f_{1\to n}f_{n\to2}$$

$$+\ \varepsilon_2(1-\varepsilon_1)(1-\varepsilon_2)(1-\varepsilon_n)^2\varepsilon_1\sigma A_1\theta_1^4 f_{1\to n}^2 f_{n\to2}^2 f_{2\to1}$$

$$+\ \varepsilon_2(1-\varepsilon_1)^2(1-\varepsilon_2)^2(1-\varepsilon_n)^3\varepsilon_1\sigma A_1\theta_1^4 f_{1\to n}^3 f_{n\to2}^3 f_{2\to1}^2 + \cdots\infty$$

$$= \frac{\varepsilon_2(1-\varepsilon_n)\varepsilon_1\sigma A_1\theta_1^4 f_{1\to n}f_{n\to2}}{1-(1-\varepsilon_1)(1-\varepsilon_2)(1-\varepsilon_n)f_{1\to n}f_{n\to2}f_{2\to1}}$$

and so the flux arriving at surface 2 from 1 via all other surfaces directly is given by

$$q_{(3)1\to2} = \sum_{i=3}^{N}\left(\frac{\varepsilon_2(1-\varepsilon_i)\varepsilon_1\sigma A_1\theta_1^4 f_{1\to i}f_{i\to2}}{1-(1-\varepsilon_1)(1-\varepsilon_2)(1-\varepsilon_i)f_{1\to i}f_{i\to2}f_{2\to1}}\right). \tag{7.25}$$

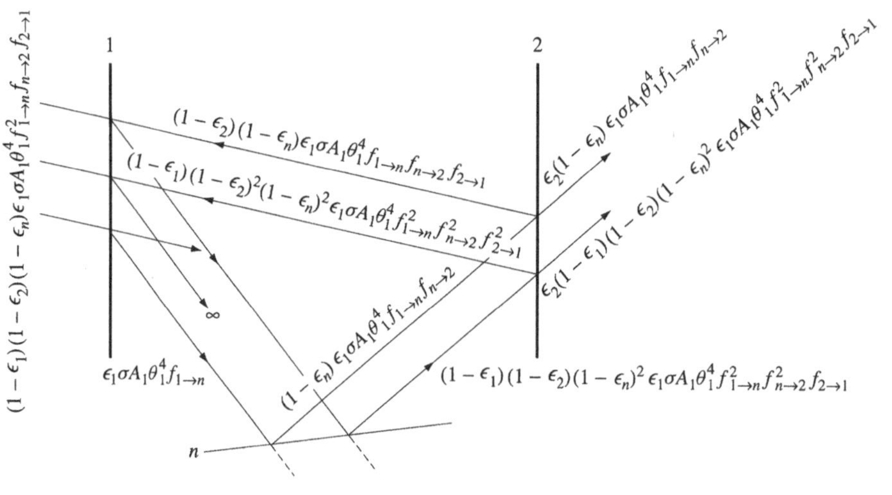

Figure 7.20: Flux absorbed at surface 2 due to three-surface interaction.

The special case of three-surface interaction is obtained when flux emitted at 1 finds its way back to 1 from another enclosure surface as shown in figure 7.21 for the case $1 \to 2 \to 1$. The flux absorbed at 1 is given by

$$q_{(3)1\to1} = \varepsilon_1(1-\varepsilon_2)\varepsilon_1\sigma A_1\theta_1^4 f_{1\to2}f_{2\to1} + \varepsilon_1(1-\varepsilon_1)(1-\varepsilon_2)^2\varepsilon_1\sigma A_1\theta_1^4 f_{1\to2}^2 f_{2\to1}^2$$

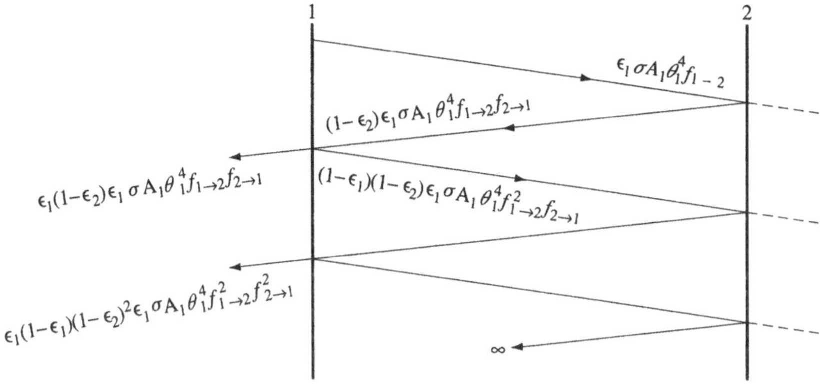

Figure 7.21: Flux absorbed at surface 1 due to reflection at surface 2.

$$+\ \varepsilon_1(1-\varepsilon_1)^2(1-\varepsilon_2)^3\varepsilon_1\sigma A_1\theta_1^4 f_{1\rightarrow 2}^3 f_{2\rightarrow 1}^3 + \cdots \infty$$

$$= \frac{\varepsilon_1(1-\varepsilon_2)\varepsilon_1\sigma A_1\theta_1^4 f_{1\rightarrow 2} f_{2\rightarrow 1}}{1-(1-\varepsilon_1)(1-\varepsilon_2)f_{1\rightarrow 2}f_{2\rightarrow 1}}\ . \tag{7.26}$$

This process can be continued indefinitely to consider four interacting surfaces and so on; each addition representing the next significant contribution in the recursive stages of table 7.15.

The addition of eqns (7.24) and (7.25) gives the approximation to the total flux absorbed by surface 2 (because of two- and three-surface diffuse reflections to infinity) due to the emission at surface 1:

$$q_{1\rightarrow 2} = \frac{\varepsilon_2\varepsilon_1\sigma A_1\theta_1^4 f_{1\rightarrow 2}}{1-(1-\varepsilon_1)(1-\varepsilon_2)f_{1\rightarrow 2}f_{2\rightarrow 1}} + \sum_{i=3}^{N}\frac{\varepsilon_2(1-\varepsilon_i)\varepsilon_1\sigma A_1\theta_1^4 f_{1\rightarrow i}f_{i\rightarrow 2}}{1-(1-\varepsilon_1)(1-\varepsilon_2)(1-\varepsilon_i)f_{1\rightarrow i}f_{i\rightarrow 2}f_{2\rightarrow 1}} \tag{7.27}$$

and, by transposing signs, the approximate flux absorbed by surface 1 due to emission at 2 is given by

$$q_{2\rightarrow 1} = \frac{\varepsilon_1\varepsilon_2\sigma A_2\theta_2^4 f_{2\rightarrow 1}}{1-(1-\varepsilon_1)(1-\varepsilon_2)f_{1\rightarrow 2}f_{2\rightarrow 1}} + \sum_{i=3}^{N}\frac{\varepsilon_1(1-\varepsilon_i)\varepsilon_2\sigma A_2\theta_2^4 f_{2\rightarrow i}f_{i\rightarrow 1}}{1-(1-\varepsilon_1)(1-\varepsilon_2)(1-\varepsilon_i)f_{2\rightarrow i}f_{i\rightarrow 1}f_{1\rightarrow 2}}\ . \tag{7.28}$$

The approximate net flux exchange between surface 2 and 1 is found from $q_{2,1}$ = [flux absorbed at 1 from 2 (i.e. $q_{2\rightarrow 1}$)] - [flux absorbed at 2 from 1 (i.e. $q_{1\rightarrow 2}$)]—i.e. eqn (7.28) - eqn (7.27), or

$$q_{2,1} = \frac{\varepsilon_1\varepsilon_2\sigma(A_2\theta_2^4 f_{2\rightarrow 1} - A_1\theta_1^4 f_{1\rightarrow 2})}{1-(1-\varepsilon_1)(1-\varepsilon_2)f_{1\rightarrow 2}f_{2\rightarrow 1}} + \sum_{i=3}^{N}\frac{\varepsilon_1(1-\varepsilon_i)\varepsilon_2\sigma A_2\theta_2^4 f_{2\rightarrow i}f_{i\rightarrow 2}}{1-(1-\varepsilon_1)(1-\varepsilon_2)(1-\varepsilon_i)f_{2\rightarrow i}f_{i\rightarrow 1}f_{1\rightarrow 2}}$$

$$-\sum_{i=3}^{N} \frac{\varepsilon_2(1-\varepsilon_i)\varepsilon_1\sigma A_1\theta_1^4 f_{1\to i} f_{i\to 2}}{1-(1-\varepsilon_1)(1-\varepsilon_2)(1-\varepsilon_i)f_{1\to i}f_{i\to 2}f_{2\to 1}} \,. \tag{7.29}$$

Note that for the special case of infinite parallel surfaces ($f_{i\to j} = 1$), the net flux exchange is also given by a flux balance conducted at either surface so that

$$q_{2,1} = q_1 - q_{2\to 1} - q_{1\to 1}$$

$$= \varepsilon_1\sigma A_1\theta_1^4 - \text{eqn (7.27)} - \text{eqn (7.26)}$$

$$= q_2 - q_{1\to 2} - q_{2\to 2}$$

$$= \varepsilon_2\sigma A_2\theta_2^4 - \text{eqn (7.27)} - [\text{eqn (7.26) with signs transposed}].$$

Eqn (7.29) defines the net transfer of energy between two surfaces separated by a non-absorbing medium and due to one other participating surface. For the example of table 7.13 accuracy is reasonably good—in excess of 94.9% of the energy emitted at surface 1 is accounted for if all two- and three-surface combinatorial pairs are considered. Since the flux magnitude decays rapidly it is reasonable to account for the missing energy (5.1% in this case) by determining the difference between the computed value and the initial emission and to then apportion this difference between surfaces on the basis of a view factor weighting. However, this will become inaccurate in the case of low emissivity surfaces for which the full recursive or matrix treatment is recommended. Alternatively, an analytical solution can be preserved by adding four interacting surface terms and so on. Figure 7.22 details, for each level (4 participating surfaces, 5 participating surfaces etc), the flowpaths then considered by the foregoing analytical technique, and table 7.14 shows the recursive technique applied in a low surface emissivity context. The inquiring reader is left to consider the analytical rigour required to satisfy this atypical problem.

7.5.2 View factor determination

The determination of the view factor relationship is a non-trivial task in the case of real enclosures with the possibility of inter-surface obstructions, re-entrant angles, surface openings and specular reflections (Moore and Numan 1983). A number of approaches to view factor assessment are possible, ranging from rigorous numerical methods to published solutions for standard geometries.

Consider figure 7.23, which shows a ray of intensity I, contained within a solid angle $d\Omega$, corresponding to an elemental area dA, propagated in a direction Ω, and making an angle θ with the normal, n, to the surface element. The emitted flux contained within $d\Omega$ is

$$dq = I\cos\theta d\Omega \tag{7.30}$$

with the flux contained in the solid angle over the entire hemisphere obtained by integration:

$$q = \int I\cos\theta d\Omega \,. \tag{7.31}$$

Now since $d\Omega = \sin\theta d\theta d\phi$, where ϕ is an azimuthal angle as illustrated in figure 7.23, substitution in eqn (7.31) gives

$$q = \int_{\phi=0}^{2\pi}\int_{\theta=0}^{\pi/2} I\cos\theta\sin\theta\, d\theta d\phi = \pi I \,.$$

Table 7.14: Application of the recursive techniques to low emissivity surfaces.

- 1:367.4
  - 2:12.3
    - 1:29.4
      - 2:0.25
        - 1:0.59
        - 3:0.15
        - 4:0.07
      - 3:0.49
        - 1:0.52
        - 2:0.07
        - 4:0.07
      - 4:0.25
        - 1:0.59
        - 2:0.07
        - 3:0.15
    - 3:7.4
      - 1:7.8
        - 2:0.07
        - 3:0.13
        - 4:0.07
      - 2:0.98
        - 1:2.4
        - 3:0.59
        - 4:0.29
      - 4:0.98
        - 1:2.4
        - 2:0.29
        - 3:0.59
    - 4:3.7
      - 1:8.8
        - 2:0.07
        - 3:0.15
        - 4:0.07
      - 2:1.1
        - 1:2.7
        - 3:0.66
        - 4:0.33
      - 3:2.2
        - 1:2.4
        - 2:0.29
        - 4:0.29
  - 3:24.5
    - 1:26.1
      - 2:0.22
        - 1:0.46
        - 3:0.06
        - 4:0.12
      - 3:0.44
        - 1:0.52
        - 2:0.07
        - 4:0.07
      - 4:0.22
        - 1:0.46
        - 2:0.12
        - 3:0.06
    - 2:3.3
      - 1:7.8
        - 2:0.09
        - 3:0.13
        - 4:0.09
      - 3:2.0
        - 1:2.1
        - 2:0.26
        - 4:0.26
      - 4:0.98
        - 1:2.4
        - 2:0.59
        - 3:0.29
    - 4:3.3
      - 1:7.8
        - 2:0.07
        - 3:0.13
        - 4:0.07
      - 3:2.0
        - 1:2.1
        - 2:0.26
        - 4:0.26
      - 4:0.98
        - 1:2.4
        - 2:0.59
        - 3:0.29
  - 4:12.2
    - 1:29.4
      - 2:0.25
        - 1:0.59
        - 3:0.15
        - 4:0.07
      - 3:0.49
        - 1:0.52
        - 2:0.07
        - 4:0.07
      - 4:0.25
        - 1:0.59
        - 2:0.07
        - 3:0.15
    - 2:3.7
      - 1:8.8
        - 2:0.07
        - 3:0.15
        - 4:0.07
      - 3:2.2
        - 1:2.4
        - 2:0.29
        - 4:0.29
      - 4:1.1
        - 1:2.7
        - 2:0.33
        - 3:0.66
    - 3:7.4
      - 1:7.8
        - 2:0.07
        - 3:0.13
        - 4:0.07
      - 2:0.95
        - 1:2.4
        - 3:0.59
        - 4:0.29
      - 4:0.95
        - 1:2.4
        - 2:0.29
        - 3:0.59

Surface i	θ (K)	Area (m^2)	ε	$f_{i->j}$ for j = 1	2	3	4
1	300	1	0.8	0	0.33	0.33	0.33
2	290	1	0.1	0.33	0	0.33	0.33
3	290	1	0.2	0.33	0.33	0	0.33
4	290	1	0.1	0.33	0.33	0.33	0

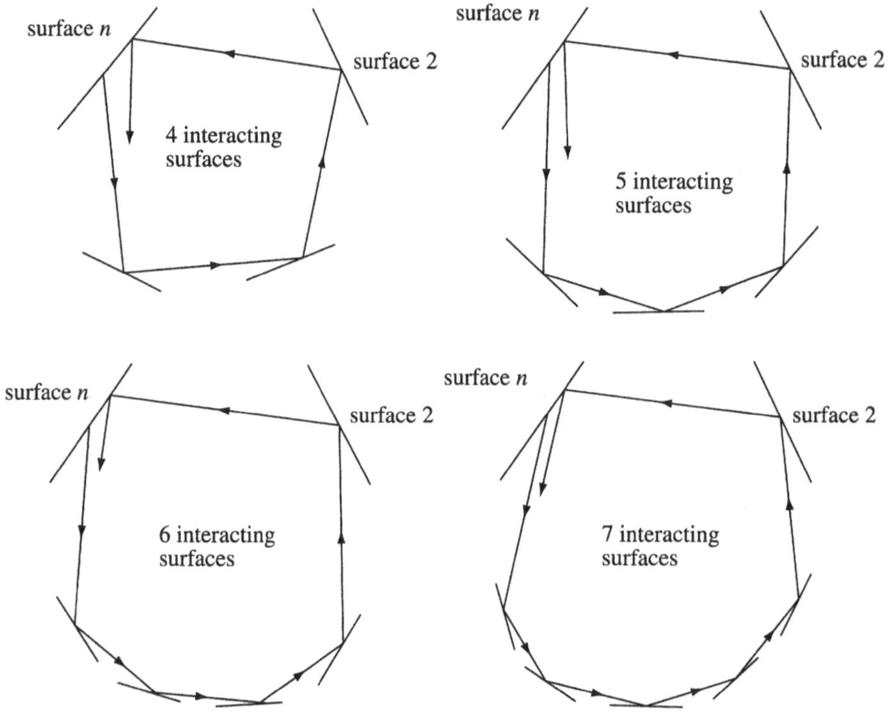

Figure 7.22: Flux absorbed at surface 2 from surface n and due to four or more interacting surfaces.

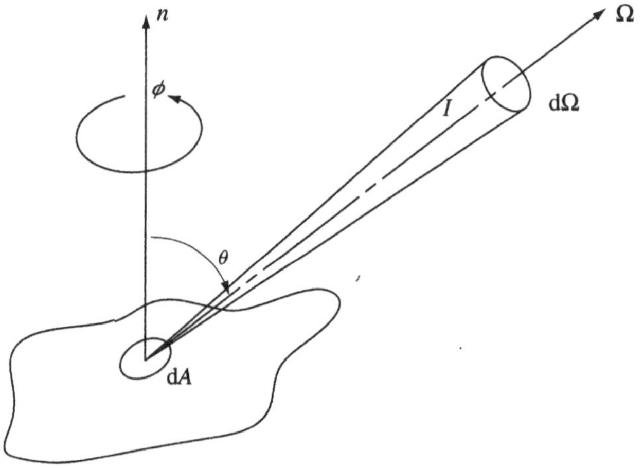

Figure 7.23: View factor prediction.

Consider figure 7.24, which shows two planar elemental areas, dA_i and dA_j, separated by a distance r and with polar angles, θ_i and θ_j, between r and normals n_i and n_j respectively. The radiative flux leaving dA_i that arrives at dA_j, as given by eqn (7.30), is

$$dq_i = I_i \cos\theta_i d\Omega_{ij}$$

and since $d\Omega_{ij} = dA_j \cos\theta_j / r^2$, it follows that

$$dq_i = I_i \cos\theta_i dA_j \cos\theta_j / r^2 \ .$$

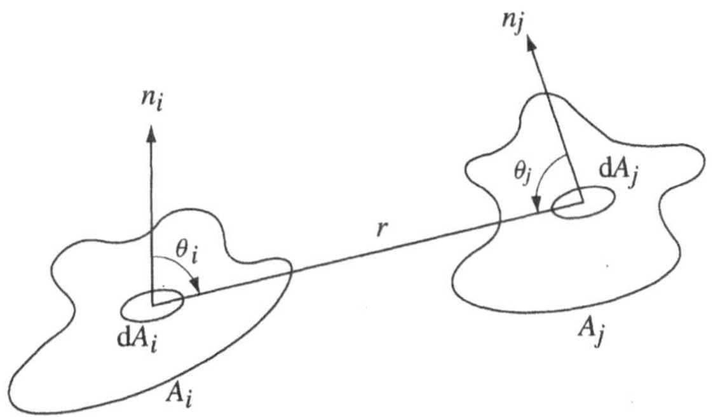

Figure 7.24: Two communicating elemental areas.

Introducing the elemental view factor, $df_{dAi \to dAj}$, which is the ratio of the radiative flux leaving dA_i that arrives directly at dA_j to the total radiative flux leaving dA_i:

$$df_{dAi \to dAj} = dq_i / q$$

$$= \cos\theta_i dA_j \cos\theta_j / \pi r^2 \ . \tag{7.32}$$

In a similar manner, the point view factor $f_{dAi \to Aj}$ can be defined as the fraction of the radiative energy leaving dA_i that is received by surface A_j:

$$f_{dAi \to Aj} = \int^{Aj} df_{dAi \to dAj}$$

$$= \int^{Aj} \frac{\cos\theta_i \cos\theta_j dA_j}{\pi r^2} \tag{7.33}$$

Now, the flux emitted by A_j is $I_j A_j$ and the amount received by dA_i is $I_j \int^{Aj} df_{dAj \to dAi} dA_j$ and therefore $f_{Aj \to dAi}$ is given by

$$f_{Aj \to dAi} = I_j \int^{Aj} \frac{df_{dAj \to dAi}}{I_j A_j} dA_j = \frac{1}{A_j} \int^{Aj} df_{dAj \to dAi} dA_j$$

and so, from eqn (7.32)

$$f_{Aj \to dAi} = \frac{dA_i}{A_j} \int^{Aj} \frac{\cos\theta_i \cos\theta_j}{\pi r^2} dA_j \ . \tag{7.34}$$

The reciprocity relationship follows by equating eqns (7.33) and (7.34):

$$dA_i f_{dAi \to Aj} = A_j f_{Aj \to dAi} \, .$$

Lastly, the flux emitted by A_i is $I_i A_i$ and the amount received by A_j is $I_i \int^{Ai} f_{dAi \to Aj} dA_i$ and therefore the area-to-area view factor, $f_{Ai \to Aj}$, is given by

$$f_{Ai \to Aj} = \frac{1}{A_i} \int^{Ai} f_{dAi \to Aj} dA_i$$

and from eqn (7.33)

$$f_{Ai \to Aj} = \frac{1}{A_i} \int^{Ai} \int^{Aj} \frac{\cos\theta_i \cos\theta_j}{\pi r^2} dA_j dA_i \qquad (7.35)$$

with, as before

$$A_i f_{Ai \to Aj} = A_j f_{Aj \to Ai} \, .$$

The determination of the view factor relationship by double integration over the surfaces of two directly communicating bodies can be achieved by a finite difference representation of eqn (7.35). One algorithmic approach (Moore and Numan 1982) starts by subdividing some surface polygon into a number of small elemental areas and, for each, establishing a unit hemisphere above the centre point representing dA_i as shown in figure 7.25.

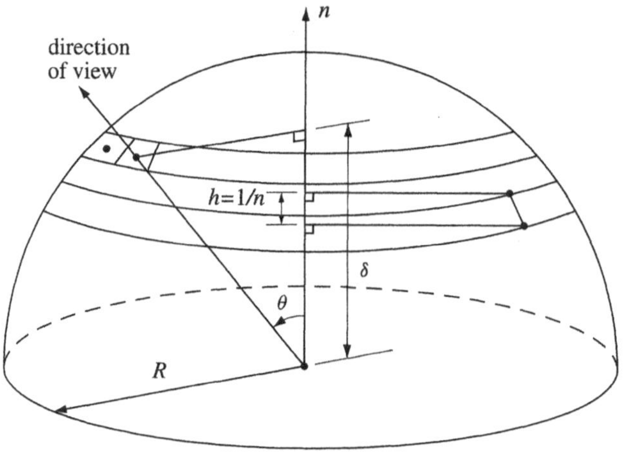

Figure 7.25: Strip and patch subdivision of a unit hemisphere.

Each unit hemisphere is then subdivided into a number of equal solid angles by 'strip and patch' subdivision as shown. Every solid angle is then projected until intersection occurs with another surface so that a 'viewed' polygon is associated with each hemispherical patch to account for all radiation emitted by dA_i. The point view factor of eqn (7.33) is then determined by summing the contributions of each viewed polygon. The view factor between two finite surfaces A_i and A_j is then found from

$$f_{Ai \to Aj} = \frac{1}{A_i} \int^{Ai} f_{dAi \to Aj} dA_i$$

to give the equivalent of eqn (7.35).

Such an algorithm therefore guarantees that $\sum^{j} f_{i \to j} = 1$ and that reciprocity will prevail. As proof of this consider eqn (7.33), which can be rewritten

$$f_{dAi \to Aj} = \int^{\Omega_i} \frac{\cos\theta_i}{\pi} d\Omega \, .$$

Now, for $f_{dAi \to Aj} = 1$, the integration must give

$$\int^{\Omega_i} \cos\theta_i d\Omega = \pi \, .$$

If this equation is approximated as a summation, and if all $d\Omega_n$ are the same, then

$$\sum^{n} \cos\theta_n d\Omega_n = d\Omega_n \sum^{n} \cos\theta_n \, . \qquad (7.36)$$

By the definition of solid angle, to have identical $d\Omega_n$, the areas subtended at the surface of a unit hemisphere (centred on dA_i) must be the same. Consider figure 7.25: the area of a strip of the hemisphere is given by $A_s = 2\pi Rh$. Therefore, to divide the hemisphere into strips of equal area requires having equal h. Having achieved this, patches of equal area can be created by implementing equal subdivisions of each strip. These patches must then subtend equal solid angles at the centre node located at the base of the hemisphere. In each case, the width of the patch is defined by some angle ϕ in a plane parallel to the hemisphere equator.

Note that for a unit hemisphere

$$\cos\theta = \delta$$

where δ is the projection of the radius onto the normal. If the hemisphere is divided into n strips, then the cosines will be

$$1/2n,\ 1/2n + 1/n,\ 1/2n + 2/n,\ \ldots,\ 1/2n + (n-1)/n \, .$$

and summing by the usual formula for arithmetic series gives

$$S_n = (n/2)[2a + (n-1)b]$$

where, in this case, $a = 1/2n$ and $b = 1/n$ and so it follows that

$$S_n = n/2 \, .$$

If each strip is further subdivided into m equal patches, then the cosine summation will be mn/2. Since each patch subtends a solid angle of $2\pi/mn$ steradians, eqn (7.36) becomes

$$(2\pi/mn)(mn/2) = \pi$$

and so $\sum^{j} f_{i \to j} = 1$ and reciprocity is achieved.

One algorithmic approach is to determine the point view factors by finding which viewed polygon is 'seen' through each of the equal area patches and then summing the contributions of each viewed polygon. A polygon is seen if the axis of the solid angle intersects that polygon at a point closer than it intersects any other polygon. Although all radiation is accounted for in theory, care should be taken in practice since numerical approximations (relating to mesh generation, polygon clipping, point containment testing, application of point view factors to finite mesh areas etc) may introduce errors. These can be minimised by increasing the number of strip and patch divisions applied to the unit hemisphere. Note, however, that since the problem is quadrate any accuracy improvement will be hard won.

This algorithm has been employed as the basis of a lighting simulation program in which surfaces may range from diffuse through specular reflectors (Maver *et al* 1987). Appendix F gives details on the ray tracing method employed and the mapping of the computed spectral luminance distribution to display colour.

For many simplified geometries such computational rigour may not be justified and simplifying assumptions are often acceptable. Some possibilities follow.

View factor between two small surfaces

If two communicating surfaces can be considered small compared with the remaining surfaces comprising the enclosure then an approximation to $f_{i \to j}$ is to use the elemental view factor defined by eqn (7.32).

View factor between large parallel surfaces

In this case most of the radiation from one surface will be intercepted by the other and so $f_{i \to j} = f_{j \to i} \to 1$.

View factor in the case of one surface enclosing another

Again $f_{i \to j} \to 1$, where surface i is contained, and from the reciprocity theorem $f_{j \to i} = f_{i \to j} A_i / A_j$.

View factors in the absence of geometric information

In many modelling applications—especially at an early design stage—exact geometrical details are unavailable. In such cases it is convenient to utilise one of the following two approaches.

It is possible to accept as representative one of the standard geometrical arrangements found in the literature (Hottel 1930, Chung and Samitra 1972). Many such arrangements have been evaluated and view factors established for various dimensionally representative cases.

Alternatively, simple area weighting techniques can be used, often without detectable loss of accuracy. For example, approximate view factors are given by

$$f_{i \to j} = A_j / \left(\sum A - A_i \right)$$

$$f_{j \to i} = A_i / \left(\sum A - A_i \right)$$

where the area summation is over all participating surfaces. Note that these expressions satisfy the reciprocity theorem but cannot ensure that $\sum^j f_{i \to j} = 1$ since they have no basis in angular considerations. They give exact answers only for a cube.

Exact analytical solutions

It is often possible to sub-divide two communicating surfaces in such a way that, by the application of view factor algebra (Welty 1974), the view factors can be determined from basic analytical formulations relating to parallel and perpendicular rectangles. With reference to figure

7.26, the view factor for two parallel surfaces of *equal area* may be determined as a function of the dimensions x, y and z:

$$f_{(\|)1\to 2} = \frac{2}{xy\pi}\left[x(y^2+z^2)^{\frac{1}{2}}\tan^{-1}\left(\frac{x}{(y^2+z^2)^{\frac{1}{2}}}\right)\right.$$

$$+\,y(x^2+z^2)^{\frac{1}{2}}\tan^{-1}\left(\frac{y}{(x^2+z^2)^{\frac{1}{2}}}\right) - xz\tan^{-1}\left(\frac{x}{z}\right)$$

$$\left.-\,yz\tan^{-1}\left(\frac{y}{z}\right) + \frac{z^2}{2}\ln\left(\frac{(x^2+z^2)(y^2+z^2)}{(x^2+y^2+z^2)z^2}\right)\right] = f_{(\|)}[x,y,z] \qquad (7.37)$$

and for two perpendicular surfaces, with *common edge* 'a' and 'height' dimensions b and c:

$$f_{(\perp)1\to 2} = \frac{1}{ab\pi}\left[ab\tan^{-1}\left(\frac{a}{b}\right) + ac\tan^{-1}\left(\frac{a}{c}\right) - a(b^2+c^2)^{\frac{1}{2}}\tan^{-1}\left(\frac{a}{(b^2+c^2)^{\frac{1}{2}}}\right)\right.$$

$$+\,\frac{(a^2-b^2)}{4}\ln(a^2+b^2) + \frac{(a^2-c^2)}{4}\ln(a^2+c^2)$$

$$-\,\frac{(a^2-b^2-c^2)}{4}\ln(a^2+b^2+c^2) - \frac{a^2}{4}\ln a^2 + \frac{b^2}{4}\ln b^2$$

$$\left.+\,\frac{c^2}{4}\ln c^2 - \frac{(b^2+c^2)}{4}\ln(b^2+c^2)\right] = f_{(\perp)}[a,b,c]\,. \qquad (7.38)$$

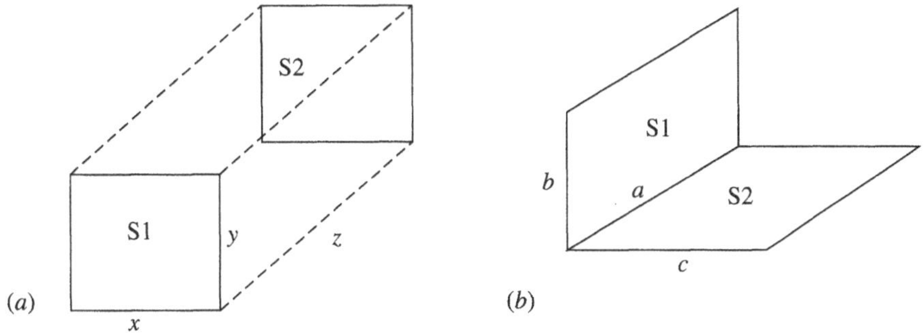

Figure 7.26: Basic view factor configurations—(a) parallel case, (b) perpendicular case.

These two basic expressions can now be combined, by shape factor algebra, to give the generalised parallel and perpendicular formulations that do not include the equal area or common edge restrictions of eqns (7.37) and (7.38).

With reference to figure 7.27, the general parallel plane view factor ($\Phi_{(\|)}$) is given by

$$\Phi_{(\|)1\to 2} = 1/4\left(f_{(\|)}[(x_{1,2}-x_{2,1}),(y_{1,2}-y_{2,1}),z_1]\right.$$

$$+\,f_{(\|)}[(x_{2,2}-x_{1,1}),(y_{1,2}-y_{2,1}),z_1] + f_{(\|)}[(x_{1,2}-x_{2,1}),(y_{2,2}-y_{1,1}),z_1]$$

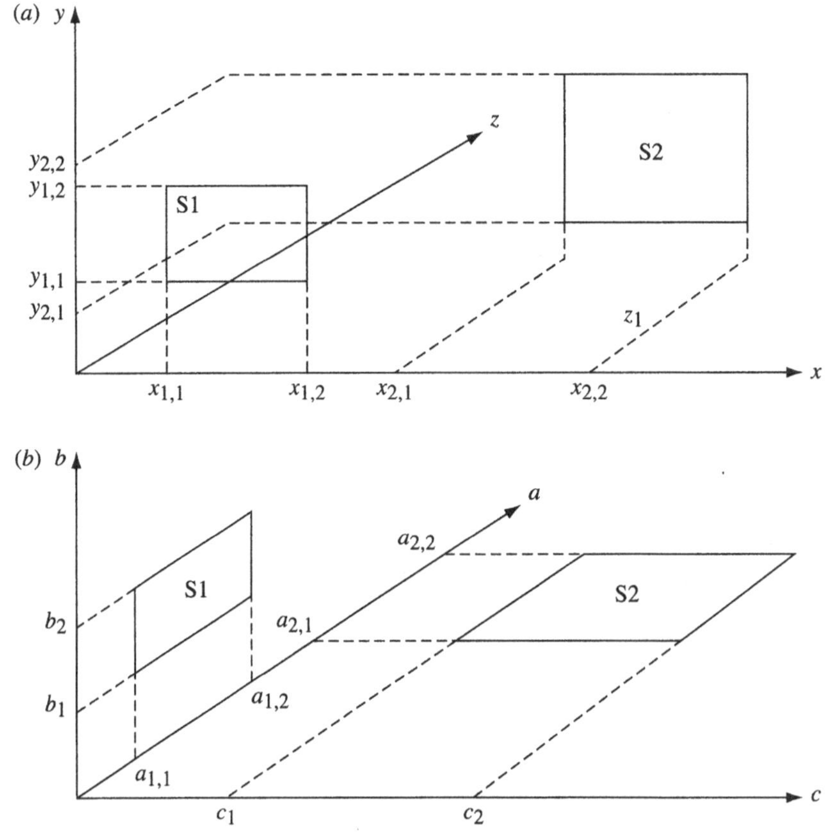

Figure 7.27: General view factor configurations—(a) parallel case, (b) perpendicular case.

$$+ f_{(||)}[(x_{2,2} - x_{1,1}), (y_{2,2} - y_{1,1}), z_1] - f_{(||)}[(x_{1,2} - x_{2,1}), (y_{1,1} - y_{2,1}), z_1]$$
$$- f_{(||)}[(x_{1,1} - x_{2,1}), (y_{1,2} - y_{2,1}), z_1] - f_{(||)}[(x_{2,2} - x_{1,1}), (y_{1,1} - y_{2,1}), z_1]$$
$$- f_{(||)}[(x_{2,2} - x_{1,2}), (y_{1,2} - y_{2,1}), z_1] - f_{(||)}[(x_{1,1} - x_{2,1}), (y_{2,2} - y_{1,1}), z_1]$$
$$- f_{(||)}[(x_{1,2} - x_{2,1}), (y_{2,2} - y_{1,2}), z_1] - f_{(||)}[(x_{2,2} - x_{1,2}), (y_{2,2} - y_{1,1}), z_1]$$
$$- f_{(||)}[(x_{2,2} - x_{1,1}), (y_{2,2} - y_{1,2}), z_1] + f_{(||)}[(x_{1,1} - x_{2,1}), (y_{1,1} - y_{2,1}), z_1]$$
$$+ f_{(||)}[(x_{2,2} - x_{1,2}), (y_{1,1} - y_{2,1}), z_1] + f_{(||)}[(x_{1,1} - x_{2,1}), (y_{2,2} - y_{1,2}), z_1]$$
$$\left. + f_{(||)}[(x_{2,2} - x_{1,2}), (y_{2,2} - y_{1,2}), z_1] \right) \tag{7.39}$$

and the general perpendicular plane value ($\Phi_{(\perp)}$) by

$$\Phi_{(\perp)1\to 2} = 1/2 \left(f_{(\perp)}[(a_{2,1} - a_{1,2}), (b_2 - b_0), (c_2 - c_0)] \right.$$
$$- f_{(\perp)}[(a_{2,1} - a_{1,2}), (b_2 - b_0), (c_1 - c_0)]$$

$$
\begin{aligned}
&+ f_{(L)}[(a_{2,2} - a_{1,1}), (b_2 - b_0), (c_2 - c_0)] \\
&- f_{(L)}[(a_{2,2} - a_{1,1}), (b_2 - b_0), (c_1 - c_0)] \\
&+ f_{(L)}[(a_{2,1} - a_{1,2}), (b_1 - b_0), (c_2 - c_0)] \\
&+ f_{(L)}[(a_{2,1} - a_{1,2}), (b_1 - b_0), (c_1 - c_0)] \\
&- f_{(L)}[(a_{2,2} - a_{1,1}), (b_1 - b_0), (c_2 - c_0)] \\
&+ f_{(L)}[(a_{2,2} - a_{1,1}), (b_1 - b_0), (c_1 - c_0)] \\
&- f_{(L)}[(a_{2,1} - a_{1,1}), (b_2 - b_0), (c_2 - c_0)] \\
&+ f_{(L)}[(a_{2,1} - a_{1,1}), (b_2 - b_0), (c_1 - c_0)] \\
&- f_{(L)}[(a_{2,2} - a_{1,2}), (b_2 - b_0), (c_2 - c_0)] \\
&+ f_{(L)}[(a_{2,2} - a_{1,2}), (b_2 - b_0), (c_1 - c_0)] \\
&+ f_{(L)}[(a_{2,1} - a_{1,1}), (b_1 - b_0), (c_2 - c_0)] \\
&- f_{(L)}[(a_{2,1} - a_{1,1}), (b_1 - b_0), (c_1 - c_0)] \\
&+ f_{(L)}[(a_{2,2} - a_{1,2}), (b_1 - b_0), (c_2 - c_0)] \\
&- f_{(L)}[(a_{2,2} - a_{1,2}), (b_1 - b_0), (c_1 - c_0)] \Big). \qquad (7.40)
\end{aligned}
$$

Eqns (7.39) and (7.40) can now be employed to evaluate the view factors for other arrangements built up from elementary parts, e.g. $f_{1\rightarrow 2}$ in figure 7.28 is given by

$$f_{1\rightarrow 2} = f_{1\rightarrow 234} - f_{1\rightarrow 3} - f_{1\rightarrow 4}$$

with each of the factors on the right-hand side determined from eqn (7.40).

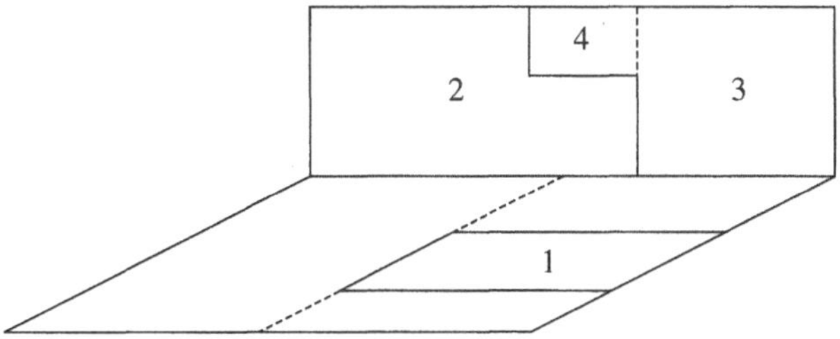

Figure 7.28: View factor relationship established from elementary surface pairings.

Once determined, these view factors can be used in the methods of §7.5.1 to assess the longwave flux exchanges, exactly or by approximation. These flux exchanges can be applied to surface nodes via the heat generation terms of the surface energy conservation equations derived in chapter 3. Alternatively, a linearised h_r value can be determined for use in forming the inter-surface cross coupling coefficients. Note that while the former approach avoids the need for linearisation, it requires that the longwave flux exchange be determined on the basis of present time-row surface temperatures, with iteration employed where high accuracy is required.

7.5.3 Linearised longwave radiation coefficients

The net longwave radiation gain at surface 1 from 2 is given by

$$q_{2,1} = h_{r2,1} A_1(\theta_2 - \theta_1) \tag{7.41}$$

where $h_{r2,1}$ is the linearised radiative heat transfer coefficient (W m^{-2}°C^{-1}).

Note that the heat flux is expressed in terms of the receiving surface area as required by the energy balance formulations of chapter 3. This is in contrast to the usual practice of expressing radiative transfer as a function of the emitter surface area. Equating eqns (7.30) and (7.41) yields the formulation for the linearised longwave radiation coefficient:

$$h_{r2,1} = \frac{\varepsilon_1 \varepsilon_2 \sigma (A_2 \theta_2^4 f_{2\to 1} - A_1 \theta_1^4 f_{1\to 2})}{A_1(\theta_2 - \theta_1)[1 - (1-\varepsilon_1)(1-\varepsilon_2) f_{1\to 2} f_{2\to 1}]}$$

$$+ \sum_{i=3}^{N} \frac{\varepsilon_1 (1-\varepsilon_i)\varepsilon_2 \sigma A_2 \theta_2^4 f_{2\to i} f_{i\to 1}}{A_1(\theta_2 - \theta_1)[1 - (1-\varepsilon_1)(1-\varepsilon_2)(1-\varepsilon_i) f_{2\to i} f_{i\to 1} f_{1\to 2}]}$$

$$- \sum_{i=3}^{N} \frac{\varepsilon_2 (1-\varepsilon_i)\varepsilon_1 \sigma A_1 \theta_1^4 f_{1\to i} f_{i\to 2}}{A_1(\theta_2 - \theta_1)[1 - (1-\varepsilon_1)(1-\varepsilon_2)(1-\varepsilon_i) f_{1\to i} f_{i\to 2} f_{2\to 1}]} .$$

Now, since reciprocity states that $f_{2\to 1} = f_{1\to 2} A_1/A_2$, and since $(\theta_2^4 - \theta_1^4) = (\theta_2^2 + \theta_1^2)(\theta_2 + \theta_1)(\theta_2 - \theta_1)$, then

$$h_{r2,1} = \frac{\varepsilon_1 \varepsilon_2 \sigma f_{1\to 2}(\theta_2^2 + \theta_1^2)(\theta_2 + \theta_1)}{[1 - (1-\varepsilon_1)(1-\varepsilon_2) f_{1\to 2}^2 A_1/A_2]} + \varepsilon_1 \varepsilon_2 \sigma A_2 (\theta_2^2 + \theta_1^2)(\theta_2 + \theta_1)$$

$$\times \sum_{i=3}^{N} \frac{(1-\varepsilon_i) f_{1\to i} f_{2\to i}}{A_i[1 - (1-\varepsilon_1)(1-\varepsilon_2)(1-\varepsilon_i) f_{1\to i} f_{i\to 2} f_{2\to 1}]} . \tag{7.42}$$

From the first law of thermodynamics it follows that

$$q_{1,2} = - q_{2,1}$$

and so

$$h_{r1,2} = h_{r2,1} A_1/A_2 . \tag{7.43}$$

Thus, when computing the radiative heat transfer coefficient for inclusion within the matrix structures of chapter 4, it is only necessary to explicitly evaluate eqn (7.42) for half the total number of combinatorial surface pairings (i.e. $1 \to 2$, $1 \to 3$, $1 \to 4$,...; $2 \to 3$, $2 \to 4$,...; $3 \to 4$, $3 \to 5$,... etc) since all self coupling is zero and the remaining half can be determined directly from the relationship of eqn (7.43). This, of course, implies that view factor information need only be established for the corresponding surface pairings.

7.5.4 Exchange between external surfaces

The net longwave radiation exchange at some exposed external building surface is given as the difference between the emitted and received flux. If the surroundings are represented by some equivalent temperature, θ_e, then the net exchange can be expressed as

$$q = A_s \varepsilon \sigma (\theta_e^4 - \theta_s^4)$$

where A_s is the surface area (m^2), ε the surface emissivity, σ the Stefan-Boltzmann constant (W $m^{-2}K^{-4}$) and θ_s is the absolute temperature of the surface (K). The equivalent temperature is a function of the temperatures of the sky, ground and surroundings:

$$\theta_e^4 = f_s\theta_{sky}^4 + f_g\theta_{grd}^4 + f_u\theta_{sur}^4$$

where f_s, f_g and f_u are the view factors to the sky, ground and surroundings respectively. Table 7.15 gives some example values. The need, then, is to estimate these temperatures from the known weather data.

Table 7.15: Representative values of sky, ground and obstructions view factors.

Location	f_s	f_g	f_u
City centre: surrounding buildings at same height, vertical surface	0.36	0.36	0.28
City centre: surrounding buildings higher, vertical surface	0.15	0.33	0.52
Urban site: vertical surface	0.41	0.41	0.18
Rural site: vertical surface	0.45	0.45	0.10
City centre: sloping roof	0.50	0.20	0.30
Urban site: sloping roof	0.50	0.30	0.20
Rural site: isolated	0.50	0.50	0.00

Sky temperature estimation

The sky temperature under non-cloudy conditions can be determined from

$$R_s = 5.31 \times 10^{-13}\theta_{sc}^6$$

where R_s is the sky radiation (W m^{-2}) and θ_{sc} the screen air temperature (K). This expression has been compared with measured data from different global locations (Swinbank 1963) and found to give reasonable accuracy. If the assumption is made that the clear sky behaves as a black body then

$$R_s = \sigma\theta_{sky}^4 \tag{7.44}$$

and so

$$\theta_{sky} = 0.05532\theta_{sc}^{1.5}\,.$$

In the presence of clouds, the mean sky temperature increases and an alternative expression has been proposed (Cole 1976):

$$R_s' = (1 - CC)R_s + CC\varepsilon_c\sigma\theta_{sc}^4 \tag{7.45}$$

where R_s' is cloudy sky radiation (W), CC the cloud cover factor (proportion of 1), and ε_c the emissivity of the cloud base given by

$$\varepsilon_c = (1 - 0.84CC)(0.527 + 0.161e^{[8.45(1-273/\theta_{sc})]} + 0.84CC)\,. \tag{7.46}$$

Substitution of eqn (7.46) in (7.45) and using eqn (7.44) gives the final expression for the effective sky temperature under cloudy conditions:

$$\theta_{sky} = \{9.365574 \times 10^{-6}(1 - CC)\theta_{sc}^6$$

$$+ \theta_{sc}^4 CC(1 - 0.84CC)[(0.527 + 0.161e^{[8.45(1-273/\theta_{sc})]}) + 0.84CC]\}^{0.25}\,.$$

Stanzel (1994) compared several empirical models for the estimation of θ_{sky} in a northern European context and concluded that the Swinbank model gave acceptable results for both the clear and cloudy sky cases. Other models may be more suited to other locations—such as the model of Berdahl and Martin (1984) formulated initially for the United States.

Ground temperature estimation

The simplest method of estimation is to use the concept of sol-air temperature so that

$$\theta_{grd} = \theta_A + (\alpha_g I_{gh} - q_{lw})/R_{so}$$

where θ_A is the air temperature (°C), α_g the ground absorptivity, I_{gh} the total solar irradiance (W m^{-2}), q_{lw} the net longwave radiation exchange (W m^{-2}) and R_{so} the combined convective/radiative ground surface layer resistance (m^2°C W^{-1}). Application of this expression will require, firstly, that the longwave exchange term be evaluated. This, in turn, will require knowledge of the temperatures of the sky and obstructions.

Alternatively, a ground nodal scheme can be introduced to the system matrix equation (of chapter 4) to allow explicit modelling of the ground exchange processes and, thereby, the removal of ground temperature from the present calculations.

Surroundings temperature estimation

In the absence of a detailed model of the surroundings for use in an explicit simulation exercise, a pragmatic approach is to evaluate the temperature of the surroundings as a weighted function of the immediate past temperatures of the surfaces of the target building.

7.6 Surface convection

In building heat transfer applications, the convective heat flux at surface layers can be evaluated in a manner analogous to that of eqn (7.41):

$$q_c = h_c A_s (\theta_A - \theta_s) \tag{7.47}$$

where h_c is the convection coefficient (W m^{-2}°C^{-1}), A_s the surface area (m^2) and θ_A, θ_s the air and surface temperatures respectively (°C).

In this form h_c is a surface-averaged value and so the apparent simplicity of eqn (7.47) is misleading since, in reality, its value is position dependent (McAdams 1954). There are two main methods for the determination of h_c values (Kreith 1973): dimensional analysis combined with experimental data and mathematical solutions of the continuity, momentum and energy equations. This section addresses the former method while §5.2 considered the latter.

7.6.1 Natural convection at internal surfaces

When a fluid comes into contact with a heated surface, heat transfer takes place by conduction and fluid temperature variations are established which give rise to density variations. Buoyancy forces then establish fluid motion to carry away the conducted heat. This process is known as natural (or buoyancy-driven) convection. The fluid flow to result will be either laminar, in which case each fluid particle follows a smooth streamline and does not interfere with adjacent streamlines, or turbulent, in which case fluid particles can cross the streamlines to increase the potential for heat transfer.

Three dimensionless groupings are of importance in natural convection estimation:

Nusselt Number (Nu)

$$Nu = h_c d/k$$

where d is some characteristic dimension (m) and k the fluid thermal conductivity (W m^{-1}°C^{-1}).

Prandtl Number (Pr)

$$Pr = C_p \mu / k$$

where C is the specific heat capacity at constant pressure (J kg^{-1}°C^{-1}) and μ the fluid viscosity (kg $m^{-1}s^{-1}$).

Grashof Number (Gr)

$$Gr = \rho^2 g \beta (\theta_s - \theta_A) d^3 / \mu^2$$

where g is the gravitational constant (m s^{-2}), β the coefficient of expansion (K^{-1}), θ_s the surface temperature (°C), θ_A the bulk fluid temperature (°C) and ρ the fluid density (kg m^{-3}).

Most data correlations are obtained from the experimental evaluation of the natural convection heat transfer from heated plates. For example, Fujii and Imura (1970), working with a 30cm × 15cm plate of arbitrary inclination, produced the following correlations between the foregoing dimensionless groupings.

Vertical plate and inclined plate facing upward or downward; laminar region

$$Nu = 0.56(GrPr \cos \theta)^{1/4} \quad ; \quad 10^5 < GrPr \cos \theta < 10^{11}$$

where, in the case of the inclined plate, the gravitational constant of the Grashof Number is adjusted to the component parallel to the surface and θ is the angle with the vertical, positive downward.

Horizontal plate facing upward, vertical plate and inclined plate facing upward; turbulent region

$$Nu = 0.13(GrPr)^{1/3} \quad ; \quad 5 \times 10^8 < GrPr \, .$$

Horizontal plate facing downward; laminar region

$$Nu = 0.58(GrPr)^{1/3} \quad ; \quad 10^6 < GrPr < 10^{11} \, .$$

Inclined plate facing upwards; transition region

$$Nu = 0.13[(GrPr)^{1/3} - (Gr_c Pr)^{1/3}] + 0.56(Gr_c Pr \cos \theta)^{1/4}$$

where Gr_c is Gr corresponding to the transition from laminar to turbulent flow.

Many other empirical formulations can be found in the literature (Jacob 1949, Min *et al* 1956, Wong 1977). These include correlations that give h_c values for building surfaces as a function of the surface-to-air temperature difference, the direction of heat flow and some characteristic dimension. To achieve this, a set of assumed conditions are imposed on the empirical equations. Care must then be taken to ensure that the correlations are not applied outwith these conditions.

Alamdari and Hammond (1983) produced a general expression for both vertical and horizontal surfaces (with upward heat flow) valid over the range, $10^4 < GrPr < 10^{12}$, which encompasses most of the flow conditions found within buildings:

$$h_c = \{[a(\Delta\theta/d)^p]^m + [b(\Delta\theta)^q]^m\}^{1/m} \tag{7.48}$$

where a, b, p, q and m are given in table 7.16 and $\Delta\theta$ is the surface-to-air temperature difference.

Table 7.16: Empirical coefficients for eqn (7.48).

Surface aspect	*a*	*b*	*p*	*q*	*m*
Vertical	1.5	1.23	1/4	1/3	6
Horizontal	1.4	1.63	1/4	1/3	6

For horizontal surfaces undergoing downward heat flow, the natural convective heat transfer coefficient is given by

$$h_c = 0.6(\Delta\theta/d^2)^{1/5} . \tag{7.49}$$

For vertical surfaces, the characteristic dimension d is given by the surface height, whereas for horizontal surfaces it is the hydraulic diameter found from

$$d = 4A/P$$

where A is the surface area (m^2) and P the perimeter length (m).

Based on an extensive literature review, Beausoleil-Morrison (2000) identified additional h_c models to create a family of correlations that cover both buoyant and mechanically induced convective regimes as found within buildings. Table 7.17 summarises these correlations. To overcome the paucity of models for mixed buoyancy and mechanically driven flows, the models of Alamdari and Hammond (1983) and Fisher (1995) were blended to obtain the models of table 7.18. These are applicable to mechanically ventilated rooms where the mechanical air supply spreads towards, and flows over, the walls, with the buoyancy force acting to augment or resist this flow. These correlations have been implemented within the ESP-r system where their assignation to surfaces may be varied over time by prescription or as a function of the changing surface conditions determined from CFD simulations (§5.2).

7.6.2 Forced convection at internal and external surfaces

When fluid motion is caused by some external force such as a fan or wind power, convection is termed forced. In this case an additional dimensionless grouping is relevant:

Reynolds Number (Re)

$$Re = Vd\rho/\mu$$

where V is the fluid velocity (m s^{-1}).

As with natural convection estimation, dimensional analysis techniques can be employed to determine some generalised expression similar to the expression for natural convection but with the Reynolds number replacing the Grashof number:

$$Nu = C\,Re^n\,Pr^m$$

with the coefficient and exponent terms being assessed from experimental observation or theoretical considerations. For example, based on a copper plate experiment using a parallel flow of air at a reference temperature of 21.1°C, McAdams (1954) produced the following expression

Table 7.17: h_c correlations for buoyancy and mechanically driven flows.

Location	*Applicability*	h_c *correlation*
Correlations from Khalifa and Marshall (1990):		
Wall	• room heated by radiator • radiator not located under window • wall surface adjacent to radiator	$1.98\ \Delta\theta^{0.32}$
Wall	• room heated by radiator • radiator located under window • wall surface adjacent to radiator	$2.30\ \Delta\theta^{0.24}$
	• room with heated walls • not applicable for heated wall	
Wall	• room with circulating fan heater • wall surface opposite fan	$2.92\ \Delta\theta^{0.25}$
Wall	• room with circulating fan heater • wall surface not opposite fan	$2.07\ \Delta\theta^{0.23}$
	• room with heated floor	
	• room heated by radiator • radiator not located under window • wall surface not adjacent to radiator	
Window	• room heated by radiator • radiator located under window	$8.07\ \Delta\theta^{0.11}$
Window	• room heated by radiator • radiator not located under window	$7.61\ \Delta\theta^{0.06}$
Ceiling	• room heated by radiator • radiator located under window	$3.10\ \Delta\theta^{0.17}$
	• room with heated walls	
Ceiling	• room with circulating fan heater	$2.72\ \Delta\theta^{0.13}$
	• room with heated floor	
	• room heated by radiator • radiator not located under window	
Correlations from Awbi and Hatton (1999):		
Wall	• heated	$\frac{1.823\ \Delta^{0.293}}{D_h^{0.121}}$
Floor	• heated	$\frac{2.175\ \Delta^{0.308}}{D_h^{0.076}}$
Correlations from Fisher (1995) and Fisher and Pederson (1997):		
Wall	• ceiling jet in isothermal room#	$-0.199 + 0.190\ (ACR)^{0.8}$
Floor	• ceiling jet in isothermal room#	$0.159 + 0.116\ (ACR)^{0.8}$
Ceiling	• ceiling jet in isothermal room#	$-0.166 + 0.484\ (ACR)^{0.8}$
Wall	• free horizontal jet in isothermal room	$-0.110 + 0.132\ (ACR)^{0.8}$
Floor	• free horizontal jet in isothermal room	$0.704 + 0.168\ (ACR)^{0.8}$
Ceiling	• free horizontal jet in isothermal room	$0.064 + 0.00444\ \frac{(ACR)^{2.8}}{\Delta T}$

(continued)

Table 7.17 (continued)

Correlations from Alamdari and Hammond (1983):
As eqns (7.48) and (7.49).

$\Delta\theta$ is the surface-to-air temperature difference, D_h the hydraulic diameter of the surface (= 4 A/P where A is the surface area and P is the perimeter length), ACR the room air changes per hour, and ΔT the temperature difference between the air supply and extract.

Some ambiguity exists in relation to these equations. Beausoleil-Morrison (2000) has compared these expressions with alternatives obtained from Fisher and Pederson (1997) and reported significant differences.

Table 7.18: h_c correlations for the mixed flows (from Beausoleil-Morrison 2000).

Location	*Applicability*	h_c *correlation*
Correlations from Beausoleil-Morrison (2000)		
Wall	• assisting forces	$\left(\left\{\left[1.5\left(\frac{\|\Delta\theta\|}{H}\right)^{1/4}\right]^6+\left[1.23\,\|\Delta\theta\|^{1/3}\right]^6\right\}^{(3\times 1/6)}+\left\{\left[\frac{\theta_s-\theta_d}{\|\Delta\theta\|}\right]\left[-0.199+0.190(ACR)^{0.8}\right]\right\}^3\right)^{1/3}$
	• opposing forces	$\max\begin{cases}\left[\left\{\left[1.5\left(\frac{\|\Delta\theta\|}{H}\right)^{1/4}\right]^6+\left[1.23\,\|\Delta\theta\|^{1/3}\right]^6\right\}^{(3\times 1/6)}-\left\{\left[\frac{\theta_s-\theta_d}{\|\Delta\theta\|}\right]\left[-0.199+0.190(ACR)^{0.8}\right]\right\}^3\right]^{1/3}\\ 80\%\text{ of }\left\{\left[1.5\left(\frac{\|\Delta\theta\|}{H}\right)^{1/4}\right]^6+\left[1.23\,\|\Delta\theta\|^{1/3}\right]^6\right\}^{1/6}\\ 80\%\text{ of }\left\{\left[\frac{\theta_s-\theta_d}{\|\Delta\theta\|}\right]\left[-0.199+0.190(ACR)^{0.8}\right]\right\}\end{cases}$
Floor	• buoyant	$\left(\left\{\left[1.4\left(\frac{\|\Delta\theta\|}{D_h}\right)^{1/4}\right]^6+\left[1.63\,\|\Delta\theta\|^{1/3}\right]^6\right\}^{(3\times 1/6)}+\left\{\left[\frac{\theta_s-\theta_d}{\|\Delta\theta\|}\right]\left[0.159+0.116(ACR)^{0.8}\right]\right\}^3\right)^{1/3}$
	• stably stratified	$\left(\left[0.6\left(\frac{\|\Delta\theta\|}{D_h}\right)^{1/5}\right]^3+\left\{\left[\frac{\theta_s-\theta_d}{\|\Delta\theta\|}\right]\left[0.159+0.116(ACR)^{0.8}\right]\right\}^3\right)^{1/3}$
Ceiling	• buoyant	$\left(\left\{\left[1.4\left(\frac{\|\Delta\theta\|}{D_h}\right)^{1/4}\right]^6+\left[1.63\|\Delta\theta\|^{1/3}\right]^6\right\}^{(3\times 1/6)}+\left\{\left[\frac{\theta_s-\theta_d}{\|\Delta\theta\|}\right]\left[-0.166+0.484(ACR)^{0.8}\right]\right\}^3\right)^{1/3}$
	• stably stratified	$\left(\left\{0.6\left(\frac{\|\Delta\theta\|}{D_h^2}\right)^{1/5}\right\}^3+\left\{\left[\frac{\theta_s-\theta_d}{\|\Delta\theta\|}\right]\left[-0.166+0.484(ACR)^{0.8}\right]\right\}^3\right)^{1/3}$

ACR is the room air changes per hour, $\Delta\theta$ the surface-to-air temperature difference, θ_s the surface temperature, θ_d the temperature of the air supplied through the ceiling diffuser, H the surface height and D_h the hydraulic diameter of the surface as before.

$$h_c = 5.678\left[a + b\left(\frac{V}{0.3048}\right)^n\right] \qquad (7.50)$$

where h_c is the forced convection coefficient (W m^{-2}°C^{-1}), a, b, n are empirical values as given in table 7.19, and V the parallel component of the flow velocity (m s^{-1}). For non-reference temperatures, a simple adjustment to the velocity term is required: $294.26V/(273.16 + \theta_n)$ where θ_n is the non-reference temperature (°C).

Table 7.19: Empirical coefficients and exponents for eqn (7.50).

	$v < 4.88$ ms^{-1}			$4.88 \leq v < 30.48$ ms^{-1}		
Nature of surface	*a*	*b*	*n*	*a*	*b*	*n*
Smooth	0.99	0.21	1	0	0.50	0.78
Rough	1.09	0.23	1	0	0.53	0.78

In any modelling application, two issues are of importance: the resolution of the prevailing wind direction into a component parallel to external surfaces, and the determination of the flow velocity at internal surfaces as generated by the HVAC system. The latter issue is addressed in §5.2, the former issue here.

Based on experimental data, Ito *et al* (1972) developed empirical relationships for the assessment of V—as required in eqn (7.50)— from the free stream velocity, V_f. The wind direction relative to some building surface is given by

$$d_r = \alpha_f - d_f \quad (\text{if } |d_r| > 180,\ d_r = 360 - |d_r|)$$

where d_r is the wind direction relative to the surface, α_f the surface azimuth, and d_f the free stream wind direction (° from north, clockwise positive).

The local surface velocity is dependent on whether the surface is windward or leeward and on the magnitude of the free stream velocity. Its value is approximated by

Windward ($|d_r| < 90$)	Leeward ($|d_r| > 90$)
If $V_f > 2$ m s^{-1} ; $V = 0.25V_f$	$V = 0.3 + 0.05V_f$
If $V_f < 2$ m s^{-1} ; $V = 0.5$	

Table 7.20 gives example values for the internal and external convection coefficients—as evaluated from eqns (7.48) and (7.50)—against a number of $\Delta\theta$ and V conditions. Note the relatively high impact of wind speed on the external h_c value and the relatively low values for h_c at internal surfaces when compared to the values embedded within U-values intended for design use.

Table 7.20: Example convection coefficients.

Natural convection equation (7.48)			*Forced convection equation (7.50)*		
$\Delta\theta$	h_{c1}	h_{c2}	V	h_{c3}	h_{c4}
0.1	0.81	0.77	0.5	7.58	8.33
0.5	1.24	1.31	1	9.53	10.47
1	1.49	1.64	5	25.17	26.68
5	2.36	2.80	10	43.21	45.81

h_{c1} vertical surface, h_{c2} horizontal surface, h_{c3} smooth surface, h_{c4} rough surface

7.7 Casual heat sources

A substantial portion of the heat injection terms of the surface and air energy conservation equations of chapter 3 will normally be due to the radiant and convective components of the heat gain from lights, occupants, equipment etc. The total radiant and convective heat gain is given by

$$Q_{RI(\xi)} = \sum_{i=1}^{M} Q_{Si(\xi)} R_i$$

$$Q_{CI(\xi)} = \sum_{i=1}^{M} Q_{Si(\xi)} C_i$$

where $Q_{RI(\xi)}$ is the total radiant flux at time ξ (W), $Q_{CI(\xi)}$ the total convective flux (W), $Q_{Si(\xi)}$ the sensible flux output for gain i (W), R_i the radiant portion of this output flux, C_i the convective portion and M the total number of casual gain sources.

The apportioning of the radiant flux between zone surfaces can be achieved by simple area weighting or, for more accuracy, on the basis of view factor data generated from a knowledge of source position and zone geometry. The convective flux is applied to the node representing the air volume that contains the heat source(s). Note that it may be necessary to process casual gains individually if the requirement is to model the position dependent injection of heat (e.g. where a CFD model is active).

Of the normal heat sources found in buildings, lighting systems normally offer the greatest potential for energy savings (Cooper and Crisp 1983). Critical sizing of fittings and switching on the basis of prevailing lighting levels can result in significant electrical energy reduction (Fontoynont 1999) and, in the case of air-conditioned buildings, reduced cooling loads. An accurate assessment of these savings will require a lamp model to determine the variation of lamp power and light output as a function of lamp wall temperature. Such a model has been developed by Sowell and Johnson (1993).

It is a common simulation exercise to moderate lighting levels, and therefore casual heat gain, as a function of the level of indoor daylight illuminance. The requirement, then, is for a model to predict the time varying internal daylight illuminance distribution that accommodates the interactions between climate, building, systems and occupants.

7.8 Daylight prediction

The total amount of natural light at any point within an enclosure depends on the size and position of the windows, the properties of the glazing, the position and reflectances of the internal finishes and fittings, the characteristics of light directing and glare control devices, the luminance distribution of the sky, and the presence of external obstructions.

The Commission Internationale de l'Eclairage (CIE) overcast and clear sky daylight factor method is often employed, with the calculation approach ranging from analytical formulations (Winkelmann and Selkowitz 1985) that are appropriate to specific cases (e.g. rectilinear geometry) to general lighting simulation programs (Szerman 1994) that can handle problems of arbitrary complexity. §7.8.2 outlines an approach of the former type.

The daylight factor method is inadequate for the modelling of problems that combine real sky types with complex problem definitions. For example, Littlefair (1992a) compared measured internal illuminance to the predictions from a variety of daylight factor based methods. The results demonstrated discrepancies of the order of 40% for some sky luminance distributions. Teregenza (1980) reported variations in the daylight factor of the order of 1:5. Another area of concern relates to luminaire control. The hourly weather data normally employed by simulation programs is incompatible with the higher frequency requirements of realistic control algorithms. It is therefore problematic to model specific control characteristics such as photocell position, field of view, time constant, switch off delay etc. The consequences of such omissions have been widely reported (Rubinstein *et al* 1989, Littlefair 1992a, Janak 1997). A more robust method is required to assess the internal daylight illuminance under realistic sky conditions, building use and luminaire control. Such a method is described in §7.8.3.

7.8.1 Sky luminance distribution

Many researchers have studied the luminance distribution of real skies and produced corresponding models (Ne'eman and Selkowitz 1983, Kittler 1985 & 1986, Littlefair 1981, Perez *et al* 1992). Typically, such models are presented as options within state-of-the-art daylight simulation programs. For the purpose of design calculations, a standard model is often employed—such as a clear or overcast sky. The CIE standard overcast sky is brightest at the zenith and approximately one-third as bright at the horizon:

$$L_\beta = L_z(1 + 2\sin\beta)/3 \tag{7.51}$$

where L_β is the luminance of the sky at an elevation of β above the horizontal (cd m^{-2}) and L_z the luminance of the sky at the zenith. Of course, in reality the luminance distribution is constantly changing due to cloud and sun movement (Tragenza 1983).

7.8.2 Internal illuminance distribution: analytical method

A common simplification is to assume that the ratio of the indoor to outdoor daylight illuminance, termed the daylight factor, is constant. While approximately true for a sky of changing brightness, an action such as adjusting a blind will clearly render the assumption invalid. The daylight factor consists of three components as follows.

The *sky component*, SC, addresses only light arriving at a point directly from the sky vault. It is the ratio of the direct illuminance at some indoor point to the direct illuminance at a point exposed to the unobstructed sky. For an unobstructed window, the projected solid angle of the window, as subtended at the internal point in question, determines the SC.

The *external reflected component*, ERC, is the ratio of the light arriving at the point, after reflection from external surfaces, to that available from the unobstructed sky.

The *internal reflected component*, IRC, is the ratio of the light arriving at the point after reflection from internal surfaces to that available from the complete and unobstructed sky.

Consider the arbitrary point P of figure 7.29. For a CIE standard overcast sky, the luminous intensity normal to the elemental area of window (dxdy) is given by

$$I_N = L_\beta \mathrm{d}x\mathrm{d}y \tag{7.52}$$

and the luminous intensity due to the elemental area in the direction of P is given by

$$I_P = I_N \cos i \tag{7.53}$$

where i is the angle of incidence as shown in figure 7.29.

The illumination on a horizontal plane at P, due to the elemental area, is found from the inverse square and cosine laws:

$$\mathrm{d}E_p = (I_P \cos\gamma)/R^2$$

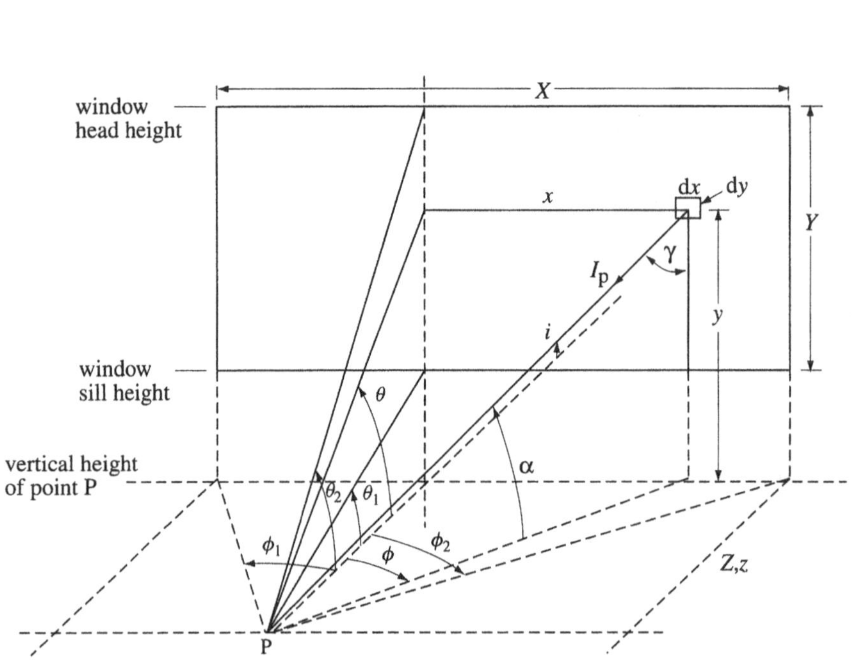

Figure 7.29: Sky component due to a vertical, rectangular window.

$$= (I_P \sin \beta)/R^2 \quad (7.54)$$

where γ is the angle between the line joining the elemental area and the point P, and the vertical at P, R the distance between the elemental area and P (m), and $\beta = (90 - \gamma)$.

Combining eqns (7.51) through (7.54) gives

$$dE_P = \frac{L_z(1 + 2\sin\beta)}{3R^2}\cos i \sin\beta dxdy$$

$$= \frac{L_z zy(1 + 2y/R)}{3R^4} dxdy$$

where z is the perpendicular distance from P to the plane of the window (m).

Now $z \tan\theta = y$, where θ is the altitude angle, and $z \tan\phi = x$, where ϕ is the azimuth angle. It follows that

$$\frac{d\theta}{dy} = \frac{1}{z\sec^2\theta} \quad \text{and} \quad \frac{d\phi}{dx} = \frac{1}{z\sec^2\phi}$$

with the illuminance at P given by

$$E_p = \frac{L_z}{3}\int_{y_1}^{y_2}\int_{x_1}^{x_2}\frac{z^2\tan\theta\left[1 + \left(2z\tan\theta)/(x^2 + y^2 + z^2)^{\frac{1}{2}}\right)\right]}{(x^2 + y^2 + z^2)^2}dxdy$$

$$= \frac{L_z}{3}\int_{\theta_1}^{\theta_2}\int_{\phi_1}^{\phi_2}\frac{\tan\theta[(1 + 2\tan\theta)/(\tan^2\theta + \tan^2\phi + 1)^{\frac{1}{2}}]\sec^2\theta\sec^2\phi}{(\tan^2\theta + \tan^2\phi + 1)^2}d\phi d\theta$$

$$= \frac{L_z}{3} \int_{\theta_1}^{\theta_2} \int_{\phi_1}^{\phi_2} \left[\frac{\tan\theta \sec^2\theta \sec^2\phi}{(\tan^2\phi + \sec^2\theta)^2} + \frac{2\tan^2\theta \sec^2\theta \sec^2\phi}{(\tan^2\phi + \sec^2\theta)^{5/2}} \right] d\phi d\theta \ .$$

Consider figure 7.30, which shows a point P under the unobstructed hemispherical sky vault. The horizontal plane illuminance at P due to the elemental ring located at any altitude β is given by

$$dE = (L_\alpha dA \sin\beta)/R^2$$

$$= \frac{L_z(1 + 2\sin\beta)}{3R^2} \sin\beta(2\pi R^2 \cos\beta d\beta)$$

$$= \frac{2\pi L_z}{3}(1 + 2\sin\beta)\sin\beta\cos\beta d\beta \ .$$

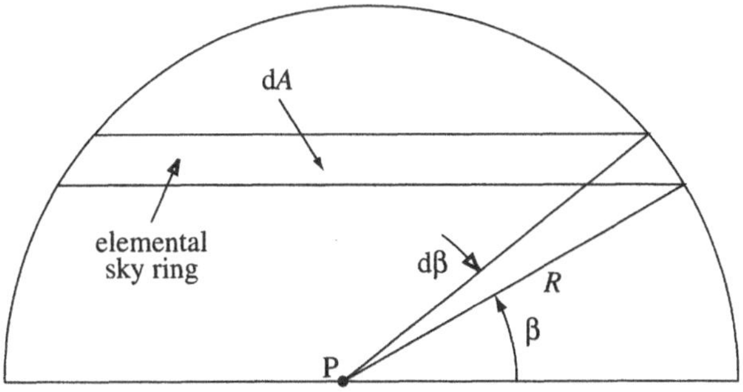

Figure 7.30: The unobstructed hemispherical sky vault.

The illuminance due to the total sky vault is therefore given by

$$E = \int_0^{\pi/2} \frac{\pi L_z}{3}(1 + 2\sin\beta)\sin 2\beta d\beta = 7\pi L_z/9 \ .$$

The sky component of the daylight factor then follows from

$$SC = \frac{E_P}{E}$$

$$= \frac{3}{7\pi} \int_{\theta_1}^{\theta_2} \int_{\phi_1}^{\phi_2} \left[\frac{\tan\theta \sec^2\theta \sec^2\phi}{(\tan^2\phi + \sec^2\theta)^2} + \frac{2\tan^2\theta \sec^2\theta \sec^2\phi}{(\tan^2\phi + \sec^2\theta)^{5/2}} \right] d\phi d\theta \ .$$

The exact analytical solution of this double integral (Clarke 1977) is

$$SC = \frac{3}{14\pi} \left\{ \frac{1}{A_1^{0.5}} \left[\tan^{-1}\left(\frac{B_2}{A_1}\right)^{0.5} - \tan^{-1}\left(\frac{B_1}{A_1}\right)^{0.5} \right] \right.$$

$$+ \frac{1}{A_2^{0.5}} \left[\tan^{-1}\left(\frac{B_1}{A_2}\right)^{0.5} - \tan^{-1}\left(\frac{B_2}{A_2}\right)^{0.5} \right]$$

$$+\frac{4}{3A_1}(A_1-1)^{0.5}\left[\left(\frac{B_2}{A_1+B_2}\right)^{0.5}-\left(\frac{B_1}{A_1+B_1}\right)^{0.5}\right]$$

$$+\frac{4}{3A_2}(A_2-1)^{0.5}\left[\left(\frac{B_1}{A_2+B_1}\right)^{0.5}-\left(\frac{B_2}{A_2+B_2}\right)^{0.5}\right]$$

$$+\frac{4}{3}\tan^{-1}\left[\frac{(A_2-1)B_2}{(A_2+B_2)}\right]^{0.5}-\frac{4}{3}\tan^{-1}\left[\frac{(A_2-1)B_1}{(A_2+B_1)}\right]^{0.5}+\frac{4}{3}\tan^{-1}\left[\frac{(A_1-1)B_1}{(A_1+B_1)}\right]^{0.5}$$

$$\left.-\frac{4}{3}\tan^{-1}\left[\frac{(A_1-1)B_2}{(A_1+B_2)}\right]^{0.5}\right\}$$

where $A_1 = \sec^2\theta_1$, $A_2 = \sec^2\theta_2$, $B_1 = \tan^2\phi_1$ and $B_2 = \tan^2\phi_2$.

This equation gives the sky component at any point due to a vertical window with sill head and edges defined by altitude angles θ_1, θ_2 and azimuth angles ϕ_1, ϕ_2. These angles can become negative if point P is to one side, above or below the window. In this case the sky component can be evaluated by the 'shape factor' algebra method as detailed in §7.5. By the same technique, the sky component for a partially obstructed window can be determined by subtracting the value for the obstructed portion from that obtained for the whole window.

External reflected component

It is common practice to assume that the relative luminance of external obstructions is one-tenth that of the sky. This allows the external reflected component to be found by dividing the sky component of the obstructed portion, when treated as unobstructed, by ten.

Internal reflected component

This component is a function of the available direct and external reflected light, the reflectance of the internal surfaces and the geometric configuration of the enclosure. There are several methods for assessing this component (Hopkinson 1963) and one of the most widespread is the empirical 'split flux' method (AJ 1954) applied to the theory of inter-reflection for an integrating sphere. This method, based on formulae proposed by Pleijel (1952), considers the flux above and below a horizontal plane passing through the mid-height of the window. Each flux component is then modified only by the receiving surfaces as illustrated in figure 7.31.

The method is derived from consideration of flux inter-reflections within a sphere as shown in figure 7.32. At the end of the first reflection, the average internal reflected component, IRC_1, is given by

$$IRC_1 = FR/4\pi r^2$$

where F is the flux entering the sphere (lm), R the average reflectance of the internal surfaces, and r the radius of the sphere (m). At the end of the second reflection, the average internal reflected component, IRC_2, is

$$IRC_2 = FR^2/4\pi r^2$$

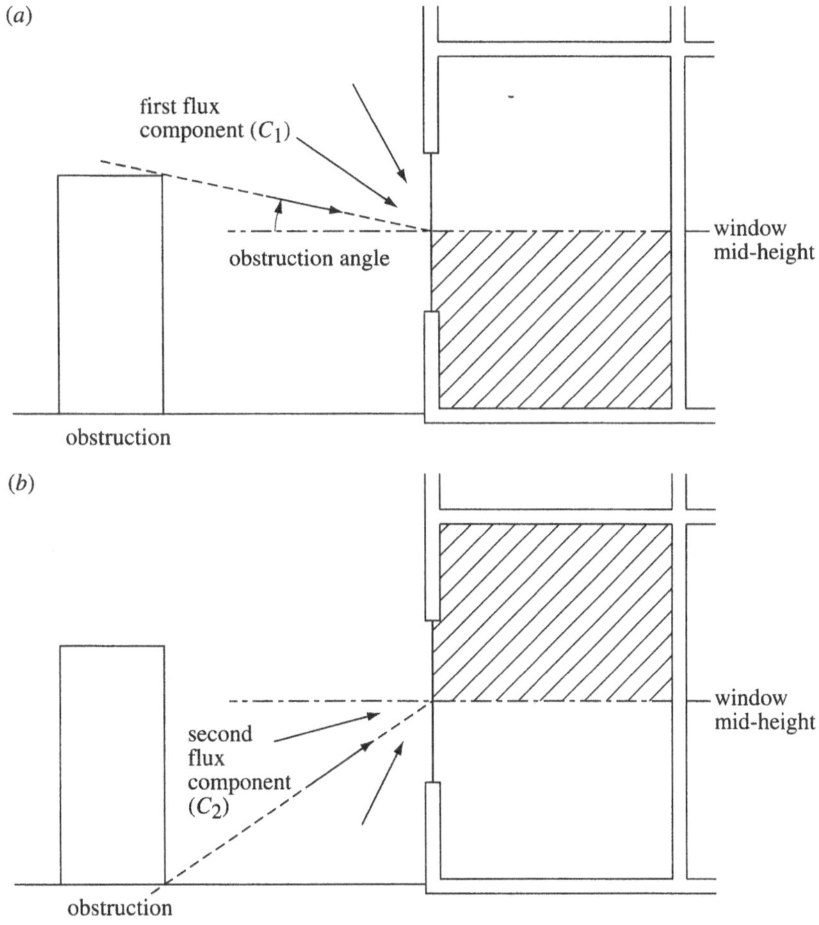

Figure 7.31: Internal reflected component, 'split flux' method—(a) upper flux component, (b) lower flux component.

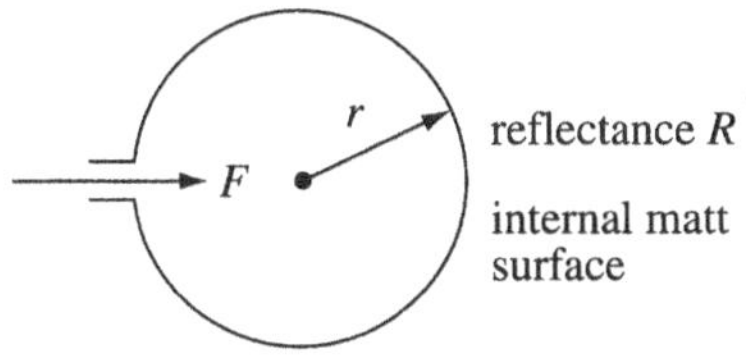

Figure 7.32: An integrating sphere.

and so on to infinity:

$$\begin{aligned} IRC &= \frac{FR}{4\pi r^2}(1 + R + R^2 + \ldots) \\ &= \frac{FR}{4\pi r^2}\left(\frac{1}{1-R}\right) \\ &= \frac{F_T}{A(1-R)} \end{aligned} \quad (7.55)$$

where IRC is the total internal reflected component, F_T the total first reflected flux (lm) and A the total internal surface area (m^2).

In applying the approach of eqn (7.55) to an actual enclosure two main assumptions are made: the interior surfaces of the enclosure are diffuse reflectors and the enclosure shape is not significantly removed from a sphere. The latter assumption is likely to contribute most to technique inaccuracy.

The total first reflected flux is given by

$$F_T = W(C_1 R_{fW} + C_2 R_{CW})$$

where W is the window area (m^2), R_{fW} the average reflectance of the working plane and those wall surfaces below the horizontal plane passing through the mid-height of the window (but excluding the window wall), and R_{CW} the average reflectance of the ceiling and those wall surfaces above the horizontal plane (again excluding the window wall).

C_1 and C_2 are functions of the flux entering the enclosure from above and below the horizontal plane. C_1 is dependent on the luminance distribution of the sky and on an angle defining the external obstructions. C_2 is dependent on the luminance of the ground and on the external obstructions below the horizontal plane. The ground luminance distribution is usually assumed to be 10% of the sky distribution. This gives a value for C_2 that is not significantly removed from 5 and, for convenience, C_2 is set equal to 5 by adjusting the value of C_1 accordingly. Eqn (7.55) can then be expressed in its final form:

$$IRC = W(CR_{fW} + 5R_{CW})/A(1 - R)$$

where C is the adjusted function of C_1 as given in table 7.21.

Table 7.21: Values of C.

Obstruction angle (°)	0	10	20	30	40	50	60	70	80
C	39	35	31	25	20	14	10	7	5

Daylight factor

The daylight factor, DF, is now obtained by adding the individual components for each window after modification by visible transmission and maintenance factors:

$$DF = \sum^{N} TM(SC + ERC + IRC)$$

where T is the visible transmittance, M a maintenance factor and N the total number of windows.

Within an energy simulation, the daylight factor is applied to the time varying external illuminance to determine the internal illuminance. This can then be used to effect luminaire control. The simulation will then include the effects of variable heat gains from the lighting installation.

7.8.3 Internal illuminance distribution: numerical method

Where it is necessary to include real world phenomena—such as varying sky luminance distribution, movable shading, light redirection and the spatial/spectral characteristics of artificial lights—it is possible to arrange that a thermal and lighting model operate in tandem at the time-step level.

Figure 7.33 shows such an arrangement involving the ESP-r and RADIANCE systems, with the former providing the overall supervisory control at simulation time.

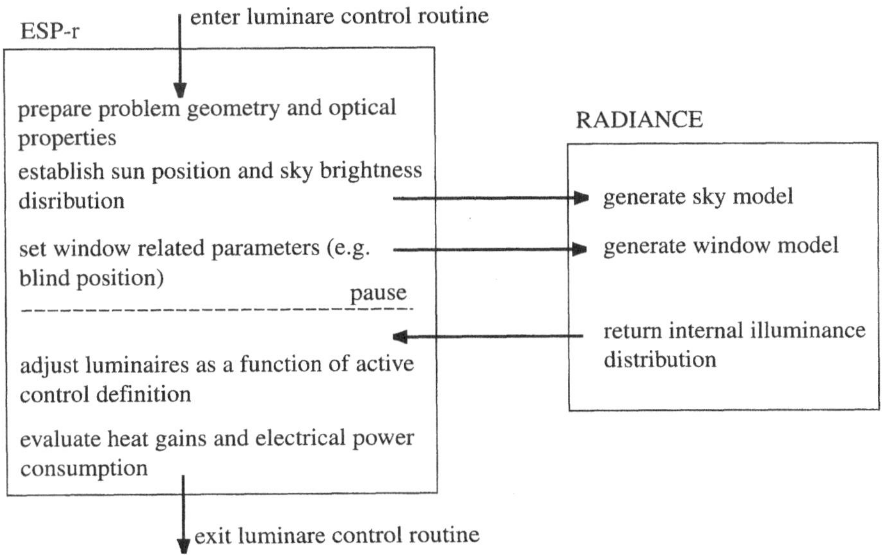

Figure 7.33: ESP-r and RADIANCE interacting at the time-step level.

At each time-step, a related ESP-r controller adjusts model parameters, such as shading device position, as a function of some active regulation law. A luminaire control algorithm then initiates the daylight simulation, co-ordinating RADIANCE to carry out several tasks as follows: (1) transfer of data defining current solar position, sky brightness and building state; (2) generation of sky model; (3) re-building of scene model; (4) calculation of internal illuminance for defined sensor locations; (5) transfer back of illuminance data to luminaire controller. The returned data are then used to determine, as a function of an active photocell algorithm, the luminaire status and hence the casual gain associated with lights at the current time-step. The theoretical basis and operational capabilities of RADIANCE are described elsewhere (Larson and Shakespeare 1998). The Perez sky luminance distribution model (Perez *et al* 1993) is used in conjunction with direct normal and diffuse horizontal solar irradiance data. Several sources exist from which these data may be obtained at higher than normal frequency: e.g. the 5 minute average data from the International Daylight Measurement Program (IDMP, Atlas 1995) stations or the use of a probability density algorithm (Skartveit and Olseth 1992) to generate high frequency irradiance data from hourly values.

7.8.4 Photocell response

Luminaire control may be imposed as a function of photocell position, vision angle, controller set-point, switch-off illumination level, switch-off delay time and minimum stop.

An integral reset controller adjusts the dimming level so that the measured photocell signal is kept at a constant reference value. This value is set during night-time photocell calibration and represents the measured signal from the artificial lighting. The dimming level within the controller's dynamic range is determined as

$$f_d^\tau = 1 - \frac{E_{d-s}^\tau}{E_{e-s}} .$$

where f_d^τ is the time varying dimming level (-), E_{d-s}^τ the time varying daylight photocell signal (lux) and E_{e-s} the artificial lighting photocell signal during night-time calibration (lux).

A closed loop proportional controller adjusts the dimming level so that it is a linear function of the difference between the photocell signal and the night-time reference level. With this controller, a day-time calibration must be performed to determine the linear control function slope:

$$m_s = \frac{E_{d-w}^\tau}{E_{d-w}^\tau E_{e-s} - E_{e-w} E_{d-s}^\tau}$$

$$f_d^\tau = \frac{1 + m_s \left(E_{d-s}^\tau - E_{e-s} \right)}{1 - m_s E_{e-s}}$$

where m_s the slope of the controller's linear function (-), E_{d-w}^τ the time varying daylight illuminance at selected workplace control points during daytime calibration (lux) and E_{e-w} the artificial lighting illuminance at selected workplace control points during night-time calibration (lux).

Assuming a closed loop proportional controller, Janak (1997) has implemented and used the method of §7.8.3 with the foregoing photocell model to investigate the influence of controller type and photocell geometry. The problem considered is shown in figure 7.34: an office of dimensions $4.5\text{m} \times 4.5\text{m} \times 3.2\text{m}$ with a combination window comprising an upper portion with integral light shelf and a lower portion with a movable blind. The office is lit by wall mounted, asymmetric luminaires designed to provide an average workplace illuminance, E_{e-w}, of 320 lux. The lamp luminous output can be regulated between 10% and 100% of the full light output. Daylight responsive control is implemented via a ceiling mounted photocell located at 2/3 of the room depth. The following control parameters were assumed in the study:

daylight sensor set-point, 320 lux	switch-off at 150% of set-point
minimum light dimming, 10% of full light output	minimum power dimming, 10% of full circuit power
switch-off delay time, 0 for 60 minutes time-step simulation, 15 minutes for 5 minute time-step simulation	blind control sensor location, vision window plane measuring vertical global irradiance
set-point for blind rotation to the shading position (45°), 300 W m^{-2}	

A sample of the 5 minute irradiance values from the IDMP station in Bratislava, Slovakia was used within a December day simulation during which the irradiance was highly variable.

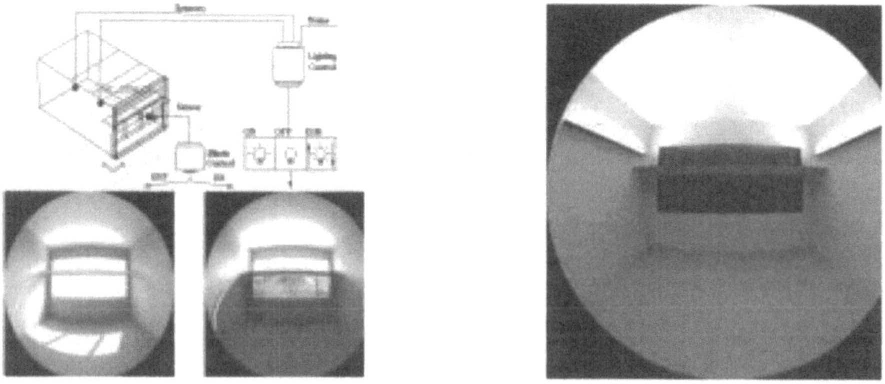

Figure 7.34: Thermal and lighting model configuration (left); office with wall mounted, asymmetric luminaires during night-time calibration (right).

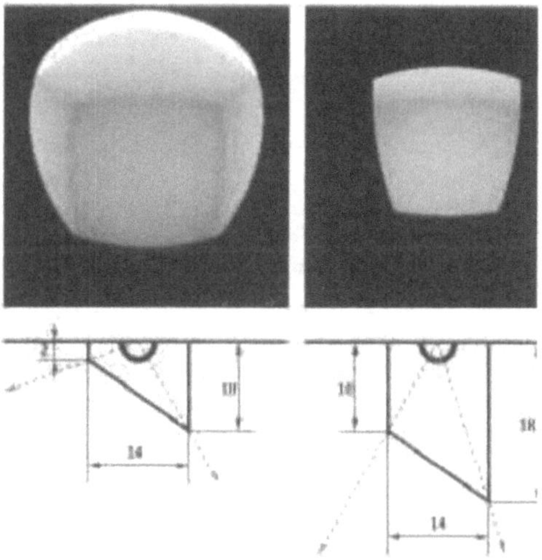

Figure 7.35: Partially and fully shading photocell geometry (top figure shows sensor's view from ceiling towards floor and walls.

Table 7.22: Comparison of predicted lighting power consumption for the different cases.

Sensor type	*Time-step (min)*	*Lighting power consumption ($Wh\ d^{-1}$)*	*Relative difference (%)*
Integral reset, partially shaded	5	1168.2	-32.5
Integral reset, fully shaded	5	997.0	-42.4
Closed loop proportional, partially shaded	5	1670.4	-3.4
Closed loop proportional, fully shaded	5	1636.0	-5.4
Closed loop proportional, partially shaded	60	1792.2	+3.6
Ideal control	60	1730.0	0.0

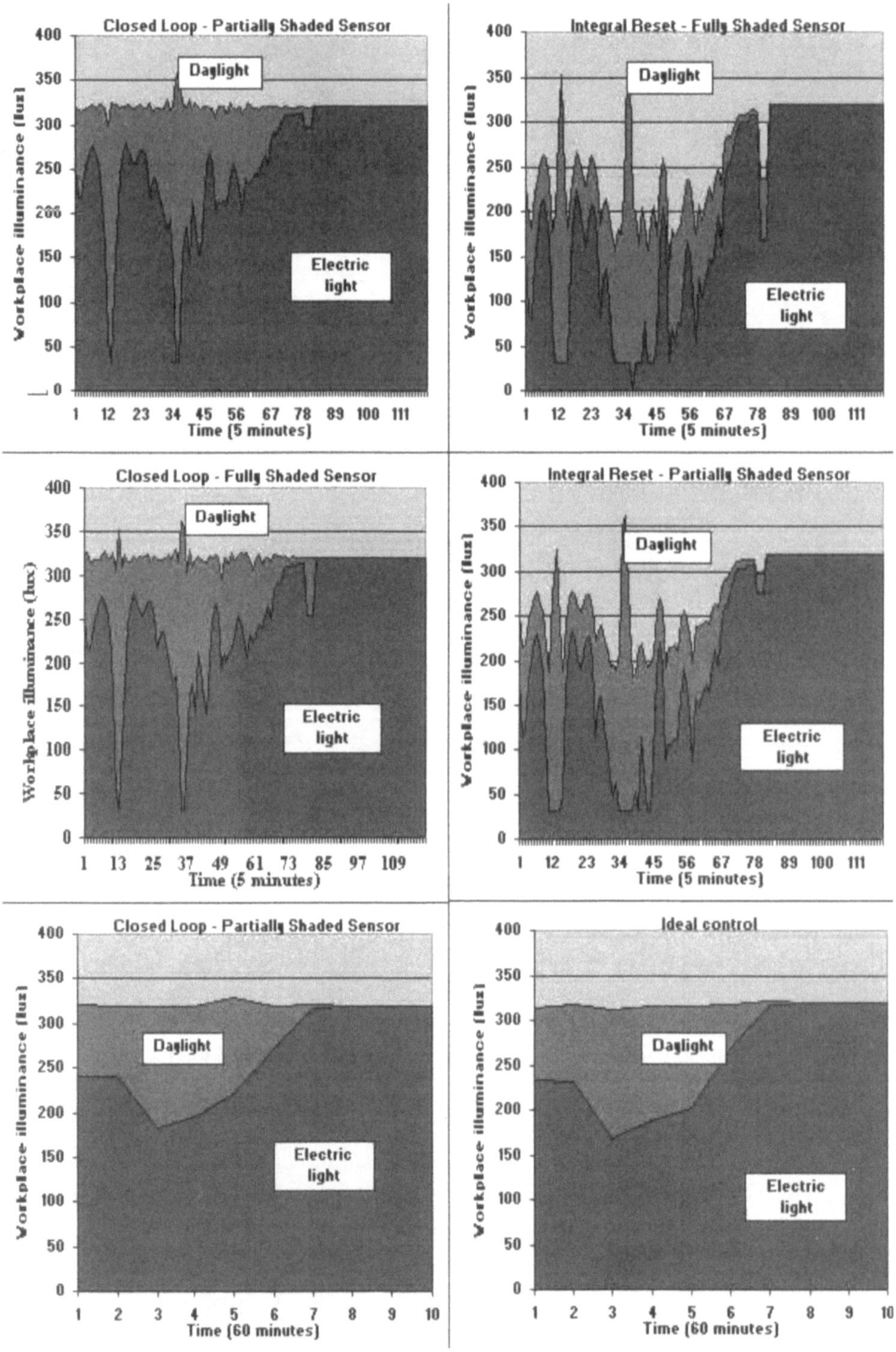

Figure 7.36: Predicted illuminance for the different cases.

Figure 7.35 shows a comparison between two photocell geometries corresponding to the partially (E_{e-s} = 44. 5 lux) and fully (E_{e-s} = 14. 1 lux) shaded cases. For the closed loop, proportional control action considered, the linear control slope m_s was set at -0.023 for the partially shaded case and -0.056 for the fully shaded case.

Table 7.22 shows the predicted power consumption for the different controller, photocell and simulation approaches studied while figure 7.36 shows the predicted illuminance variation at selected locations.

It is clear that to achieve realistic luminaire control behaviour, short-term daylight availability should be considered. It is also clear that the different controllers give rise to large differences in the estimated power consumption (e.g. compare integral reset with closed loop action). On the other hand, a well calibrated closed loop proportional controller will give rise to the same power consumption as a model based on ideal control with a 60 minute time step. With an integral reset controller, an ideal control model will fail to predict the correct dynamic behaviour and power consumption. It is clear therefore that the method of direct run-time coupling allows explicit modelling of the many important interactions that occur between the thermal and visual domains. Two modelling constructs have been pursued in an attempt to reduce the computational burden of long term simulations: daylight coefficients (Littlefair 1992) and the grouping of Perez model parameters (Perez *et al* 1993) into classes.

7.9 Mould growth

Apart from possible degradation of the building fabric, there is growing epidemiological evidence to suggest that mould infestation can present a distinct health risk to the people who live in such environments (Dales *et al* 1991a & 1991b, Platt *et al* 1989). Respiratory and/or allergenic disorders, particularly in children, have been diagnosed and a range of other symptoms, including nausea and vomiting, have been reported (Burr *et al* 1988, Martin *et al* 1989). While the precise mechanisms for these symptoms are not fully understood, toxic fungal metabolites, particularly mycotoxins produced by fungal spores and absorbed through the mucous membrane of the respiratory tract, are increasingly being implicated (Lewis et al 1989, 1994).

Clarke *et al* (1996) undertook an inter-disciplinary engineering and bioscience research project to categorise the mould types that are problematic in housing and develop a model of the limiting growth conditions for each category.

The moulds encountered in the built environment are members of the Deuteromycetes subgroup of fungi. Mould growth is essentially a surface phenomenon and will occur if fungal propagules are present on a surface and the physiological requirements for growth are met (Adan 1994). The most important factors known to control growth are relative humidity and temperature. The widely held belief that condensation has to occur before mould growth will appear is invalid.

From an extensive literature review covering over 250 publications, the principal mould species affecting dwellings were identified and their minimum growth requirements established in terms of local surface temperature and relative humidity (RH). The moulds were then assigned to one of six categories, A-F, ranging from highly xerophilic (dry loving) to highly hydrophilic (wet loving). Each category constitutes a family of mould species possessing similar growth requirements. On the basis of a representative mould from each category, growth limit curves were then established as shown in figure 7.37.

These curves define the minimum combination of local surface temperature and humidity for which mould growth will occur on building surfaces where conditions are sustained. Below any given curve, growth will not occur for moulds in that category. Examples of moulds in

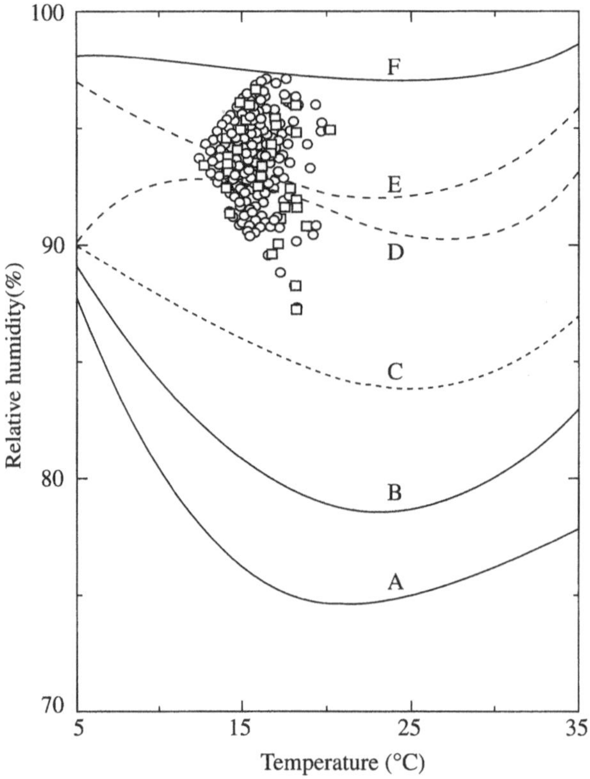

A - highly xerophilic
B - xerophilic
C - moderately xerophilic
D - moderately hydrophilic
E - hydrophilic
F - highly hydrophilic

Figure 7.37: Mould limiting growth curves.

each category include *Aspergillus repens* (A), *Aspergillus versicolor* (B), *Penicillium chysogenum* (C), *Cladosporium sphaerospermum* (D), *Ulocladium consortiale* (E) and *Stachybotrys atra* (F).

In order to verify the figure 7.37 model, samples were collected from a mould infested house and incubated at 25°C under a range of RH conditions over a 120 day period. The identified mould species, and the lowest RH value at which growth was detected, are given in table 7.23.

These data show good agreement with the growth limit curves of figure 3.37: the minimum RH values supporting growth for category A, B and C moulds, and the lack of growth below 74.5%, are consistent with the predictions of the model. Note, however, that the minimum RH for the category D and F moulds are one RH step below that predicted by the model. This can be explained, in part, by the fact that the moulds were cultivated on laboratory media that are nutritionally richer than building materials.

The curves may now be used in conjunction with the previously derived building and flow models as the basis of a mould growth prediction facility. Given a building model incorporating refined models for zone and construction moisture distribution estimation (as introduced in chapter 5), the local surface temperature and relative humidity profile may be determined. As

Table 7.23: Mould growth data.

Isolated mould	*Growth category*	*Lowest RH (%) permitting growth*	*Time (days) for growth*
Yeasts	F	94.5	85
Cladosporium sphaerospermum	D	88.5	78
Penicillium spp.	C	85.5	97
Alternaria alternata	D	88.5	90
Aureobasidium pullulans	D	88.5	103
Aspergillus versicolor	B	81.0	97
Eurotium herbariorum	A	78.5	97

No mould growth was observed at 74.5% RH after 120 days.

shown in figure 7.37, these data may then be superimposed on the growth limit curves and the concentration of plotted points relative to the various growth bands used to assess the risk of mould growth and its probable persistence over time.

7.10 References and further reading

Adan O C G 1994 On the Fungal Defacement of Interior Finishes *PhD Thesis* (Eindhoven: Eindhoven University of Technology)

AJ 1954 The Split Flux Method *Architects J.* **120** 335

Alamdari F and Hammond G P 1983 Improved Data Correlation for Buoyancy-Driven Convection in Rooms *Rep. SME/J/83/01* (Cranfield: Applied Energy Group, Cranfield Institute of Technology)

Anderson B 1974 Meteorological Data for Design of Building and Installation *Rep. 89* (Denmark: Danish Building Research Institute)

Ångström A 1930 On the Atmospheric Transmission of Sun Radiation *Geografis. Annal.* **2 & 3** 130-59

ASHRAE 1997 *Handbook of Fundamentals* (ASHRAE)

— 1985 *Weather Year for Energy Calculations* (Atlanta: American Society of Heating, Refrigeration and Air-Conditioning Engineers)

Atlas 1995 Design of a European Daylighting Atlas - Luminous Climate in Central Europe *Final report for EU project JOU2-CT 920144 CIPD-CT 925033* (Bratislava: ICA SAS)

Awbi H B and Hatton A 1999 Natural Convection from Heated Room Surfaces *Energy and Buildings* **30** 233-44

Barbaro S, Coppolino S, Leone C and Sinagra E 1979 An Atmospheric Model for Computing Direct and Diffuse Solar Radiation *Solar Energy* **22** 225

Beausoleil-Morrison I 2000 The Adaptive Coupling of Heat and Air Flow Modelling within Dynamic Whole-Building Simulation *PhD Thesis* (Glasgow: University of Strathclyde)

Berdahl P and Martin M 1984 Characteristics of Infrared Sjy Radiation in the United States *Solar Energy* **33**(3/4) 321-6

Bird R E and Hulstrom R L 1979 Application of Monte Carlo Technique to Insolation Characterisation and Prediction *Tech. Report TR-642-761* (Bolder: Solar Energy Research Institute)

Born M and Wolf E 1965 *Principles of Optics (3rd edn)* (London: Pergammon)

Burr M I, Mullins J, Merrett T G and Stott N C H 1988 Indoor Moulds and Asthma *J. Royal Society of Health* **3** 99-101

CEC 1996 *European Solar Radiation Atlas* Eds. Palz W and Greif J (Berlin: Springer-Verlag)

Chung B T F and Samitra P S 1972 Radiation Shape Factors from Plane Point Sources *J. Heat Transf.* **94** 328-30

CIBSE 2001 Guide to Weather, Solar and Illuminance Data (CDROM) (London: Chartered Institution of Building Services Engineers)

Clarke J A 1977 Environmental Systems Performance *PhD Thesis* (Glasgow: University of Strathclyde)

Clarke J A, Johnstone C M, Kelly N J, McLean R C, Anderson J G, Rowan N J and Smith J E 1999 A Technique for the Prediction of the Conditions Leading to Mould Growth in Buildings *Building and Environment* **34**(4) 515-21

Cole R J 1976 The Longwave Radiation Incident Upon the External Surface of Buildings *Building Services Engineer* **44** 195-206

Cooper I and Crisp V H C 1983 Barriers to the Exploitation of Daylighting in Building Design *Proc. Int. Daylighting Conf. (Phoenix)*

Crawley D B 1998 Which Weather Data Should You Use for Energy Simulations of Commercial Buildings? *ASHRAE Trans.* **104** 2

Dales R E, Zwanenburg H, Burnett R and Flannigan C A 1991a Respiratory Health Effects of Home Dampness and Moulds Among Canadian Children *American Journal of Epidemiology* **134** 196-206

Dales R E, Burnett R and Zwanenburg H 1991b Adverse Health Effects Among Adults Exposed to House Dampness and Moulds *American Reviews of Respiratory Disease* **143** 505-9

Davies J A, Schertzer W and Nunez M 1975 Estimating Global Solar Radiation *Boundary-Layer Meteorol.* **9** 33-52

Degelman L O 1997 Examination of the Concept of Using 'Typical-week' Weather Data for Simulation of Annualized Energy Use in Buildings *Proc. Building Simulation '97* (Prague)

Duffie J A and Beckman W A 1980 *Solar Engineering of Thermal Processes* (New York: John Wiley)

Fisher D E 1995 An Experimental Investigation of Mixed Convection Heat Transfer in a Rectangular Enclosure *PhD Thesis* (Urbana: University of Illinois)

Fisher D E and Pederson C O 1997 Convective Heat Transfer in Building Energy and Thermal Load Calculations *ASHRAE Transactions* **103**(2) 137-48

Foley J D and Van Dam A 1982 *Fundamentals of Interactive Computer Graphics* (Reading, Mass.: Addison-Wesley)

Fontoynont M (Ed) 1999 *Daylight Performance of Buildings* (James and James)

Fujii T and Imura H 1970 Natural Convection Heat Transfer from a Plate with Arbitrary Inclination *J. Heat and Mass Transf* **15** 755-67

Gringley P C, Batty W J and Probert S D 1995 Mathematical Model for Predicting the Magnitudes of Total, Diffuse and Direct-Beam Insolation *Applied Energy* **52** 89

Hay J E 1979 Calculation of Monthly Mean Solar Radiation for Horizontal and Inclined Surfaces *Solar Energy* **23** 301

Heavens O S 1955 *Optical Properties of Thin Solid Films* (London: Butterworth)

Holmes M J and Hitchin E R 1978 An Example Year for the Calculation of Energy Demand in Buildings *J. Building Services Engineer* (January)

Hopkinson R G 1963 *Architectural Physics - Lighting* (London: Her Majesty's Stationary Office)

Hottel H C 1930 Radiant Heat Transmission *J. Mech. Eng.* **52** 699-704

Howson R P, Ridge M I, Bishop C A and Cottman GP 1984 The Production of Selective Optical Coatings on Plastic Sheet *Proc. EC Seminar on Energy Saving in Buildings* 421-33 (Dordrecht: Teidel)

IHVE (Now CIBSE) 1973 Some Fundamental Data Used by Building Services Engineers *Rep of Working Group 6, Building Services Sub-Committee of CACCI* (London)

IAUC 2001 International Association for Urban Climate (http://www.geography.ohio-state.edu/UrbanClimate/index.html)

Ito N, Kimura K and Oka J 1972 A Field Experiment Study on the Convective Heat Transfer Coefficient on Exterior Surface of a Building *ASHRAE Trans.* (2225) 184-191

Iqbal M 1983 *An Introduction to Solar Radiation* (New York: Academic Press)

Jacob M 1949 *Heat Transfer* (New York: Wiley)

Janak M 1997 Coupling Building Energy and Lighting Simulation *Building Simulation '97* (Prague)

Jensen S O (Ed) 1994 Validation of Building Energy Simulation Programs *Final Report, PASSYS Model Validation Subgroup* EUR 15115 EN (European Commission)

Khalifa A J N and Marshall R H 1990 Validation of Heat Transfer Coefficients on Interior Building Surfaces Using a Real-Sized Indoor Test Cell *Int. J. Heat Mass Transfer* **33**(10) 2219-36.

Kimura K and Stephenson D G 1969 Solar Radiation on Cloudy Days *ASHRAE Transactions Part I* 227-233

Kittler R 1985 Luminance Distribution Characteristics of Homogeneous Skies: A Measurement and Prediction Strategy *Lighting Research and Technology* **17**(4) 183-8

— 1986 Luminance Models of Homogeneous Skies for Design and Energy Performance Prediction *Proc. Int Daylighting Conf.* (Long Beach)

Klucher T M 1979 Evaluation of Models to Predict Insolation on Tilted Surfaces *J. Solar Energy* **23**(2) 111-14

Kreith F 1973 *Principles of Heat Transfer* (3rd edn) (New York: Harper and Row)

Lacis A A and Hansen J E 1974 A Parameterisation for the Absorption of Solar Radiation in the Earth's Atmosphere *J. Atmos. Sci.* **31** 118-32

Larson G W and Shakespeare R 1998 *Rendering with Radiance - The Art and Science of Lighting Visualization* (San Francisco: Morgan Kaufmann)

Lewis C W, Anderson J G, Smith J E, Morris G P and Hunt S M 1989 The Incidence of Moulds Within 525 Dwellings in the United Kingdom *International Journal of Environmental Studies* **25** 105-12

Lewis C W, Smith J E, Anderson J G and Murad Y M 1994 The Presence of Mycotoxin-Associated Fungal Spores Isolated from the Indoor Air of the Damp Domestic Environment and Cytotoxic to Human Cells *Indoor Environment* **3** 323-30

Littlefair 1981 The Luminance Distribution of an Average Sky *Lighting Research and Technology* **13**(4) 192-8

Littlefair P J 1992 Daylight Coefficients for Practical Computation of Internal Illuminance *Lighting Research and Technology* **24**(3) 127-35

Littlefair P J 1992a Modelling Daylight Illuminance in Building Environmental Performance Analysis *J. of the Illuminating Eng. Soc.* 25-34

Liu B Y H and Jordan R C 1960 The Inter-relationship and Characteristic Distribution of Direct, Diffuse and Total Solar Radiation *Solar Energy* **1**

Lund H 1976 Requirements and Recommendations for the Test Reference Year *Proc. CIB Symp.* (Vienna)

Ma C C Y and Iqbal M 1983 Statistical Comparison of Models for Estimating Solar Radiation on Inclined Surfaces *J. Solar Energy* **31**(2) 313-17

Martin C J, Platt S D and Hunt S M 1989 Housing conditions and ill health *British Medical Journal* **294** 1125-7

Maver T W, Clarke J A, Stearn D D and Kim J J 1987 Lighting Simulation in Building Design *Proc. Electronic Imaging '87* (Boston: Institute for Graphics Communication)

Markus T A, Clarke J A and Morris E N 1984 Climatic Severity *Technical Rep., Dept. Architecture and Build. Sci.* (Glasgow, University of Strathclyde)

McAdams W H 1954 *Heat Transmission* (New York: McGraw-Hill)

Meteonorm 2000 http://www.meteotest.ch/products/meteonorm/

Min T C, Schutrum L F, Parmelee G V and Vouris J D 1956 Natural Convection and Radiation in a Panel-Heated Room *ASHRAE Trans.* **62** 337-58

Moore G and Numan M Y 1982 *Form Factors: The Problem of Partial Obstruction* (Cambridge: University of Cambridge, Martin Centre Report)

— 1983 *Specular Reflection of Radiation* (Cambridge: University of Cambridge, Martin Centre Report)

Muneer T 1997 *Solar Radiation & Daylight Models for the Energy Efficient Design of Buildings* (London: Architectural Press) ISBN 0 7506 2495 7

— 1990 Solar Radiation Model for Europe *BSER&T* **11** 153

Muneer T, Gul M, Kambezedis H and Alwinkle S 1996 An All-Sky Solar Radiation Model Based on Meteorological Data *Proc. CIBSE/ASHRAE Joint Annual Conf.* (Harrogate)

NCDC 1976 Test Reference Year *Reference Manual TD-9706* (North Carolina: National Climatic Data Center)

— 1981 Typical Meteorological Year *Reference Manual TD-9734*

— 1993 Solar and Meteorological Surface Observation Network, 1961-90

Ne'eman E and Selkowitz S 1983 Daylight Availability as a Function of Atmospheric Conditions *Proc. Atmospheric Radiation Conf.* (Baltimore)

NREL 1995 TMY2 User's Manual National Report *Rep. NREL/SP-463-7668* (Golden, Colorado: National Renewable Energy Laboratory)

Optical Society of America 1978 *Handbook of Optics* (New York: McGraw-Hill)

Perez R, Michalsky J and Seals R 1992 Modelling Sky Luminance Angular Distribution for Real Sky Conditions: Experimental Evaluation of Existing Algorithms *IES J.* **21** 84-92

Perez R, Seals R, Michalsky J (1993) All-Weather Model for Sky Luminance Distribution-Preliminary Configuration and Validation *Solar Energy* **50**(3) 235-43

Pilkington 1973 *Thermal Transmission of Windows* (St Helens: Pilkington Glass Environmental Advisory Service)

Platt S D, Martin C J, Hunt S M and Lewis C W 1989 Damp Housing, Mould Growth and Symptomatic Health State *British Medical Journal* **298** 1673-8

Pleijel G 1952 *J. Sci. Instrum.* **29** 137

Rogers D F and Adams J A 1976 *Mathematical Elements for Computer Graphics* (New York: McGraw-Hill)

Rousseau and Mathieu 1973 *Problems in Optics* (London: Pergammon)

Rubinstein F, Ward G, Verderber R 1989 Improving the Performance of Photo-Electrically Controlled Lighting Systems *J. of the Illuminating Eng. Soc.* 70-94

Saito H and Matsuo Y 1974 Standard Weather Data for SHASE Computer Program *Proc. 2nd Symp. Use of Computers for Environ. Eng. related to Build.* (Paris)

Skartveit A and Olseth J A 1992 The Probability Density and Autocorrelation of Short-Term Global and Beam Irradiance *Solar Energy* **49**(6) 477-87

Sowell E F and Johnson K 1993 Improved Fluorescent Lighting Models for Building Energy Programs *Proc. Building Simulation '93* (Adelaide)

Stamper E 1977 Weather Data *ASHRAE J.* **Feb.** 47.

Stanzel B 1994 External Longwave Radiation *Chapter 5 in Final Report, PASSYS Model Validation Subgroup* EUR 15115 EN (European Commission)

Steven M D and Unsworth M H 1980 The Angular Distribution and Interception of Diffuse Solar Radiation Below Overcast Skies *Q. J. Royal Met. Soc.* **105** 593

Swinbank W C 1963 Longwave Radiation from Clear Skies *J. R. Meteorol. Soc.* **89** 339-48

Szerman M 1994 Auswirkung der Tageslichtnutzung auf das Energetische Verhalten von Burogebauden *PhD Thesis* (University of Stuttgart)

Temps R C and K L Coulson 1977 Solar Radiation Incident on Slopes of Different Orientations *J. Solar Energy* **19**(2) 179-84

Thekaekara M P 1973 Solar Energy Outside the Earth's Atmosphere *J. Solar Energy* **14** 109-27

Tragenza P R 1980 The Daylight Factors and Actual Illuminance Ratios *Lighting Research and Technology* **12**(2) 64-8

Tragenza P R 1983 Predicting Daylighting from Skies of Random Cloudiness *Proc. Int. Daylighting Conf.* (Phoenix)

Vasicek A 1960 *Optics of Thin Films* (Amsterdam: North-Holland)

Welty J R 1974 *Engineering Heat Transfer* (London: Wiley)

Winkelmann F C and Selkowitz S 1985 Daylighting Simulation in the DOE-2 Building Energy Analysis Program *Energy and Buildings* **8** 271-86

Wong H Y 1977 *Heat Transfer for Engineers* (London: Longman)

Yallop B D 1992 *Technical Note* (Cambridge: Royal Greenwich Observatory)

8

Use in practice

Although most practitioners will be aware of building simulation, few as yet are able to claim expertise in its application. This situation is poised to change given society's aspirations towards sustainable practices and the drive to deploy new and renewable energy systems within the built environment at all scales. Simulation provides a way to assess the benefits of particular schemes, to improve life cycle performance, to enhance design quality, to appraise climate change mitigation measures, to undertake scenario based energy planning, to link energy and health, to enable inter-organisation partnerships, and so on. How else might such aspirations be achieved?

Applying simulation to the design and management of the built environment represents a paradigm shift of unparalleled scale. For the first time, the construction industry has the means to address the underlying thermodynamic complexities and undertake integrated performance appraisals of options at reasonable cost. Table 8.1 places the emergence of simulation within the spectrum of issues facing the evolution of computer-aided building design in general. Clearly, progress is required on all fronts. What is the point of developing powerful tools without putting in place the means to train and support users? And what is the point of deploying advanced IT methods within an outdated approach to work-flow management?

While there is little doubt that numerical techniques can be used to construct programs of high integrity (vis-à-vis the real world), accuracy (vis-à-vis building physics) and flexibility (vis-à-vis designer requirements), several barriers to uptake remain. These may best be overcome by action on five fronts. First, there is a need to establish a validation method for routine application as models are developed and refined. By demonstrating that simulation can be used to predict performance accurately, practitioners will come to realise the inadequacy of traditional methods. Second, there is a need to establish user interfaces that are able to protect users from the complexity of simulation while giving them access to its considerable power. Third, there is a need to establish a performance assessment method to guide those designers who wish to employ a computational approach to design. Fourth, there is a need to allow for the many uncertainties that impact on the problem description and appraisal process. And, last, there is a need to establish support mechanisms for those wishing to apply simulation in the real time, real scale context of design practice. The following sections address these disparate needs in turn.

Table 8.1: Computer-aided building design—the spectrum of issues.

Issue	*Time Scale*		
	Now	*Short-term*	*Long-term*
Technology	• Partial CABD • Early expert systems	• Full CABD • Expert systems	• Natural language interface • IKBS
Applications	• Mono-variate appraisal • 3D visualisation • Regulations	• Multi-variate appraisal • Life cycle analysis	• Post occupancy applications • Client-oriented CABD
Training	• In-depth postgraduate training	• Advanced undergraduate & CPD training	• Computer assisted design
Research	• VR systems • Web-enabled systems	• Systems for computer-naive designers	• Non-traditional communication • Design optimisation
Impact	• Skills shortage	• Breakdown of professional boundaries	• De-professionalisation • Improved product performance

CABD = Computer-Aided Building Design
IKBS = Intelligent, Knowledge-Based Systems
VR = Virtual Reality

8.1 Validation

Ultimately, a program's predictive accuracy can only be assessed by comparing its outputs with results from buildings in use. This task is frustrated by the difficulties associated with data acquisition and the shortcomings inherent in even the most sophisticated program that render it impossible to model reality exactly. Empirical validation is therefore expensive and time consuming and can only be pursued within well resourced projects. A recurring message from such projects is that it is likely that compensating errors result in different programs fortuitously giving similar outputs. To address this issue, additional testing procedures are required.

Inter-program comparison allows new programs to be tested against well established ones. This is a particularly useful device where the input models can be established to stress a particular aspect known to be well handled by one of the programs. Analytical solutions are a powerful verification device where the assumptions made to permit the solution can be equivalently imposed on a program. Sensitivity analysis allows the influence of input parameters on outputs to be determined. This information can then be used to direct program refinement.

Such devices have been applied with varying levels of success within several validation projects addressing building simulation programs: IEA Annex 1 (DOE 1981), IEA Annex 4 (BRE 1984), IEA Annex 10 (Lebrun and Liebecq 1988), IEA Annex 21 (Lomas *et al* 1994), IEA Task 8 (Morck 1986), EC Passive Solar Project (Dupagne 1983), EPSRC Validation Project (Bloomfield 1988), ETSU Applicability Study (Lomas and Eppel 1992), EC PASSYS Project (Jensen 1994) and BRE/EdF Validation Project (Bloomfield 1999).

Strachan (2000) has summarised the outcome from these and other projects from the perspective of the ESP-r system. One issue is raised here: while ESP-r performed well when exercised in a certain way in one project, it sometimes failed when the same or similar test was repeated in a later project. While the causes of this are largely known (e.g. new code development), it gives rise to a key requirement: for the integrated simulation approach to be reliable, validation checks must be made on a regular basis. This is best achieved by encapsulating relevant quality assurance checks within the programs. For example, IEA Annex 21 developed a set of diagnostic tests known collectively as BESTEST (Judkoff and Neymark 1995), which could be included within a program package to allow users to determine the impact of version updates. Likewise, there exist standard analytical tests (Bloomfield 1988) and sensitivity analysis techniques (Lomas and Eppel 1992) that can also be included to provide on-line quality assurance.

8.2 User interface

As the power of the integrated simulation approach has evolved, so the demands on the user interface have increased. In establishing an interface, there is a balance to be struck between allowing full access to a program's capabilities, and protecting users from unnecessary data inputs and advanced technical features. To add to the problem, this balance is dynamic because it depends on the user's (evolving) capabilities, the design stage/issue being addressed and the technical requirements of the project. This suggests the use of an adaptive interface that is able to change its style of interaction as a function of in-built user and process models. Section 9.1 elaborates such an approach while this section considers the requirements, however implemented, for the definition and maintenance of the related data model. Figure 8.1 summarises the integrated data model as required to support the domains addressed in previous chapters†. Such a decomposition has three principal advantages as follows.

1. A problem can be composed of one or more domains. For example, initial simulations might focus only on the building domain in order to maximise the use of natural resources (e.g. natural ventilation, daylight, solar power etc). As required, an air flow network can be added to allow possible natural ventilation approaches to be explored in detail (e.g. night purge, atrium induced etc). Where necessary, a plant/control system might be added to allow HVAC system sizing and the evaluation of alternative approaches to control. At a later stage, a CFD domain might be established to ensure adequate indoor thermal comfort and air quality. This implies that any interface should be flexible enough to support the incremental and random order definition of domains.

2. Any single domain can have several levels of abstraction. For example, initially a building model might be established as a small number of representative zones, each attributed with standard occupancy and infiltration profiles. While such a model is easy to construct, it is nevertheless a powerful aid to decision-making in relation to form and fabric and issues such as overheating potential. Likewise, an ideal controller might be used to represent the design capabilities of an entire HVAC system, which is only explicitly defined at a later stage. In other words, the interface must support the substitution of abstractions as the performance appraisal issues evolve.

3. Portions of a domain can be selectively enhanced to accommodate special cases. For example, the discretisation level of a single multi-layered construction might be increased to support the assessment of condensation risk and mould growth potential, or an electrical network model added to a given zone to support the study of daylight responsive luminaire switching. In other words, the interface must support targeted improvements in modelling resolution in order to allow specialist simulations to proceed within the overall scheme.

† The specific data underlying this model is described elsewhere (ESRU 2001).

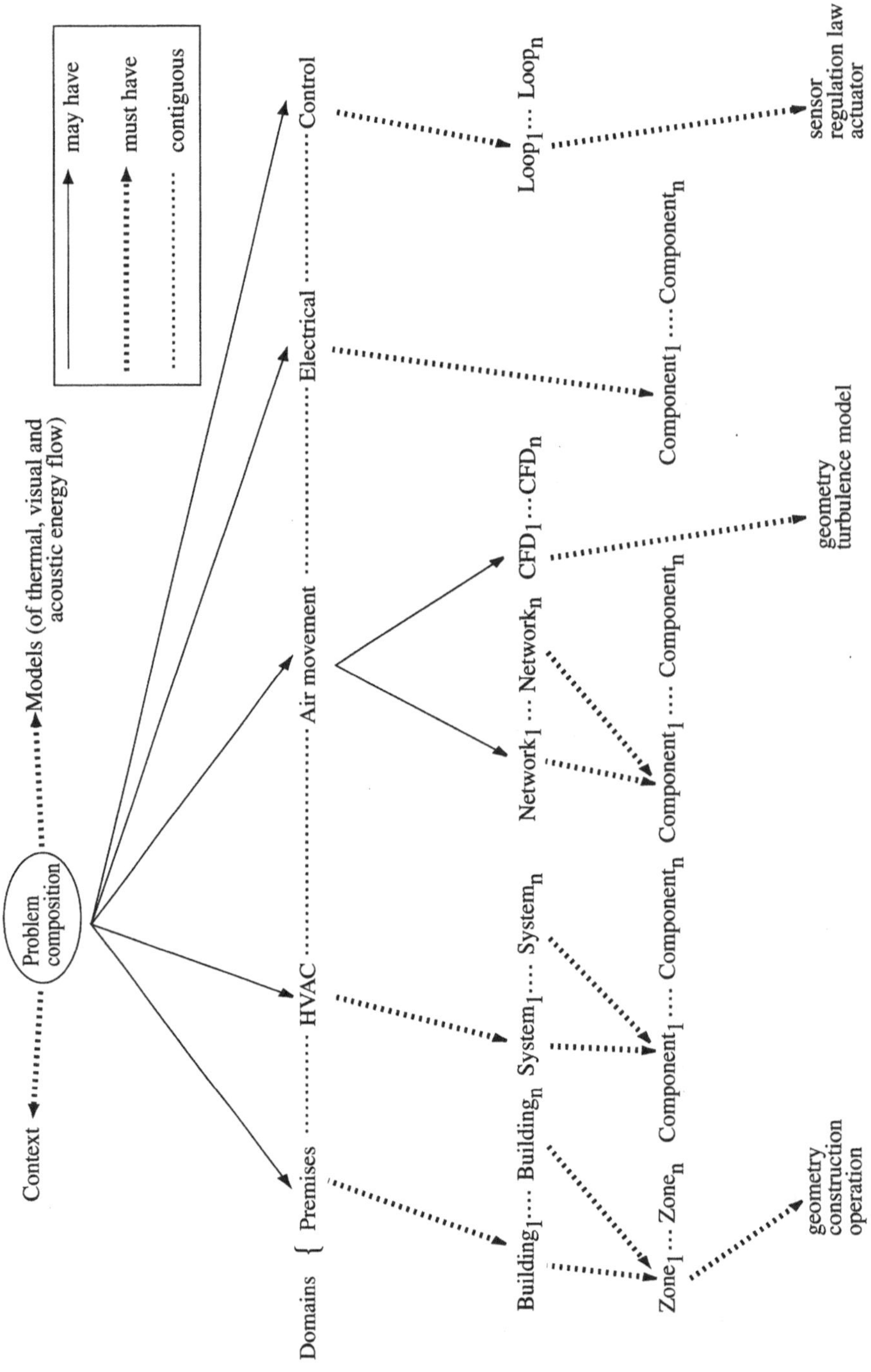

Figure 8.1: Integrated data model.

By arranging that the entities underlying each domain be contained within standard databases (e.g. material properties and plant/flow/electrical/control components) and ensuring that these entities are available at the point of domain definition, it is possible to establish an interface that embodies the above principles. The success of such an interface will depend on the user's ability to conceptualise a problem and co-ordinate the related data model as new aspects come into focus and old ones change. Such an interface (Hand 2000), based on the X Windows System open standard (http://www.x.org/) has been established as part of the ESP-r system and subjected to critical review within educational, commercial and research contexts. The versatility of the underlying data model has been demonstrated by its interfacing to several programs including TSBI3 (Johnsen 2001), RADIANCE (Larson and Shakespeare 1998) and AutoCAD (Autodesk 1989). Appendix G summarises the ESP-r system and gives details on accessing the source code which corresponds to the theories detailed in chapters 3 through 7 and includes the above mentioned user interface.

8.3 Performance assessment method

Within a simulation based approach, and as summarised in table 8.2 in relation to the ESP-r system, behaviour follows description (or, in other words, reward follows effort). This means that significant decision support can often be achieved for little input effort. It also means that more detail can be added to a model as the design hypothesis progresses and the complexity of the domain interactions grow. It is a paradox that simulation is at its most powerful when used in conjunction with uncertain data. This is because it can be used to explore the impact of expected ranges in the design parameters and so inform the decision-making process.

Table 8.2: Mapping of problem description to model behaviour.

Cumulative model description	*Typical behaviour enabled*
pre-existing databases	simple performance indicators (e.g. material behaviour);
+ geometry	visualisation, photomontage, shading, insolation etc;
+ constructional attribution	material quantities, embodied energy, time constants etc;
+ operational attribution	casual gains, electricity demands etc;
+ boundary conditions	photo-realistic imaging, illuminance distribution, no-systems thermal and visual comfort levels etc;
+ special materials	photovoltaics and switchable glazings evaluation etc;
+ control system	daylight utilisation, energy use, system response etc;
+ flow network	ventilation and heat recovery evaluation etc;
+ HVAC network	psychrometric analysis, component sizing etc;
+ CFD domain	indoor air quality, thermal comfort etc;
+ electrical power network	renewable energy integration, load control etc;
+ enhanced geometrical resolution	thermal bridging etc;
+ moisture network	local condensation, mould growth and health.

Consider the following scenario†, the purpose of which is to highlight the integrated appraisal process and, by implication, indicate the nature of a possible future design activity.

† This scenario employs the ESP-r system when its underlying data model is cumulatively refined according to the process of table 8.2.

A Project Manager module (Hand 1998) gives access to support databases, a simulation engine, performance assessment tools and a variety of third party applications for CABD, visualisation, report generation etc. Its function is to co-ordinate problem definition and give/receive the data model to/from the support applications. Most importantly, it supports an incremental evolution of the descriptive process and gives access to the simulation engine's corresponding functionality at each stage.

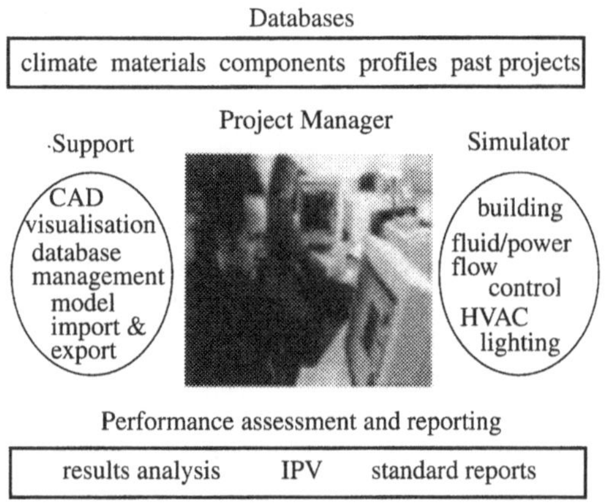

The starting point for a new project is to scrutinise and make ready the support databases. These include hygro-thermal, embodied energy and optical properties for construction elements and composites, typical occupancy profiles, pressure coefficient sets for use in problems involving air flow modelling, plant components for use in HVAC systems modelling, mould species data for use with predicted local surface conditions to assess the risk of mould growth, and climate collections representing different locations and severities.

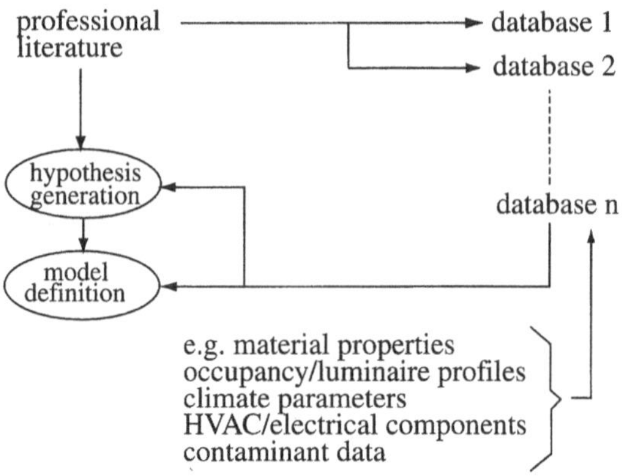

Embedded within such databases is knowledge that might usefully support the design conceptualisation process. As an example, the construction elements database will contain sets of fundamental hygro-thermo and optical properties for a range of construction materials, and derived properties from which behaviour may be deduced (e.g. the use of thermal diffusivity to characterise the rate of response or the thermal transmittance to characterise the rate of heat loss).

fundamental:
- conductivity
- density
- specific heat
- transmissivity
- reflectivity
- vapour diffusivity
- surface absorptivity
- surface emissivity

derived:
- thermal transmissivity
- thermal diffusivity
- thermal effusivity

Although the procedure for problem definition is largely a matter of personal preference, it is not uncommon to commence the process with the specification of a building's geometry using a CABD tool. ESP-r can interoperate with the AutoCAD (Autodesk 1989) and XZIP (Stearn 1993) programs, either of which can be used to create a building model of arbitrary complexity. This model can then be imported to the Project Manager where the attribution process is enabled.

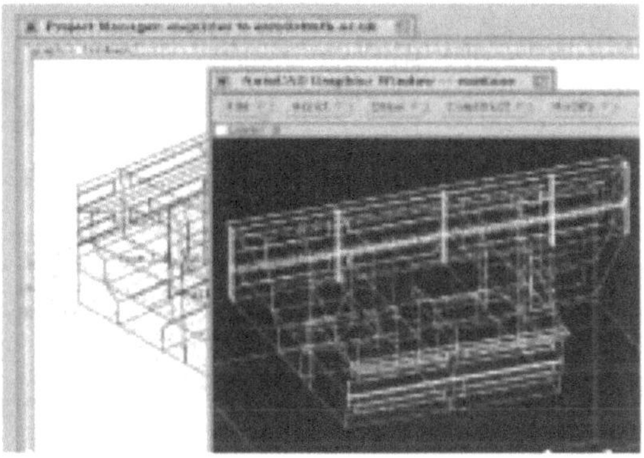

Simple wireline or false coloured images can then be generated as an aid to the communication of design intent or the study of solar/daylight access. The Project Manager provides wireframe photomontages via the Viewer system (Parkins 1977) and coloured, textured images via the Radiance system (Larson and Shakespeare 1998), automatically generating the required input models, driving these two applications and receiving back their outputs.

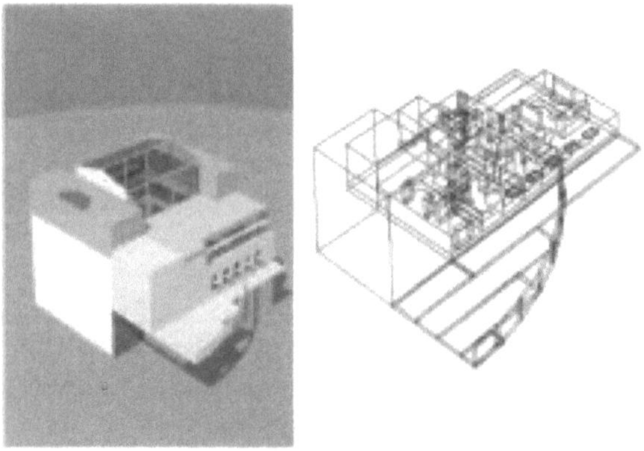

Constructional and operational attribution is now achieved by selecting products (e.g. wall constructions) and entities (e.g. occupancy profiles) from the support databases and associating these with the geometrical entities previously defined. It is at this stage that the simulation novice will appreciate the importance of a well conceived problem abstraction, which achieves an adequate resolution while minimising the number of entities requiring attribution, simulation processing and performance appraisal. Problem abstraction is an acquired skill that develops over time.

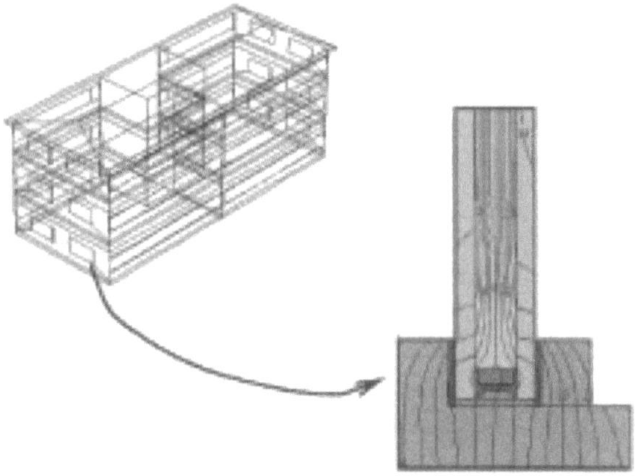

Temperature, wind, radiation and luminance boundary conditions of the required severity are now associated with the model to enable an appraisal of environmental performance—e.g. thermal and visual comfort levels throughout the year—and to gain an insight into the extent of any required remedial action. As appropriate, these boundary conditions can be modified to represent extreme weather events or micro-climate phenomena such as wind hollows and vegetative cooling.

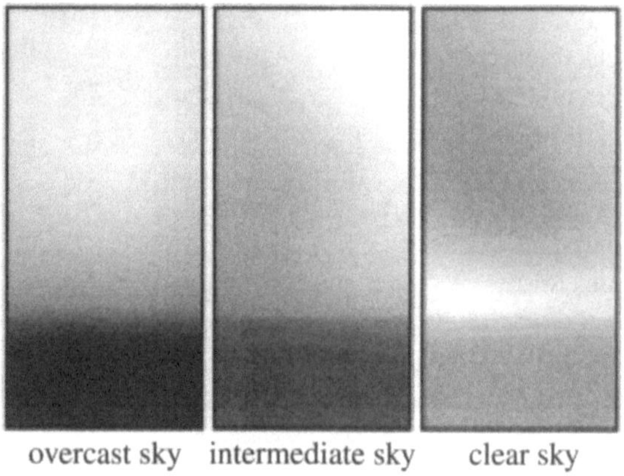

As required, geometrical, constructional or operational changes can be applied to the model in order to determine their impact on performance. For example, alternative constructional systems might be investigated, different occupancy loadings might be imposed, or different approaches to daylight utilisation assessed along with the extent and location of glare as shown here for the case of an office with added light shelf. The possibilities are limited only by the designer's imagination.

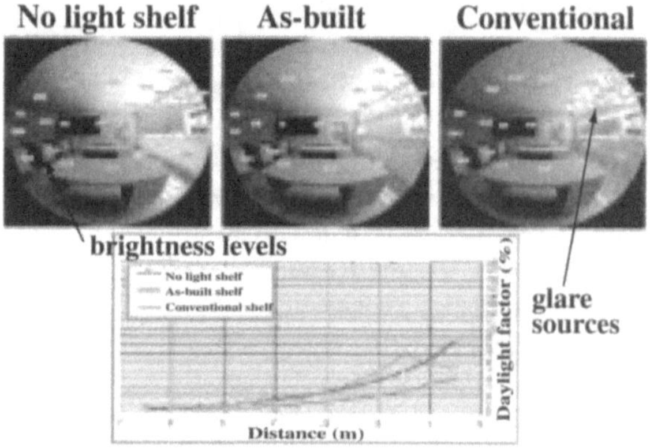

Special facade systems might now be considered: photovoltaic (PV) components to transform part of the solar power spectrum into electricity (and heat); transparent insulation to capture passively and process solar energy, or adaptable glazings (electro-, photo- or thermo-chromic) to control glare and/or solar gain. In each case, the contribution to improved environmental performance and reduced energy use can be determined. It is even possible to study ways to eliminate conflict as in the case of a PV facade reducing daylight penetration to the interior.

To access the energy displacement potential of this daylight, a luminaire control system might now be introduced, comprising one or more photocells linked to a circuit switch or luminaire dimming device. Subsequent simulations can then be undertaken to optimise the parameters of this control system in an attempt to maximise the displacement of the electricity required for lighting purposes. In this way, the conflict that exists between the beneficial aspects of daylight capture and the detrimental effects of the reduced heat gain from lights (on heating load) can be studied.

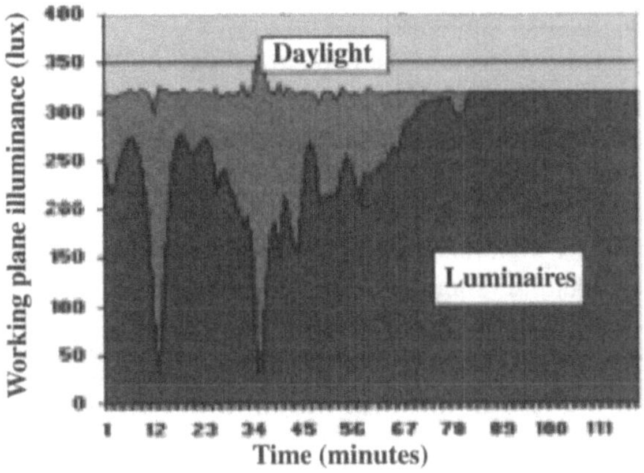

The issue of integrated environmental control can now be explored by establishing a control system conceived as a collection of open or closed loops. Some of these loops will dictate the availability of heating, cooling, ventilation, lighting etc, while others serve only to resolve conflict between these delivery systems. Previous aspects of the model may now be revisited in order to change the building's dynamic response to accommodate the intended control action.

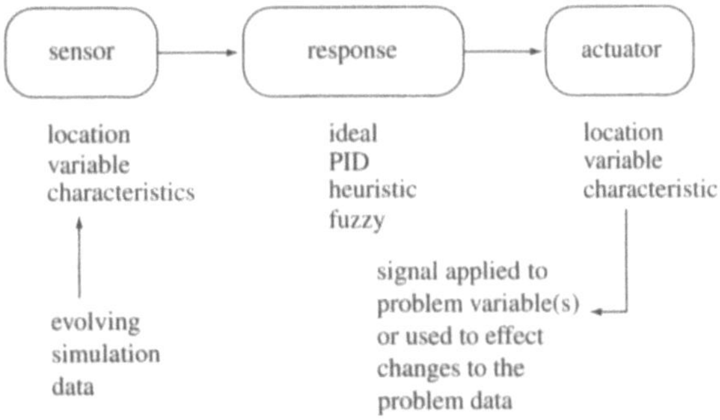

To study the feasibility of natural ventilation, a flow network can be associated with the building model so that the dynamic interactions are explicitly represented. The control definition may then be extended to include actions applied to the components of this network—e.g. to emulate window opening or flow damper control. This model can be used to examine the impact on air flow of alternative assumptions applied to the distributed leakage paths. In this way robust natural or mixed mode ventilation schemes may be designed.

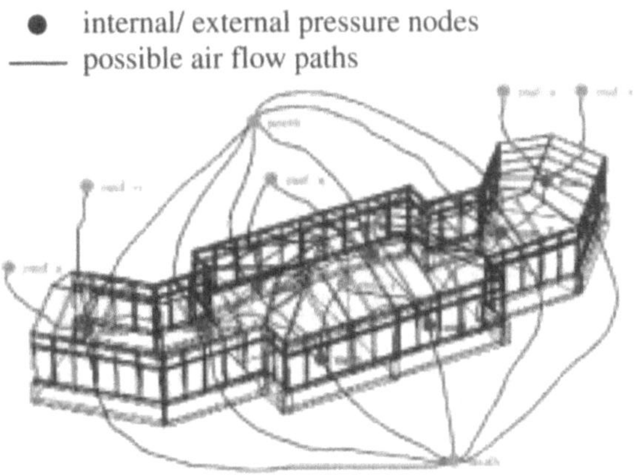

Where mechanical intervention is required, a component network can be defined to represent the HVAC system for association with both the building model and any active flow network. The control definition previously established may be further extended to provide internal component control and link the room states to the supply condition. Such a model can be used to study the operational characteristics of individual plant components as well as the behaviour of the overall system.

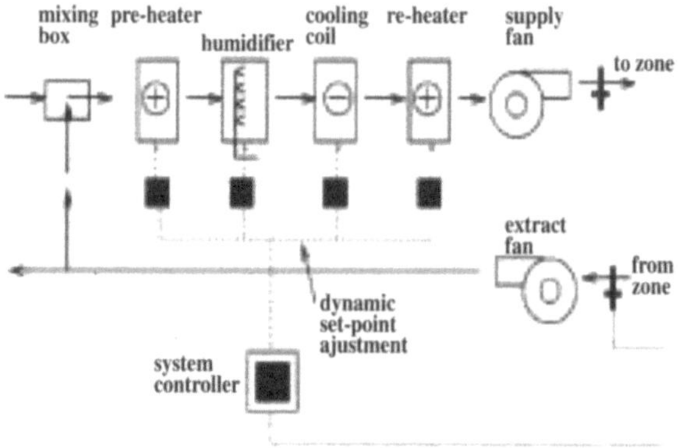

In order to examine indoor air quality, one or more spaces within the building model can be further discretised to enable the application of computational fluid dynamics in order to evaluate the intra-space air movement and the distribution of temperature, humidity and species concentration. These data may then be combined to determine the comfort levels and air quality at different points within the space. A useful determinant of indoor air quality is the distribution of the mean age of air.

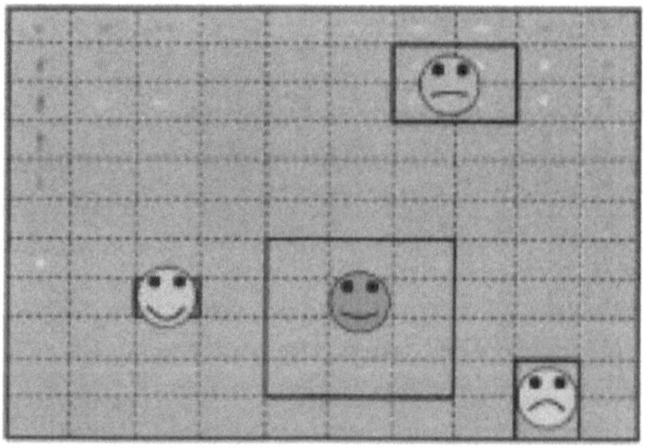

While the components of a model—the building, flow and HVAC networks, and the CFD domain—may be processed independently, it is usual to subject them to an integrated assessment whereby the dynamic interactions are explicitly represented. In the example shown here, a house model has been assigned a flow network to represent natural ventilation, an HVAC network to represent a ventilation heat recovery system, a CFD domain to enable the detailed analysis of air quality and a moisture flow model to allow an accurate assessment of humidity distribution.

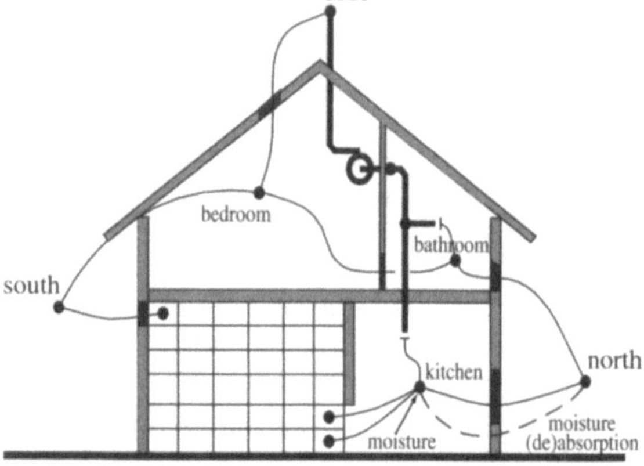

A further network might now be added to represent the building's electrical power circuits. This can be used in conjunction with the previously established models for facade-integrated photovoltaics, luminaire control, HVAC and flow networks to study scenarios for the local utilisation of the outputs from building-integrated renewable energy components, the co-operative switching with the public electricity supply and the shedding of load as an energy efficiency measure. Other technologies, such as combined heat and power plant and fuel cells, can also be assessed.

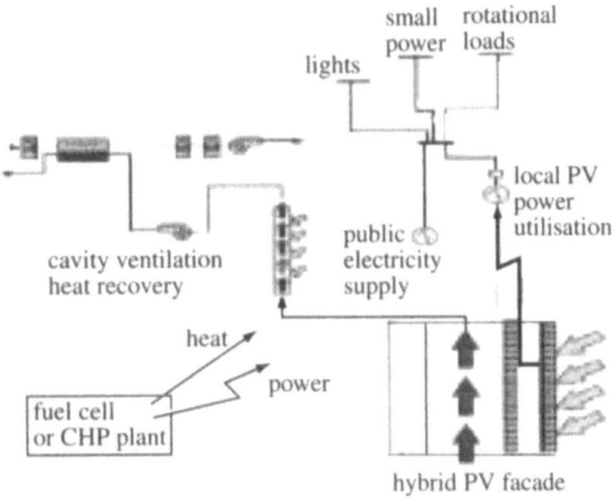

For specialist applications, the resolution of parts of the model can now be selectively enhanced to allow the detailed study of particular issues. For example, a portion of a multi-layered constructions might be finely discretised to enable the study of the behaviour of a thermal bridge or innovative building component. A moisture flow network might then be added to support an assessment of the potential for interstitial or surface condensation.

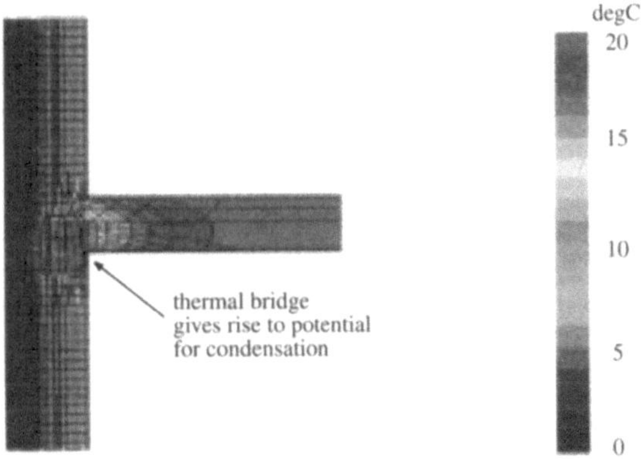

By associating the time series pairs of near surface temperature and relative humidity (to emerge from the integrated building, CFD and network air/moisture flow models) with the growth limit data as held in the mould species database, it is possible to determine the risk of mould growth and explore the different possible remedial actions—from eliminating moisture at source to modifying the constructional material or arrangement in order to prevent optimum growth conditions from occurring in the first place.

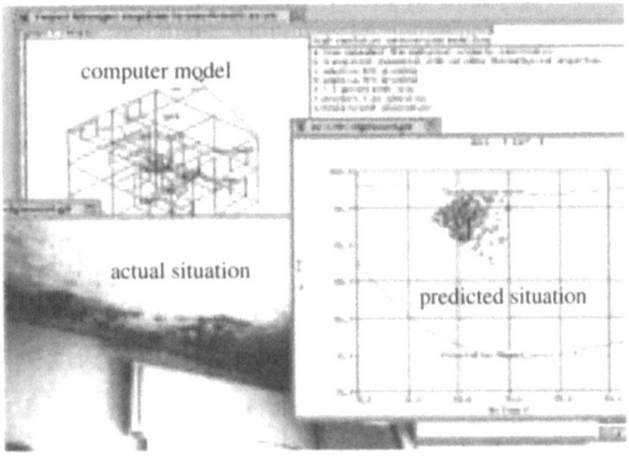

The Project Manager requires that a record be kept of the problem composition and to this end is able to store and manipulate text and images that document the problem and any special technical features. At any stage it is possible to bring together the results for the different aspects of performance and to present these in the form of an Integrated Performance View in order to summarise issues such as seasonal fuel use, environmental emissions, thermal/visual comfort, daylight utilisation, risk of condensation, renewable energy contribution etc.

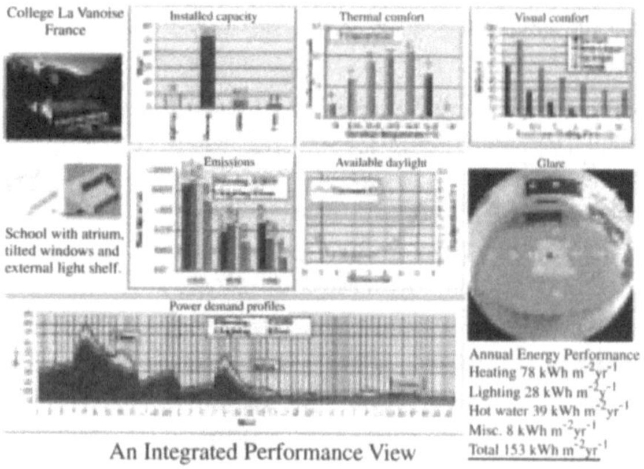

An Integrated Performance View

The core message is that any problem—from a single space with simple control and prescribed ventilation, to an entire building with systems, distributed control and enhanced resolutions—can be passed to the Simulator where its multi-variate performance is assessed and made available to inform the process of design evolution. By integrating the different technical domains, the approach supports the identification of trade-offs. This, in turn, nurtures sustainable approaches to building design and operation.

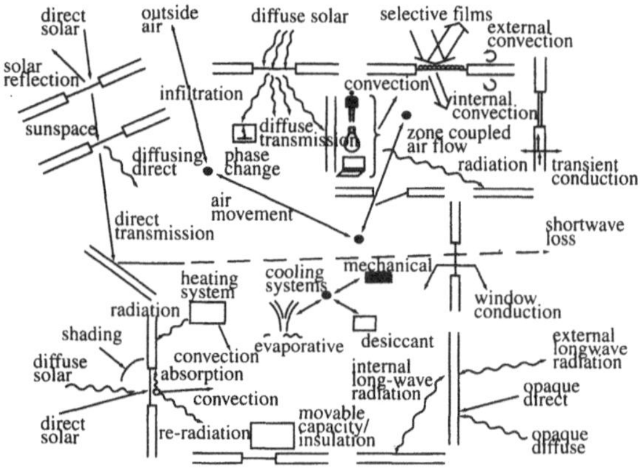

Importantly, integrated modelling supports team working because it provides a mechanism whereby the different professional viewpoints can come together and contribute equally to the eventual outcome. Moreover, given the electronic form of the underlying model, and the possibility of efficiently updating this model as the design hypothesis evolves, the different members of the team may operate from different locations and within different time zones. Such an interdisciplinary approach is likely to give rise to more sustainable solutions.

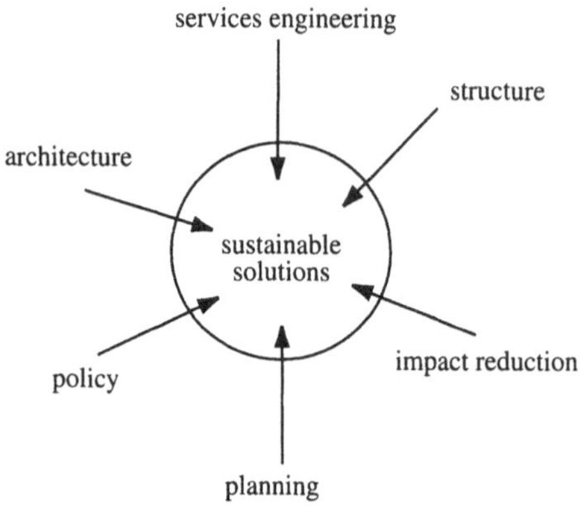

Citherlet (2001) has extended the foregoing integrated performance modelling approach in two significant respects: assessing the acoustic performance of indoor spaces and accounting for the energy use and environmental emissions throughout the building's life cycle. In the former respect, the reverberation time is used to characterise the sound quality within a space. In the latter respect, a life cycle impact assessment (LCIA) procedure has been added to the ESP-r system (Citherlet *et al* 2001). This supports the assessment of the energy use and environmental emissions relating to the manufacture, transport, assembly, maintenance and disposal of construction materials in addition to those associated with building environmental control. Four environmental impact indicators are used to quantify the overall impact: global warming potential, acidification potential, ozone depletion potential and the use of non-renewable energy. Such impacts may be directly estimated from the predicted energy demand given that suitable conversion factors are available. These factors will depend on the fuel types and mix, and on the energy conversion technologies used (Liscom 1984).

To support the LCIA procedure an environmental database has been constructed from a number of sources. This has a significant overhead in terms of the number of parameters required to define a multi-layer construction. The reason for this is illustrated in figure 8.2: an environmental impact assessment requires the complete description of the construction while the other modelling domains require only a sub-set.

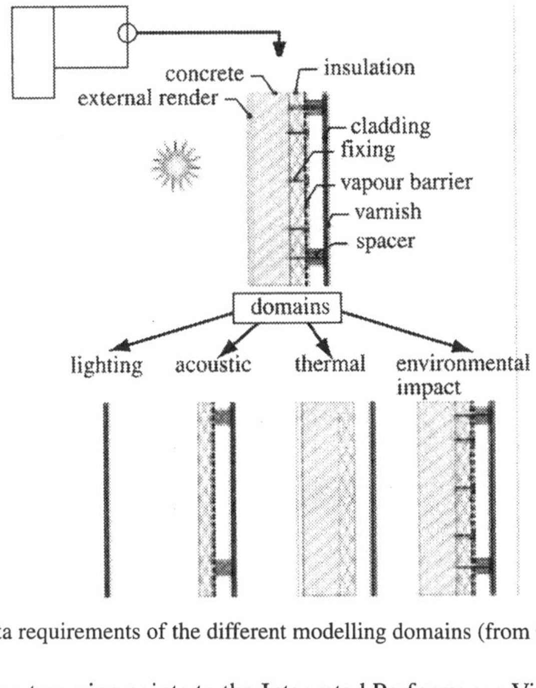

Figure 8.2: Data requirements of the different modelling domains (from Citherlet 2001).

The addition of these two viewpoints to the Integrated Performance View (IPV) concept provides useful support for decision makers concerned with sustainable development. Figure 8.3, for example, shows the outcome of a simulation of an office development in Lausanne, Switzerland. This building incorporates a number of passive features including an external light shelf and central atrium.

To achieve effective application in practice, a performance assessment method (PAM) is required to direct the user's line of inquiry. Table 8.3 shows the stages of a generic PAM in

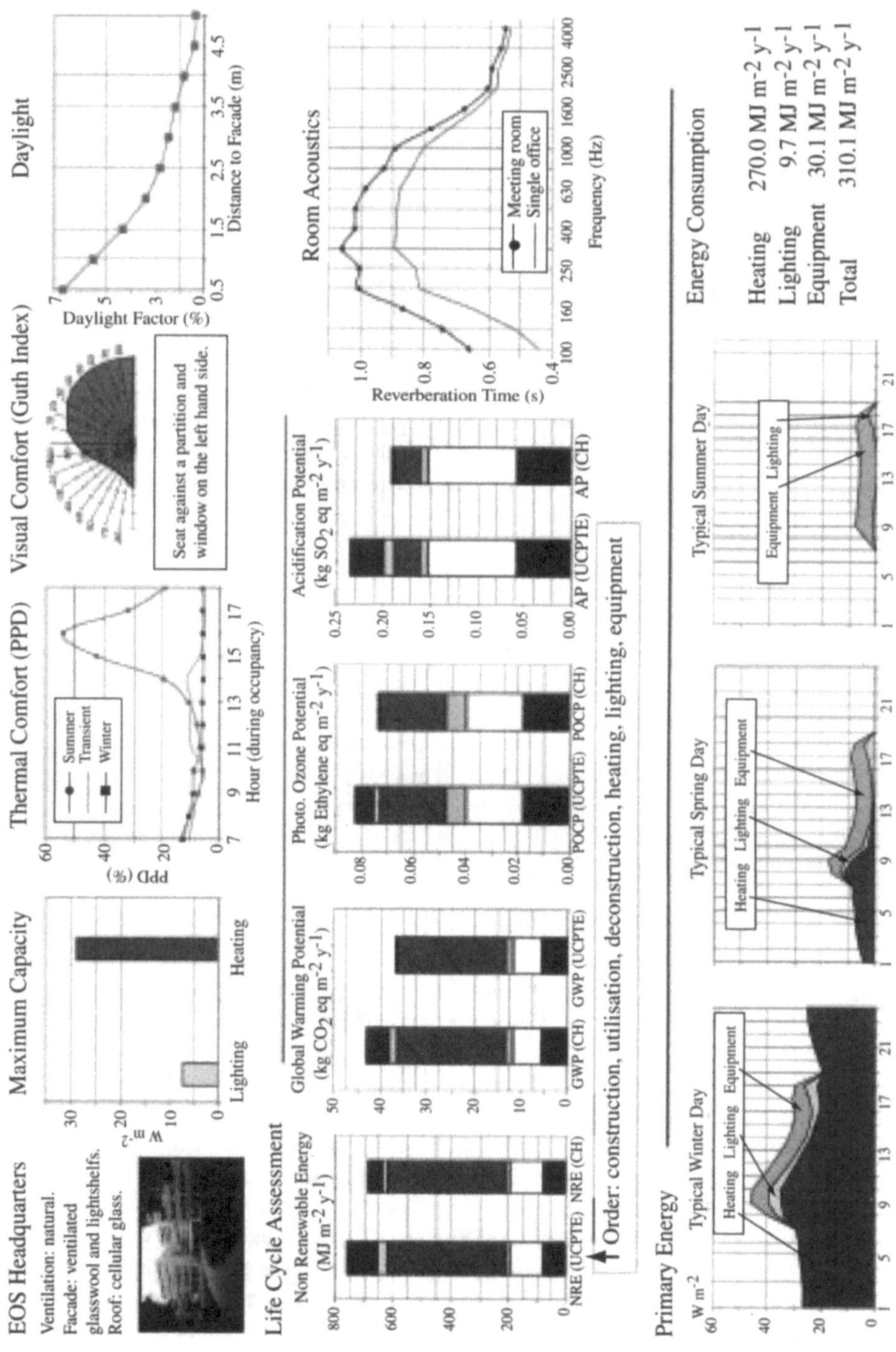

Figure 8.3: Example of an IPV for sustainable development support.

which the action required at each stage is underlined and the knowledge required to implement this action is shown in *italics*.

Table 8.3: A generic PAM for building energy simulation.

Stage	*Activity*
1	Establish a computer representation corresponding to a *base case design*.
2	Calibrate this model using *reliable techniques*.
3	Locate representative boundary conditions of *appropriate severity*.
4	Undertake integrated simulations using *suitable applications*.
5	Express multi-variate performance in terms of suitable *criteria*.
6	Identify problem areas as a function of *criteria acceptability*.
7	Analyse simulation results to identify *cause of problems*.
8	Postulate remedies by associating problem causes with *appropriate design options*.
9	For each postulate, establish a reference model to a *justifiable level of resolution*.
10	Iterate from step 4 until the overall performance is *satisfactory*.
11	Repeat from step 3 to establish replicability for other *weather conditions* (where applicable).

Such a PAM can be attributed with alternative knowledge instances depending on the user's viewpoint, the application topic(s) and the program's capabilities. To illustrate the approach, consider the embedding of renewable energy systems within the Lighthouse Building in Glasgow (Clarke *et al* 2000). This project employed the integrated modelling approach to determine the best possible match between energy demand and the local renewable energy resource without compromising power quality.

A base case model, compliant with best practice, was established and its multi-variate performance determined against relevant weather conditions. A number of energy efficiency measures were then applied to the model (initially independently and then, where warranted, jointly) to determine their potential to reduce energy demand and alter the demand profile to accommodate the integration of active renewable components.

Figure 8.4 shows the cumulative impact of several measures: advanced glazing, daylight responsive luminaire control, facade integrated transparent insulation, efficient lighting and dynamic heating set-point temperature control. When compared with the original design, these measures resulted in a 68% reduction in annual energy demand (corresponding to a 58% reduction in heating and an 80% reduction in lighting). More significantly, the final demand profiles were better matched to the output from locally deployed renewable energy systems: a photovoltaic (PV) component operating in hybrid mode to provide both power and heat; and ducted wind turbines (DWT) with an integral photovoltaic aerofoil section to increase the output power density. Also shown in figure 8.4 is the predicted power outputs from these two RE technologies superimposed on the most favourable demand profile.

As shown in figure 8.5, the hybrid PV component was subsequently incorporated within a south-facing facade of the Lighthouse building, while the DWTs were mounted on the south-

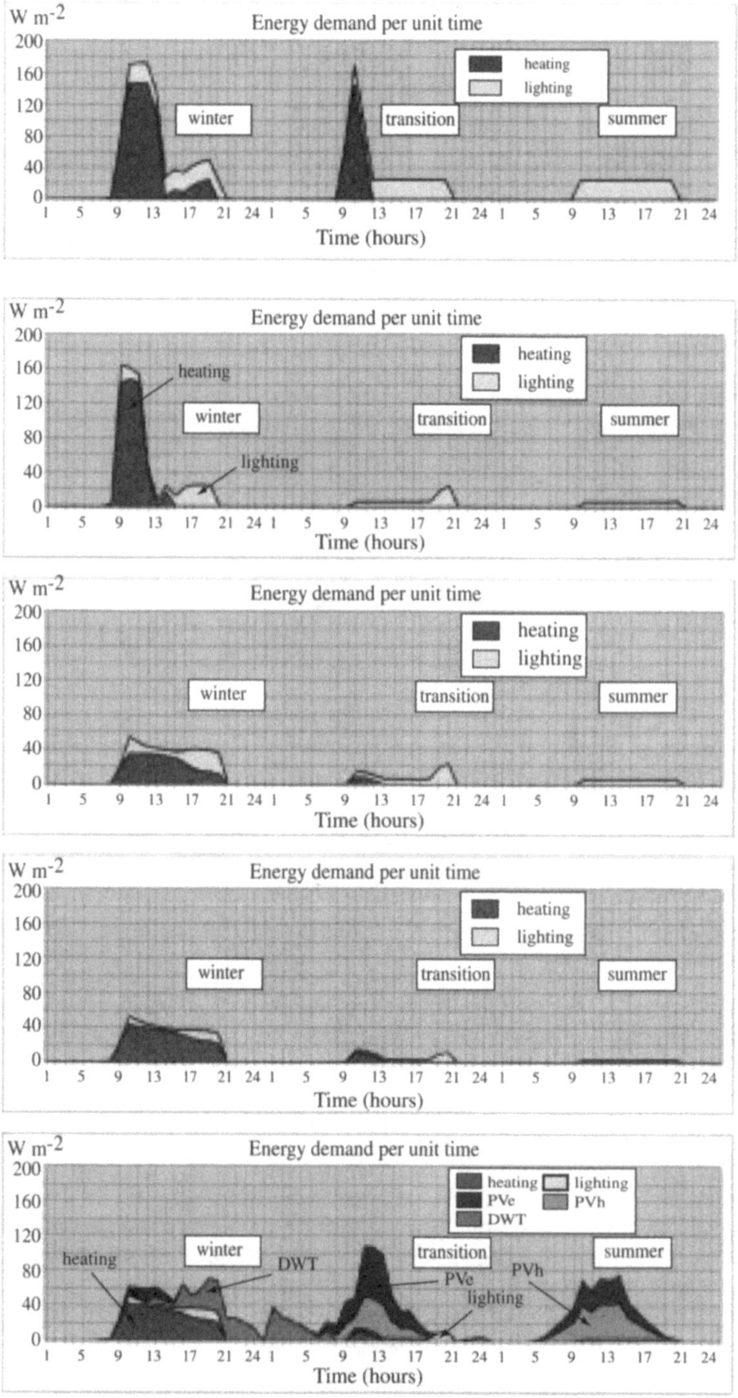

Base Case

Annual Energy Demands

Heating 118.3 kWh $m^{-2}y^{-1}$
Lighting 100.1 kWh $m^{-2}y^{-1}$
Total 218.4 kWh $m^{-2}y^{-1}$

As above
+ advanced glazing

Annual Energy Demands

Heating 64.5 kWh $m^{-2}y^{-1}$
Lighting 41.6 kWh $m^{-2}y^{-1}$
Total 106.1 kWh $m^{-2}y^{-1}$

As above
+ transparent insulation
+ daylight utilisation

Annual Energy Demands

Heating 38.2 kWh $m^{-2}y^{-1}$
Lighting 41.6 kWh $m^{-2}y^{-1}$
Total 79.8 kWh $m^{-2}y^{-1}$

As above
+ efficient lighting
+ responsive heating

Annual Energy Demands

Heating 49.0 kWh $m^{-2}y^{-1}$
Lighting 20.0 kWh $m^{-2}y^{-1}$
Total 69.0 kWh $m^{-2}y^{-1}$

As above
+ active renewables

Annual Energy Demands

Heating 49.0 kWh $m^{-2}y^{-1}$
Lighting 20.0 kWh $m^{-2}y^{-1}$
Total 69.0 kWh $m^{-2}y^{-1}$

DWT 25.0 kWh $m^{-2}y^{-1}$
PVe 33.8 kWh $m^{-2}y^{-1}$
PVh 41.0 kWh $m^{-2}y^{-1}$

Figure 8.4: Energy efficiency measures facilitating the introduction of renewable energy components.

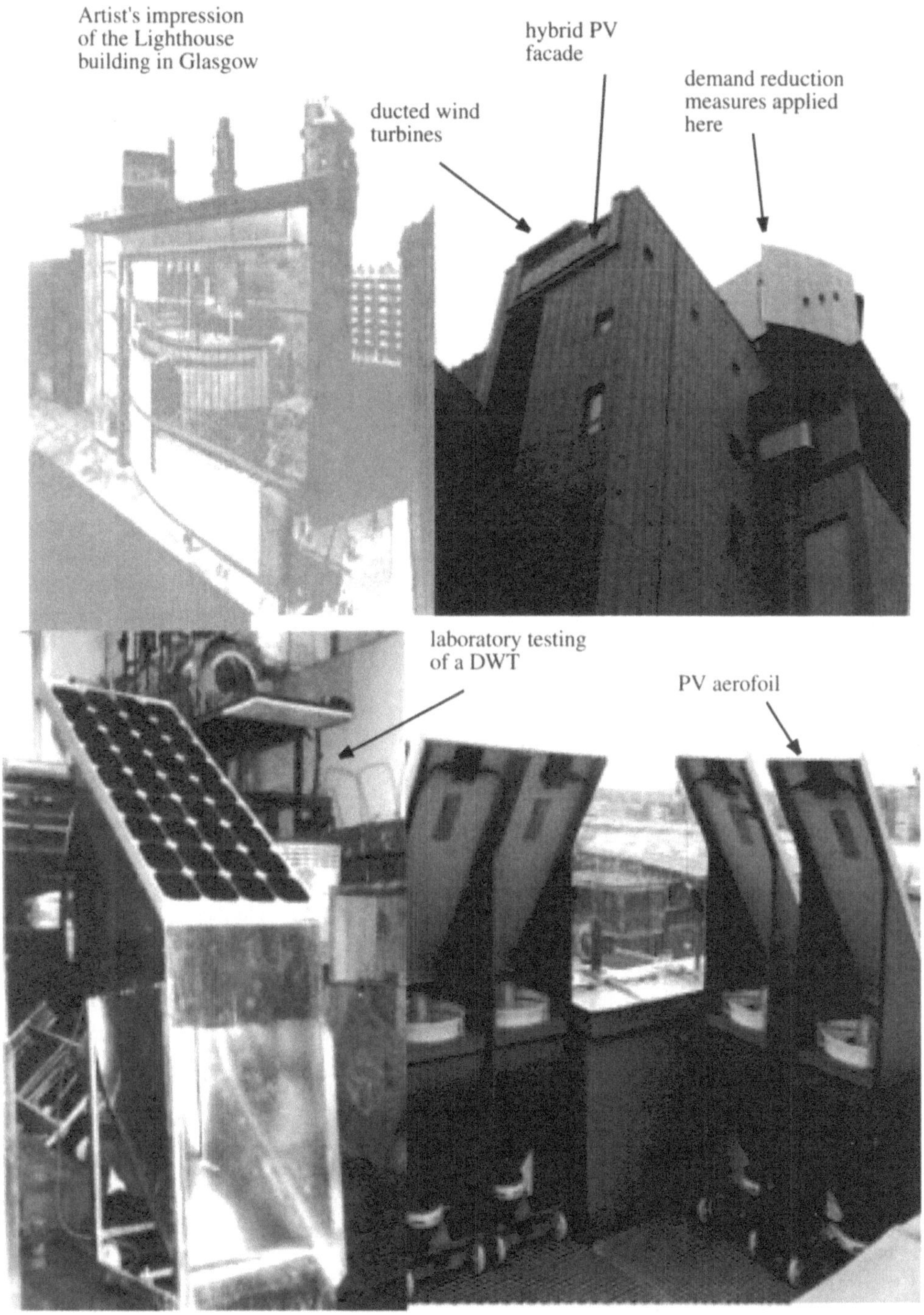

Figure 8.5: Renewable energy components as incorporated in the Lighthouse building in Glasgow.

and west-facing edges of the roof.

Section 8.7 gives a number of brief application examples to further illustrate the applicability of the integrated simulation approach as a comprehensive design and planning tool.

8.4 Uncertainty

Of the issues confronting design tool refinement, one of the most problematic is how to quantify and process the uncertainties that are inherent in the problem description and appraisal process. Such uncertainties occur in relation to the building's dimensional, constructional and operational specification, in relation to climate and simulation parameter definition, and in relation to the assumptions inherent in a program's internal algorithms. To make effective use of simulation, designers require information on performance robustness, in the light of these uncertainties, rather than performance quantification with no account taken of uncertainty (De Wit 2001, Macdonald *et al* 1999).

When using simulation, it is important to try to express the input parameters as confidence intervals so that the resulting outputs may be used to express performance in probabilistic terms. At the present time the treatment of uncertainty is left to a user's common sense. To exercise this common sense requires an appreciation of the issues involved.

Climate

As a stochastic system, the weather is inherently uncertain. It is therefore important to subject models to a range of boundary conditions in order to test the robustness of a given design hypothesis.

Micro-climate phenomena—such as wind hollows, urban canyons and vegetation related cooling—can also have a significant impact on performance. Either location specific data should be used or reference data modified to emulate the anticipated conditions.

Lighting

Atmospheric phenomena, especially clouds and pollution, will impact on the sky luminance distribution. The use of simulation to undertake realistic appraisals of the performance of daylight responsive luminaires, switchable glazings, daylight capture devices and the like, will require that a range of sky types be processed. Note however that changes in the indoor luminance of less than 30% will be imperceptible.

Indoor surface finishes, photocell response characteristics and lamp emissions each have an associated uncertainty which derives from the manufacturing process and calibration/maintenance considerations. Furniture will also modify daylight distribution and, because actual types and positions are largely unknown at simulation time, it is usual to assume optimistic and pessimistic scenarios in order to ensure an adequate lighting provision.

Glazing

The thermo-optical properties of glazings may exhibit a significant variation both within a given sample (e.g. from centre to edge) and between samples (e.g. uncertainty in convective heat exchange, particularly with solar shading devices).

Context factors—such as frame conduction, urban air pollution, window maintenance, shading device robustness etc—can have a significant effect on performance related properties. Where this is likely, a higher resolution should be adopted.

Form and Fabric

The translation from design intent to on-site realisation gives rise to many uncertainties in relation to final dimensions, construction composition and tightness, and fenestration operation. Each of these factors can impact significantly on the final performance of the design. In addition, the operation and use of a building is likely to change during its lifetime.

Contemporary concepts, such as breathable facades, PV-integrated facades, transparent insulation, switchable glazings etc, will increase the level of uncertainty inherent in the problem definition process and associated with the selection of the thermophysical properties of materials.

Ventilation

The quantification of a building's leakage distribution is an inherently uncertain task and there exists scant guidance on the likely ranges to be found in practice.

Likewise, the determination of surface pressure distribution is dependent on the ameliorative effect of local wind shelter phenomena that are the subject of considerable uncertainties. An otherwise sophisticated program may well provide results that are inaccurate and inapplicable. A sensitivity analysis, giving the variation of the results when input data are changed, is usually required.

Monitoring

All instruments have an intrinsic accuracy limit associated with their design and calibration. Further, an instrument's suitability of purpose is an important determinant of accuracy, e.g. the use of PRTs rather than thermocouples where accuracy of temperature measurement is an important consideration.

Errors are also associated with each of the steps following data acquisition: averaging, filtering and filling. These errors can accumulate so that the inherent uncertainties render the empirical data useless for program validation or calibration. Furthermore, monitoring campaigns must be comprehensive and detailed if resulting data are to be used for comparison with predicted data.

Occupant Interactions

The physiological and psychological processes that give rise to particular occupant responses to their environment are not well understood and few models exists for use in predicting how people interact with ventilation, lighting and heating/cooling systems.

The levels of heat and moisture production can vary significantly, both between individuals and as a function of the context.

Electrical Systems

All electrical systems are characterised by fundamental parameters (impedance, capacitance, inductance etc) that are affected by variations in temperature, moisture and demand. As a stochastic process, demand, in particular, gives rise to a significant uncertainty.

Many electrical systems—such as PV components, power electronics and rotary generators—are characterised by parameters that relate to standard test conditions (STC). As conditions depart from STC so the applicability of the parameter values, and hence the uncertainty, will grow.

Modelling

Many techniques exist for the modelling of the overall impact of uncertainty where the individual constituent uncertainties can be defined, and these techniques are beginning to be incorporated within contemporary modelling systems (Macdonald and Strachan 2001).

While procedures have been defined to allow the association of uncertainties with model input parameters, and to automate the determination of the effects of these uncertainties on the outputs, there is a paucity of data to assist users in defining the confidence limits associated with the input parameters, and of procedures for presenting the results to designers.

8.5 Large scale considerations

At the large scale, the problem is less to do with imminent depletion of energy resources and more to do with public acceptability of exploitation options. With energy demand growing each year, and the development of lower carbon energy technologies being supported by governments, integrated simulation provides a valuable tool for the scenario-based planning of new supply options and demand reduction measures. Simulation offers a practical way to plan a modern energy economy in which demand and supply are integrated at all levels.

The essential element in promoting the rational use of energy is that decision-makers be given access to relevant sources of information. These include the temporal and spatial patterns of energy demand, the potential impact of energy saving measures, and the characteristics of sources of energy supply, including renewables. It has been observed that the comprehensive picture needed to support informed planning decisions is elusive: fuel use data lacks resolution or is incomplete, and certain key information is commercially sensitive and so cannot be easily accessed. The contribution that modelling and simulation can make to fill these gaps is becoming increasingly apparent.

Consider the scheme of figure 8.6, which shows a decision-support system comprising three modules: a fuel use database to support the storage and analysis of historical fuel consumption data; an energy systems simulation package to appraise demand reduction measures and assess supply potentials; and a Geographical Information System (GIS) to support the integration of energy data with other technical and policy factors (such as site access and community proximity respectively).

Such a system would offer substantial new functionality:

Quantification of energy demand by fuel type, sector and time.
Ranking of alternative building energy efficiency measures.
Assessment of autonomous (e.g. photovoltaic facade) or grid connected (e.g. wind farm) renewable energy schemes from a standpoint of technical feasibility and planning acceptability.
Imposition of environmental and socio-economic impact considerations.
Correlation of energy supply with demand, on a spatial and temporal basis, for a single entity or at the institutional/regional scale.

Such systems are beginning to emerge (Clarke *et al* 1997, Evans 2001, Born 2001) and are seen as an aid to the local and regional planning/management process where objective, subjective, social and political considerations must be balanced. Essentially, the system links this process to the building design activity where specific decisions on energy conservation and renewable energy integration are taken.

Within such a system, simulation may be used in situations where energy demand data are

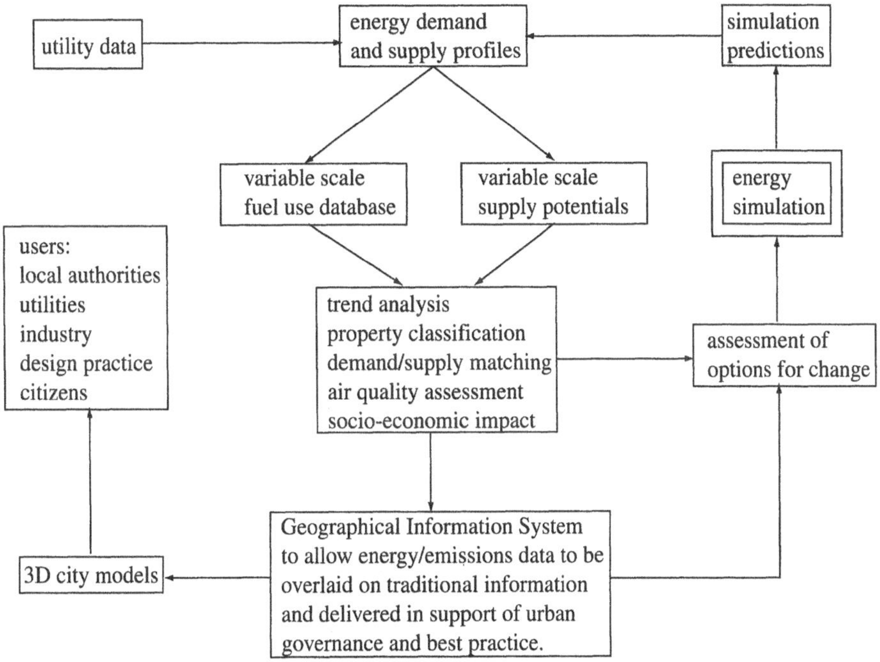

Figure 8.6: A decision-support system for fuel use management.

sparse to create 'virtual meters' that generate data at the required temporal and spatial resolution or to appraise specific options for change.

Such a system would support a scenario-based approach to planning and would provide an essential feedback loop between intervention intent and actual outcomes. In this way the process of sustainable development can be put on a rational basis. This, of course, would require significant changes to current work practices, not least the adoption of a 'digital city' approach to planning and a computational approach to intervention impact appraisal.

8.6 Support mechanisms

Many designers are reaching the stage where they wish to try simulation. What then? The following pragmatic advice is offered.

Research your options. There are many useful sources of information on program availability and application. The following three are particularly relevant.

The US Department of Energy's Web site <http://www.eren.doe.gov/buildings/tools_directory/> gives information on the many programs available from vendors and research groups worldwide. While there are many excellent commercial systems, it is important to realise that the best system in its class is sometimes part of the open source community and therefore available free of charge.

The CIBSE/BEPAC application manual, *Building Energy and Environmental Modelling* (ISBN 0 900953 85 3), provides an excellent introduction to the issues surrounding practical modelling.

The long established International Building Performance Simulation Association (IBPSA)

<http://www.mae.okstate.edu/ibpsa/> has regional affiliates who organise seminars and software demonstrations, publish support documents, arrange licence discounts and provide low cost training.

Be patient. A recent survey of program developers and, independently, of users (Crawley *et al* 1997) revealed that a program's capabilities were thought to be more important than the elegance of its interface. This finding is substantiated by the all too common occurrence of a well researched technical feature being accessible only via an obtuse interface.

Start simple. Acquire the software on a trial basis and apply it to familiar tasks—the assessment of overheating potential or winter heat loss under design weather conditions, for example. What's the point of turning on modelling features just because they're there, and getting hopelessly lost in the process. Rely on simplifying assumptions as you have always done. For example, use design air change rates rather than succumb to the temptation to use that fashionable, but pernicious, air flow prediction feature. By initially staying on familiar ground, your competence and confidence will grow apace, forming a solid foundation upon which to build.

Get support. The simulation of energy/environmental systems is part art, part science. No one person is an expert in all aspects, even the software authors! Team work is required. In some cases it may be possible to form an in-house team comprising individuals with complementary skills in the different technical and design aspects. Where this is not possible, strategic alliances with the vendor organisation or a local university might be appropriate. Some organisations are able to provide substantial support. For example, IBPSA Scotland <http://www.sesg.strath.ac.uk/> offers a Supported Technology Deployment service (McElroy and Clarke 1999) whereby in-house support is given by application specialists seconded to the design team for part of a project. Where required, a fully configured computer can also be made available. By gaining low-cost, risk-free access to modelling in the context of live projects and normal work practices, a company is better able to identify the financial and human resource barriers to routine program use. Further, by placing specialists within an active design team, a two way flow of information is supported: modelling know-how is acquired by practitioners, while the application specialists are exposed to the inadequacies of their software when applied to real design issues.

Get training. Following on from trial deployments, an informed decision on whether or not to acquire a program can be taken and, where the answer is yes, the level of staff training should be increased. This is a key point: a serious investment in software and training is only made after the benefits of the program has been demonstrated in a commercial setting. Where possible, training should be a mix of intensive short courses, literature review (e.g. Bartholomew *et al* 1997, Hand and Crawley 1997, Hensen *et al* 1997) and work related self study with application support. Remember that while the rewards from the successful adoption of a computational approach to design can be great, the pitfalls are many and varied. Take care, be cautious.

Pace your progress. After you have acquired the basic skills, which will happen sooner than you might think, you will be ready to delve deeper into the exciting and rewarding world of integrated performance simulation. This step is better taken in a focused and systematic manner rather than through a scattergun approach. Select a topic—air flow, control or renewable energy system modelling for example—and explore the program's related data requirements and analysis capabilities. If possible, attend a specialist course to ensure that you understand the technical details (a little knowledge is a dangerous thing). Within a short period you will be ready to try out your new skill on a real project. As long as you resist the temptation to over elaborate, and ensure that specialist help is to hand if needed, the outcome will be rewarding. In this way you will be able to incrementally develop your know-how in one topic area before moving on to another. The deployment of modelling in practice is less about software licencing costs and more about practical staff training and appropriate adjustments to current work practices. That this goal can best be pursued within a partnership between the design and vendor organisations should be seen as good news for the economic prospects of both.

Reflect on your learning achievements. Against the maxim that reflection helps consolidation, you should take every opportunity to discuss with colleagues your approach to modelling and the usefulness of the results you generate. The aim is to evolve a performance assessment method that fits well with your company's procedures, improves staff productivity and adds value to the product delivered to your client. Remember that different people will progress at different rates. Perhaps your destiny is to manage the modelling process rather than push the buttons yourself. The industry would benefit immensely from a larger pool of experienced managers.

Now, let's jump forward a few years. You've served your time and mastered the art of simulation (or at least some parts of it). You are able to provide others with relevant information on issues such as energy use, environmental impact, component performance, comfort levels, indoor air quality and the like and so help to improve the overall performance of the related product, process or system. Significantly, you are able to do this within the real time, real scale, resource constrained context of design practice. Can you think of a more apt contribution to sustainable development?

In the longer term, when computing power permits, simulation might well give rise to an exciting new prospect: the design of energy systems in real time through the on-screen manipulation of component parts. Such a virtual design environment (as elaborated in §9.2) would enable the rapid prototyping of solutions that strike the correct balance between product performance and sustainable development. Imagine that!

8.7 Example applications

It is clear then that the integrated simulation approach both deepens and broadens the appraisal capability of practitioners. Consider the following short descriptions of typical applications of integrated simulation (IS). Additional examples of the application of the IS approach are available as on-line exemplars within the ESP-r system.

Resource allocation

Many government agencies own housing which dates from the middle of the last century and much of this stock has fallen due for upgrading. IS has been used to help establish the most cost-beneficial upgrading strategy.

A sample of houses in any estate are simulated, firstly in their original form, and subsequently with a range of design modifications formulated on the basis of the results to emerge from the initial simulations. Typically, the analyses will focus on issues relating to construction, fenestration, ventilation and condensation.

In one case the major heat flowpath was identified as being through the suspended timber floor. This occurred in a refurbishment programme for which substantial wall insulation was proposed. As a result the upgrading proposal was reformulated and the client's investment put to better use.

Building conversion

IS has been used by local authorities to appraise options for the conversion of existing buildings. one study entailed an investigation of diversity of heating demand as affected by a variety of proposed zoning strategies, plant operating schedules and building modifications. It was observed that the total cost incurred in the modelling exercise was approximately half the cost incurred in a parallel manual exercise which involved only the straightforward calculation of individual zone peak heating loads under steady state conditions.

The graphical presentation of results was also considered invaluable as a means of conveying information to the design team. See Foley and Van Dam (1982) for a detailed discussion on the role of interactive graphics in design.

Passive solar architecture

In an attempt to achieve energy savings, practitioners are increasingly attempting to harness solar energy as a supplementary and renewable fuel source. The technical problems inherent in such passive solar design are often considerable and require sophisticated modelling methods. IS provides a solution to this problem.

Simulation can be used to incorporate passive solar elements such as direct gain systems, sunspaces and Trombe-Michel walls within a design. Other devices—such as shading control, movable insulation and advanced glazings—can then be evaluated and optimised (e.g. Clarke *et al* 1998).

In one case an architectural practice was commissioned to examine the performance of sunspaces when applied to UK houses. A simulation model was established in which thermostatically controlled vents operated to transmit energy from the sunspace to the house. When overheating occurred, the sunspace was vented to ambient. The passive solar design was simulated and the results compared with an identical simulation applied to the original house. It was shown that the sunspace would reduce the annual energy demand by ~30%.

Innovation in design

IS is particularly well suited to design innovation. One application involved the design of a solar wall construction as part of a new laboratory complex.

The need to move large quantities of air had suggested to the design team a scheme in which this air could be used to capture solar energy by passing it over the south-facing facade when contained within an outer glass skin. This 'solar wall' was simulated in order to determine the heat recovery potential. The modelling task was complicated by the inclusion of ducts within the air space which caused facade shading and convective heat pick-up.

Load diversity

The conversion of a building to house a computer facility and ancillary activities was already well advanced when the opportunity arose to use IS. The decision to use simulation resulted from a need to alter the building envelope from a lightweight cladding system to brickwork and increase the glazing from 25% to 40%. The effect of these changes on the air conditioning system had, as a matter of urgency, to be determined.

Among other things, the simulation results showed that the cooling loads in different areas occurred at significantly different times of the day. While the individual VAV terminal box duties had to be increased, diversity considerations meant that no significant modifications were required to be made to the central plant.

Atrium air flow

Accompanying improved living standards is an increased demand for leisure and shopping facilities. Many of the proposed designs incorporate atria. Because of the pressure and temperature induced air movement, the prediction of likely environmental conditions will require the simultaneous treatment of heat and air flow.

In one application the intention was to create a large shopping, office and recreation complex within an outer glass construction. A model was constructed and simulations undertaken to assist the design team in relation to issues such as overheating, shading control, smoke extraction, condensation prevention and local comfort maintenance.

Electrical storage heaters

In some applications it is necessary to explicitly model electrical systems. In one project fan-assisted electrical storage units were to be incorporated within a building conversion to small workshop units. A computer model of the building and storage units was established and used to investigate issues such as local comfort, unit control, charge/discharge characteristics, weather anticipation and running cost. It was shown that the units would only perform well if fitted with relatively sophisticated control of heat output.

Parametric studies

A parametric analysis of building performance can lead to the identification of effective energy saving measures. In one case IS was used to study the cost-benefit associated with different window designs when placed in Scandinavian and UK climates. Simulations were performed for combinations of orientation, window size, window type and building structural mass. The results identified those design parameters that were best suited to the two climates.

Plant simulation

Simultaneous building and plant modelling supports the study of HVAC design and control. In one case the design team wished to compare a conventional air conditioning design with a system based on displacement ventilation using a sub-floor plenum. Issues to be studied included the effects on energy consumption and comfort conditions of a number of alternative control options.

Detailed models were established for both systems and combined with a multi-zone building model. The models were then subjected to various control regimes including intermittent and continuous plant operation, solar reset of the perimeter heating, and the night purging of the plenum with outside air in anticipation of the next day's cooling demand. In each case the impact on central plant energy consumption was determined and assessed relative to the level of comfort achieved in the occupied zones.

Future possibilities

With the proliferation of powerful, low cost computing, it is possible to incorporate IS within a building's control system. Weather data and spatial requirements are fed to an explicit building/plant model which then anticipates the consequences of any proposed control action. In this way the control system is endowed with a predictive capability and so is able to take corrective action in advance.

As described in §8.3, IS is uniquely able to support the deployment of micro power systems within the built environment as an apt response to the challenge of sustainable development.

8.8 References and further reading

Autodesk Ltd 1989 *AutoCAD Release 10 Reference Manual* (Exeter: Autodesk Ltd)

Bartholomew D, Hand J, Irving S, Lomas K, McElroy L B, Parand F, Robinson D and Strachan P 1997 An Application Manual for Building Energy and Environmental Modelling *Proc. Building Simulation '97* (Prague)

Bloomfield D P 1988 An Investigation into Analytical and Empirical Validation Techniques for Dynamic Thermal Models of Buildings *EPSRC Project Final Report* (Garston: Building Research Establishment)

Bloomfield D P 1999 BRE Office Empirical Validation Study *Reports v2-breoff-5a, v2-bre18a and v2-brelisses-3a* (Garston: Building Research Establishment)

BRE 1984 Glasgow Commercial Building Monitoring Project *Final Report for IEA Annex 4* (Garston: Building Research Establishment)

Born F 2001 Aiding Renewable Energy Integration through Complementary Demand-Supply Matching *PhD Thesis* (Glasgow: University of Strathclyde)

Citherlet S 2001 Towards the Holistic Assessment of Building Performance Based on an Integrated Simulation Approach *PhD Thesis* (Lausanne: Swiss Federal Institute of Technology)

Citherlet S, Clarke J A and Hand J 2001 Integration in Building Physics Simulation *Energy and Buildings* **33**(5) 451-61

Clarke J A, Evans M S, Grant A D, Kelly N (1997) Simulation Tools for the Exploitation of Renewable Energy in the Built Environment: The EnTrack-GIS System *Proc. Building Simulation '97* (Prague)

Clarke J A, Janak M and Ruyssevelt P 1998 Assessing the Overall Performance of Advanced Glazing Systems *Solar Energy* **63**(4) 231-41

Clarke J A, Johnstone C M, Macdonald I A, French P, Gillan S, Glover C, Tatton D, Devlin J and Mann R 2000 The Deployment of Photovoltaic Components Within the Lighthouse Building in Glasgow *Proc. 16th European Photovoltaic Solar Energy Conference* (Glasgow)

Crawley D B, Lawrie L K, Winkelmann F C, Buhl W F, Erdem A E, Pedersen C O, Liesen R J and Fisher D E 1997 What Next for Building Energy Simulation—A Glimpse of the Future *Proc. Building Simulation '97* (Prague)

De Wit S de 2001 Uncertainty in Predictions of Thermal Comfort in Buildings *PhD Thesis* (Technische Universiteit Delft)

DOE 1981 IEA Annex 1: Comparison of Load Determination Methodologies for Building Analysis Programs *Report DOE/CE/20184-1* (Washington: Department of Energy)

Dupagne A 1983 Comparison of Passive Solar Design Methods *Proc. Conf. on Design Methods for Passive Solar Buildings* (London: International Solar Energy Society)

Evans M 2001 Energy Management in Diverse Estates *PhD Thesis* (Glasgow: University of Strathclyde)

ESRU 2001 Data Requirements of the ESP-r System *ESRU Technical Report* (Glasgow: University of Strathclyde)

Foley J D and Van Dam A 1982 *Fundamentals of Interactive Computer Graphics* (New York: Addison-Wesley)

Hand 1998 Removing Barriers to the Use of Simulation in the Building Design Professions *PhD Thesis* (Glasgow: Department of Mechanical Engineering, University of Strathclyde)

Hand J W and Crawley D B 1997 Forget the Tool When Training New Simulation Users *Proc. Building Simulation '97* (Prague)

Hensen J, Janak M, Kaloyanov N and Rutten P 1997 Introducing IT Based Energy Simulation Courses in Central/Eastern Europe *Proc. Building Simulation '97* (Prague)

Jensen S O (Ed) 1994 Validation of Building Energy Simulation Programs *Final Report, PASSYS Model Validation Subgroup* EUR 15115 EN (European Commission)

Johnsen K 2001 *TSBI3 User Manual* (Danish Building Research Institute)

Judkoff R and Neymark J 1995 International Energy Agency Building Energy Simulation Test (BESTEST) and Diagnostic Method *Report TP-472-6231 (Goldev, CO.: National Renewable Energy Laboratory)*

Larson G W and Shakespeare R 1998 *Rendering with Radiance - The Art and Science of Lighting Visualization* (San Francisco: Morgan Kaufmann)

Lebrun J and Liebecq G 1988 Annex 10 Systems Simulation Synthesis Report *Report AN10 881020-RF* (University of Liege)

Liscom W 1984 *Energy Data Conversion Handbook: How to Combine and Compare International Energy Data* (New York: McGraw-Hill)

Lomas K J and Eppel H 1992 Sensitivity Analysis Techniques for Building Thermal Simulation Programmes *Energy and Buildings*, **19** 21-44

Lomas K J, Eppel H, Martin C and Bloomfield D 1994 Empirical Validation of Thermal Building Simulation Programs Using Test Room Data *Final Report for IEA Annex 21/ Task 12* (Leicester: De Montfort University)

Macdonald I A, Clarke J A and Strachan P A 1999 Assessing Uncertainty in Building Simulation *Proc. Building Simulation '99* (Kyoto)

Macdonald I A and Strachan P A 2001 Practical Application of Uncertainty Analysis *Energy and Buildings* **33**(3) 219-27

McElroy L B and Clarke J A 1999 Embedding Simulation Within Energy Sector Businesses *Proc. Building Simulation '99* (Kyoto)

Morck O C (ed) 1986 Simulation Model Validation Using Test Cell Data *Report for IEA Task 8* (Lyngby: Technical University of Denmark)

Parkins R P 1977 VIEWER Users Manual (Glasgow: Department of Architecture, University of Strathclyde)

Stearn D 1993 XZIP Users Manual (Glasgow: Department of Architecture, University of Strathclyde)

Strachan P A 2000 ESP-r: Summary of Validation Studies Technical Report (Glasgow, University of Strathclyde, Energy Systems Research Unit)

9

Future trends

One trend is clear: as the technology is more widely applied, the demands on simulation programs will grow. While this is welcome, in that demand fuels development, it is also problematic because the underlying issues are complex. Although contemporary programs are equipped to deliver an impressive array of performance assessments, there are barriers to their routine use in practice, not least the absence of a standard building product model whereby inter-program operability can be supported. Figure 9.1 summarises the three principal devices available to overcome these barriers: technology transfer via specialists, intelligent user interfaces and software engineering constructs to enable energy simulation functionality to be encapsulated within other applications such as CABD. The previous chapter gave summary details on the first device, while this chapter considers the last two.

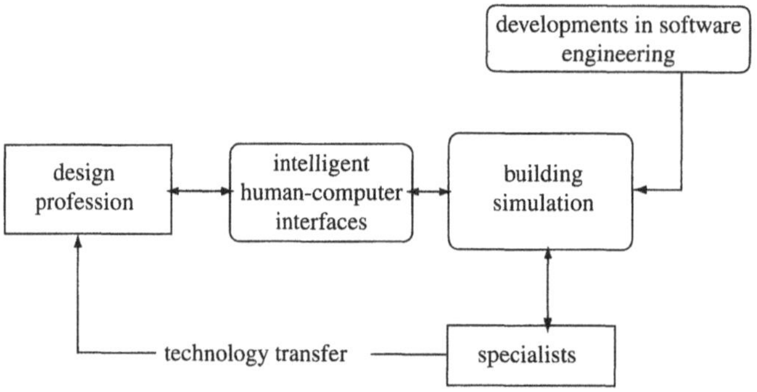

Figure 9.1: Routes to advanced simulation.

9.1 Design process integration

To bring real benefit, building performance modelling must be fully integrated within the design process—see figure 1.1b. In such a computer-supported design environment (CSDE), the practitioner evolves the design hypothesis in such a way that the tools are able to automatically access the data describing the design and give feedback on all aspects of performance and cost in meaningful terms. The attainment of a CSDE is a non-trivial task involving the development of *integrated product models* and *intelligent interfaces*. In the case of integrated product models, the complexity stems from the temporal aspects of the design process, the diversity of the applications to be served, and the different professional viewpoints and vocabularies. In the case of intelligent interface the complexity stems from the nature of the transactions (human-to-human, human-to-application and application-to-application) that require to be managed.

9.1.1 Integrated product models

Bjork (1991) defines an integrated product model (IPM) as the "description of a product, capable of structuring all the information necessary for the design, manufacturing and use of that product". By this definition, and in the context of a building, an IPM must be comprehensive to accommodate the needs of the different technical domains as each evolve throughout the building's life cycle. No such IPM exists at the present time. There are two principal factors that have contributed to its absence. First, a generic building IPM is necessarily extensive because it must encapsulate many aspects: geometry, construction, operation, systems, finance, performance, environmental impact, disassembly and so on. As a consequence, it has proved difficult to obtain an overall representation consensus. For example, while STEP (Standards for the Exchange of Product Data; ISO 1989) Part 42 provides a consistent approach to the representation of geometry, no similar approach has yet emerged for material hygro-thermal, visual and acoustic properties or for aspects relating to cost and performance. Second, the maintenance of IPM integrity can become intractable where the transactions are concurrent, where the problem domain, as here, is large, or where there is a high level of data redundancy (e.g. as the design progresses from concept to detail design).

Citherlet (2001) has suggested the building life cycle decomposition as shown in Table 9.1.

Table 9.1: Phases of the building life cycle.

Major phase	*Elementary phase*
Material production	extraction transport manufacture
Building construction	transport of material to site assembly
Building use	operation maintenance replacements
Building disassembly	deconstruction disposal/ re-use management transport to disposal site

To achieve a robust IPM that is capable of accommodating such a range of activities will require substantial further developments. Such developments are underway at the present time. The International Standards Organisation is developing standards for the exchange of product data within all industry sectors, including construction (ISO 1994), while the International Alliance for Interoperability is developing International Foundation Classes that relate directly to building simulation (Bazjanec and Crawley 1997).

Within the EC's COMBINE project (Augenbroe 1992), the objective was to evolve an IPM that could satisfy the needs of a co-operating set of design tools for CAD, layout planning, energy/lighting analysis and regulations compliance checking. The IPM, or strictly speaking its computer implemented counterpart termed the Data Exchange System (DES), is then able to receive from, store and deliver data to these design tools in a manner which ensures that the data are semantically unambiguous. The key issue was the structure of the IPM to ensure efficient exchange and allow future extension as additional or more powerful design tools become available. The decision to base the DES implementation on the object-oriented paradigm and to contain the product model within an object-oriented database was seen as the best way to achieve these goals.

The approach adopted within COMBINE was to firstly capture the aspect models of the individual applications and then to merge these to form the IPM. These aspect models were constructed using an ATLIAM graphic representation (similar to a NIAM model with extensions; Nijssen and Halpin 1989), with a mapping to the EXPRESS language (Spiby 1991) to support DES to application transactions. In relation to the ESP-r system, for example, this gave rise to 54 ATLIAM diagrams of which 48 related to building description and 6 to building performance (Clarke *et al* 1995).

Of the difficulties encountered when constructing the ESP-r aspect model, the representation of performance proved to be the most challenging. An initial attempt led to a simple two level representation: the first encapsulating the time-step temperatures, flows and fluxes relating to any finite volume within the discretised domain; the second holding derived data corresponding to commonly requested information (e.g. on energy consumption, installed plant capacities, overheating risk and indoor air quality).

Although these two entities accurately describe the data that can be obtained from ESP-r, they do not directly lead to a performance assessment because they must firstly be interpreted. This consideration gives rise to an alternative approach. First, the data that may be considered to underlie the answer to a particular performance evaluation is encapsulated. Second, methods that serve to integrate these data and compare the result to appropriate benchmarks are embedded. Third, a meta-entity is created to weight the outputs from the different performance entities that comprise a given integrated performance view (§8.3).

To elaborate the approach, consider the case of an overheating risk assessment for which the following data and methods would be required. The first step is to rank order the occupied zones within the building according to their peak operative temperature over some period. This allows problematic zones to be located. For these zones a profile of the operative temperature is constructed over some suitable period. This allows the severity of the problem to be quantified. Where unacceptable, an energy balance is formed. This allows the cause of the problem to be determined. For known causes, possible remedial measures are suggested.

After several such performance assessments have been completed—for comfort, energy consumption, environmental emissions, indoor air quality etc—the next stage would be to weight them. Such a weighting might be problem specific, e.g. where sustainable development considerations mitigate against energy use and environmental emissions, or they might be context specific, e.g. where overheating is severe, its alleviation might be achieved (high weighting factor) even if at the expense of increased energy use. Such considerations may be expected to appear in next generation simulation programs.

9.1.2 Intelligent interfaces

An intelligent interface (II) is a device to manage the disparate transactions that can occur within the building design process. Such transactions include supporting design concurrency (designer-to-designer and designer(s)-to-application(s) inter-communication);

preserving audit trail (who did what, when and why?);

supporting a constructive user dialogue (style of interaction, feedback and tutoring);

evolving the product model (incremental problem definition and intelligent defaulting);

and handling application semantics (application-to-product model and application-to-application).

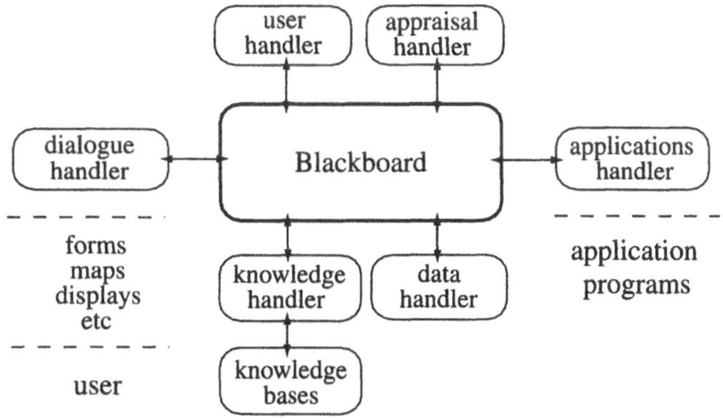

Figure 9.2: Architecture of the COMBINE IIBDS.

To address these issues the COMBINE project established an intelligent, integrated building design system (IIBDS; Clarke *et al* 1995, 1995a) as shown in figure 9.2. The IIBDS, which was constructed using an intelligent front-end authoring system (Clarke and Mac Randal 1989, 1991) comprises the following elements.

A Blackboard to serve as a communication centre for its various clients. By this means concurrency can be supported and traceability achieved through the collection, organisation and storage of the session chronicle.

An Application Handler to control the various design tools, pass them their data and receive their returns.

A Knowledge Handler to control design tool access to the product model and the communication with the designer (verification of entries, supplemental inferencing and feedback/guidance).

A Dialogue Handler to converse with the user by means of appropriate interface tools employing acceptable concepts that relate to the different user types and levels of expertise.

A User Handler to track the user's progress and ensure that the system responds in an appropriate manner when difficulties are encountered.

An Appraisal Handler to hold the design tool control syntax against standard performance assessment methodologies.

A Data Handler to extract an application's data from the Blackboard and organise these data in the required format.

In dealing with the design process the IIBDS must support the flow of data/information between work-steps (WS) and handle events in terms of starting and stopping the design tools (DT). The mechanism adopted to handle these issues within the IIBDS was as follows.

The required sequencing of DTs (i.e. the process model) is captured in the form of a Petri-Net† (Javor 1993) and is then transformed into a file of Prolog facts. This gives the basis of the process as a formal description. This Petri-Net File (PNF) is then read dynamically into a knowledge base associated with the Knowledge Handler—the Application Knowledge Handler (AKH)—where it is used by an inference engine to animate the process.

By modifying this process knowledge base it is possible to control the rigidity of the system, its handling of parallelism etc. Consider the following three examples.

DT invocation is not sequenced nor functionally constrained so that the designer is able to invoke the DTs in any order and activate their internal functions as required. That is, the PNF is used only for DT access control.

The DTs are sequenced but not functionally constrained so that DT selection is prescribed while function invocation is not. That is, the PNF controls DT ordering but DT use is opportunistic; concurrency is allowed.

The DTs are both sequenced and functionally constrained so that the system, not the designer, controls the order of DT selection and the invocation of the WS. The user remains in control of the process and whether the outcome of a WS is acceptable: the PNF enforces rigid DT use but no concurrency is allowed.

The process model corresponding to the first case relates to 'shallow' control whereby DT transactions are managed, while the model corresponding to the last case relates to 'deep' control whereby knowledge is introduced in relation to design purpose. The PNF can be changed in mid-process should it become necessary to adapt the rigidity of the process in response to events. The external Petri-Net description and the dynamic loading makes it possible to change the process being enacted.

Each node in the Petri-Net corresponds to a WS and triggers a knowledge predicate that "knows" what should be done at this point in the process. This is where the problematic issue of concurrency is handled. The knowledge base has access to the Blackboard (i.e. the design process state) and to the DES (i.e. the state of the problem description). This knowledge base will either be established to react only to the Petri-Net (when in prescriptive mode of operation), or to react to the design process state (when in reactive mode of operation). After deciding to carry out a WS, the knowledge base ensures that a) the data required for the task is available, b) starts the appropriate DT and then c) hands control to the user. It then monitors what the DT is doing and finally ensures that the results of the WS are captured and propagated via the Blackboard. While the DES is responsible for the handling of the data, the AKH is responsible for driving the process (i.e triggering state changes in the Petri-Net), propagating information to other knowledge bases and keeping track of the design status and history.

The process and design tool knowledge bases are event driven and operate asynchronously. This enables concurrency. Event driven controllers can handle any amount of concurrency, subject only to their ability to 'understand' what the other controllers are 'saying'. In practice,

† Petri-Nets provide a formal technique for the representation and analysis of concurrent, discrete events such as found (in this context) within work groups engaged in building design activities. Further information is available via the discussion group located at <PetriNets@daimi.au.dk>.

unconstrained concurrency is of little value as it is inherently unstable and unpredictable. The DT knowledge base is therefore made subordinate to the process knowledge base, which activates/de-activates the former as appropriate. By activating more than one process knowledge base at a time, Petri-Net handling can effectively move from single token passing to coloured token based. Furthermore, DT knowledge bases can be forced to listen only to the Petri-Net, giving a slavish compliance to the specified process, or encouraged to react to other DTs giving a more dynamic, context sensitive system.

Figure 9.3 shows a Petri-Net comprising an arrangement of DTs corresponding to the intermediate case process model (DT invocation sequenced but use unconstrained). On entering the design session the user is required to use AutoCAD or MicroStation to create a new problem geometry. On exiting the CAD DT, the ATTRIBUTE DT is used to complete the site, composition and operational characteristics of the problem. When attribution is complete, the user is presented with a choice of compliance checking or thermal/lighting performance appraisal. In the case of compliance checking, the conclusions provided will influence a user's choice to modify the problem geometry (via the CAD DT), its composition (via the ATTRIBUTE DT) or invoke either an energy/comfort assessment (ESP-r) or a lighting/visual evaluation (RADIANCE). Finally, the user can either revisit the CAD or ATTRIBUTE DTs or exit the session. Although this process model supports a co-operative dialogue between the user and the active DTs, the session possesses no knowledge about design purpose.

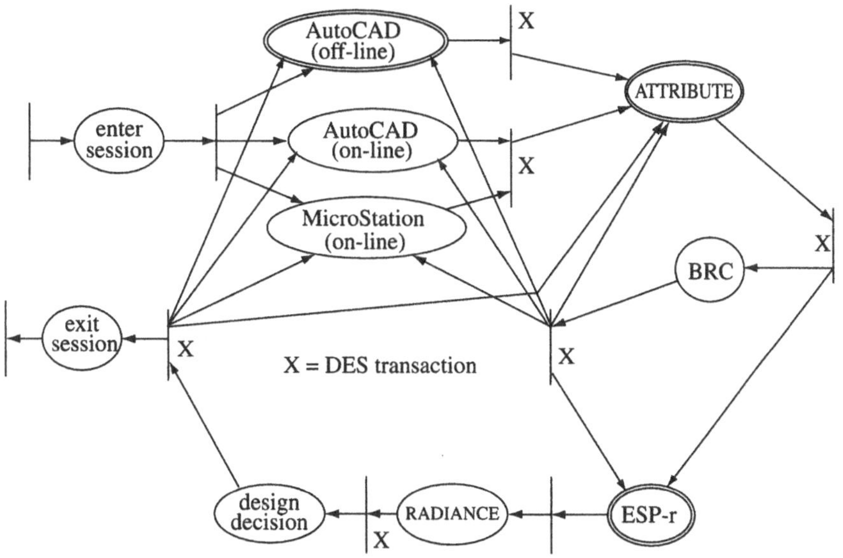

Figure 9.3: Petri-Net for an intermediate case process model.

Figure 9.4 shows the corresponding internal structure of the IIBDS. There are two knowledge handlers, relating to user and application control, communicating through the Blackboard as indicated.

The message passing between the user and application domains has been isolated within a "transaction" area of the Blackboard. The aim of introducing knowledge about design purpose is further supported by the addition of a "journal" area on the Blackboard. This is a repository for the aggregate log of transactions within the system and is used to feed the Prolog predicates of the design session knowledge base. In particular, the nature of the DTs presented to the

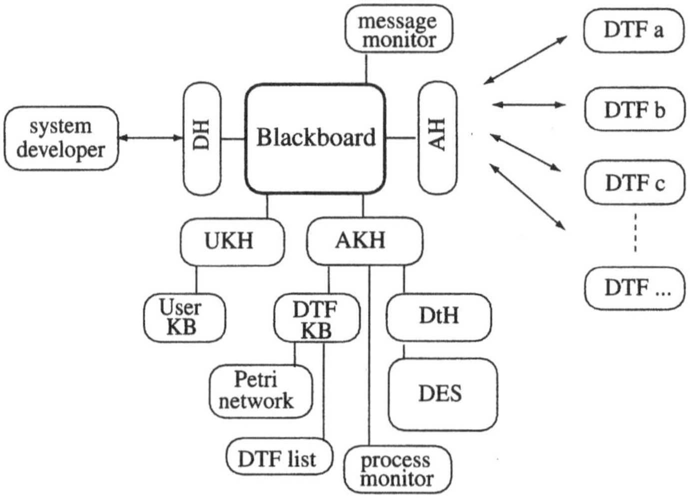

Figure 9.4: Internal structure of the IIBDS.

user, and how they are sequenced and constrained, are supported by the addition of design process knowledge to the AKH. This has been achieved by arranging for the AKH to load the Petri-Net representations as implied by the user's choice of design session.

Between the AKH and the Data Handler/DES resides the Process Monitor, which presents the current position of the token in the Petri-Net and the passing of STEP compliant files (ISO 1994) to and from the DES. Also shown is the Transaction Monitor (TM), which observes the transactions between the knowledge handlers, the DTs and DES. The TM is used to observe and analyse an active design session.

In order to explain the working of the IIBDS, a series of snap-shots follow that record a user's progress when the active design session corresponds to the intermediate level process model previously outlined. Within these snap-shots the arrows show the potential flows of information: a single arrow indicates a notification, while a double arrow indicates sending and listening. The "user_dialog" area of the Blackboard is reserved for user interaction transactions, while the "application_dialog" area is reserved for transactions related to the DTs. The "journal" area receives messages from the various knowledge handlers and organises these for subsequent analysis and process control.

Figure 9.5 defines the state of the system after the user has selected a design tool function (DTF) and the corresponding actions have been triggered: a message passes to the Dialogue Handler (DH) indicating the requested interaction; the DH passes the message to the "user_dialog" area; and the User Knowledge Handler (UKH) tells the Blackboard to "start DTF_x".

Figure 9.6 is the state of the Blackboard after the UKH has issued a message "application_dialog start DTF_x" to the transaction area. The Application Knowledge Handler (AKH) finds the actual application and posts the message "new_application DTF_x" to the "application_dialog" area. The Data Handler (DtH) then queries the DES, "get_data_for DTF_x", and the DES returns the appropriate data for design tool 'x'.

In figure 9.7, after the DES issues "data_for DTF_x file", the DtH posts "new_application application_parameters" to the application dialog area, the Application Handler (AH) starts the DT and establishes a pipe to receive the performance return(s) and, when the application is complete, the AH records this and sends "closed DTF_x revised_data_file" to the

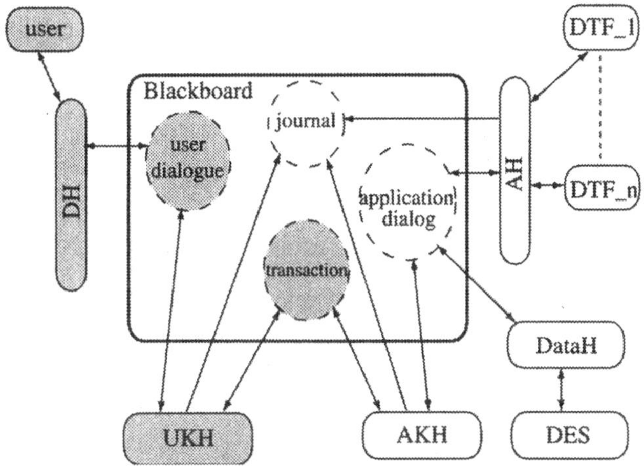

Figure 9.5: State of the IIBDS after user action.

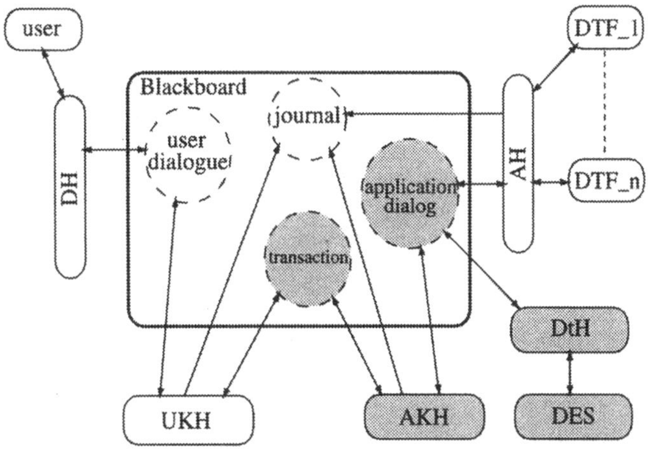

Figure 9.6: User knowledge handler requesting a DTF.

application_dialog area. The AKH then posts "closed DTF_x" to the transaction area, which is received by the UKH for transmission (not shown) back to the user.

Consider now a session that incorporates knowledge in relation to design purpose. With reference to the Petri-Net shown in figure 9.8, the initial portion of the design session might proceed as in the previous unconstrained case (in terms of geometric specifications and attribution). For those users who enter the summer overheating design session with an existing model there is a direct path to the overheating assessment.

Simulation environments, such as ESP-r, provide facilities to enable direct access to their internal functions. In the current example some of these functions are accessed, with the user involved as the arbitrator of acceptability or otherwise of the performance returns. Here, the process model involves the determination of the climate patterns that would constitute an acceptable test of summer overheating risk. While such a decision is implicit in most simulation based studies, here it has been made explicit. Next, the focus is shifted to a simulation of

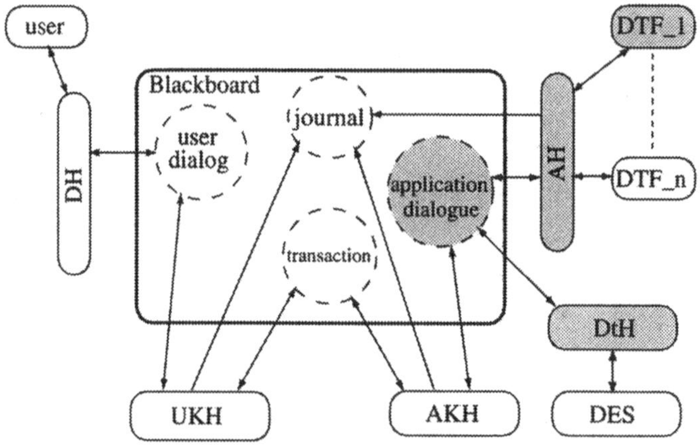

Figure 9.7: DES returns STEP file and DTF_x starts.

the current model and then determination of what constitutes the worst spaces in terms of overheating. The search rules operate on the basis of the highest operative temperature in an occupied space as determined from a series of inquiries of the database of simulation results. It is quite possible to have alternative rules governing this DT function.

Assuming that overheating has been detected, two presentations are made to the user. First, a frequency binning of temperatures and a graph of temperatures in the worst zone is displayed. This sets the context that violated the 'rules' of the assessment. Second, the process model calls for the presentation of information on the likely causal factors. For example, if high internal gains were the cause of the overheating then only this information would be presented. The user can then either exit the design session, select a different climate condition or return to the CAD or ATTRIBUTE DTs. A typical design session is as shown in figure 9.9.

While systems such as the IIBDS are entirely possible, their emergence as robust, useful tools will only occur if the construction industry adopts a simulation approach to design/management of the built environment and seeks to do so within a multi-disciplinary framework. It is likely that such a viewpoint will be fostered more by the emergence of a ubiquitous Internet than by altruistic considerations of best practice.

9.2 Virtual construction

Looking to the long term, it is possible to envisage a modelling environment where the time delay between problem description and performance feedback is effectively eliminated. By endowing the computational objects that are used to describe a problem with behaviour, it is possible to generate useful performance information as soon as relevant object groupings are created. For example, a sun, wall and obstruction object would give shading information as each is moved relative to the other. The prerequisite of such an environment is the existence of a set of objects that encapsulate both description and behaviour. The creation of such an object set, and the means to manipulate it, was the aim of the Energy Kernel System (EKS) project funded by the UK Engineering and Physical Sciences Research Council (Wright *el al* 1992, Hammond and Irving 1992, Clarke *et al* 1992). Essentially, the project set out to identify the real and abstract (mathematical) entities underlying building energy modelling, research the feasibility of adopting the object-oriented (OO) programming paradigm in the representation of

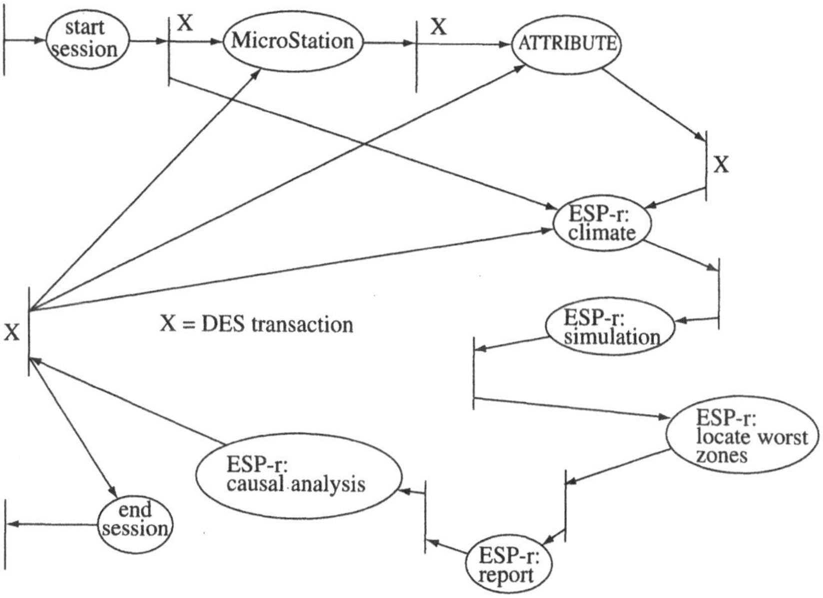

Figure 9.8: Summer overheating Petri-Net.

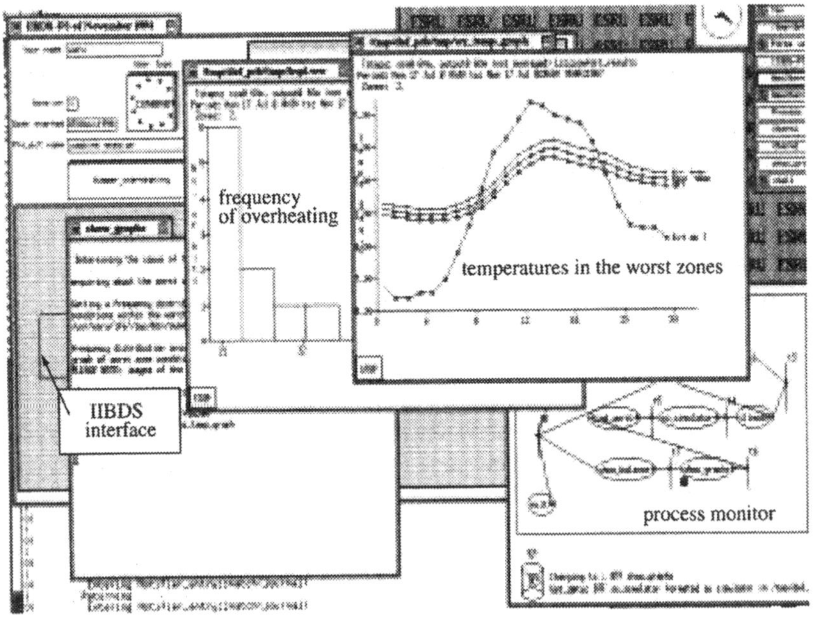

Figure 9.9: Summer overheating design session.

these entities, and establish a demonstration system. The essence of the OO approach is that a program can be composed of independent objects communicating via messages. The main challenge was therefore to decompose building modelling into classes and subclasses and define the properties of these classes in terms of their data members, behaviour and inter-relationship.

It has been stated that "the most important concept in the object-oriented approach is data abstraction" (Kim and Lochovsky 1990). Since the EKS was intended to be a construction environment for building simulation programs, rather than a model of a building, the "data" was here viewed as the required behaviour and so a functional decomposition approach was adopted (Clarke *et al* 1989). The important point about the approach is that it eliminates all aspects of the problem that are not related to the target domain, ensures maximum class reuse and extensibility, and provides a mechanism to deal with the physical/abstract mix so dominant in thermodynamic modelling.

The building energy modelling domain was decomposed into its primitive functionality, such as sun position tracking, conduction, convection, radiation, equation solving, polygon operations and the like. These are the functions, albeit at different levels of abstraction, to be found within all contemporary building modelling systems. The initial research task was therefore to undertake a comprehensive review of existing models for building, plant and control system simulation (Tang 1990). The functions so identified are described elsewhere (Wright *et al* 1990, Clarke *et al* 1990).

A minimum level of data requirement was then associated with each function. For example, a one dimensional layer conduction function will require a set of thermophysical properties irrespective of the underlying mathematical model and so its minimum data requirement (in EKS terms) is a "Material"† object (or strictly speaking a pointer to such an object) and a "Dimension" object. Likewise, a matrix inversion function requires the matrix topology and coefficient values irrespective of the inversion technique to be used. It is also worth noting that only building related functionality was considered. For example, when considering solar processes the sun position is relevant while its angular rotation is not.

Each functions is now associated with a physical class which, logically, knows the context of the function. For example, "Construction" logically knows about thermal resistance, while "Room" logically knows about shortwave response. More contentious perhaps, "Construction" may know about reference U-values (which have prescribed surface resistances), while only "Building" may know about actual U-values because to evaluate this parameter requires knowledge of the properties and thermodynamic state of different entities (air volumes, surfaces, constructions, exposures and so on). It follows that class functions must relate only to the intrinsic data and properties of a class and not require the existence of, or assume data or properties of, another class. This ensures that a class will encapsulate only data which is pertinent to that class and that the data members of each class, as required to support its functions, can be guaranteed to be available at run-time to the object made from the class.

Abstract classes are now identified by gathering together related functionality as implied by the functions of the physical classes. For example, the convection function of "Room" requires surface area, hydraulic diameter and heat flow direction, all of which are geometrical entities and so are gathered together into an abstract class "Polygon". This ensures that classes will not possess functionality where that functionality could be made more generally available by encapsulation within another class.

Where a class has several domain theory functions (e.g. convection, shortwave response and

† Henceforth quoted and capitalised names signify classes or their instances, termed objects.

occupant behaviour in the case of the "Room" class or different matrix solution techniques in the case of the "Solver" class), these functions are implemented as links to abstract classes containing the required functionality of the domain theory. The different formulations of any given domain theory can then be handled by classes derived from these abstract classes: this facilitates the handling of the multiplicity of domain theories, without incurring combinatorial explosion in the parent classes.

By utilising the C++ language (Stroustrup 1987) and an OO database (ONTOS 1989), a class taxonomy was constructed from which models of different functionality could be built. The classes were organised into 'used by' and 'derived from' hierarchies and placed under the control of an instantiation mechanism. This means that programs offering different modelling capabilities can be automatically constructed by merely selecting the required class variants.

Figure 9.10 shows the EKS 'used by' hierarchy. This specifies how the classes interrelate and defines the information flows. Orthogonal to this plane is the EKS inheritance hierarchy used to represent the alternative domain theories, e.g. the alternative conduction theories used by the "Layer" class. The dilemma between extensibility (i.e. the ability of an existing class to use, or be used by, a new class) and security (i.e. the guarantee that classes are compatible) is solved by the use of meta-classes. These define the behaviour of each EKS class and have knowledge of other dependent classes. By insisting that programs can only be built from meta-classes, it is possible to have an extensible and secure system without paying the performance penalty of run-time type checking.

EKS classes are of three types: base, principal and intrinsic.

Base classes provide the interface to the principal classes and are 'generic' in that they cannot be instantiated because their functions are virtual, being implemented only in the principal classes derived from them. They exist to allow the future extension of the class taxonomy into other domains without the need to carry the thermodynamic functionality of the existing implementation. Also, they guarantee that future derived classes will interoperate with the other EKS classes.

Principal classes are selected by a program builder to define the capabilities of some program. Each principal class is derived from a corresponding base class and offers a particular implementation of the virtual functions of its parent—that is, they represent the physical or thermodynamic states of the entities they represent. In the case of the domain theory classes, alternative principal classes will exist to represent the different possible theoretical approaches.

Intrinsic classes are the internal work horses of the EKS, serving to transport data between the principal classes, control the class selection process at template specification time, contain support data such as climate time series and dimension all properties of state. Because they are not selected by the program builder, intrinsic classes are not shown in the figure 9.10 taxonomy.

The separation of the underlying functions in this way gives maximum flexibility and code reuse when creating new programs. The base classes provide a common semantic for the principal classes and permit classes derived from them to be reliably used in different contexts. Derived classes are subtypes of their parent—that is, a class may be transparently replaced (even at run-time) by any class derived from it. Thus, as shown in figure 9.11, if the functionality of "DynamicConduction" is required, either finite difference or response function approaches can be selected whereas, if only the lesser functionality of "Conduction" is called for, both steady state or dynamic conduction classes can be selected. It is this facility that gives the EKS its flexibility: in terms of model building (support of theoretical variants); in terms of run-time features such as dynamic model substitution; in terms of support for program validation and in terms of program maintenance. Where different program architectures handle the

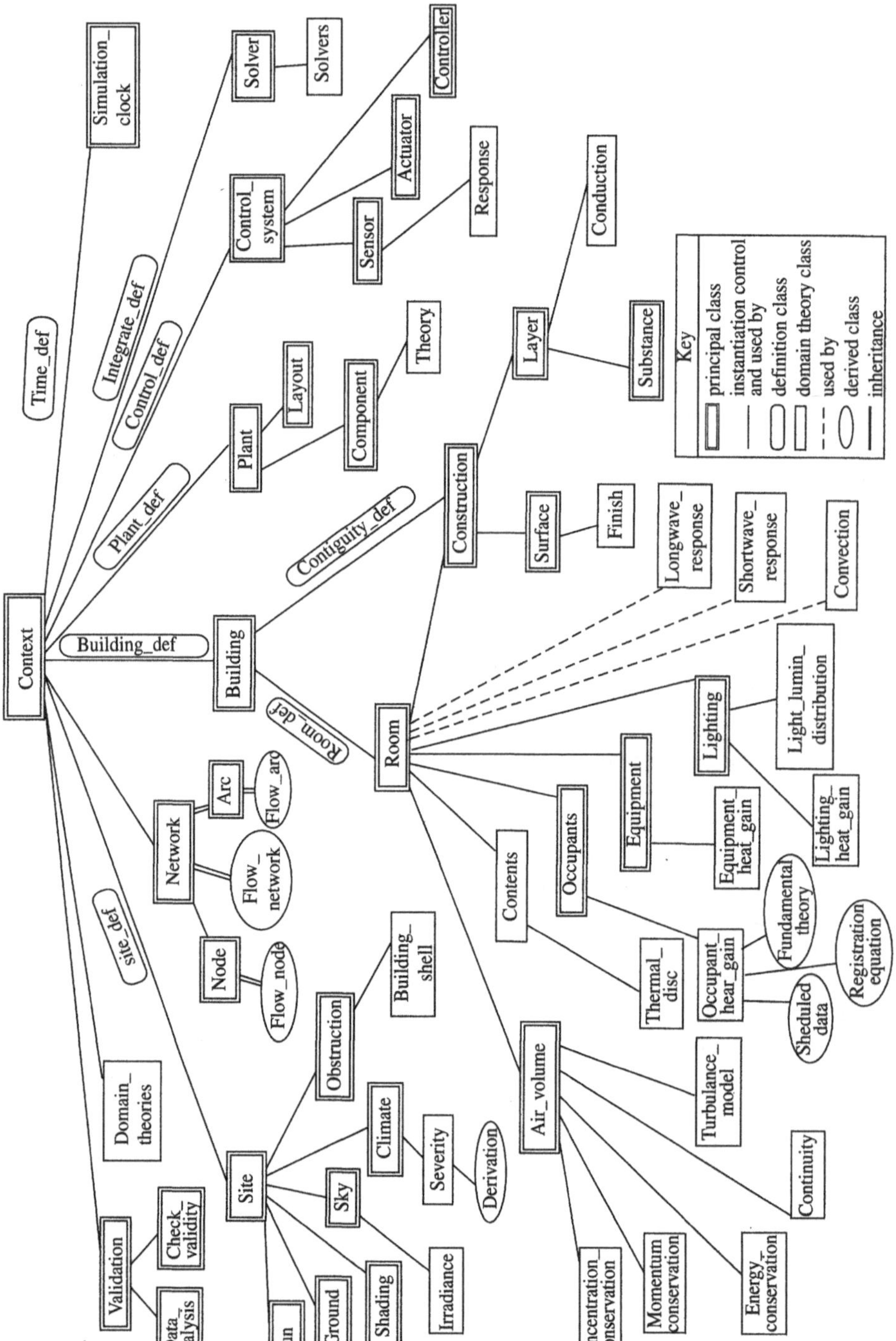

Figure 9.10: EKS 'used by' class hierarchy.

same task in different ways, derived classes provide a means to incorporate that functionality while ensuring minimal impact on the rest of the system.

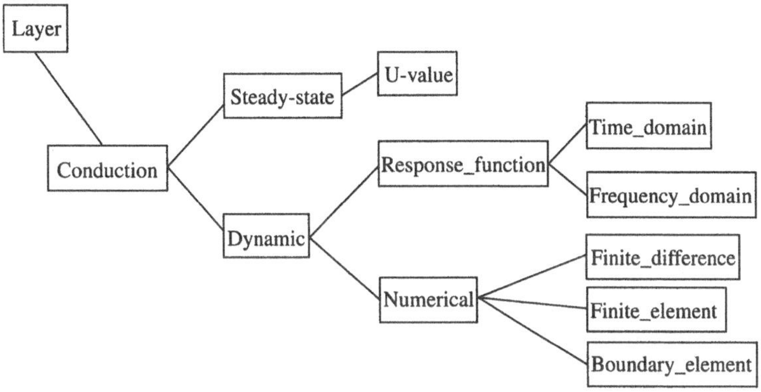

Figure 9.11: Use of derived classes.

Behavioural inheritance also reduces coding and improves reliability since new classes gain access to their parent's code and so need only add the code for the extra functionality they provide. Extensibility is also assisted because new classes and variants of existing classes can be added without requiring any changes to existing classes.

Actual program building is carried out by a special template class. Given that the required program capability is known, a "Template" is constructed as a collection of meta-class instances. Since the "Template" is the program specification, there is no need to generate an executable program at this stage. Instead, once the problem data is available, the "Template" uses its meta-class objects to create a program tailored to the problem being addressed.

As a simulation model development platform, it is expected that the EKS will need to support a wide range of systems represented by different mathematical categories, e.g. hyperbolic partial differential models for aerodynamic systems, parabolic partial differential models for diffusion systems, elliptic partial differential models for wave and vibration systems, ordinary differential models for lumped parameter systems and so on. One way to achieve this would be to provide representation schemes for all the different possible approaches—a finite difference "Layer" and a finite difference "Conduction"; a steady-state "Layer" and a steady-state "Conduction" and so on. Apart from being inelegant, this would give rise to the problems of interface complexity and combinatorial explosion (Clarke *et al* 1990), result in duplication of code/functionality and create intractable maintenance difficulties.

Instead, the EKS employs a theory representation based on a vectorised state-space equation method, implemented using sparse matrix techniques. Using this technique, it is possible to represent most of the equation-based theories in the domain of building energy modelling in a consistent manner as vectorised state-equations. The details of this method are given elsewhere (Tang 1990, Clarke *et al* 1992). Theory, in vectorised form, is encapsulated within four special transport classes: "Equation_set", "Equation", "Coefficient" and "State_variable".

As shown in figure 9.12, an OO database lies at the heart of the EKS and has three roles. First, it holds persistent objects relating to entities such as climate and material properties. Second, it holds the problem description and results as objects, enabling the interfaces to be separated from the body of the performance prediction engine. Last, it holds the "Template" and the meta-class objects. The programs surrounding the database are used to specify a particular program architecture.

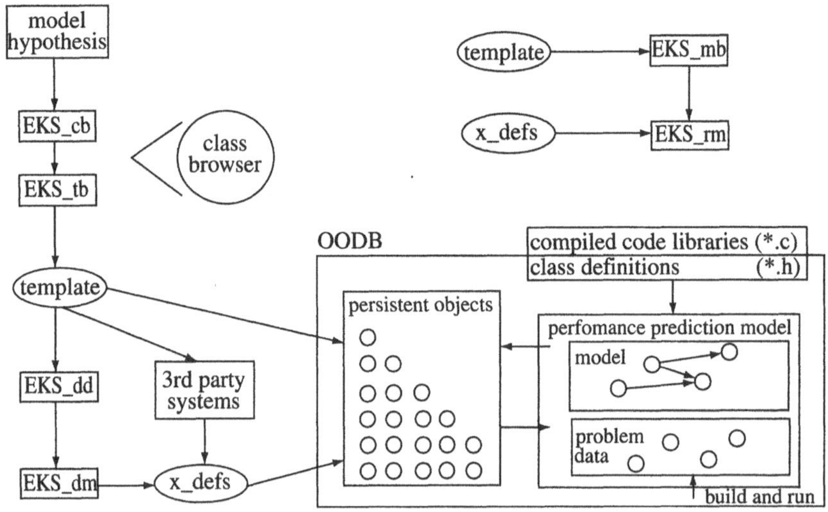

Figure 9.12: EKS elements.

EKS_cb builds the problem context in terms of a definition of the features required of a program.

EKS_tb builds a given "Template" by allowing class variants to be selected as required. As each class is selected, the corresponding meta-class defines the dependent classes so that the process is automated.

EKS_dd takes a "Template" as input and outputs the corresponding OO product model.

EKS_dm supports the user definition of the product model and holds the data in the form of "X_def" ('something_definition') objects. Typically, a program will require several "X_def" objects to define building geometry, construction, operation etc.

EKS_mb builds a program on the basis of a given "Template".

EKS_rm controls a simulation by associating the program with its corresponding "X_def" objects.

An environment such as the EKS opens up the prospect of a radical change to the design support process. Imagine, for example, a computer system that allowed the definition of a design hypothesis by the graphical selection of component parts representing a site topology, walls, windows, radiators, shading devices, sensors, a sun and so on. If these components were related to EKS classes then the user would effectively be constructing, in real time, a model that is matched to the envisaged design. Given the functionality of the EKS, it would then be possible to arrange that the object instances externalise their behaviour immediately on selection. By bringing together hypothesis manipulation and performance appraisal, a real-time computer-supported design environment is enabled. This, in turn, would enable the application of simulation at the conceptualisation stage of the design process where the potential benefits are greatest.

9.3 Concluding remark

I would like to conclude this book by quoting Tom Maver, Professor of CAD at the University of Strathclyde:

> *"Today's justifiable concern with the energy conscious design of buildings is a recurring echo of a fundamental theme dating back to the first human settlements: the need for shelter from an inclement environment. Of the factors which determine the degree of effective shelter provided by a building, none are more important, or more worthy of the designer's consideration, than its form, fabric and systems; but these are also the determinants, for better or worse, of the entire range of attributes which make up the quality of the built environment—from the life-cycle cost of the building to its visual impact on the site.*
>
> *The exciting prospect which emerges, then, is of an integrated computer-based system in which the exploration of form, fabric and operation leads to a design solution embodying a balance between cost and performance, between investment and return, between need and aspiration.*
>
> *The new generation of integrated energy models exemplify the benefits which a sustained intellectual commitment can bring to the quality of the built environment. It is to be hoped that this endeavour will itself be a model, and an inspiration, for future developments in computer-aided design."*

9.4 References and further reading

Augenbroe G 1992 Integrated Building Performance Evaluation in the Early Design Stages *Building and Environment* **V27**(2) 149-61

Bazjanac V and Crawley D B 1997 The Implementation of Industry Foundation Classes in Simulation Tools for the Building Industry *Proc. Building Simulation '97* Prague

Citherlet S 2001 Towards the Holistic Assessment of Building Performance Based on an Integrated Simulation Approach *PhD Thesis* (Lausanne: Swiss Federal Institute of Technology)

Clarke J A, K James and D Tang 1989 Simulation Methods Concerning Building Performance Prediction: A General Review *EKS Project Report* (Glasgow: Energy Systems Research Unit, University of Strathclyde)

— 1990 EKS Prototype Status Review and Issues Arising *EKS Project Report* (Glasgow: Energy Systems Research Unit, University of Strathclyde)

Clarke J A and Mac Randal D F 1989 The Application of Intelligent Knowledge Based Systems in Building Design *Final Report for Grant GR/E/18018* (Swindon: Engineering and Physical Science Research Council)

— 1991 An Intelligent Front-End for Computer-Aided Building Design *Int. J. Artificial Intelligence in Engineering* **61** 36-45

Clarke J A, D Tang, K James and D F Mac Randal 1992 Energy Kernel System *Final Report for Grant GR/F/07880* (Swindon: Science and Engineering Research Council)

— 1995, The Development of an Intelligent Integrated Building Design System Within the European COMBINE Project *Building Simulation '95* (Madison)

Clarke J A, Hand J W, Mac Randal D F and Strachan P A 1995 Data Decomposition of the ESP-r System, Development of a DES Interface and the Prototyping of an Intelligent,

Integrated Building Design System *Final Report for the COMBINE II Project* (European Commission)

— 1995a, The Development of an Intelligent Integrated Building Design System Within the European COMBINE Project *Building Simulation '95* (Madison)

Hammond G and Irving I 1992 Validation Studies Appropriate to the Development of the Energy Kernel System for Building Environmental Analysis *Final Grant Report* (Swindon: Science and Engineering Research Council)

ISO 1989 External Representation of Product Representation Data *ISO 10303* (International Standards Organisation)

ISO 1994 Industrial Automation Systems and Integration - Product Data Representation and Exchange. Part 1: Overview and Fundamental Principles *ISO TC 184/SC4* (International Standards Organisation)

Javor A 1993 Petri Nets in Simulation *EUROSIM - Simulation News Europe* November 6-7

Kim W and F H Lochovsky 1990 *Objected-Oriented Concepts, Databases, and Applications* (New York: ACM Press)

Nijssen G M and Halpin T A 1989 Conceptual Schema and Relational Database Design - A Fact Oriented Approach (New York: Prentice Hall)

ONTOS 1989 *Object Database Documentation* (Burlington, MA: Ontologic Inc)

Stroustrup B 1987 *The C++ Programming Language* (Addison-Wesley)

Spiby P (Ed) 1991 EXPRESS Language Reference Manual *ISO TC184/SC4/WG5 Document N14.*

Tang D 1990 The EKS Theory Representation *EKS Project Report* (Glasgow: Energy Systems Research Unit, University of Strathclyde)

Wright A J, Charlesworth P, Clarke J A, Hammond G P, Irving A, James K A, Mac Randal D and Tang D 1990 The use of Object-Oriented Programming Techniques in the UK Energy Kernel System for Building Simulation *Proc. European Simulation Multiconference* Nuremberg 548-52

Wright A J, Wiltshire T J and Lockley S 1992 *EKS Project Final Report* (Swindon: Science and Engineering Research Council)

Appendix A

Thermophysical properties

This appendix presents the partial outcome from a review of existing datasets of thermophysical properties of building materials and an assessment of data reliability in terms of the underlying test procedures (Clarke *et al* 1990). Scrutiny of the data gave rise to the following points.

The range of properties for which values are quoted is frequently limited to thermal conductivity, density and vapour resistivity as required for simple steady state heat loss and condensation calculations.

The sources of much of the data are not identified, and little information is given on the underlying experimental conditions or procedures. As a consequence, it is often impossible to check compatibility between different values.

Many of the data values are derived from work carried out with non-standard apparatus or from a date that precedes modern standards of equipment and operation.

Much of the agreement that does exist between different data-sets may be attributable to historical 'borrowing'. This may lead, erroneously, to an optimistic assessment of the inherent uncertainty.

No guidance is given on the variation in properties such as density and internal structure inherent in the production of many building materials, and there is no agreement on the procedure for determining the thermal conductivity of materials as the moisture content varies. Such variations can lead to very large differences in reported material properties.

There is little consensus on the manner in which materials are grouped for presentation of data to designers. What is needed is a common system such as the CIB Master list of materials (Eldridge 1974) which integrates thermal properties within a broad material classification system.

Many values are quoted without any statement as to whether they correspond to single or multiple measurements. A random inspection of several referenced works would suggest that values are usually derived from the work of a single researcher on the basis of a small sample size.

There is no agreement on the procedure for the determination of the thermal conductivity of materials in the moist state or guidance given on the variation of density inherent in the production process.

There is tacit agreement that the uncertainty within the data is use-context dependent. The traditional calculation methods are clearly expected to yield no more than ball-park estimates of real conditions.

The work distinguished between two contexts in which the data might be used. The first is in studies where the aim is to compare different buildings made of ostensibly the same materials. In such cases, the absolute accuracy of data is not paramount and 'reference' data may be used. The second is in the calculation of real building performance where the variations of properties with moisture content, and the inherent uncertainties in the manufacture and use of building materials, are of key importance.

The following material properties were examined: conductivity (W $m^{-1}°C^{-1}$), density (kg m^{-3}), specific heat (J $kg^{-1}°C^{-1}$), surface emissivity (-), surface shortwave absorptivity (-) and vapour resistivity or resistance (MNs $kg^{-1}m^{-1}$ or MNs kg^{-1}). From consideration of the use context of the materials, and the reliability/scope of their underlying test procedures, the data were classified as follows.

Category 1 - Impermeables

Materials which act as a barrier to water in the vapour and/or liquid states and do not alter their hygro-thermal properties by absorbing or being wetted by water.

Category 2 - Non-Hygroscopic

Lightweight insulations, such as mineral wools and foamed plastics, which display water vapour permeability, zero hygroscopic water content and an apparent thermal conductivity, and which operate under conditions of air-dry equilibrium normally protected from wetting by rain.

Category 3 - Inorganic-Porous

Masonry and related materials that are inorganic, porous and may contain significant amounts of water due to hygroscopic absorption from the air or wetting by rain, which affects their hygro-thermal properties and their thermal conductivity in particular.

Category 4 - Organic-Hygroscopic

Organic materials such as wood and wood based products which are porous and strongly hygroscopic and display a highly non-linear water vapour permeability characteristic.

Further information is available in the literature on the underlying test methods and their accuracy (Shirtcliffe and Tye 1985, Bomberg and Solvason 1985, Hager 1985), on the correlation between conductivity and density (Jakob 1949, Ball 1968, Arnold 1970, Van Geem and Fiorato 1983), on the correlation between conductivity and moisture content/temperature (Jespersen 1953, Billington 1952, Valore 1980, Stuckes and Simpson 1986), on the correlation between conductivity and thickness (ASHRAE 1985, Siviour 1985), on the determination of specific heat capacity (Hens 1984), on the determination of vapour permeability (MacLean and Galbraith 1988) and on the determination of surface properties (Holden and Greenland 1951).

A.1 Category One: Impermeables

Material	*Conductivity* *(W m^{-1}°C^{-1})*	*Density* *(kg m^{-3})*	*Specific Heat* *(J kg^{-1}°C^{-1})*
Asphalt			
poured	1.20	2100.	920.
reflective coat	1.20	2300.	1700.
roofing	1.15	2330.	840.
Bitumen			
flooring	0.85	2400.	1000.
insulation	0.20	1000.	1700.
Ceramics			
glazed	1.40	2500.	840.
Glass			
cellular sheet	0.048	140.	840.
foamed	0.052	140.	840.
4mm clear float	1.05	2500.	750.
6mm antisun	1.05	2500.	750.
block	0.70	3500.	840.
ceramic	1.40	2500.	840.
plate	0.76	2710.	840.
brick	1.40	2500.	840.
mirror	2.80	2500.	840.
Linoleum			
regular	0.19	1200.	1470.
Metal			
aluminium	203.	2700.	880.
brass	110.	8500.	390.
bronze	64.	8150.	-
copper	384.	8600.	390.
duraluminium	160.	2800.	580.
iron	72.	7900.	530.
cast iron	56.	7500.	530.
lead	35.	11340.	130.
steel	45.	7800.	480.
stainless steel	29.	7850.	480.
tin	65.	7300.	240.
zinc	113.	7000.	390.
PVC			
regular	0.16	1380.	1000.
tiles	0.19	1200.	1470.

Material	*Conductivity* *(W m^{-1}°C^{-1})*	*Density* *(kg m^{-3})*	*Specific Heat* *(J kg^{-1}°C^{-1})*
Roof covering			
felt	0.19	960.	840.
Rubber			
regular	0.17	1500.	1470.
hard	0.15	1200.	1000.
expanded board	0.032	70.	1680.
tiles	0.30	1600.	2000.

A.2 Category Two: Non-Hygroscopic

Material	*Conductivity* *(W m^{-1}°C^{-1})*	*Density* *(kg m^{-3})*	*Specific Heat* *(J kg^{-1}°C^{-1})*
Carpet			
cellular rubber underlay	0.10	400.	1360.
synthetic	0.06	160.	2500.
Foam			
phenol	0.04	30.	1400.
phenol, rigid	0.035	110.	1470.
polyisocyanate	0.03	45.	1470.
polyurethane	0.028	30.	1470.
polyvinyl chloride	0.035	37.	1470.
urea formaldehyde foam	0.054	14.	1470.
Glass Fibre			
quilt	0.04	12.	840.
slab	0.035	25.	1000.
strawboard	0.085	300.	2100.
wool	0.04	12.	840.
Insulating Board			
mineral fibre, wet felted	0.051	290.	800.
mineral fibre, wet moulded	0.061	370.	590.
mineral fibre, resin binder	0.042	240.	710.
Loose Fill			
cellulosic insulation	0.042	43.	1380.
exfoliated vermiculite	0.069	260.	880.
glass, granular	0.07	180.	840.
gravel	0.36	1840.	840.
perlite, expanded	0.051	100.	1090.
roof gravel or slag	1.44	880.	1680.

Material	*Conductivity* (*W* m^{-1}°C^{-1})	*Density* (*kg* m^{-3})	*Specific Heat* (*J* kg^{-1}°C^{-1})
sand	1.74	2240.	840.
stone chippings	0.96	1800.	1000.
dry render	0.50	1300.	1000.
Mineral Wool			
regular	0.038	140.	840.
fibrous	0.043	96.	840.
resin bonded	0.036	99.	1000.
rock wool	0.033	100.	710.
Miscellaneous			
acoustic tile	0.057	290.	1340.
cratherm board	0.05	176.	840.
perlite	0.046	65.	840.
perlite panel	0.055	170.	840.
vermiculite	0.058	350.	840.
vermiculite panel	0.082	350.	840.
felt sheathing	0.19	960.	950.
mineral board, preformed	0.042	240.	760.
mineral fibre	0.04	100.	1800.
mineral fibre slab	0.035	30.	1000.
polystyrene, rigid	0.036	16.	1210.
silicon	0.18	700.	1000.
plastic tiles	0.50	1050.	840.
polyisocyanurate board	0.02	32.	920.
Polystyrene			
extruded	0.035	25.	1470.
expanded	0.035	23.	1470.
expanded PVC	0.04	100.	750.
Polyurethane			
cellular board	0.023	24.	1590.
expanded	0.023	24.	1590.
unfaced	0.023	32.	1590.

A.3 Category Three: Inorganic-Porous

Material	*Conductivity* (*W* m^{-1}°C^{-1})	*Density* (*kg* m^{-3})	*Specific Heat* (*J* kg^{-1}°C^{-1})
Asbestos			
cement	1.02	1750.	840.
cement board	0.58	1920.	1010.
cement decking	0.36	1500.	1050.

Material	*Conductivity* *(W m^{-1}°C^{-1})*	*Density* *(kg m^{-3})*	*Specific Heat* *(J kg^{-1}°C^{-1})*
Brick			
aerated	0.30	1000.	840.
breeze block	0.44	1500.	650.
burned	0.75	1300.	840.
inner leaf	0.62	1800.	840.
outer leaf	0.96	2000.	650.
paviour	0.96	2000.	840.
reinforced	1.10	1920.	840.
Cement & Plaster			
cement, regular	0.72	1860.	840.
cement blocks	0.33	520.	2040.
cement fibreboard	0.082	350.	1300.
cement mortar	0.93	1900.	840.
limestone mortar	0.70	1600.	840.
plaster	1.50	1900.	840.
plaster/sand aggregate	0.72	1860.	840.
cement screed	1.40	2100.	650.
plaster	0.51	1120.	960.
plasterboard	0.17	800.	1090.
plaster, dense	0.50	1300.	1000.
plaster, lightweight	0.16	600.	1000.
rendering	0.79	1330.	1000.
plaster, vermiculite	0.20	720.	840.
Ceramics			
tiles	1.20	2000.	850.
Clay			
tile	0.85	1900.	840.
tile, burnt	1.30	2000.	840.
tile, hollow	0.623	1120.	840.
tile, paver	1.803	1920.	840.
Concrete			
heavyweight	1.30	2000.	840.
lightweight	0.20	620.	840.
medium lightweight	0.32	1050.	840.
very lightweight	0.14	370.	840.
no fines	0.96	1800.	840.
aerated, cellular	0.70	1000.	840.
aerated roofing slab	0.16	500.	840.
block, lightweight	0.64	1660.	840.
block, mediumweight	0.86	1970.	840.
block, heavyweight	1.31	2240.	840.
block, aerated	0.24	750.	1000.
block, hollow, lightweight	0.58	720.	840.

Material	*Conductivity* *(W m^{-1}°C^{-1})*	*Density* *(kg m^{-3})*	*Specific Heat* *(J kg^{-1}°C^{-1})*
block, hollow, mediumweight	0.86	930.	840.
block, hollow, heavyweight	1.35	1220.	840.
block, partially filled, lightweight	0.67	1090.	840.
block, partially filled, mediumweight	0.85	1260.	840.
block, partially filled, heavyweight	1.35	1570.	840.
block, perlite filled, lightweight	0.17	770.	840.
block, perlite filled, mediumweight	0.20	900.	840.
cast	1.28	2100.	1010.
cast, dense	1.40	2100.	840.
cast, dense, not reinforced	1.70	2400.	840.
cast, dense, reinforced	1.90	2500.	840.
cast, lightweight	0.38	1200.	1000.
cement or lime based, aerated	0.21	580.	840.
cinder	0.69	1410.	840.
foamed	0.08	400.	920.
glass reinforced	0.90	1950.	840.
refractory insulating	0.25	10.	840.
tiles	1.10	2100.	840.
vermiculite aggregate	0.17	450.	840.
Masonry			
heavyweight	0.90	1850.	840.
lightweight	0.22	750.	840.
mediumweight	0.32	1050.	840.
quarry stones, calcareous	1.40	2200.	840.
semi-heavy blocks	0.60	1350.	840.
very light blocks	0.19	470..	840.
Miscellaneous			
calcium silicate brick	1.50	2000.	840.
dried aggregate	1.31	2240.	840.
granolithic	0.87	2085.	840.
lime stone	1.80	2420.	840.
siporex	0.12	550.	1000.
thermalite	0.19	750.	840.
Roofing			
tile	0.84	1900.	800.
terracotta	81.	1700.	840.
Soil			
earth, common	1.28	1460.	880.
earth, gravel	0.52	2050.	180.
alluvial clay, 40% sands	1.21	1960.	840.

Material	*Conductivity* *(W m^{-1}°C^{-1})*	*Density* *(kg m^{-3})*	*Specific Heat* *(J kg^{-1}°C^{-1})*
Stone			
basalt	3.5	3000.	840.
gneiss	3.49	2880.	840.
granite	3.49	2880.	840.
limestone	2.9	2750.	840.
marble	2.9	2750.	840.
porphyry	3.49	2880.	840.
red granite	2.9	2650.	900.
sandstone	1.3	2150.	840.
sandstone tiles	1.2	2000.	840.
slate	1.72	2750.	840.
slate shale	2.1	2700.	840.
white calcareous stone	2.09	2350.	840.
white marble	2.	2500.	880.

A.4 Category Four: Organic-Hygroscopic

Material	*Conductivity* *(W m^{-1}°C^{-1})*	*Density* *(kg m^{-3})*	*Specific Heat* *(J kg^{-1}°C^{-1})*
Cardboard/Paper			
bitumen impregnated	0.06	1090.	1000.
laminated	0.072	480.	1380.
Cloth, Carpet, Felt			
bitumen/felt layers	0.50	1700.	1000.
carpet, simulated sheep wool	0.06	200.	1360.
carpet, wilton	0.06	190.	1360.
felt, semi-rigid, organic bonded	0.035	48.	710.
jute fibre	0.067	330.	1090.
wool felt underlay	0.04	160.	1360.
Cork			
board	0.04	160.	1890.
expanded	0.044	150.	1760.
expanded and impregnated	0.043	150.	1760.
tiles	0.08	530.	1800.
Straw			
board	0.057	310.	1300.
fibre board or slab	0.10	300.	2100.
thatch	0.07	240.	180.

Material	*Conductivity (W m^{-1}°C^{-1})*	*Density (kg m^{-3})*	*Specific Heat (J kg^{-1}°C^{-1})*
Miscellaneous			
afzelia, minunga, meranti	0.29	850.	2070.
expanded ebonite	0.035	100.	1470.
expanded perlite board, organic bonded	0.052	16.	1260.
glass fibre board, organic bonded	0.036	100.	960.
thermalite turbo block	0.11	480.	1050.
weatherboard	0.14	650.	2000.
Wood			
fir, pine	0.12	510.	1380.
maple, oak	0.16	720.	1260.
beech, ash, walnut, meranti	0.23	650.	3050.
spruce, sylvester	0.12	530.	1880.
willow, birch	0.14	520.	2280.
softwood	0.17	550.	1880.
hardwood	0.05	90.	2810.
Wood Derivatives			
cellulosic insulation, loose fill	0.042	43.	1380.
chipboard	0.066	350.	1260.
chipboard, bonded with P.F.	0.12	650.	2340.
chipboard, bonded with melamine	0.12	630.	2260.
hardboard	0.13	900.	2000.
particle board	0.12	800.	1300.
plywood	0.15	700.	1420.
sawdust	0.065	180.	1380.
wood chippings, cement bonded	0.15	530.	1470.
woodwool	0.10	500.	1000.

A.5 Absorptivity and Emissivity

These data are grouped into 4 categories (one null) as follows.

Category/ Material	*Absorptivity*	*Emissivity*
Impermeables		
aluminium (polished)	0.10-0.40	0.03-0.06
aluminium (dull, rough polish)	0.40-0.65	0.18-0.30
aluminium (anodised)	-	0.72
aluminium surfaced roofing	-	0.216
asphalt (new)	0.91-0.93	-
asphalt (block)	0.85-0.98	0.90-0.98
asphalt (weathered)	0.82-0.89	-
asphalt (pavement)	0.852-0.928	-
bitumen felt/roofing sheets	0.86-0.89	0.91

Category/ Material	*Absorptivity*	*Emissivity*
bitumen (parking lot)	0.86-0.89	0.90-0.98
brass (polished)	0.30-0.50	0.03-0.05
brass (dull)	0.40-0.065	0.20-0.30
brass (anodised)	-	0.59-0.61
bronze	0.34	-
ceramics		
copper (polished)	0.18-0.50	0.02-0.05
copper (dull)	0.40-0.065	0.20-0.30
copper (anodised)	0.64	0.60
duraluminium	-	-
glass	-	0.88-0.937
iron (unoxidised)	-	0.05
iron (polished/bright)	0.40-0.65	0.20-0.377
iron (oxidised)	-	0.736-0.74
iron (red rusted)	-	0.61-0.65
iron (heavily rusted)	0.737	0.85-0.94
iron, cast (unoxidised/polished)	-	0.21-0.24
iron, cast (oxidised)	-	0.64-0.78
iron, cast (strongly oxidised)	-	0.95
iron, galvanised (new)	0.64-0.66	0.22-0.28
iron, galvanised (aged/very dirty)	0.89-0.92	0.89
lead (unoxidised)	-	0.05-0.075
lead (old/oxidised)	0.77-0.79	0.28-0.281
rubber (hard/glossy)	-	0.945
rubber (grey/rough)	-	0.859
steel (unoxidised/polished/stainless)	0.20	0.074-0.097
steel (oxidised)	0.20	0.79-0.82
tin (highly polished/unoxidised)	0.10-0.40	0.043-0.084
paint, Aluminium	0.30-0.55	0.27-0.67
paint, Zinc	0.30	0.95
PVC	-	0.90-0.92
tile (light)	0.3-0.5	0.85-0.95
varnishes	-	0.80-0.98
zinc (polished)	0.55	0.045-0.053
zinc (oxidised)	0.05	0.11-0.25

Non-Hygroscopic

Because insulants and the like are never used as surface finishes, there are no entries under this heading.

Inorganic-Porous

asbestos board	-	0.96
asbestos paper	-	0.93-0.94
asbestos cloth	-	0.90
asbestos cement (new)	0.61	0.95-0.96
asbestos cement (very dirty)	0.83	0.95-0.96

Category/ Material	*Absorptivity*	*Emissivity*
brick (glazed/light)	0.25-0.36	0.85-0.95
brick (light)	0.36-0.62	0.85-0.95
brick (dark)	0.63-0.89	0.85-0.95
cement mortar, screed	0.73	0.93
clay tiles (red/brown)	0.60-0.69	0.85-0.95
clay tiles (purple, dark)	0.81-0.82	0.85-0.95
concrete and plain concrete tile	0.65-0.80	0.85-0.95
concrete block	0.56-0.69	0.94
plaster	0.30-0.50	0.91
stone, granite (red)	0.55	0.90-0.93
stone, limestone	0.33-0.53	0.90-0.93
stone, marble	0.44-0.592	0.90-0.93
stone, sandstone	0.54-0.76	0.90-0.93
stone, slate	0.79-0.93	0.85-0.98
stone, quartz	-	0.90
Organic Hygroscopic		
paper	-	0.091-0.94
paper (white, bond)	0.25-0.28	-
cloth, cotton, black	0.67-0.98	-
cloth, cotton, deep blue	0.82-0.83	-
cloth, cotton, red	0.562	-
cloth, wool, black	0.87-0.88	-
cloth, wool, black	0.749	-
cloth, felt, black	0.775-0.861	-
cloth, all fabrics	-	0.89-0.92
wood, Beach	-	0.94
wood, Oak	-	0.89-0.90
wood, Spruce	-	0.82
wood, Walnut	-	0.83

A.6 Vapour Resistivity

Vapour resistivity is grouped into the same 4 categories as follows.

Category/ Material	*Vapour Resistivity (MNs $g^{-1}m^{-1}$)*
Impermeables	
asphalt (laid)	∞
bitumen roofing sheets	2,000-60,000
bituminous felt	15,000
glass, cellular	∞
glass, sheet/mirror/window	∞
glass, expanded/foamed	∞
glass brick	∞

Category/ Material	*Vapour Resistivity (MNs $g^{-1}m^{-1}$)*
linoleum (1200 kg m^{-3})	9,000
metals and metal cladding	∞
paint, Gloss (vapour resistant)	40-200
plastics, PVC sheets on tile	800-1,300
plastics, hard	45,000
rubber (1200-1500 kg m^{-3})	4500
rubber Tiles (1200-1500 kg m^{-3})	∞
tiles, Ceramic	500-5,000
tiles, Glazed ceramic	∞
plastics, PVC sheets on tile	800-1,300
plastics, hard	45,000
Non-Hygroscopic	
mineral fibre, glass fibre/wool	5-7
mineral fibre/wool	5-9
mineral fibre, rock wool	6.5-7.5
phenolic (closed cell)	150-750
phenol formaldehyde	19-20
polystyrene, expanded	100-750
polystyrene, extruded	600-1,500
polystyrene, extruded without skin	350-400
polyethylene foam	20,000
polyurethylene foam	115-1,000
PVC foam (rigid)	40-1,300
urea formaldehyde foam	5-20
Inorganic-Porous	
asbestos cement (800 kg m^{-3})	70
asbestos cement, sheeting, substitutes (1600-1900 kg m^{-3})	185-1000
brick, blast furnace slag (1000-2000 kg m^{-3})	350-500
brick, calcium silicate (<1400 kg m^{-3})	25-50
brick, calcium silicate (>1400 kg m^{-3})	75-125
brick, dense (>2000 kg m^{-3})	100-250
brick, heavyweight (>1700 kg m^{-3})	45-70
brick, lightweight (<1000 kg m^{-3})	25-50
brick, mediumweight (>1300 kg m^{-3})	23-45
brick, sand lime (<1400 kg m^{-3})	25-50
brick, sand lime (>1500 kg m^{-3})	75-200
concrete, cellular (450-1300 kg m^{-3})	9-50
concrete, cast (<1000 kg m^{-3})	14-33
concrete, cast (>1000 kg m^{-3})	30-80
concrete, cast (>1900 kg m^{-3})	115-1,000
concrete, expanded clay (500-1,000 kg m^{-3})	25-33
concrete, expanded clay (1,000-1,800 kg m^{-3})	33-75
concrete, foamed steam hardened (400-800 kg m^{-3})	25-50
concrete, natural pumice (500-1,400 kg m^{-3})	25-75

Category/ Material	*Vapour Resistivity (MNs $g^{-1}m^{-1}$)*
concrete, no fines (1800 kg m^{-3})	20
concrete, polystyrene foamed (400 kg m^{-3})	80-100
concrete, porous aggregate (1,000-2,000 kg m^{-3})	15-50
concrete, porous aggregate without quartz sand	25-75
concrete, close textured	350-750
concrete, slag and Rhine sand (1,500-1,700 kg m^{-3})	50-200
concrete, insulating	23-26
concrete blocks (very light)	15-150
plaster/ Mortar, cement based (1900-2000 kg m^{-3})	75-205
plaster/ Mortar, lime based (1600-1800 kg m^{-3})	45-205
plaster/ Mortar, gypsum, gypsum plasterboard	30-60
stone, Basalt, porphory, bluestone	∞
stone, Granite, marble	150-∞
stone, Slate	150-450
stone, slate shale	>3,000
stone, Limestone, firm	350-450
stone, Limestone, soft	130-160
stone, Limestone, soft tufa	25-50
stone, Sandstone	75-450
stone, Clay	75
tiles, clay tile, ceramic	750-1,500
tiles, floor tile, ceramic	115
tiles, terracotta roof tile	180-220
Organic-Hygroscopic	
carpet, normal backing	7-20
carpet, foam backed or foam underlay	100-300
chipboard	230-500
chipboard, bonded with cement	19-50
chipboard, bonded with U.F.	200-700
chipboard, bonded with melanine	300-500
chipboard, bonded with P.F.	250-750
corkboard	50-200
cork insulation	25-50
cork, expanded	23-50
cork, expanded and impregnated 45-230	
cork, expanded with bitumous binding	45-230
hardboard	230-1000
fibreboard	150-375
fibreboard, hard wood fibres	350
fibreboard, porous wood fibres	25
fibreboard, bitumened	25
fibreboard, cement based	19-50
mineral and vegetable fibre insulation	5
multiplex (800 kg m^{-3})	200-2000

Category/ Material	*Vapour Resistivity (MNs $g^{-1}m^{-1}$)*
multiplex, light pine	80
multiplex, North Canadian Gaboon	80
multiplex, red pine	875-250
paper	500
particle board, soft wood	25
plywood	150-2000
plywood, decking	1000-6000
plywood, marine	230-375
plywood, sheathing	144-1000
strawboard	45-70
triplex - Multiplex (700 kg m^{-3})	200-500
wood, ash	200-1850
wood, balsa	45-265
wood, beech	200-1850
wood, beech, soft	90-700
wood, birch	90-700
wood, fir	45-1850
wood, gaboon, North Canadian	45-1850
wood, oak	200-1850
wood, pine	45-1850
wood, pine, Northern red; Oregon	90-200
wood, pitch pine	200-1850
wood, spruce	45-1850
wood, teak	185-1850
wood, walnut	200-1850
wood, willow	45-1850
wood wool slabs	15-40
wood wool/ cement slabs	15-50
wood wool/ magnesia slabs	19-50
wood lath	4

A.7 References and further reading

Arnold P J 1970 Thermal Conductivity of Masonry Materials *BRS Current Paper CP 1/70*

ASHRAE 1985 Thermal Insulation and Water Vapor Retarders *Handbook of Fundamentals Chapter 20*

Ball E F 1968 Measurements of Thermal Conductivity of Building Materials *JIHVE 36* 51-6

Billington N S 1952 *Thermal Properties of Buildings* (London: Cleaver-Hume Press Ltd)

Bomberg M and Solvason K R 1985 Discussion of Heat flow Meter Apparatus and Transfer Standards Used for Error Analysis *ASTM Publication 879* 140-53 (Philadelphia American Society for Testing and Materials)

Clarke J A, Yaneske P P and Pinney A A 1990 The Harmonisation of Thermal Properties of Building Materials *BEPAC Research Report* (http://www.bepac.dmu.ac.uk)

Eldridge H J 1974 *Properties of Building Materials* (Lancaster: Medical and Technical Publishing Co.)

Hager N E 1985 Recent Developments with the Thin-Heater Thermal Conductivity Apparatus *ASTM Publication 879* 180-90 (Philadelphia: American Society for Testing and Materials)

Hens H 1984 Kataloog van Hygrothermische Eigenschappen van Bouwen Isolatiematerialen *Technical Report 1.* (Catholic University of Louvain: Laboratory for Building Physics)

Holden T S and Greenland J J 1951 The Coefficients of Solar Absorptivity and Low Temperature Emissivity of Various Materials - a Review of the Literature *Report R.6* (CSIRO: Division of Building Research)

Jakob M 1949 *Heat Transfer, Part 1* (London: Chapman and Hall)

Jespersen H B 1953 Thermal Conductivity of Moist Materials and its Measurement *JIHVE* **21** 157-74

MacLean R C and Galbraith G H 1988 Interstitial Condensation: Applicability of Conventional Vapour Permeability Values *Building Serv. Eng. Res. & Technol.* **9**(1) 29-34

Shirtcliffe C J and Tye R P (Eds) 1985 Guarded Hot Plate and Heat Flow Meter Methodology *ASTM Publication 879* 140-53 (Philadelphia: American Society for Testing and Materials)

Siviour J B 1985 Thermal Performance of Mineral Fibre Insulation *Building Serv. Eng. Res. and Technol.* **6**(2) 91-2

Stuckes A D and Simpson A 1986 Moisture Factors and the Thermal Conductivity of Aerated Concrete *Building Serv. Eng. Res. & Technol* **7**(2) 73-7

Valore R C 1980 Calculation of U-values of Hollow Concrete Masonry *Concrete International* **2**(2) 40-63

Van Geem M G and Fiorato A E 1983 Thermal Properties of Masonry Materials for Passive Solar Design - a State-of-the-Art Review' *Report DOE/CE/30739* (Skokie, IL: Construction Technology Laboratories)

Appendix B

Deficiencies of simplified methods

Following on from the introduction of the 1978 Building Regulations for England and Wales, a study was undertaken (ABACUS and Valtos 1979) to examine the consequences of non-compliance with the 'deemed-to-satisfy' provisions.† Regulation FF3 addressed the conservation of fuel and power and stated:

> "A building or part of a building to which this part applies shall be so designed and constructed that the enclosing structure provides adequate resistance to the passage of heat the loss of which from the building or part would entail the consumption of fuel or power to enable temperature conditions normal for the proposed use of the building or part to be maintained."

Two approaches to complying with FF3 were set out in the provisions of FF4:

> "Walls, floors and roofs of a building must be designed and constructed to meet prescribed U-values and the total percentage areas of the openings provided for windows and roof lights in these walls and roofs must not exceed prescribed limits.
>
> A wall, floor or roof may have a higher U-value provided the total rate of heat loss through all the walls, floors and roofs does not exceed that which would have resulted if the first approach had been adopted. Similarly the limits on openings for windows and roof lights may be exceeded provided that the total rate of heat loss through the glazed areas does not exceed that which would have resulted had the limits been observed, for example through the use of double or triple glazing."

The study team felt that it was important to draw a distinction between prescriptive and performance requirements. The deemed-to-satisfy provisions of FF4, by focusing on heat loss rate per square meter of fabric, prescribed allowable construction in large measure, thus precluding innovatory facade treatment. More worryingly, it was entirely possible to satisfy the provisions

† Were the study to be repeated for the current regulations, the outcome would be similar although it should be noted that current regulations are less prescriptive.

with a design which, in terms of geometry, thermal mass, insulation, orientation, plant control etc, was energy profligate. Had the provisions dealt directly with performance—maximum annual energy consumption based on typical occupancy and operational statistics—the onus would be on the designer to present a design solution, together with appropriate evidence of its energy behaviour. The issue then, if such a performance concept is accepted, is one of modelling accuracy and flexibility.

As part of the study, a multi-storey hotel complex was simulated using the (then) ESP-r system and the annual energy requirements determined for alternative glazing scenarios. Figure B.1 gives the results (the full curve) and demonstrates that areas of glazing greater than the deemed-to-satisfy limit (25% single glazing in this case) can offer a significant saving in cost-in-use terms (point C-4 compared with point R-1, the latter obtained by subjecting the regulation limit scheme to the same simulation). In other words, the provisions by their apparent exclusion of building geometry, orientation, thermal inertia, shading, weather variability etc, may limit a designer to a solution which is not optimum in terms of energy consumption and comfort.

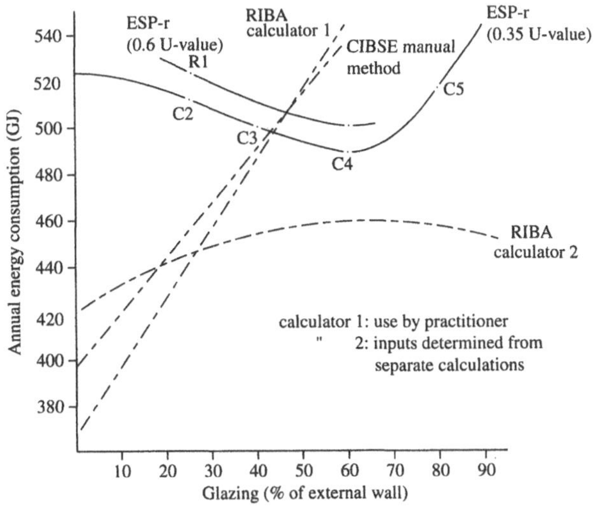

Figure B.1: Energy consumption predictions by alternative methods.

Also shown in figure B.1 (the chain curves) are the results obtained from two of the (then) commonly employed techniques: the RIBA calculator method and manual calculation methods taken from the CIBSE Guide. These results raise an additional point: if significant energy saving can only be achieved by going beyond the constraints of the regulations then the designer may not be able to rely on simplified methods to provide the necessary evidence of performance since the results so obtained may be inaccurate and therefore misleading. This problem is further compounded if additional technical complexity is introduced through the incorporation of passive solar components, advanced control systems or renewable energy technologies.

References and further reading

ABACUS and Valtos 1979 Deemed to Satisfy? *Architects Journal* October 345-55

Appendix C

Fourier heat equation and construction time constant

C.1 Fourier heat equation

Consider a closed surface S comprising a solid of volume V within which the temperature θ varies as a function of position x, y, z and time t. The total heat content of V is given by

$$\int^{V} \rho C \theta dV$$

where ρ is the density (kg m^{-3}) of V and C the specific heat capacity (J $kg^{-1}{}^{\circ}C^{-1}$).

For a small surface element, δS, with normal direction n and conductivity k_n (W $m^{-1}{}^{\circ}C^{-1}$) in the normal direction, the rate of heat flow out of δS is given by

$$-k_n \frac{\partial \theta}{\partial n} \delta S$$

where the minus sign indicates that the heat flow takes place in the direction of decreasing temperature. Since the rate of change of heat content in V is equal to the rate of heat flow out of S, equating these equations gives

$$\frac{\partial}{\partial t} \int^{V} \rho C \theta dV = \int^{S} k_n \frac{\partial \theta}{\partial n} dS + q \qquad \text{(C.1)}$$

where q (W) has been added to take account of any internal heat generation (e.g. from a solar flux injection where the material is transparent).

If the material is homogeneous and isotropic then ρ, C and k_n may be assumed constant. Setting $k_n = k$ in eqn (C.1) gives

$$\frac{\rho C}{k} \int^{V} \frac{\partial \theta}{\partial t} dV = \int^{S} \frac{\partial \theta}{\partial n} dS + \frac{q}{k} . \qquad \text{(C.2)}$$

If V is an infinitesimal parallelepiped of dimensions δx, δy and δz as shown in figure C.1 then

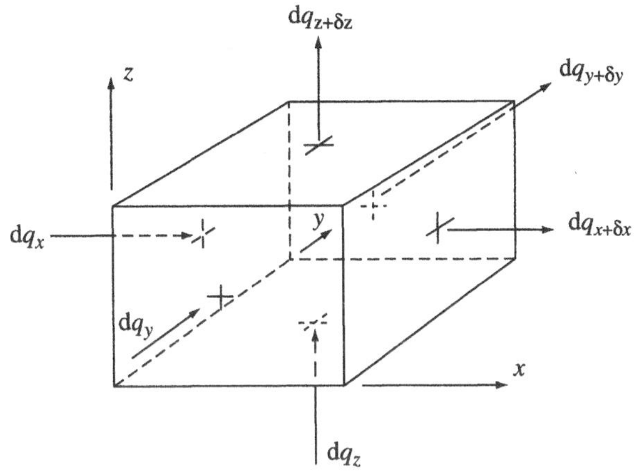

Figure C.1: A homogeneous, isotropic material.

$$\int^{V} \frac{\delta\theta}{\delta t}\, dV \approx \frac{\delta\theta}{\delta t}\, \delta x\, \delta y\, \delta z \tag{C.3}$$

and the surface integral $\int^{S} \frac{\partial\theta}{\partial n}\, dS$ will have 6 parts. For the x direction:

$$\left[\frac{\partial\theta(x+\delta x)}{\partial x} - \frac{\partial\theta(x)}{\partial x}\right] \delta y\, \delta z$$

$$= \frac{\partial}{\partial x}\left[\frac{\theta(x+\delta x) - \theta(x)}{\delta x}\right] \delta x\, \delta y\, \delta z$$

$$\approx \frac{\partial^2\theta}{\partial x^2}\, \delta x\, \delta y\, \delta z\,.$$

Taking account of the contributions from the other two pairs of faces, the surface integral becomes

$$\int^{S} \frac{\partial\theta}{\partial n}\, dS = \left[\frac{\partial^2\theta}{\partial x^2} + \frac{\partial^2\theta}{\partial y^2} + \frac{\partial^2\theta}{\partial z^2}\right] \delta x\, \delta y\, \delta z\,. \tag{C.4}$$

Substituting eqns (C.3) and (C.4) in (C.2) gives the three-dimensional, time dependent Fourier heat equation for a homogeneous, isotropic material.

$$\frac{\rho C}{k} \frac{\partial\theta}{\partial t} = \frac{\partial^2\theta}{\partial x^2} + \frac{\partial^2\theta}{\partial y^2} + \frac{\partial^2\theta}{\partial z^2} + \frac{q}{k\,(\delta x\, \delta y\, \delta z)}\,. \tag{C.5}$$

If the material has reached its steady state, i.e. $\frac{\partial\theta}{\partial t} = 0$, and there is no internal heat generation then equation (C.5) reduces to the Laplace equation:

$$\frac{\partial^2\theta}{\partial x^2} + \frac{\partial^2\theta}{\partial y^2} + \frac{\partial^2\theta}{\partial z^2} = 0$$

Finally, the Fourier equation in one space dimension and with heat generation is given by

$$\frac{\partial\theta}{\partial t} = \alpha\frac{\partial^2\theta}{\partial x^2} + \frac{q}{\rho C}$$

where α is the thermal diffusivity (m^2s^{-1}) as discussed in chapter 1.

C.2 Construction time constant

The time constant, τ, of a multi-layered construction may be evaluated from a method by Mackey and Wright (1946):

$$\tau = \frac{(k\rho C)_o\left(R_o - 0.1R_i - 0.1\sum_{j=1}^{M} R_j\right) + 1.1\left((\rho Cx)_i + \sum_{j=1}^{M}(\rho Cx)_j\right)}{U}$$

where k is the conductivity ($W\ m^{-1}{}^\circ C^{-1}$), ρ the density ($kg\ m^{-3}$), C the specific heat ($J\ kg^{-1}{}^\circ C^{-1}$), R the thermal resistance ($m^2{}^\circ C\ W^{-1}$), x the thickness (m), and o, i, j refer to the outermost, innermost and M central layers respectively.

Within a simulation, the start-up period (required to eliminate the effects of the arbitrarily assigned initial conditions) can then be empirically related to the maximum time constant occurring within a given model.

C.3 References and further reading

Mackey C O and Wright L T 1946 Periodic Heat Flow—Composite Walls or Roofs *Heating, Piping and Air Conditioning* **18** 107-10

Appendix D

Admittance method: worked example

Using relevant external data sources, this appendix demonstrates the use of the admittance method (as described in §2.3) to calculate the internal environmental temperature likely to occur at 15h00 during a warm, sunny day in August in a south facing office as described by the following data.

Construction Element	*U-Value* ($W\ m^{-2}{}^{\circ}C^{-1}$)	*Admittance* ($W\ m^{-2}{}^{\circ}C^{-1}$)	*Decrement* *(-)*	*Time Lag* *(h)*
External wall: 220mm brickwork, 25mm cavity, 25mm insulation, 10mm plasterboard	0.59	0.91	0.3	8
Window: double glazed, 12mm air gap, normal exposure (ignore frame)	2.9	2.9		
Internal walls: 220mm brickwork, 13mm light plaster	1.9	3.6		
Floor: 25mm wood block, 50mm screed, 150mm cast concrete	1.5	2.9		
Ceiling: as floor but reversed	1.5	6.0		

Latitude: 51.7°N.
Internal dimensions: 6m × 5m × 3m high.
External wall: 5m × 3m, dark external finish.
Window: 3m × 2m, unshaded, open during day, closed at night.
Occupancy: 5 persons for 8 hours at 85W (sensible) per person.
Lighting: 25W m^{-2} of floor area, ON 08h00-18h00.

Mean solar heat gain

$$Q'_s = S'\ I'_T\ A_g = 0.64 \times 175 \times (3 \times 2) = 672W$$

where Q'_s is the mean solar gain (W), A_g the sunlit area of glazing (m^2), I'_T the mean total solar irradiance (W/m^2; see Table A8.1†) and S' the mean solar gain factor (see Table A8.2).

Mean casual gains

$$Q'_c = \frac{(g_{c1} \times t_1) + (g_{c2} \times t_2) + \cdots}{24} = \frac{(5 \times 85 \times 8) + (6 \times 5 \times 25 \times 10)}{24} = 454\text{W}$$

where Q'_c is the mean casual gain (W), g_{c1}, g_{c2}, ... the instantaneous casual gains (W) and t_1, t_2, ... the durations of g_{c1}, g_{c2}, ... (hours).

Total mean heat gain

$$Q'_t = 672 + 454 = 1126\text{W}$$

Mean internal environmental temperature

$$Q'_t = (\textstyle\sum A_g U_g + C_v)(\theta'_{ei} - \theta'_{ao}) + \sum A_f U_f(\theta'_{ei} - \theta'_{eo}) \qquad \text{(D.1)}$$

where g and f refer to glazed and opaque surfaces respectively and C_v is the ventilation conductance evaluated from:

$$\frac{1}{C_v} = \frac{1}{0.33NV} + \frac{1}{4.8\sum A}$$

where N is the effective ventilation rate (hr^{-1}), V the room volume (90 m^3) and $\sum A$ the total internal surface area (120 m^2). From Table A8.4, N = 3 hr^{-1} and so $C_v = 78.3$ W/K.

From Table A8.3, the mean outside air temperature, θ'_{ao}, is 16.5°C and the mean sol-air temperature, θ'_{eo}, is 23°C. Therefore from eqn (D.1):

$$1126 = [2.9 \times 6 + 78.3][\theta'_{ei} - 16.5] + [0.59 \times 9][\theta'_{ei} - 23]$$

$$\Rightarrow \theta'_{ei} = 28°\text{C}$$

Swing in effective solar heat gain

$$\tilde{Q}_s = S_a A_g (I_p - I') = 0.56(3 \times 2)(490 - 175) = 1058.4\text{W}$$

where $\tilde{Q}_s$ is the swing in effective heat gain due to solar radiation (W), S_a the alternating solar gain factor (see Table A8.6) and I_p the peak intensity of solar radiation (W/m^2) which here is 490 W/m^2 (i.e. the value at 14h00 allowing for a 1 hour time lag).

Structural gain

$$\tilde{Q}_s = fAU(\theta_{eo} - \theta'_{eo}) = 0.3 \times 5.31(15.5 - 23) = -12\text{W}$$

where $\tilde{Q}_f$ is the swing in effective heat input due to structural gain (W), f the decrement factor, θ_{eo} the sol-air temperature at time of peak less time lag (i.e. 07h00 (15h00 - 8) = 15.5°C) and θ'_{eo} the mean sol-air temperature (= 23°C).

Casual gain

$$\tilde{Q}_c = Q_c - Q'_c = 5 \times 85 + 30 \times 25 - 454 = 721\text{W}$$

† Refers to table in the CIBSE Guide.

where Q_c is the casual gain value at the peak hour (= $g_{c1} + g_{c2} + ...$).

Swing in gain, air-to-air

$$\tilde{Q}_a = (\sum A_g U_g + C_v)\tilde{\theta}_{ao} = [(17.4 + 77.2] \times 5 = 478.5W$$

where $\tilde{Q}_a$ is the swing in the effective heat input due to swing in outside air temperature and $\tilde{\theta}_{ao}$ the swing in outside air temperature (from Table A8.3 = 21.5-16.5 = 5°C).

Total swing

$$\tilde{Q}_t = 1058.4 - 12 + 721 + 478.5 = 2246W$$

Swing in internal environmental temperature

$$\tilde{Q}_t = (\sum AY + C_v)\tilde{\theta}_{ei}$$

with the sum of the product of surface area and admittance given by $\sum AY = 476.19$ W/K:

Element	A	Y	$\sum AY$
ext wall	9	.91	8.19
window	6	2.9	17.4
int walls	51	3.6	183.6
floor	30	2.9	87
ceiling	30	6	180
			476.19

$$=> \tilde{\theta}_{ei} = \frac{2246}{(476.19 + 78.3)} = 4.1°C$$

Peak internal environmental temperature

$$\theta''_{ei} = \theta'_{ei} + \tilde{\theta}_{ei} = 28 + 4.1 = 32.1°C$$

Appendix E

Point containment algorithm

This Fortran 77 algorithm determines if a point X,Y is within or outwith a polygon with NV vertices defined by points Xp,Yp.

```
      subroutine point(X,Y,NV,Xp,Yp,inside)
      dimension Xp(NV),Yp(NV),C(2*NV+2)
      logical inside
      inside=.false.
      do 10 I=1,NV
        II=2*I-1
        C(II)=Xp(I)
        II=II+1
        C(II)=Yp(I)
10    continue
      last1=2*NV+1
      last2=2*NV+2
      C(last1)=Xp(1)
      C(last2)=Yp(1)
      LN=0
      IP=NV+1
      do 20 I=2,IP
        N=2*I
        IF((C(N-2)-Y)*(Y-C(N)))20,1,2
 2      IF((Y-C(N-2))*(C(N-1)-C(N-3))/(C(N)-C(N-2))+C(N-3)-X)20,3,4
 1      IF(C(N-2)-C(N))5,6,7
 6      IF((C(N-3)-X)*(X-C(N-1)))20,3,3
 5      IF((Y-C(N-2))*(C(N-1)-C(N-3))/(C(N)-C(N-2))+C(N-3)-X)20,3,8
 7      IF((Y-C(N-2))*(C(N-1)-C(N-3))/(C(N)-C(N-2))+C(N-3)-X)20,3,9
 9      LN=LN-2
 8      LN=LN-1
 4      LN=LN+2
20    continue
      IF((LN/4)*4.NE.LN)goto 3
      goto 11
 3    inside=.true.
11    return
      end
```

Appendix F

Radiosity based lighting simulation

The technique of §7.5.2 can be used to establish a computationally efficient lighting simulation program. This may be implemented as follows.

Zone discretisation

A finite element grid is applied to each polygon comprising the zone and its various contents. The unit hemisphere of figure 7.25 is generated and placed over each grid cell in turn so that the lines connecting a cell's centre point with the centroids of each hemispherical patch represents the cell's radiosity to a given resolution. Each ray is then projected to locate the point of intersection and establish the associated grid cell located on the other polygon. In this way each projection gives rise to a ray, and each ray has a source and sink cell.

The technique makes allowance for the relationship between the source/sink separation distance and the spread of light due to the solid angle effect. Because of this spread, the number of illuminated cells arising from one exit ray increases as the distance to the point of intersection increases. The radiosity technique represents this phenomenon while minimising the number of rays for processing. For example, 10 polygons, each divided into 100 grid cells, and using a unit hemisphere with 50 patches, gives rise to 50,000 rays. If, instead, each polygon cell is joined to all the others, the total number of combinatorial pairs is 450,000 rays! (i.e. $n(n-1)100^2/2$, where n is the number of polygons). The latter approach is computationally unacceptable because it involves many redundant rays: there will be many rays with identical point view factors that could be substituted by one representative parent ray.

All un-hit cells are now grouped and marked as family members of some parent ray. These "secondary" rays are never processed at ray tracing time but, instead, inherit the illuminance properties of their parent. To determine family members, the following technique may be employed. An un-hit cell is connected back to the centre point of the initial polygon cell. This gives the point of intersection with the hemispherical surface and indicates an associated patch. All back projections associated with the same patch are then marked as family members of the source/sink ray that was initially processed for this patch and source polygon cell.

At the end of the discretisation process a number of principal rays exist, each one with a distinct source and sink cell. These comprise the discrete model of the zone and its contents, and are organised into a stack for subsequent processing. The secondary rays, grouped together as family members, are withheld from the stack and are not processed at simulation time. They are used only during results recovery to determine the finite element patches that are actually illuminated within the resolution of the numerical processing scheme.

Modelling of light sources

Three entities are of importance in light source representation: the geometry of the source, its intensity distribution and its spectral power distribution.

Geometry may be differentiated into point, linear and area sources. The spatial distribution of luminous intensity, as defined from manufacturer's intensity distribution data, can be made discrete and held as a collection of vectors.

The energy emitted from a light source at each wavelength in the visible spectrum is given by a spectral energy distribution curve. This spectrum may be conveniently modelled as a number of mono-chromatic wavebands. For the case of daylight, a window may be treated as an area light source with a luminous intensity distribution superimposed as a function of the prevailing sky type (overcast, clear or intermediate).

Ray tracing

The luminous flux of a light source is radiant flux in the visible range (390 nm to 780 nm) weighted according to a visual response as specified by a spectral luminous efficiency function. The unit of luminous flux is the lumen (lm) defined as a radiant flux of $\frac{1}{683}$ W at a wavelength of 555 nm in air. The elemental luminous flux associated with a spectral radiant flux $\phi_{e\lambda}$ over an elemental range $d\lambda$ is given by

$$d\lambda_v = 683V(\lambda)\phi_{e\lambda}d\lambda$$

and the total luminous flux, ϕ_v, is obtained by integrating over the visible range:

$$\phi_v = 683\int V(\lambda)\phi_{e\lambda}d\lambda\ .$$

The luminous intensity, that is the luminous flux per steradian, emitted in a given direction is given by

$$I = \frac{d\phi}{d\omega}\ .$$

The spectral luminous intensity is determined from the spectral energy distribution function:

$$I(\lambda) = I\,F(\lambda) \tag{E.1}$$

where $F(\lambda)$ is the spectral weighting factor, defined as the fraction of luminous energy contained in a mono-chromatic band $\delta\lambda$ at some wavelength λ. It is found from

$$F(\lambda) = \frac{\sum S(\lambda)\delta\lambda}{\sum \delta\lambda}$$

where $S(\lambda)$ is the relative spectral energy distribution function.

The direct illuminance of a sink cell of elemental area dA due to a light source is given by

$$E_d = I\frac{d\omega}{dA}$$

and since $d\omega = \frac{dA \cos\theta}{r^2}$ then

$$E_d = I \frac{\cos\theta}{r^2} . \tag{E.2}$$

Combining eqns (E.1) and (E.2) gives the final expression for the direct illuminance due to a mono-chromatic waveband:

$$\delta E(\lambda) = \frac{I F(\lambda) \cos\theta}{r^2}$$

with the total illuminance for the visible spectrum given by

$$E = \sum \frac{I F(\lambda) \cos\theta}{r^2} .$$

In this way, and for each waveband, the direct illuminance of each polygon cell may be determined and the light source vectors 'locked' onto the discrete ray model of the zone. As each ray is processed, a surface reflection model is invoked to determine the reflections at the sink surface cell. This gives rise to one or several exit rays depending on the degree of specularity of the surface. The light intensity of these rays is written to an output database while the exit rays themselves are appended to the ray stack for onward processing. The next ray in the stack is then processed, the reflection point detected, the surface reflection algorithm invoked, the reflected intensities recorded and the new exit rays appended to the stack. The process continues until new exit rays can be discarded because their intensity has diminished below some threshold value. In this way the stack is processed until exhausted. The next mono-chromatic waveband or light source vector is then processed.

Surface reflection models

The reflection of light at a surface interface occurs through two mechanisms. The first, termed specular, is due to a primary reflection at the boundary surface; the second, termed diffuse, is due to multiple inter-reflections arising from the surface roughness. Most building materials exhibit a mix of both specular and diffuse reflection. For fully diffusing surfaces, the reflected rays are equal in number to the number of initial hemispherical patches, with an intensity obtained directly from the average wavelength dependent reflectance assigned to the polygon cell in question. With specular and off-specular surfaces, the number of exit rays is greatly reduced.

The biangular reflectance technique of Torrance and Sparrow (1965) allows a surface to approach a diffuse distribution at short wavelengths and a specular distribution at long wavelengths. The technique that the requires surface reflectance be known as a function of the angle of incidence of the incoming light.

The light flux falling on an opaque surface is partly absorbed and partly reflected. The magnitude of this reflected flux depends on a property of the surface, termed reflectance:

$$\phi_r = \phi_i r$$

where ϕ_r is the reflected flux, ϕ_i the incoming flux and r the surface reflectance.

With materials which are neither perfect diffusers nor perfect specular reflectors, the determination of the spatial distribution of this reflected flux is based on the concept of biangular reflectance. This is defined as the reflected radiance in the direction (θ_r, ξ_r) divided by the incident radiant flux from the direction (θ_i, ξ_i), where θ and ξ are the azimuth and altitude angles in polar coordinates. Thus, the biangular reflectance, ρ, is given by

$$\rho(\theta_i, \xi_i; \theta_r, \xi_r) = \frac{I_r(\theta_r, \xi_r)}{I_i(\theta_i, \xi_i) \cos \theta d\omega_i} .$$

Biangular reflectance data can be obtained by experiment or analytically. In the former case the directional reflectance characteristics are measured using either gonionic reflectometers or luminance meters. The decomposition of the specular and diffuse components can also be made through the use of polarising glass on luminance meters. But because of the large number of combinations of incoming and outgoing directions, a major problem exists with the size of the data set for even a single material. For this reason an analytical approach is often favoured. This entails the theoretical determination of material reflectance based on geometrical optics. This requires that the biangular reflectance be expressed in terms of specular (ρ_s) and diffuse (ρ_d) components:

$$\rho = k_s \rho_s + k_d \rho_d$$

where $k_s + k_d = 1$.

While the diffuse reflectance data are generally available as measured values, the specular component requires a mathematical model:

$$\rho_s = \frac{F D G}{\pi \cos \theta_l \cos \theta_v}$$

where F is the Fresnel reflectance, D a distribution function for the micro-facets comprising the surface, G a geometric attenuation factor arising from micro-facet shadowing and masking, θ_l the angle of the incoming light vector and θ_v the angle of the exit vector.

The Fresnel reflectance is found from

$$F = \tfrac{1}{2}\left[\frac{\sin^2(\phi - \theta)}{\sin^2(\phi + \theta)} + \frac{\tan^2(\phi - \theta)}{\tan^2(\phi + \theta)}\right]$$

where $\sin \theta = \dfrac{\sin \phi}{\eta}$ and η is the index of refraction of the reflecting surface.

The geometric attenuation factor is a function of the angular relationship between the incident light and the surface facet geometry:

$$G = 1 - \frac{m}{l} .$$

where m is the root mean square of the micro-facets.

Torrence and Sparrow (1965) proposed the following equation for the distribution function

$$D = c_1 e^{-(\delta/m)^2}$$

where c_1 is an arbitrary constant.

Results integration

The reflected flux distribution, held separately for each polygon cell, light source vector and mono-chromatic waveband can now be integrated to provide the usual engineering quantities (daylight factors, surface illuminance, contrast ratios etc). Alternatively, the data may be used to provide a coloured perspective image. For a given eye and focus point, a perspective transformation is applied to each polygon cell before terminal display. When combined with a visibility priority algorithm (e.g. z-depth sorting), such a 'scan-cell' algorithm results in a screen image whose resolution depends on the initial polygon gridding scheme.

The reproduction of surface colour from the computed spectral luminance distribution of room surfaces may be achieved using the CIE chromaticity system (Greenberg and Joblove 1980), which allows the specification of colour on a physical basis. Using the method, the relative fraction of each of the theoretical primary colours (red, green and blue) can be mathematically derived from the spectral distribution of luminous power. The first step is to weight the calculated surface spectral luminance distribution. This is done by using three CIE colour-matching functions, which represent the contribution of a unit amount of energy at each wavelength. This gives rise to a tri-stimulus value:

$$\delta X = \hat{x}(\lambda)L(\lambda)\delta\lambda$$

$$\delta Y = \hat{y}(\lambda)L(\lambda)\delta\lambda$$

$$\delta Z = \hat{z}(\lambda)L(\lambda)\delta\lambda$$

where $\hat{x}(\lambda)$, $\hat{y}(\lambda)$ and $\hat{z}(\lambda)$ are the CIE colour matching functions for the theoretical primaries X, Y and Z, and $L(\lambda)$ the luminance of a surface at a mono-chromatic band $\delta\lambda$.

The final theoretical primaries are obtained by integrating the elemental tri-stimulus values over the entire visible spectrum, i.e. $X = \sum \delta X$ and so on. The CIE chromaticity coordinates of a colour comprised of X, Y and Z primaries is then derived by normalising the tri-stimulus values such that

$$x = \frac{X}{X+Y+Z}$$

$$y = \frac{Y}{X+Y+Z}$$

$$z = \frac{Z}{X+Y+Z}$$

and since $x + y + z = 1$ then only x and y values need be plotted on the Chromaticity Diagram.

The relationship between these theoretical colour coordinates x, y and z and the red, green and blue primaries of a colour terminal is found by solving the following system of simultaneous colour matching equations.

$$X = x_r R + x_g G + x_b B$$

$$Y = y_r R + y_g G + y_b B$$

$$Z = z_r R + z_g G + z_b B$$

where the three sets of coefficients, (x_i, y_i, z_i), $i = r, g, b$, are the x, y and z coordinates of the red, green and blue primaries of a given display monitor.

The luminance range of typical monitors—of the order of 1 cd/m^2—is somewhat inadequate when compared to the luminance variation encountered in the real world (e.g. between 1 and several thousand cd/m^2). It is therefore necessary to scale the foregoing luminances down to the range manageable by the monitor. In this context it is normal to assume that when the brightness ratio is held constant, the perception of relative brightness in the model will approximate to the real world. First, the computed surface luminance values are transformed into apparent brightness values as perceived by the human eye:

$$b_i = k(L - L_o)^c \tag{E.3}$$

where b_i is the brightness at location i, L the surface luminance, L_o the threshold luminance of the human eye and c, k are correlation coefficients.

The maximum monitor brightness can also be determined from eqn (E.3):

$$b'_{max} = k(L'_{max} - L_o)^c$$

where b'_{max} is the maximum monitor brightness and L'_{max} the maximum monitor luminance.

Second, the luminance of the surface in terminal space is determined from

$$b'_i = (b_i/b_{max})b'_{max} \; .$$

Finally, from the terminal brightness, the luminance of the terminal is determined through the inverse of the luminance/brightness function.

References and further reading

Greenberg D and Joblove G H 1980 Colour Spaces for Computer Graphics *Proc. SIGGRAPH 1980* 20-5

Torrance K E and Sparrow E M 1965 Biangular Reflectance of an Electric Nonconductor as a Function of Wavelength and Surface Roughness *Journal of Heat Transfer* 283-92

Appendix G

The ESP-r system

Figure G.1 shows the principal components of the ESP-r system: a central project manager, an integrated simulator, support databases, a performance appraisal tool and support utilities for CAD and visualisation.

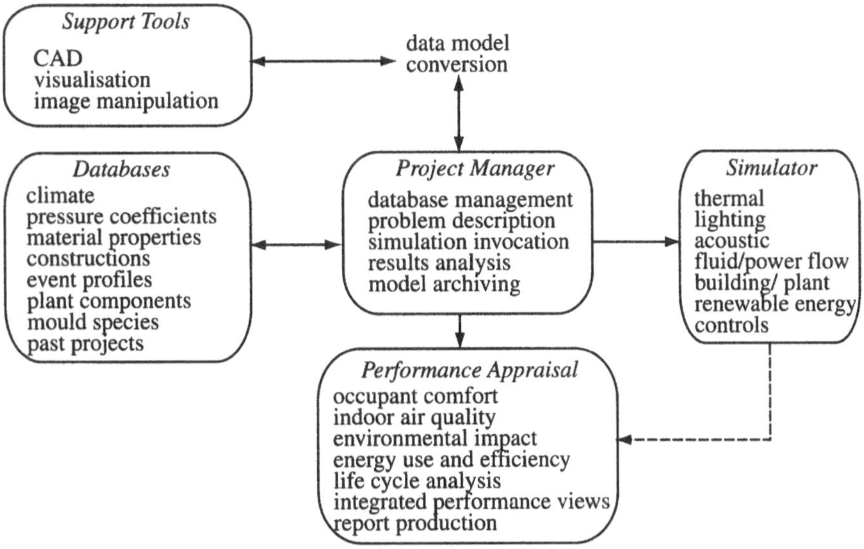

Figure G.1: The ESP-r system.

To facilitate system evolution the simulator source code is partitioned into parts relating to the different thermo-fluid domains—radiation heat exchange, network fluid flow, CFD, HVAC, control etc, and user interface constructs.

ESP-r employs the X Windows System open standard and is designed to run under UNIX operating systems (Bourne 1982) such as Linux and Solaris. The system source code, corresponding to the theories described in this book, is made available and regular updates are provided. The installation procedure entails the downloading of the source code from an ftp site, followed by local installation using a supplied script.

The internal architecture and data requirements of ESP-r are described elsewhere (Clarke 1982, 1985). While this description corresponds to an earlier version, it nevertheless provides a useful insight into the design of a building simulation program in relation to theory organisation, user interface elements, data management and processor/storage requirements. Similar information for the current version of ESP-r may be obtained directly from the source code which is available from <http://www.esru.strath.ac.uk> along with information on installation procedures, third party tool compatibility, training and support.

References and further reading

Bourne J R 1982 *The UNIX System* (New York: Addison-Wesley)

Clarke J A 1982 Documentation of the ESP-r System *Final Grant Report to SERC (now EPSRC)* (Available from: ESRU, University of Strathclyde)

Clarke J A 1985 *Energy Simulation in Building Design (1st Edition)* (Bristol: Adam Hilger)

Index